W9-AMB-127

PROTECTIVE GROUPS IN ORGANIC SYNTHESIS

PROTECTIVE GROUPS IN ORGANIC SYNTHESIS

Second Edition

THEODORA W. GREENE
The Rowland Institute for Science, Inc.

and

PETER G. M. WUTS
The Upjohn Company

A WILEY-INTERSCIENCE PUBLICATION

JOHN WILEY & SONS, INC.

New York / Chichester / Brisbane / Toronto / Singapore

***Library of Congress Cataloging in Publication Data*:**

Greene, Theodora W., 1931-
Protective groups in organic synthesis / Theodora W. Greene and Peter G. M. Wuts.—2nd ed.
p. cm.
Includes index.
ISBN 0-471-62301-6
1. Organic compounds—Synthesis. 2. Protective groups (Chemistry)
I. Wuts, Peter G. M.
QD262.G665 1991
547.2—dc20 90-45708
CIP

Printed in the United States of America

10 9 8 7 6 5 4 3

Printed and bound by Courier Companies, Inc.

PREFACE TO THE SECOND EDITION

Since publication of the first edition of this book in 1981, many new protective groups and many new methods of introduction or removal of known protective groups have been developed: 206 new groups and approximately 1500 new references have been added. Most of the information from the first edition has been retained. To conserve space, generic structures used to describe Formation/Cleavage reactions have been replaced by a single line of conditions, sometimes with explanatory comments, especially regarding selectivity. Some of the new information has been obtained from Online Searches of Chemical Abstracts, which have limitations. For example, Chemical Abstracts indexes a review article about protective groups only if that word appears in the title of the article. References are complete through 1989. Some references, from more widely circulating journals, are included for 1990.

Two new sections on the protection for indoles, imidazoles, and pyrroles, and protection for the amide —NH are included. They are separated from the regular amines because their chemical properties are sufficiently different to affect the chemistry of protection and deprotection. The Reactivity Charts in Chapter 8 are identical to those in the first edition. The chart number appears beside the name of each protective group when it is first discussed.

A number of people must be thanked for their contributions and help in completing this project. I am grateful to Gordon Bundy, who loaned me this card file, which provided many references that the computer failed to find, and to Bob Williams, Spencer Knapp, and Tohru Fukuyama for many references on amine and amide protection. I thank Theo Greene who checked and rechecked the manuscript for spelling and consistency and for the herculean task of checking all the references to make sure my 3s and 8s and 7s and 9s were not interchanged, all without a single complaint. I thank Fred Greene, who read the manuscript and provided valuable suggestions for its improvement. My wife Lizzie was a major contributor to getting this project finished by looking up and xeroxing references, turning on the computer in an evening ritual, and retyping many sections of the original book, which made the changes and additions much easier. Without her understanding and encouragement it probably would never have been completed.

PETER G. M. WUTS

Kalamazoo, Michigan
November 1990

PREFACE TO THE FIRST EDITION

The selection of a protective group is an important step in synthetic methodology, and reports of new protective groups appear regularly. This book presents information on the synthetically useful protective groups (~500) for five major functional groups: —OH, —NH, —SH, —COOH, and >C=O. References through 1979, the best method(s) of formation and cleavage, and some information on the scope and limitations of each protective group are given. The protective groups that are used most frequently and that should be considered first are listed in Reactivity Charts, which give an indication of the reactivity of a protected functionality to 108 prototype reagents.

The first chapter discusses some aspects of protective group chemistry: the properties of a protective group, the development of new protective groups, how to select a protective group from those described in this book, and an illustrative example of the use of protective groups in a synthesis of brefeldin. The book is organized by functional group to be protected. At the beginning of each chapter are listed the possible protective groups. Within each chapter protective groups are arranged in order of increasing complexity of structure (e.g., methyl, ethyl, *t*-butyl, · · ·, benzyl). The most efficient methods of formation or cleavage are described first. Emphasis has been placed on providing recent references, since the original method may have been improved. Consequently, the original reference may not be cited; my apologies to those whose contributions are not acknowledged. Chapter 8 explains the relationship between reactivities, reagents, and the Reactivity Charts that have been prepared for each class of protective groups.

This work has been carried out in association with Professor Elias J. Corey, who suggested the study of protective groups for use in computer-assisted synthetic analysis. I appreciate his continued help and encouragement. I am grateful to Dr.

J. F. W. McOmie (Ed., *Protective Groups in Organic Chemistry*, Plenum Press, New York and London, 1973) for his interest in the project and for several exchanges of correspondence, and to Mrs. Mary Fieser, Professor Frederick D. Greene, and Professor James A. Moore for reading the manuscript. Special thanks are also due to Halina and Piotr Starewicz for drawing the structures, and to Kim Chen, Ruth Emery, Janice Smith, and Ann Wicker for typing the manuscript.

THEODORA W. GREENE

Harvard University
September 1980

CONTENTS

ABBREVIATIONS

PROTECTIVE GROUPS

In some cases, several abbreviations are used for the same protective group. We have listed the abbreviations as used by an author in his original paper, including capital and lowercase letters. Occasionally, the same abbreviation has been used for two different protective groups. This information is also included.

Ac	acetyl
Acm	acetamidomethyl
Ad	1-adamantyl
Adoc	1-adamantyloxycarbonyl
Adpoc	1-(1-adamantyl)-1-methylethoxycarbonyl
Alloc or AOC	allyloxycarbonyl
AOC or Alloc	allyloxycarbonyl
p-AOM	*p*-anisyloxymethyl or (4-methoxyphenoxy)methyl
BDT	1,3-benzodithiolan-2-yl
Bic	5-benzisoxazolylmethoxycarbonyl
Bmpc	2,4-dimethylthiophenoxycarbonyl
Bmpm	1,1-bis(4-methoxyphenyl)-1′-pyrenylmethyl
Bn	benzyl
BOC	*t*-butoxycarbonyl
BOM	benzyloxymethyl
Bpoc	1-methyl-1-(4-biphenyl)ethoxycarbonyl
Bum	*t*-butoxymethyl

t-Bumeoc	1-(3,5-di-*t*-butylphenyl)-1-methylethoxycarbonyl
Bz	benzoyl
Cam	carboxamidomethyl
Cbz or Z	benzyloxycarbonyl
Cee	1-(2-chloroethoxy)ethyl
Coc	cinnamyloxycarbonyl
CPTr	4,4′,4″-tris(4,5-dichlorophthalimido)triphenylmethyl
CTMP	1-[(2-chloro-4-methyl)phenyl]-4-methoxypiperidin-4-yl
Cys	cysteine
DAM	di(4-methoxyphenyl)methyl
Bd-*t*-BOC	1,1-dimethyl-2,2-dibromoethoxycarbonyl
DBD—Tmoc	2,7-di-*t*-butyl[9-(10,10-dioxo-10,10,10,10-tetrahydrothioxanthyl)]methylcarbonyl
DBN	*p,p*′-dinitrobenzhydryl
DEIPS	diethylisopropylsilyl
Dim	[2-(1,3-dithianyl)methyl]
DMIPS	dimethylisopropylsilyl
Dmoc	dithianylmethoxycarbonyl
Dmp	dimethylphosphinyl
DMPM	3,4-dimethoxybenzyl
DMTr	di(*p*-methoxyphenyl)phenylmethyl
DNB	*p,p*′-dinitrobenzhydryl
DNMBS	4-(4′,8′-dimethoxynaphthylmethyl)benzenesulfonyl
DNP	2,4-dinitrophenyl
Dobz	*p*-(dihydroxyboryl)benzyloxycarbonyl
DPM or Dpm	diphenylmethyl
DPMS	diphenylmethylsilyl
Dpp	diphenylphosphinyl
Dppe	2-(diphenylphosphino)ethyl
Dppm	(diphenyl-4-pyridyl)methyl
DTBMS	di-*t*-butylmethylsilyl
DTBS	di-*t*-butylsilylene
Dts	dithiasuccinimidyl
EE	1-ethoxyethyl
Fcm—NR_2	*N*-ferrocenylmethyl
Fcm—SR	*S*-ferrocenylthiomethyl
Fm	9-fluorenylmethyl
Fmoc	9-fluorenylmethoxycarbonyl
GUM	guaiacolmethyl
hZ (or homo Z)	homobenzyloxycarbonyl
IDTr	3-(imidazol-1-ylmethyl)-4′, 4″-dimethoxytriphenylmethyl
iMds	2,6-dimethoxy-4-methylbenzenesulfonyl
Ipaoc	1-isopropylallyloxycarbonyl
IPDMS	isopropyldimethylsilyl

Lev	levulinoyl
LevS	levulinoyldithioacetal ester [4,4-(ethylenedithio)pentanoyl]
MBE	1-methyl-1-benzyloxyethyl
MBF	2,3,3a,4,5,6,7,7a-octahydro-7,8,8-trimethyl-4,7-methanobenzofuran-2-yl
MBS or Mbs	*p*-methoxybenzenesulfonyl
Mds	2,6-dimethyl-4-methoxybenzenesulfonyl
MEM	2-methoxyethoxymethyl
MeOZ or Moz	*p*-methoxybenzyloxycarbonyl
MMTr	*p*-methoxyphenyldiphenylmethyl
MOM	methoxymethyl
Moz or MeOZ	*p*-methoxybenzyloxycarbonyl
MPM or PMB	*p*-methoxyphenylmethyl or *p*-methoxybenzyl
Mps	*p*-methoxyphenylsulfonyl
Mpt	dimethylthiophosphinyl
Ms	methanesulfonyl (mesylate)
Msib	4-(methylsulfinyl)benzyl
Msz	4-methylsulfinylbenzyloxycarbonyl
Mtb	2,4,6-trimethoxybenzenesulfonyl
Mte	2,3,5,6-tetramethyl-4-methoxybenzenesulfonyl
MTHP	4-methoxytetrahydropyranyl
MTM	methylthiomethyl
MTMB	4-(methylthiomethoxy)butyryl
MTMECO	2-(methylthiomethoxy)ethoxycarbonyl
MTMT	2-(methylthiomethoxymethyl)benzoyl
Mtpc	4-methylthiophenoxycarbonyl
Mtr	2,3,6-trimethyl-4-methoxybenzenesulfonyl
Mts	2,4,6-trimethylbenzenesulfonyl (mesitylenesulfonyl)
Noc	4-nitrocinnamyloxycarbonyl
NPS or Nps	2-nitrophenylsulfenyl
Npys	3-nitro-2-pyridinesulfenyl
ONB	*o*-nitrobenzyl
Peoc	2-phosphonioethoxycarbonyl or 2-(triphenylphosphonio)ethoxycarbonyl
Pet	2-(2′-pyridyl)ethyl
Phenoc	4-methoxyphenacyloxycarbonyl
Pixyl	9-(9-phenyl)xanthenyl
PMB or MPM	*p*-methoxybenzyl or *p*-methoxyphenylmethyl
PMBM	*p*-methoxybenzyloxymethyl
Pmc	2,2,5,7,8-pentamethylchroman-6-sulfonyl
Pme	pentamethylbenzenesulfonyl
PMS	*p*-methylbenzylsulfonyl
POM	4-pentenyloxymethyl or pivaloyloxymethyl

Ppoc	2-triphenylphosphonioisopropoxycarbonyl
Ppt	diphenylthiophosphinyl
Psec	2-(phenylsulfonyl)ethoxycarbonyl
PTM	phenylthiomethyl
Pv	pivaloyl
Pyoc	2-(2′- and 4′-pyridyl)ethoxycarbonyl
Scm	*S*-carboxymethylsulfenyl
SEM	2-(trimethylsilyl)ethoxymethyl
SES	β-trimethylsilylethanesulfonyl
SMOM	(phenyldimethylsilyl)methoxymethyl
Snm	*S*-(*N*′-methyl-*N*′-phenylcarbamoyl)sulfenyl
STABASE	1,1,4,4-tetramethyldisilylazacyclopentane
Tacm	trimethylacetamidomethyl
TBDMS or TBS	*t*-butyldimethylsilyl
TBDPS	*t*-butyldiphenylsilyl
TBDS	tetra-*t*-butoxydisiloxane-1,3-diylidene
TBMPS	*t*-butylmethoxyphenylsilyl
TBS or TBDMS	*t*-butyldimethylsilyl
TBTr	4,4′,4″-tris(benzyloxy)triphenylmethyl
TcBoc	1,1-dimethyl-2,2,2-trichloroethoxycarbonyl
Tcroc	2-(trifluoromethyl)-6-chromonylmethylenecarbonyl
Tcrom	2-(trifluoromethyl)-6-chromonylmethylene
TDS	thexyldimethylsilyl
Teoc	2-(trimethylsilyl)ethoxycarbonyl
TES	triethylsilyl
TFA	trifluoroacetyl
Thexyl	2,3-dimethyl-2-butyl
THF	tetrahydrofuranyl
THP	tetrahydropyranyl
TIPDS	1,3-(1,1,3,3-tetraisopropyldisiloxanylidene)
TIPS	triisopropylsilyl
TLTr	4,4′,4″-tris(levulinoyloxy)triphenylmethyl
TMPM	trimethoxyphenylmethyl
TMS	trimethylsilyl
TMSEC	2-(trimethylsilyl)ethoxycarbonyl
TMTr	tri(*p*-methoxyphenyl)methyl
Tos or Ts	*p*-toluenesulfonyl
TPS	triphenylsilyl
Tr	triphenylmethyl (trityl)
Tritylone	9-(9-phenyl-10-oxo)anthryl
Troc	2,2,2-trichloroethoxycarbonyl
Ts or Tos	*p*-toluenesulfonyl
Tse	2-(*p*-toluenesulfonyl)ethyl
Voc	vinyloxycarbonyl
Z or Cbz	benzyloxycarbonyl

REAGENTS

9-BBN	9-borabicyclo[3.3.1]nonane
BOP reagent	benzotriazol-1-yloxytris(dimethylamino)phosphonium hexafluorophosphate
BOP—Cl	*N,N*-bis(2-oxo-3-oxazolidinyl)phosphinic chloride
Bt	benzotriazol-1-yl or 1-benzotriazolyl
CAN	ceric ammonium nitrate
CMPI	2-chloro-1-methylpyridinium chloride
CSA	camphorsulfonic acid
DABCO	1,4-diazabicyclo[2.2.2]octane
DBN	1,5-diazabicyclo[4.3.0]non-5-ene
DBU	1,8-diazabicyclo[5.4.0]undec-7-ene
DCC	dicyclohexylcarbodiimide
DDQ	2,3-dichloro-5,6-dicyano-1,4-benzoquinone
DEAD	diethyl azodicarboxylate
DIAD	diisopropyl azodicarboxylate
DIBAL—H	diisobutylaluminum hydride
DIPEA	diisopropylethylamine
DMAP	4-*N,N*-dimethylaminopyridine
DME	1,2-dimethoxyethane
DMF	*N,N*-dimethylformamide
DMPU	1,3-dimethyl-3,4,5,6-tetrahydro-2(1*H*)-pyrimidinone
DMS	dimethyl sulfide
DMSO	dimethyl sulfoxide
EDCI	1-ethyl-3-[3-(dimethylamino)propyl]carbodiimide
EDTA	ethylenediaminetetraacetic acid
HMDS	1,1,1,3,3,3-hexamethyldisilazane
HMPA	hexamethylphosphoramide
HMPT	hexamethylphosphorous triamide
HOBT	1-hydroxybenzotriazole
Im	imidazol-1-yl or 1-imidazolyl
MAD	methylaluminum bis(2,6-di-*t*-butyl-4-methylphenoxide)
MCPBA	*m*-chloroperoxybenzoic acid
MoOPH	oxodiperoxymolybdenum(pyridine)hexamethylphosphoramide
MSA	methanesulfonic acid
NBS	*N*-bromosuccinimide
$Ni(acac)_2$	nickel acetylacetonate
PCC	pyridinium chlorochromate
$PdCl_2(tpp)_2$	dichlorobis[tris(2-methylphenyl)phosphine]palladium
$Pd_2(dba)_3$	tris(dibenzylideneacetone)dipalladium
PG	protective group
PPTS	pyridinium *p*-toluenesulfonate
Pyr	pyridine
ScmCl	methoxycarbonylsulfenyl chloride

Su	succinimidyl
TEA	triethylamine
TEBAC	benzyltriethylammonium chloride
Tf	trifluoromethanesulfonate
TFA	trifluoroacetic acid
TFAA	trifluoroaetic anhydride
TfOH	trifluoromethanesulfonic acid
THF	tetrahydrofuran
THP	tetrahydropyran
TMEDA	*N*,*N*,*N'*, *N'*-tetramethylethylenediamine
Ts	toluenesulfonyl

PROTECTIVE GROUPS IN ORGANIC SYNTHESIS

1

THE ROLE OF PROTECTIVE GROUPS IN ORGANIC SYNTHESIS

Properties of a Protective Group

When a chemical reaction is to be carried out selectively at one reactive site in a multifunctional compound, other reactive sites must be temporarily blocked. Many protective groups have been, and are being, developed for this purpose. A protective group must fulfill a number of requirements. It must react selectively in good yield to give a protected substrate that is stable to the projected reactions. The protective group must be selectively removed in good yield by readily available, preferably nontoxic reagents that do not attack the regenerated functional group. The protective group should form a crystalline derivative (without the generation of new stereogenic centers) that can easily be separated from side products associated with its formation or cleavage. The protective group should have a minimum of additional functionality to avoid further sites of reaction.

Historical Development

Since a few protective groups cannot satisfy all these criteria for elaborate substrates, a large number of mutually complementary protective groups are needed and, indeed, are becoming available. In early syntheses the chemist chose a standard derivative known to be stable to the subsequent reactions. In a synthesis of callistephin chloride the phenolic —OH group in **1** was selectively protected as an acetate.[1] In the presence of silver ion the aliphatic hydroxyl group in **2** displaced

$HO-C_6H_4-C(O)CH_2OH$ (**1**) $\xrightarrow{NaOH}$ $\xrightarrow{CH_3COCl}$ $AcO-C_6H_4-C(O)CH_2OH$ (**2**)

the bromide ion in a bromoglucoside. In a final step the acetate group was removed by basic hydrolysis. Other classical methods of cleavage include acidic hydrolysis (eq. 1), reduction (eq. 2), and oxidation (eq. 3):

(1) ArO—R → ArOH
(2) $RO-CH_2Ph \rightarrow ROH$
(3) $RNH-CHO \rightarrow [RNHCOOH] \rightarrow RNH_3^+$

Some of the original work in the carbohydrate area in particular reveals extensive protection of carbonyl and hydroxyl groups. For example, a cyclic diacetonide of glucose was selectively cleaved to the monoacetonide.[2] A summary[3] describes the selective protection of primary and secondary hydroxyl groups in a synthesis of gentiobiose, carried out in the 1870s, as triphenylmethyl ethers.

Development of New Protective Groups

As chemists proceeded to synthesize more complicated structures, they developed more satisfactory protective groups and more effective methods for the formation and cleavage of protected compounds. At first a tetrahydropyranyl acetal was prepared,[4] by an acid-catalyzed reaction with dihydropyran, to protect a hydroxyl group. The acetal is readily cleaved by mild acid hydrolysis, but formation of this acetal introduces a new stereogenic center. Formation of the 4-methoxytetrahydropyranyl ketal[5] eliminates this problem.

Catalytic hydrogenolysis of an *O*-benzyl protective group is a mild, selective method introduced by Bergmann and Zervas[6] to cleave a benzyl carbamate ($>NCO-OCH_2C_6H_5 \rightarrow >NH$) prepared to protect an amino group during peptide syntheses. The method also has been used to cleave alkyl benzyl ethers, stable compounds prepared to protect alkyl alcohols; benzyl esters are cleaved by catalytic hydrogenolysis under neutral conditions.

Three selective methods to remove protective groups are receiving much attention: "assisted," electrolytic, and photolytic removal. Four examples illustrate "assisted removal" of a protective group. A stable allyl group can be converted to a labile vinyl ether group (eq. 4);[7] a β-haloethoxy (eq. 5)[8] or a β-silylethoxy (eq. 6)[9] derivative is cleaved by attack at the β-substituent; and a stable *o*-nitrophenyl derivative can be reduced to the *o*-amino compound, which undergoes cleavage by nucleophilic displacement (eq. 7)[10]:

(4) $ROCH_2CH{=}CH_2 \xrightarrow{t\text{-BuO}^-} [ROCH{=}CHCH_3] \xrightarrow{H_3O^+} ROH$

(5) $RO{-}CH_2{-}CCl_3 + Zn \rightarrow RO^- + CH_2{=}CCl_2$

(6) $RO{-}CH_2{-}CH_2{-}SiMe_3 \xrightarrow{F^-} RO^- + CH_2{=}CH_2 + FSiMe_3$

R = alkyl, aryl, R′CO—, or R′NHCO—

(7)

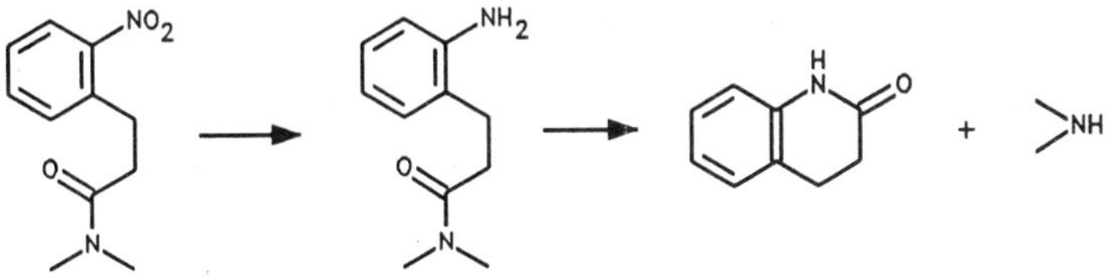

The design of new protective groups that are cleaved by "assisted removal" is a challenging and rewarding undertaking.

Removal of a protective group by electrolytic oxidation or reduction can be very satisfactory. The equipment required ranges from a minimum of two electrodes, a potentiostat, and a source of DC current to quite sophisticated systems. A suitable electrolyte–solvent system is needed, and the deprotected product must not undergo further electrochemistry under the experimental conditions. The use and subsequent removal of chemical oxidants or reductants (e.g., Cr or Pb salts; Pt— or Pd—C) are eliminated. Reductive cleavages have been carried out in high yield at −1 to −3 V (vs. SCE) depending on the group; oxidative cleavages in good yield have been realized at 1.5–2 V (vs. SCE). For systems possessing two or more electrochemically labile protective groups, selective cleavage is possible when the half-wave potentials, $E_{1/2}$, are sufficiently different; excellent selectivity can be obtained with potential differences on the order of 0.25 V. Protective groups that have been removed by electrolytic oxidation or reduction are described at the appropriate places in this book; a review article by Mairanovsky[11] discusses electrochemical removal of protective groups.[12]

Photolytic cleavage reactions (e.g., of *o*-nitrobenzyl, phenacyl, nitrophenylsulfenyl derivatives) take place in high yield on irradiation of the protected compound for a few hours at 254–350 nm. For example, the *o*-nitrobenzyl group, used to protect alcohols,[13] amines,[14] and carboxylic acids,[15] has been removed by irradiation. Protective groups that have been removed by photolysis are described at the appropriate places in this book; in addition, the reader may wish to consult five review articles.[16-20]

One widely used method of formation of protected compounds involves polymer-supported reagents,[21-27] with the advantage of simple workup by filtration and automated syntheses, especially of polypeptides. Polymer-supported reagents are used to protect a terminal —COOH group as a polymer-bound ester (RCOOR′—Ⓟ) during peptide syntheses,[21] to protect primary alcohols as Ⓟ—trityl ethers,[28] and to protect 1,2- and 1,3-diols as Ⓟ—phenyl boronates.[29] Monoprotection of symmetrical dialdehydes and diacid chlorides has been reported;[23] some diprotection occurs with diols and diamines.[23]

Internal protection, used by van Tamelen in a synthesis of colchicine, may be appropriate:[30]

1. DCC
2. CH_2N_2

Selection of a Protective Group from This Book

To select a specific protective group, the chemist must consider in detail all the reactants, reaction conditions, and functionalities involved in the proposed synthetic scheme. First he or she must evaluate all functional groups in the reactant to determine those that will be unstable to the desired reaction conditions and require protection. The chemist should then examine reactivities of possible protective groups, listed in the Reactivity Charts, to determine compatibility of protective group and reaction conditions. The protective groups listed in the Reactivity Charts (see Chapter 8) have been used most widely; consequently, considerable experimental information is available for them. The chemist should consult the complete list of protective groups in the relevant chapter and consider their properties. It will frequently be advisable to examine the use of one protective group for several functional groups (i.e., a 2,2,2-trichloroethyl group to protect a hydroxyl group as an ether, a carboxylic acid as an ester, and an amino group as a carbamate). When several protective groups are to be removed simultaneously, it may be advantageous to use the same protective group to protect different functional groups (e.g., a benzyl group, removed by hydrogenolysis, to protect an alcohol and a carboxylic acid). When selective removal is required, different classes of protection must be used (e.g., a benzyl ether, cleaved by hydrogenolysis but stable to basic hydrolysis, to protect an alcohol, and an alkyl ester cleaved by basic hydrolysis but stable to hydrogenolysis, to protect a carboxylic acid).

If a satisfactory protective group has not been located, the chemist has a number of alternatives: rearrange the order of some of the steps in the synthetic scheme so that a functional group no longer requires protection or a protective group that was reactive in the original scheme is now stable; redesign the synthesis, possibly making use of latent functionality[31] (i.e., a functional group in a precursor form; e.g., anisole as a precursor of cyclohexanone). Or it may be necessary to include the synthesis of a new protective group in the overall plan.

A number of standard synthetic reference books are available.[32,33] A review article by Kössell and Seliger[34] discusses protective groups used in oligonucleotide syntheses, including protection for the phosphate group, which is not included in this book, and a series of articles[35] describe various aspects of protective group chemistry.

Total Synthesis of Palytoxin Carboxylic Acid: An Example of the Selection, Introduction, and Removal of Protective Groups

Palytoxin carboxylic acid, $C_{123}H_{213}NO_{53}$ (Figure 1, R^1–R^8 = H), derived from palytoxin, $C_{129}H_{223}N_3O_{54}$, contains 41 hydroxyl groups, one amino group, one ketal, one hemiketal, and one carboxylic acid, in addition to some double bonds and ether linkages.

The total synthesis[36] was achieved through the synthesis of eight different segments, each requiring extensive use of protective group methodology, followed by the appropriate coupling of the various segments in their protected forms.

The choice of what protective groups to use in the synthesis of each segment was based on three aspects: (a) the specific steps chosen to achieve the synthesis of each segment, (b) the methods to be used in coupling the various segments, and (c) the conditions needed to deprotect the 42 blocked groups in order to liberate palytoxin carboxylic acid in its unprotected form. (These conditions must be such that the functional groups already deprotected are stable to the successive deblocking conditions.) Kishi's synthesis employed only eight different protective groups for the 42 functional groups present in the fully protected form of palytoxin carboxylic acid (Figure 1, **1**). A few additional protective groups were used for "end-group" protection in the synthesis and sequential coupling of the eight different segments. The synthesis was completed by removal of all the groups by a series of five different methods. The selection, formation, and cleavage of these groups are described below.

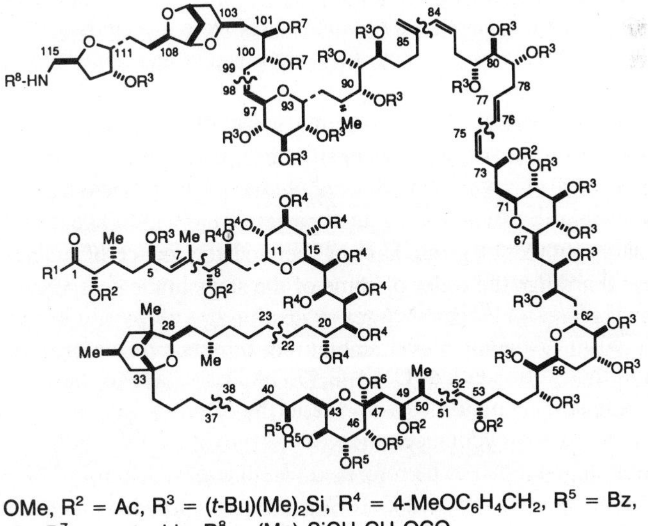

1: R^1 = OMe, R^2 = Ac, R^3 = (*t*-Bu)(Me)$_2$Si, R^4 = 4-MeOC$_6$H$_4$CH$_2$, R^5 = Bz, R^6 = Me, R^7 = acetonide, R^8 = (Me)$_3$SiCH$_2$CH$_2$OCO

2: Palytoxin carboxylic acid: R^1 = OH, R^2–R^8 = H

Figure 1. Palytoxin carboxylic acid. (Structure kindly provided by Professor Yoshito Kishi.)

For the synthesis of the C.1–C.7 segment, the C.1 carboxylic acid was protected as a methyl ester. The C.5 hydroxyl group was protected as the *t*-butyldimethylsilyl (TBS) ether. This particular silyl group was chosen because it improved the chemical yield and stereochemistry of the Ni(II)/Cr(II)-mediated coupling reaction of segment C.1–C.7 with segment C.8–C.51. Nine hydroxyl groups were protected as *p*-methoxyphenylmethyl (MPM) ethers, a group that was stable to the conditions used in the synthesis of the C.8–C.22 segment. These MPM groups were eventually cleaved oxidatively by treatment with 2,3-dichloro-5,6-dicyano-1,4-benzoquinone (DDQ).

The C.2 hydroxyl group was protected as an acetate, since cleavage of a *p*-methoxyphenylmethyl (MPM) ether at C.2 proved to be very slow. An acetyl group was also used to protect the C.73 hydroxyl group during synthesis of the right-hand half of the molecule (C.52–C.115). Neither a *p*-methoxyphenylmethyl (MPM) nor a *t*-butyldimethylsilyl (TBS) ether was satisfactory at C.73: dichlorodicyanobenzoquinone (DDQ) cleavage of a *p*-methoxyphenylmethyl (MPM) ether at C.73 resulted in oxidation of the *cis–trans* dienol at C.78–C.73 to a *cis–trans* dienone. When C.73 was protected as a *t*-butyldimethylsilyl (TBS) ether, Suzucki coupling of segment C.53–C.75 (in which C.75 was a vinyl iodide) to segment C.76–C.115 was too slow. In the synthesis of segment C.38–C.51, the C.49 hydroxyl group was also protected at one stage as an acetate, to prevent benzoate migration from C.46. The C.8 and C.53 hydroxyl groups were protected as acetates for experimental convenience. A benzoate ester, more electron-withdrawing than an acetate ester, was used to protect the C.46 hydroxyl group to prevent spiroketalization of the C.43 and C.51 hydroxyl groups during synthesis of the C.38–C.51 segment. Benzoate protection of the C.46 hydroxyl group also increased the stability of the C.47 methoxy group (part of a ketal) under acidic cleavage conditions. Benzoates rather than acetates were used during the synthesis of the C.38–C.51 segment since they were more stable and better chromophores in purification and characterization.

Several additional protective groups were used in the coupling of the eight different segments. A tetrahydropyranyl (THP) group was used to protect the hydroxyl group at C.8 in segment C.8–C.22 and a *t*-butyldiphenylsilyl (TBDPS) group for the hydroxyl group at C.37 in segment C.23–C.37. The TBDPS group at C.37 was later removed by $Bu_4N^+F^-$/THF in the presence of nine *p*-methoxyphenylmethyl (MPM) groups. After the coupling of segment C.8–C.37 with segment C.38–C.51, the C.8 THP ether was hydrolyzed with pyridinium *p*-toluenesulfonate (PPTS) in methanol–ether, 42°, in the presence of the bicyclic ketal at C.28–C.33 and the cyclic ketal at C.43–C.47. (As noted above, the resistance of this ketal to these acidic conditions was due to the electron-withdrawing effect of the benzoate at C.46.) A cyclic acetonide (a 1,3-dioxane) at C.49–C.51 was also removed by this step and had to be reformed (acetone/PPTS) prior to the coupling of segment C.8–C.51 with segment C.1–C.7. After coupling of these segments to form segment C.1–C.51, the new hydroxyl group at C.8 was protected as an acetate, and the acetonide at C.49–C.51 was again removed without alteration of the bicyclic ketal at C.28–C.33 or the cyclic ketal at C.43–C.47, still stabilized by the benzoate at C.46.

The synthesis of segment C.77–C.115 from segments C.77–C.84 and C.85–C.115 involved the liberation of an aldehyde at C.85 from its protected form as a dithioacetal, $RCH(SEt)_2$, by mild oxidative deblocking ($I_2/NaHCO_3$, acetone, water) and the use of the *p*-methoxyphenyldiphenylmethyl (MMTr) group to protect the hydroxyl group at C.77. The C.77 MMTr ether was subsequently converted to a primary alcohol (PPTS/MeOH–CH_2Cl_2, rt) without affecting the 19 *t*-butyldimethylsilyl (TBS) ethers or the cyclic acetonide at C.100–C.101.

The C.100–C.101 diol group, protected as an acetonide, was stable to the Wittig reaction used to form the *cis* double bond at C.98–C.99, and to all the conditions used in the buildup of segment C.99–C.115 to fully protected palytoxin carboxylic acid (Figure 1, **1**).

The C.115 amino group was protected as a trimethylsilylethyl carbamate ($Me_3SiCH_2CH_2OCONHR$), a group that was stable to the synthesis conditions and cleaved by the conditions used to remove the *t*-butyldimethylsilyl (TBS) ethers.

Thus the 42 functional groups in palytoxin carboxylic acid (39 hydroxyl groups, one diol, one amino group, and one carboxylic acid) were protected by eight different groups:

1 methyl ester	—COOH
5 acetate esters	—OH
20 *t*-butyldimethylsilyl (TBS) ethers	—OH
9 *p*-methoxyphenylmethyl (MPM) ethers	—OH
4 benzoate esters	—OH
1 methyl "ether"	—OH of a hemiketal
1 acetonide	1,2-diol
1 $Me_3SiCH_2CH_2OCO$	$—NH_2$

The protective groups were then removed in the following order by the five methods listed below:

(1) To cleave *p*-methoxyphenylmethyl (MPM) ethers: DDQ (dichlorodicyanobenzoquinone)/*t*-BuOH–CH_2Cl_2–phosphate buffer (pH 7.0), 4.5 h.

(2) To cleave the acetonide: 1.18 *N* $HClO_4$–THF, 25°, 8 days.

(3) To hydrolyze the acetates and benzoates: 0.08 *N* LiOH/H_2O–MeOH–THF, 25°, 20 h.

(4) To remove *t*-butyldimethylsilyl (TBS) ethers and the carbamoyl ester ($Me_3SiCH_2CH_2OCONHR$): $Bu_4N^+F^-$, THF, 22°, 18 h → THF–DMF, 22°, 72 h.

(5) To hydrolyze the methyl ketal at C.47, no longer stabilized by the C.46 benzoate: HOAc–H_2O, 22°, 36 h.

This order was chosen so that DDQ (dichlorodicyanobenzoquinone) treatment would not oxidize a deprotected allylic alcohol at C.73, and so that the C.47 hemiketal would still be protected (as the ketal) during basic hydrolysis (step 3).

And so the skillful selection, introduction, and removal of a total of 12 different protective groups have played a major role in the successful total synthesis of palytoxin carboxylic acid (Figure 1, **2**).

1. A. Robertson and R. Robinson, *J. Chem. Soc.*, 1460 (1928).
2. E. Fischer, *Ber.*, **28**, 1145–1167 (1895); see p. 1165.
3. B. Helferich, *Angew. Chem.*, **41**, 871 (1928).
4. W. E. Parham and E. L. Anderson, *J. Am. Chem. Soc.*, **70**, 4187 (1948).
5. C. B. Reese, R. Saffhill, and J. E. Sulston, *J. Am. Chem. Soc.*, **89**, 3366 (1967).
6. M. Bergmann and L. Zervas, *Chem. Ber.*, **65**, 1192 (1932).
7. J. Cunningham, R. Gigg, and C. D. Warren, *Tetrahedron Lett.*, 1191 (1964).
8. R. B. Woodward, K. Heusler, J. Gosteli, P. Naegeli, W. Oppolzer, R. Ramage, S. Ranganathan, and H. Vorbruggen, *J. Am. Chem. Soc.*, **88**, 852 (1966).
9. P. Sieber, *Helv. Chim. Acta*, **60**, 2711 (1977).
10. I. D. Entwistle, *Tetrahedron Lett.*, 555 (1979).
11. V. G. Mairanovsky, *Angew. Chem., Int. Ed. Engl.*, **15**, 281 (1976).
12. See also: M. F. Semmelhack and G. E. Heinsohn, *J. Am. Chem. Soc.*, **94**, 5139 (1972).
13. S. Uesugi, S. Tanaka, E. Ohtsuka, and M. Ikehara, *Chem. Pharm. Bull.*, **26**, 2396 (1978).
14. S. M. Kalbag and R. W. Roeske, *J. Am. Chem. Soc.*, **97**, 440 (1975).
15. L. D. Cama and B. G. Christensen, *J. Am. Chem. Soc.*, **100**, 8006 (1978).
16. V. N. R. Pillai, *Synthesis*, 1–26 (1980).
17. P. G. Sammes, *Quart. Rev., Chem. Soc.*, **24**, 37–68 (1970); see pp. 66–68.
18. B. Amit, U. Zehavi, and A. Patchornik, *Isr. J. Chem.*, **12**, 103–113 (1974).
19. V. N. R. Pillai, "Photolytic Deprotection and Activation of Functional Groups," *Org. Photochem.*, **9**, 225–323 (1987).
20. V. Zehavi, "Applications of Photosensitive Protecting Groups in Carbohydrate Chemistry," *Adv. Carbohydr. Chem. Biochem.*, **46**, 179–204 (1988).
21. R. B. Merrifield, *J. Am. Chem. Soc.*, **85**, 2149 (1963).
22. P. Hodge, *Chem. Ind. (London)*, 624 (1979).
23. C. C. Leznoff, *Acc. Chem. Res.*, **11**, 327 (1978).
24. E. C. Blossey and D. C. Neckers, Eds., *Solid Phase Synthesis*, Halsted, New York, 1975; P. Hodge and D. C. Sherrington, Eds., *Polymer-Supported Reactions in Organic Synthesis*, Wiley-Interscience, New York, 1980. A comprehensive review of polymeric protective groups by J. M. J. Fréchet is included in this book.
25. D. C. Sherrington and P. Hodge, *Syntheses and Separations Using Functional Polymers*, Wiley-Interscience, New York, 1988.
26. J. M. Stewart and J. D. Young, *Solid Phase Peptide Synthesis*, 2nd ed., Pierce Chemical Company, Rockford, IL, 1984.

27. E. Atherton and R. C. Sheppard, *Solid Phase Peptide Synthesis. A Practical Approach*, Oxford-IRL Press, New York, 1989.
28. J. M. J. Fréchet and K. E. Haque, *Tetrahedron Lett.*, 3055 (1975).
29. J. M. J. Fréchet, L. J. Nuyens, and E. Seymour, *J. Am. Chem. Soc.*, **101,** 432 (1979).
30. E. E. van Tamelen, T. A. Spencer, Jr., D. S. Allen, Jr., and R. L. Orvis, *Tetrahedron*, **14,** 8 (1961).
31. D. Lednicer, *Adv. Org. Chem.*, **8,** 179–293 (1972).
32. J. F. W. McOmie, Ed., *Protective Groups in Organic Chemistry*, Plenum, New York and London, 1973.
33. *Organic Syntheses*, Wiley-Interscience, New York, *Collect. Vols. I–VII*, 1941–1990, **65–68**, 1987–1989; W. Theilheimer, Ed., *Synthetic Methods of Organic Chemistry*, S. Karger, Basel, Vols. 1–43, 1946–1988; E. Müller, Ed., *Methoden der Organischen Chemie* (Houben-Weyl), G. Thieme Verlag, Stuttgart, Vols. 1–19b, 1958–1989; *Spec. Period Rep.: General and Synthetic Methods*, **1–10** (1978–1988); S. Patai, Ed., *The Chemistry of Functional Groups*, Wiley-Interscience, New York, Vols. 1–34, 1964–1990.
34. H. Kössell and H. Seliger, *Prog. Chem. Org. Nat. Prod.*, **32,** 297–366 (1975).
35. J. F. W. McOmie, *Chem. Ind. (London)*, 603–609; E. Haslam, *Chem. Ind. (London)*, 610–617 (1979); P. M. Hardy, *Chem. Ind. (London)*, 617–624 (1979); P. Hodge, *Chem. Ind. (London)*, 624–627 (1979).
36. R. W. Armstrong, J.-M. Beau, S. H. Cheon, W. J. Christ, H. Fujioka, W.-H. Ham, L. D. Hawkins, H. Jin, S. H. Kang, YOSHITO KISHI, M. J. Martinelli, W. W. McWhorter, Jr., M. Mizuno, M. Nakata, A. E. Stutz, F. X. Talamas, M. Taniguchi, J. A. Tino, K. Ueda, J.-i. Uenishi, J. B. White, and M. Yonaga, *J. Am. Chem. Soc.*, **111,** 7530–7533 (1989). See also: *Idem.*, *ibid.*, **111,** 7525–7530 (1989).

2

PROTECTION FOR THE HYDROXYL GROUP, INCLUDING 1,2- AND 1,3-DIOLS

ETHERS

Ethers are among the most used protective groups in organic synthesis. They vary from the simplest, most robust, methyl ether to the more elaborate, substituted, trityl ethers developed for use in nucleotide synthesis. They are formed and removed under a wide variety of conditions. Some of the ethers that have been used to protect alcohols are included in Reactivity Chart 1.[1]

1. See also: C. B. Reese, "Protection of Alcoholic Hydroxyl Groups and Glycol Systems," in *Protective Groups in Organic Chemistry*, J. F. W. McOmie, Ed., Plenum, New York and London, 1973, pp. 95–143; H. M. Flowers, "Protection of the Hydroxyl Group," in *The Chemistry of the Hydroxyl Group*, S. Patai, Ed., Wiley-Interscience,

New York, 1971, Vol. 10/2, pp. 1001–1044; C. B. Reese, *Tetrahedron*, **34,** 3143–3179 (1978), see pp. 3145–3150; V. Amarnath and A. D. Broom, *Chem. Rev.*, **77,** 183–217 (1977) (see pp. 184–194); M. Lalonde and T. H. Chan, "Use of Organosilicon Reagents as Protective Groups in Organic Synthesis," *Synthesis*, 817 (1985).

1. Methyl Ether: ROMe (Chart 1)

Formation

1. Me_2SO_4, NaOH, $Bu_4N^+I^-$, org. solvent, 60–90% yield.[1]
2. CH_2N_2, silica gel, 0–10°, 100% yield.[2]

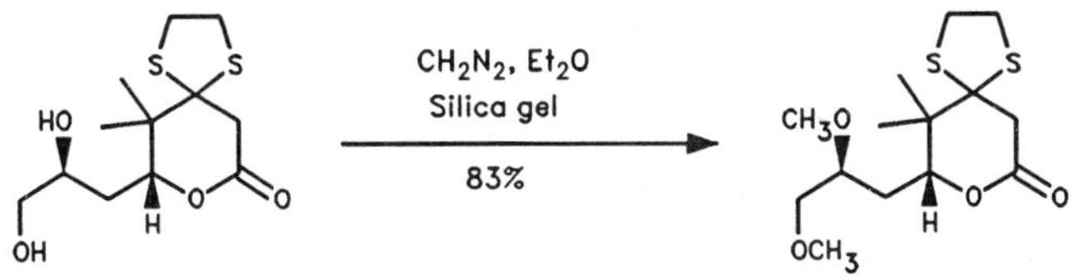

CH_2N_2, Et_2O; Silica gel; 83% — Ref. 3

3. CH_2N_2, HBF_4, CH_2Cl_2, Et_3N, 25°, 1 h, 95% yield.[4,5]
4. MeI, solid KOH, DMSO, 20°, 5–30 min, 85–90% yield.[6]
5. $(MeO)_2POH$, cat. TsOH, 90–100°, 12 h, 60% yield.[7]
6. $Me_3O^+BF_4^-$, 3 days, 55% yield.[8]
7. CF_3SO_3Me, CH_2Cl_2, Pyr, 80°, 2.5 h, 85–90% yield.[9]
8. Because of the increased acidity and reduced steric requirement of the carbohydrate hydroxyl, *t*-BuOK can be used as a base to achieve ether formation.[10]

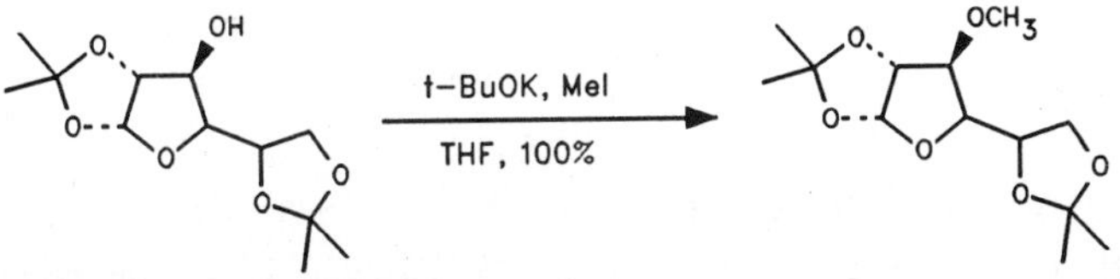

t-BuOK, MeI; THF, 100%

9. MeI, Ag_2O, 93% yield.[11]

MeI, Ag_2O; 93%

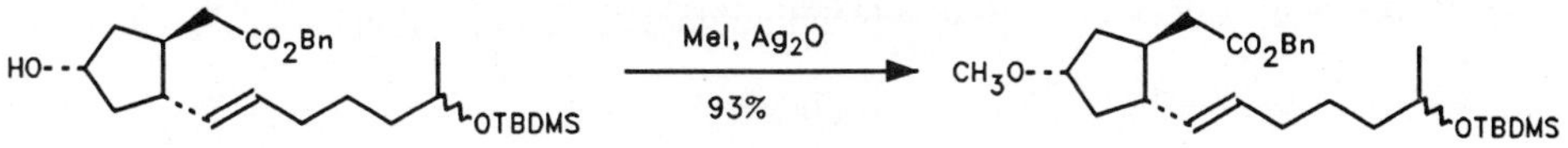

10. Me_2SO_4, DMSO, DMF, $Ba(OH)_2$, BaO, rt, 18 h, 88% yield.[12]

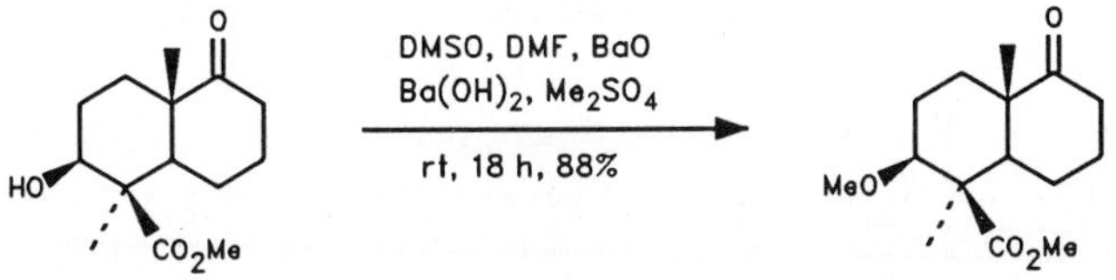

DMSO, DMF, BaO; $Ba(OH)_2$, Me_2SO_4; rt, 18 h, 88%

11. MeI or Me_2SO_4,[13] NaH or KH, THF. This is the standard method for introducing the methyl ether function onto hindered and unhindered alcohols.

Cleavage

1. Me_3SiI, $CHCl_3$, 25°, 6 h, 95% yield.[14] A number of methods have been reported in the literature for the *in situ* formation of Me_3SiI[15] since Me_3SiI is somewhat sensitive to handle. This reagent also cleaves many other ether-type protective groups, but selectivity can be maintained by control of the reaction conditions and the inherent rate differences between functional groups.
2. BBr_3, NaI, 15-crown-5.[16] Methyl esters are not cleaved under these conditions.[17]
3. BBr_3, EtOAc, 1 h, 95% yield.[18]
4. BBr_3, CH_2Cl_2, high yields.[19]

CH3O OAc CH3O2C OAc CO2CH3 — BBr_3, CH_2Cl_2, −78° -> 12° → O O OAc H OAc CO2CH3

This method is probably the most commonly used method for the cleavage of ethers because it generally gives excellent yields with a variety of structural types. BBr_3 will cleave ketals.

5. $BF_3 \cdot Et_2O$, $HSCH_2CH_2SH$, HCl, 15 h, 82% yield.[20,21]
6. $MeSSiMe_3$ or $PhSSiMe_3$, ZnI_2, $Bu_4N^+I^-$.[22] In this case the 6-*O*-methyl ether was cleaved selectively from permethylated glucose.
7. $SiCl_4$, NaI, CH_2Cl_2, CH_3CN, 80–100% yield.[23]
8. AlX_3 (X = Br, Cl), EtSH, 25°, 0.5–3 h, 95–98% yield.[24]
9. *t*-BuCOCl or AcCl, NaI, CH_3CN, 37 h, rt, 84% yield.[25] In this case the methyl ether is replaced by a pivaloate or acetate group that can be hydrolyzed with base.
10. Ac_2O, $FeCl_3$, 80°, 24 h.[26] In this case the methyl ether is converted to an acetate. The reaction proceeds with complete racemization.
11. AcCl, NaI, CH_3CN.[27]

1. A. Merz, *Angew. Chem., Int. Ed. Engl.*, **12,** 846 (1973).
2. K. Ohno, H. Nishiyama, and H. Nagase, *Tetrahedron Lett.*, 4405 (1979).
3. T. Nakata, S. Nagao, N. Mori, and T. Oishi, *Tetrahedron Lett.*, **26,** 6461 (1985).
4. M. Neeman and W. S. Johnson, *Org. Synth., Collect. Vol. V*, 245 (1973).
5. A. B. Smith, III, K. J. Hale, L. M. Laakso, K. Chen, and A. Riera, *Tetrahedron Lett.*, **30,** 6963 (1989).

6. R. A. W. Johnstone and M. E. Rose, *Tetrahedron*, **35,** 2169 (1979).
7. Y. Kashman, *J. Org. Chem.*, **37,** 912 (1972).
8. H. Meerwein, G. Hinz, P. Hofmann, E. Kroning, and E. Pfeil, *J. Prakt. Chem.*, **147,** 257 (1937).
9. J. Arnarp and J. Lönngren, *Acta Chem. Scand. Ser. B*, **32,** 465 (1978).
10. P. G. M. Wuts and S. R. Putt, unpublished results.
11. A. E. Greene, C. L. Drian, and P. Crabbe, *J. Am. Chem. Soc.*, **102,** 7583 (1980).
12. J. T. A. Reuvers and A. de Groot, *J. Org. Chem.*, **51,** 4594 (1986).
13. M. E. Jung and S. M. Kaas, *Tetrahedron Lett.*, **30,** 641 (1989).
14. M. E. Jung and M. A. Lyster, *J. Org. Chem.*, **42,** 3761 (1977).
15. M. E. Jung and T. A. Blumenkopf, *Tetrahedron Lett.*, 3657 (1978); G. A. Olah, A. Husain, B. G. B. Gupta, and S. C. Narang, *Angew. Chem., Int. Ed. Engl.*, **20,** 690 (1981); T.-L. Ho and G. Olah, *Synthesis*, 417 (1977). For a review on the uses of Me_3SiI, see: A. H. Schmidt, *Aldrichimica Acta*, **14,** 31 (1981).
16. H. Niwa, T. Hida, and K. Yamada, *Tetrahedron Lett.*, **22,** 4239 (1981).
17. M. E. Kuehne and J. B. Pitner, *J. Org. Chem.*, **54,** 4553 (1989).
18. H. Shimomura, J. Katsuba, and M. Matsui, *Agric. Biol. Chem.*, **42,** 131 (1978).
19. M. Demuynck, P. De Clercq, and M. Vandewalle, *J. Org. Chem.*, **44,** 4863 (1979); P. A. Grieco, M. Nishizawa, T. Oguri, S. D. Burke, and N. Marinovic, *J. Am. Chem. Soc.*, **99,** 5773 (1977).
20. G. Vidari, S. Ferrino, and P. A. Grieco, *J. Am. Chem. Soc.*, **106,** 3539 (1984).
21. M. Node, H. Hori, and E. Fujita, *J. Chem. Soc., Perkin Trans.*, ***I,*** 2237 (1976).
22. S. Hanessian and Y. Guindon, *Tetrahedron Lett.*, **21,** 2305 (1980); R. S. Glass, *J. Organomet. Chem.*, **61,** 83 (1973); I. Ojima, M. Nihonyangi, and Y. Nagai, *J. Organomet. Chem.*, **50,** C26 (1973).
23. M. V. Bhatt and S. S. El-Morey, *Synthesis*, 1048 (1982).
24. M. Node, K. Nishide, M. Sai, K. Ichikawa, K. Fuji, and E. Fujita, *Chem. Lett.*, 97 (1979).
25. A. Oku, T. Harada, and K. Kita, *Tetrahedron Lett.*, **23,** 681 (1982).
26. B. Ganem and V. R. Small, Jr., *J. Org. Chem.*, **39,** 3728 (1974).
27. T. Tsunoda, M. Amaike, U. S. F. Tambunan, Y. Fujise, S. Ito, and M. Kodama, *Tetrahedron Lett.*, **28,** 2537 (1987).

Substituted Methyl Ethers

2. Methoxymethyl Ether (MOM Ether): CH_3OCH_2–OR (Chart 1)

Formation

1. CH_3OCH_2Cl, NaH, THF, 80% yield.[1]
2. CH_3OCH_2Cl, *i*-Pr_2NEt, 0°, 1 h → 25°, 8 h, 86% yield.[2] This is the most commonly employed procedure for introduction of the MOM group. The reagent chloromethyl methyl ether is reported to be carcinogenic.
3. $CH_2(OMe)_2$, cat. P_2O_5, $CHCl_3$, 25°, 30 min, 95% yield.[3]
4. $CH_2(OMe)_2$, Me_3SiI or $CH_2{=}CHCH_2SiMe_3$, I_2, 76–95% yield.[4]

5. $CH_2(OMe)_2$, Nafion H.[5]
6. $CH_2(OMe)_2$, TsOH, LiBr, 9 h, rt, 71–100% yield.[6] This method is suitable for the formation of primary, secondary, allylic, and propargylic MOM ethers. Tertiary alcohols fail to give complete reaction. 1,3-Diols give methylene acetals (89% yield).
7. $CH_2(OMe)_2$, $CH_2 = CHCH_2SiMe_3$, Me_3SiOTf, P_2O_5, 93–99% yield.[7] This method was used to protect the 2′-OH of ribonucleosides and deoxyribonucleosides as well as the hydroxyl groups of several other carbohydrates bearing functionality such as esters, amides, and acetonides.
8. $CH_2(OEt)_2$, montmorillonite clay (H^+), 72–80% for nonallylic alcohols, 56% for a propargylic alcohol.[8]
9. Selective formation of MOM ethers has been achieved in a diol system.[9]

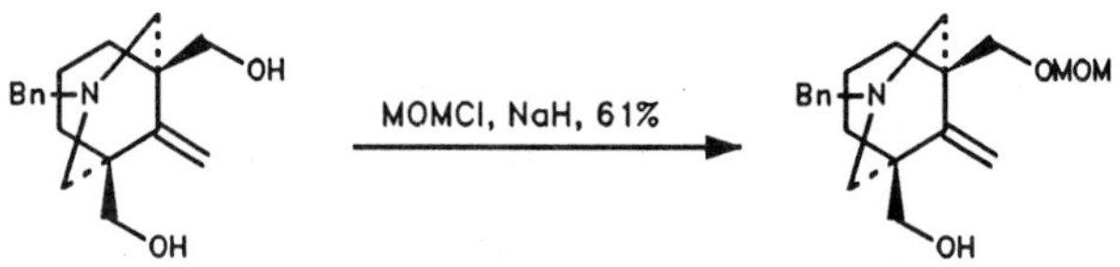

Bn–N, OH, OH — MOMCl, NaH, 61% → Bn–N, OMOM, OH

Mono MOM derivatives of diols can be prepared from the ortho esters by diisobutylaluminum hydride reduction (46–98% yield). In general, the most hindered alcohol is protected.[10]

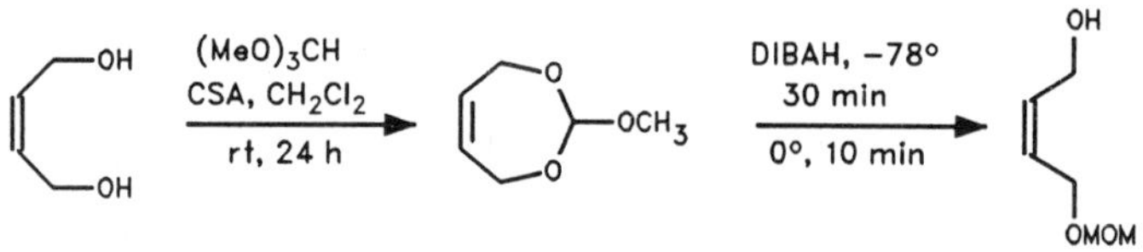

OH, OH — $(MeO)_3CH$, CSA, CH_2Cl_2, rt, 24 h → O, O, OCH_3 — DIBAH, −78° 30 min, 0°, 10 min → OH, OMOM

Cleavage

1. Trace concd. HCl, MeOH, 62°, 15 min.[11]
2. 6 *M* HCl, aq. THF, 50°, 6–8 h, 95% yield.[12] An attempt to cleave the MOM group with acid in the presence of a dimethyl acetal resulted in the cleavage of both groups, probably by intramolecular assistance.[13]
3. 50% AcOH, cat. H_2SO_4, reflux, 10–15 min, 80% yield.[14]
4. PhSH, $BF_3 \cdot Et_2O$, 98% yield.[15]
5. $Ph_3C^+BF_4^-$, CH_2Cl_2, 25°.[16]
6. Pyridinium *p*-toluenesulfonate, *t*-BuOH or 2-butanone, reflux, 80–99% yield.[17] This method is recommended for allylic alcohols. MEM ethers are also cleaved under these conditions.
7. 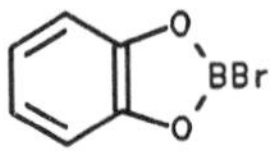 O, O, BBr

Catechol boron halides, particularly the bromide, are effective reagents for the cleavage of MOM ethers. The bromide also cleaves the following groups in the order: MOMOR ≈ MEMOR > *t*-BOC > Cbz ≈ *t*-BuOR > BnOR > allylOR > *t*-BuO_2CR ≈ 2° alkylOR > BnO_2CR > 1° alkylOR >> $alkylO_2CR$. The *t*-butyldimethylsilyl (TBDMS) group is stable to this reagent. The chloride is less reactive and thus may be more useful for achieving selectivity in multifunctional substrates. Yields are generally > 83%.[18]

8. $(i\text{-PrS})_2BBr$, MeOH, 94% yield.[19] This method has the advantage that 1,2- and 1,3-diols do not give formyl acetals as is sometimes the case in cleaving MOM groups with neighboring hydroxyl groups.[20] The reagent also cleaves MEM groups and under basic conditions affords the *i*-$PrSCH_2OR$ derivatives.

TIPSO ... OH OMOM ... CO_2Me —[$(i\text{-PrS})_2BBr$, MeOH, 94%]→ OH OH

TIPS = triisopropylsilyl

9. Me_2BBr, CH_2Cl_2, −78°, then $NaHCO_3/H_2O$, 87–95% yield.[21] This reagent also cleaves the MEM, MTM, and acetal groups. Esters are stable to this reagent.
10. Me_3SiBr, CH_2Cl_2, 0°, 8–9 h, 80–97% yield.[22] This reagent also cleaves the acetonide, THP, trityl, and *t*-$BuMe_2Si$ groups. Esters, methyl and benzyl ethers, *t*-butyldiphenylsilyl ethers, and amides are reported to be stable.

OR ... OCH_2CCl_3 ... OTBDPS, R = MOM —[TMSBr, −30°, 25 min]→ R = H

11. CF_3COOH, CH_2Cl_2, > 85% yield.[23]

MOMO ... OBn —[TFA, CH_2Cl_2, >85%]→ HO ... OBn

12. $LiBF_4$, CH_3CN, H_2O, 72°, 100% yield.[24]

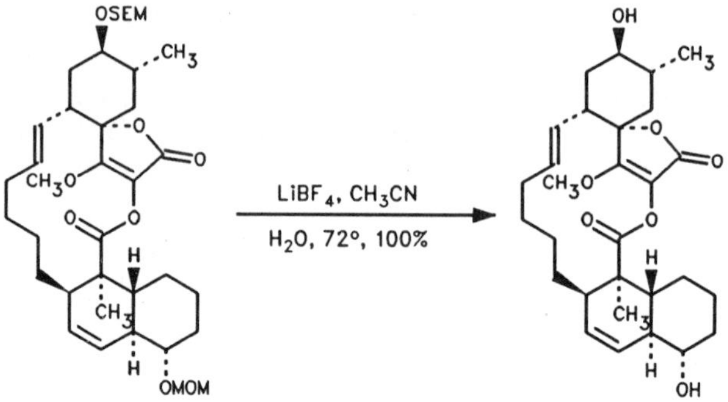

1. A. F. Kluge, K. G. Untch, and J. H. Fried, *J. Am. Chem. Soc.*, **94,** 7827 (1972).
2. G. Stork and T. Takahashi, *J. Am. Chem. Soc.*, **99,** 1275 (1977).
3. K. Fuji, S. Nakano, and E. Fujita, *Synthesis*, 276 (1975).
4. G. A. Olah, A. Husain, and S. C. Narang, *Synthesis*, 896 (1983).
5. G. A. Olah, A. Husain, B. G. B. Gupta, and S. C. Narang, *Synthesis*, 471 (1981).
6. J.-L. Gras, Y.-Y. K. W. Chang, and A. Guerin, *Synthesis*, 74 (1985).
7. S. Nishino and Y. Ishido, *J. Carbohydr. Chem.*, **5,** 313 (1986).
8. U. A. Schaper, *Synthesis*, 794 (1981).
9. M. Ihara, M. Suzuki, K. Fukumoto, T. Kametani, and C. Kabuto, *J. Am. Chem. Soc.*, **110,** 1963 (1988).
10. M. Takasu, Y. Naruse, and H. Yamamoto, *Tetrahedron Lett.*, **29,** 1947 (1988)
11. J. Auerbach and S. M. Weinreb, *J. Chem. Soc., Chem. Commun.*, 298 (1974).
12. A. I. Meyers, J. L. Durandetta, and R. Munavu, *J. Org. Chem.*, **40,** 2025 (1975).
13. M. L. Bremmer, N. A. Khatri, and S. M. Weinreb, *J. Org. Chem.*, **48,** 3661 (1983).
14. F. B. Laforge, *J. Am. Chem. Soc.*, **55,** 3040 (1933).
15. G. R. Kieczykowski and R. H. Schlessinger, *J. Am. Chem. Soc.*, **100,** 1938 (1978).
16. T. Nakata, G. Schmid, B. Vranesic, M. Okigawa, T. Smith-Palmer, and Y. Kishi, *J. Am. Chem. Soc.*, **100,** 2933 (1978).
17. H. Monti, G. Léandri, M. Klos-Ringquet, and C. Corriol, *Synth. Commun.*, **13,** 1021 (1983).
18. R. K. Boeckman, Jr., and J. C. Potenza, *Tetrahedron Lett.*, **26,** 1411 (1985).
19. E. J. Corey, D. H. Hua, and S. P. Seitz, *Tetrahedron Lett.*, **25,** 3 (1984).
20. B. C. Barot and H. W. Pinnick, *J. Org. Chem.*, **46,** 2981 (1981).
21. Y. Guindon, H. E. Morton, and C. Yoakim, *Tetrahedron Lett.*, **24,** 3969 (1983).
22. S. Hanessian, D. Delorme, and Y. Dufresne, *Tetrahedron Lett.*, **25,** 2515 (1984). For *in situ* prepared TMSBr, see: R. B. Woodward and 48 co-workers, *J. Am. Chem. Soc.*, **103,** 3213 (note 2) (1981).
23. R. B. Woodward and 48 co-workers, *J. Am. Chem. Soc.*, **103,** 3210 (1981).
24. R. E. Ireland and M. D. Varney, *J. Org. Chem.*, **51,** 635 (1986).

3. Methylthiomethyl Ether (MTM Ether): CH_3SCH_2OR (Chart 1)

4. *t*-Butylthiomethyl Ether: *t*-$BuSCH_2OR$

Methylthiomethyl ethers are quite stable to acidic conditions. Most ethers and 1,3-dithianes are stable to the neutral mercuric chloride used to remove the MTM group. One problem with the MTM group is that it is sometimes difficult to introduce.

Formation

1. NaH, DME, CH_3SCH_2Cl, NaI, 0°, 1 h → 25°, 1.5 h, >86% yield.[1]
2. CH_3SCH_2I, DMSO, Ac_2O, 20°, 12 h, 80–90% yield.[2]
3. DMSO, Ac_2O, AcOH, 20°, 1–2 days, 80%.[3]
4. CH_3SCH_2Cl, $AgNO_3$, Et_3N, benzene, 22–80°, 4–24 h, 60–80% yield.[4]
5. DMSO, molybdenum peroxide, benzene, reflux, 7–20 h, ≈60% yield.[5] This method was used to monoprotect 1,2-diols. The method is not general because oxidation to α-hydroxy ketones and diketones occurs with some substrates. On the basis of the mechanism and the results it would appear that overoxidation has a strong conformational dependence.
6. MTM ethers can be prepared from MEM and MOM ethers by treatment with Me_2BBr to form the bromomethyl ether that is trapped with MeSH and $(i\text{-}Pr)_2NEt$. This method may have some advantage since the preparation of MTM ethers directly is not always simple.[6]
7. CH_3SCH_3, CH_3CN, $(PhCO)_2O_2$, 0°, 2 h, 75–95% yield.[7] Acetonides, THP ethers, alkenes, ketones, and epoxides all survive these conditions.
8. $(COCl)_2$, DMSO, −78° → −50°; Et_3N, −78° → −15°.[8]

Cleavage

1. $HgCl_2$, CH_3CN, H_2O, 25°, 1–2 h, 88–95% yield.[1] If 2-methoxyethanol is substituted for water, the MTM ether is converted to a MEM ether. Similarly, substitution with methanol affords a MOM ether.[9] If the MTM ether has an adjacent hydroxyl, it is possible to form the formylidene acetal as a byproduct of cleavage.[10]
2. $AgNO_3$, THF, H_2O, 2,6-lutidine, 25°, 45 min, 88–95% yield.[1]
3. MeI, acetone, H_2O, $NaHCO_3$, heat for a few hours, 80–95% yield.[3]
4. Electrolysis: applied voltage = 10 V, AcONa, AcOH; K_2CO_3, MeOH, H_2O, 80–95% yield.[11]
5. Me_3SiCl, Ac_2O, 90%.[12] These conditions were used to cleave the related *t*-butylthiomethyl ether.

t-BuSCH2O — TMSCl, Ac2O or Hg(OAc)2, CH3CN → AcOCH2O

6. $Ph_3C^+BF_4^-$, CH_2Cl_2, 5–30 min, 80–95% yield.[13]

OMTM — $Ph_3C^+BF_4^-$, CH_2Cl_2 → OH

In this case the use of $HgCl_2$, $AgNO_3$, and MeI gave extensive decomposition.

7. $Hg(OTf)_2$, CH_2Cl_2, H_2O, Na_2HPO_4.[14]

1. E. J. Corey and M. G. Bock, *Tetrahedron Lett.*, 3269 (1975).
2. K. Yamada, K. Kato, H. Nagase, and Y. Hirata, *Tetrahedron Lett.*, 65 (1976).
3. P. M. Pojer and S. J. Angyal, *Aust. J. Chem.*, **31,** 1031 (1978).
4. K. Suzuki, J. Inanaga, and M. Yamaguchi, *Chem. Lett.*, 1277 (1979).
5. Y. Masuyama, M. Usukura, and Y. Kurusu, *Chem. Lett.*, 1951 (1982).
6. H. E. Morton and Y. Guindon, *J. Org. Chem.*, **50,** 5379 (1985).
7. J. C. Medina, M. Salomon, and K. S. Kyler, *Tetrahedron Lett.*, **29,** 3773 (1988).
8. D. R. Williams, F. D. Klinger, and V. Dabral, *Tetrahedron Lett.*, **29,** 3415 (1988).
9. P. K. Chowdhury, D. N. Sarma, and R. P. Sharma, *Chem. Ind. (London)*, 803 (1984).
10. M. P. Wachter and R. E. Adams, *Synth. Commun.*, **10,** 111 (1980).
11. T. Mandai, H. Yasunaga, M. Kawada, and J. Otera, *Chem. Lett.*, 715 (1984).
12. D. N. Sarma, N. C. Barua, and R. P. Sharma, *Chem. Ind. (London)*, 223 (1984).
13. P. K. Chowdhury, R. P. Sharma, and J. N. Baruah, *Tetrahedron Lett.*, **24,** 4485 (1983).
14. G. E. Keck, E. P. Boden, and M. R. Wiley, *J. Org. Chem.*, **54,** 896 (1989).

5. (Phenyldimethylsilyl)methoxymethyl Ether (SMOM–OR): $C_6H_5(CH_3)_2SiCH_2OCH_2OR$

Formation

1. SMOMCl, *i*-$PrEt_2N$, CH_3CN, 3 h, 40°, 87–91% yield.[1]

Cleavage

1. AcOOH, KBr, AcOH, NaOAc, 1.5 h, 20°, 82–92% yield.[1] The SMOM group is stable to $Bu_4N^+F^-$; NaOMe/MeOH; 4 *N* NaOH/dioxane/methanol; *N*-iodosuccinimide, cat. trifluoromethanesulfonic acid.

1. G. J. P. H. Boons, C. J. J. Elie, G. A. van der Marel, and J. H. van Boom, *Tetrahedron Lett.*, **31,** 2197 (1990).

6. Benzyloxymethyl Ether (BOM–OR): $PhCH_2OCH_2OR$

Formation

1. $PhCH_2OCH_2Cl$, $(i\text{-}Pr)_2NEt$, 10–20°, 12 h, 95% yield.[1]

Cleavage

1. Na, NH_3, EtOH.[1]
2. Li, NH_3.[2]
3. PhSH, $BF_3 \cdot Et_2O$, CH_2Cl_2, −78°, 95% yield.[3]

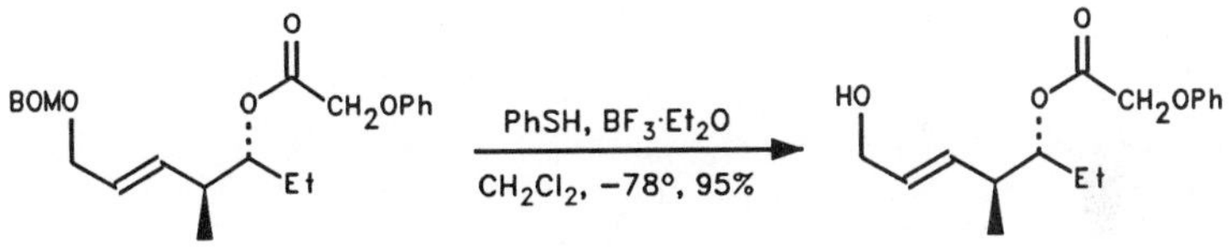

BOMO … CH_2OPh, Et —[PhSH, $BF_3 \cdot Et_2O$ / CH_2Cl_2, −78°, 95%]→ HO … CH_2OPh, Et

4. H_2, 1 atm, Pd–C, EtOAc–hexane, 68% yield.[4]

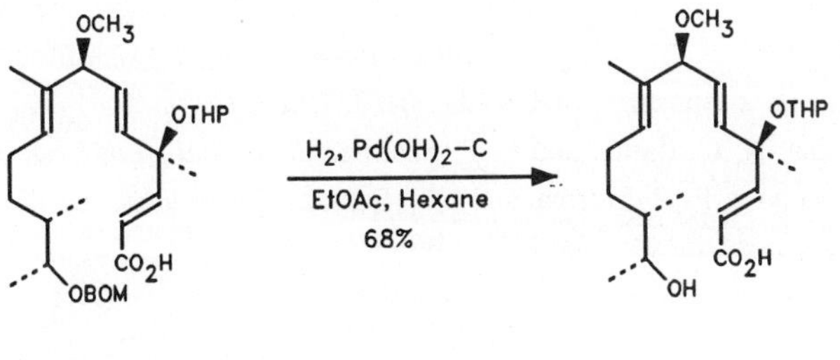

OCH_3, OTHP, CO_2H, OBOM —[H_2, $Pd(OH)_2$–C / EtOAc, Hexane / 68%]→ OCH_3, OTHP, CO_2H, OH

5. H_2, 1 atm, 10% Pd–C, 0.01 *N* $HClO_4$, in 80% THF/H_2O, 25°.[5]

1. G. Stork and M. Isobe, *J. Am. Chem. Soc.*, **97,** 6260 (1975).
2. H. Nagaoka, W. Rutsch, G. Schmid, H. Iio, M. R. Johnson, and Y. Kishi, *J. Am. Chem. Soc.*, **102,** 7962 (1980).

3. K. Suzuki, K. Tomooka, E. Katayama, T. Matsumoto, and G.-P. C. Tsuchihashi, *J. Am. Chem. Soc.*, **108,** 5221 (1986).
4. D. Tanner and P. Somfai, *Tetrahedron*, **43,** 4395 (1987).
5. D. A. Evans, S. L. Bender, and J. Morris, *J. Am. Chem. Soc.*, **110,** 2506 (1988).

7. *p*-Methoxybenzyloxymethyl Ether (PMBM–OR): *p*-$MeOC_6H_4CH_2OCH_2OR$

Formation

1. *p*-$MeOC_6H_4CH_2OCH_2Cl$, (*i*-$Pr)_2NEt$ (DIPEA), CH_2Cl_2, 78–100% yield.[1]

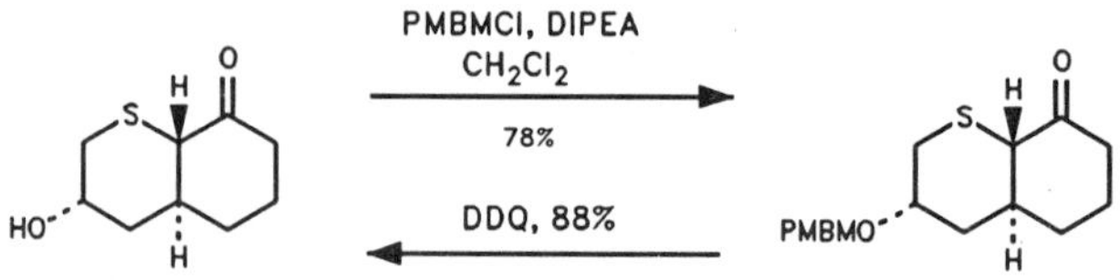

PMBMCl, DIPEA
CH_2Cl_2
78%
DDQ, 88%

Cleavage

1. DDQ, H_2O, rt, 1–10 h, 63–96% yield.[1]
2. 3:1 THF-6 *M* HCl, 50°, 6 h.[1]

1. A. P. Kozikowski and J.-P. Wu, *Tetrahedron Lett.*, **28,** 5125 (1987).

8. (4-Methoxyphenoxy)methyl Ether (*p*-AOM–OR), (*p*-Anisyloxymethyl Ether): $ROCH_2OC_6H_4$-4-OCH_3

Formation[1]

1. *p*-AOMCl, $PhCH_2N^+Et_3Cl^-$, CH_3CN, 50% NaOH, rt, 46–91% yield.
2. *p*-AOMCl, DIPEA, CH_2Cl_2, reflux.
3. *p*-AOMCl, DMF, 18-crown-6, K_2CO_3, rt.

Cleavage

1. CAN, CH_3CN, H_2O, 0°, 0.5 h, 60–98% yield.[1] In some cases the addition of pyridine improves the yields.

1. Y. Masaki, I. Iwata, I. Mukai, H. Oda, and H. Nagashima, *Chem. Lett.*, 659 (1989).

9. Guaiacolmethyl Ether (GUM-OR): 2-$MeOC_6H_4OCH_2OR$

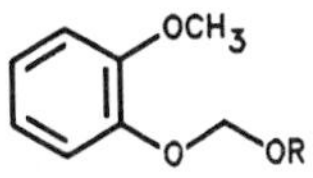

Formation/Cleavage[1]

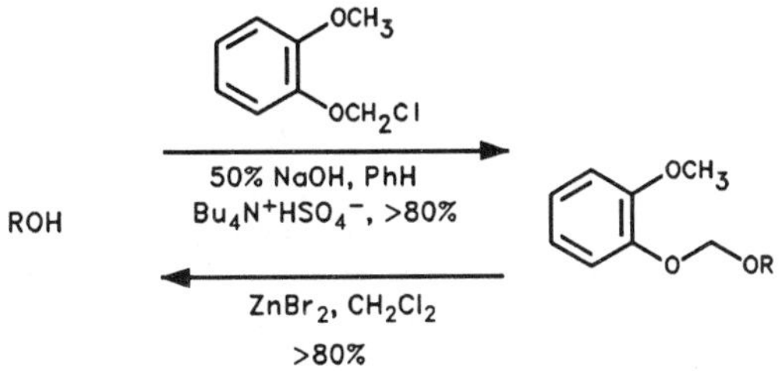

ROH → 2-$MeOC_6H_4OCH_2Cl$, 50% NaOH, PhH, $Bu_4N^+HSO_4^-$, >80% → 2-$MeOC_6H_4OCH_2OR$

← $ZnBr_2$, CH_2Cl_2, >80%

It is possible to introduce this group selectively onto a primary alcohol in the presence of a secondary alcohol. The derivative is stable to $KMnO_4$, *m*-chloroperoxybenzoic acid, $LiAlH_4$, and CrO_3–Pyr. Since this derivative is similar to the *p*-methoxyphenyl ether it should also be possible to remove it oxidatively. The GUM ethers are less stable than the MEM ethers in acid but have stability comparable to that of the SEM ethers. It is possible to remove the GUM ether in the presence of a MEM ether.

1. B. Loubinouz, G. Coudert, and G. Guillaumet, *Tetrahedron Lett.*, **22,** 1973 (1981).

10. *t*-Butoxymethyl Ether: *t*-$BuOCH_2OR$

Formation

1. *t*-$BuOCH_2Cl$,[1] Et_3N, −20° → 20°, 3 h, 54–80% yield.[2]

Cleavage

1. CF_3COOH, H_2O, 20°, 48 h, 85–90% yield.[2] The *t*-butoxymethyl ether is stable to hot glacial acetic acid; aqueous acetic acid, 20°; and anhydrous trifluoroacetic acid.

1. For an improved preparation of this reagent, see: J. H. Jones, D. W. Thomas, R. M. Thomas, and M. E. Wood, *Synth. Commun.*, **16,** 1607 (1986).
2. H. W. Pinnick and N. H. Lajis, *J. Org. Chem.*, **43,** 3964 (1978).

11. 4-Pentenyloxymethyl Ether (POM–OR): $CH_2{=}CHCH_2CH_2CH_2OCH_2OR$

Formation

1. POMCl, $(i\text{-Pr})_2EtN$, CH_2Cl_2.[1] The related pentenyl glycosides (an example is shown below), prepared by the usual methods, were used to protect the anomeric center.[2]

Cleavage

1. NBS, CH_3CN, H_2O, 62–90% yield.[1,2] The POM group has been selectively removed in the presence of an ethoxyethyl ether, TBDMS ether, benzyl ether, *p*-methoxybenzyl ether, an acetate, and an allyl ether. Because the hydrolysis of a pentenyl 2-acetoxyglycoside was so much slower than a pentenyl 2-benzyloxyglycoside, the 2-benzyl derivative could be cleaved selectively in the presence of the 2-acetoxy derivative.[3] The POM group is stable to 75% AcOH, but is cleaved by 5% HCl.

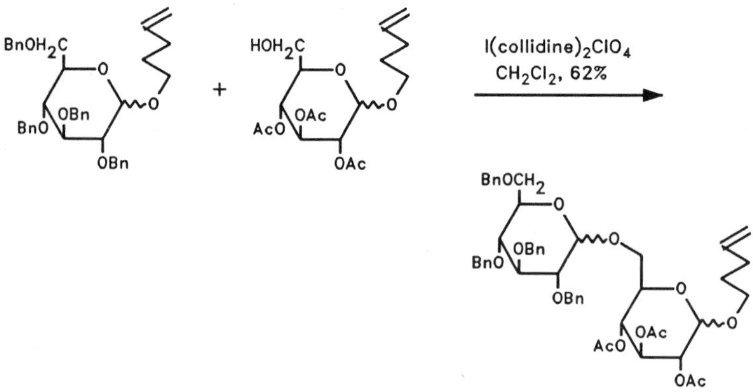

BnOH₂C / BnO / OBn / OBn + HOH₂C / AcO / OAc / OAc — $I(collidine)_2ClO_4$, CH_2Cl_2, 62% → BnOCH₂ / BnO / OBn / OBn / AcO / OAc / OAc

1. Z. Wu, D. R. Mootoo, and B. Fraser-Reid, *Tetrahedron Lett.*, **29,** 6549 (1988).
2. D. R. Mootoo, V. Date, and B. Fraser-Reid, *J. Am. Chem. Soc.*, **110,** 2662 (1988).
3. D. R. Mootoo, P. Konradsson, U. Udodong, and B. Fraser-Reid, *J. Am. Chem. Soc.*, **110,** 5583 (1988).

12. Siloxymethyl Ether: RR'_2SiOCH_2OR'', R′ = Me, R = *t*-Bu; R = Thexyl, R′ = Me; R = *t*-Bu, R′ = Ph

Formation

1. RR'_2SiOCH_2Cl, diisopropylethylamine, CH_2Cl_2, 73–92% yield.[1]

Cleavage

1. $Bu_4N^+F^-$, THF, 70–80% yield.[1]
2. $Et_4N^+F^-$, CH_3CN, rt, 64–75% yield.[1]
3. AcOH, H_2O.[1]

1. L. L. Gundersen, T. Benneche, and K. Undheim, *Acta Chem. Scand.*, **43**, 706 (1989).

13. 2-Methoxyethoxymethyl Ether (MEM Ether): $CH_3OCH_2CH_2OCH_2OR$ (Chart 1)

Formation[1]

1. NaH or KH, MEMCl, THF or DME, 0°, 10–60 min, >95% yield.
2. $MEMN^+Et_3Cl^-$, CH_3CN, reflux, 30 min, >90% yield.
3. MEMCl, $(i\text{-Pr})_2NEt$ (DIPEA), CH_2Cl_2, 25°, 3 h, quant.
4. The MEM group has been introduced on one of two sterically similar but electronically different alcohols in a 1,2-diol.[2]

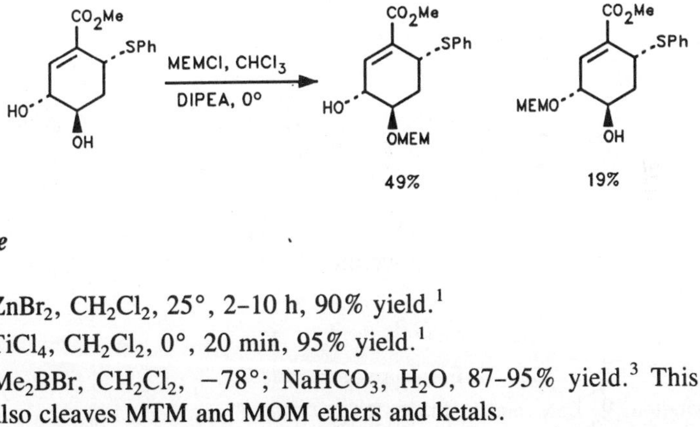

Cleavage

1. $ZnBr_2$, CH_2Cl_2, 25°, 2–10 h, 90% yield.[1]
2. $TiCl_4$, CH_2Cl_2, 0°, 20 min, 95% yield.[1]
3. Me_2BBr, CH_2Cl_2, −78°; $NaHCO_3$, H_2O, 87–95% yield.[3] This method also cleaves MTM and MOM ethers and ketals.
4. $(i\text{-PrS})_2BBr$, DMAP; K_2CO_3, H_2O.[4] In this case the MEM ether is converted into the $i\text{-PrSCH}_2$—ether, which can be cleaved using the same conditions used to cleave the MTM ether.
5. Pyridinium *p*-toluenesulfonate, *t*-BuOH or 2-butanone, heat, 80–99% yield.[5] This method also cleaves the MOM ether and has the advantage that it cleanly cleaves allylic ethers that could not be cleaved by Corey's original procedure.
6. Me_3SiCl, NaI, CH_3CN, −20°, 79%.[6] Allylic and benzylic ethers tend to form some iodide as a byproduct, but less iodide is formed than when Me_3SiI is used directly.

7. 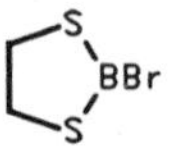 2 eq., −78°, CH_2Cl_2.[7]

Benzyl, allyl, methyl, THP, TBDMS, and TBDPS ethers are all stable to these conditions. A primary MEM group could be selectively removed in the presence of a hindered secondary MEM group.

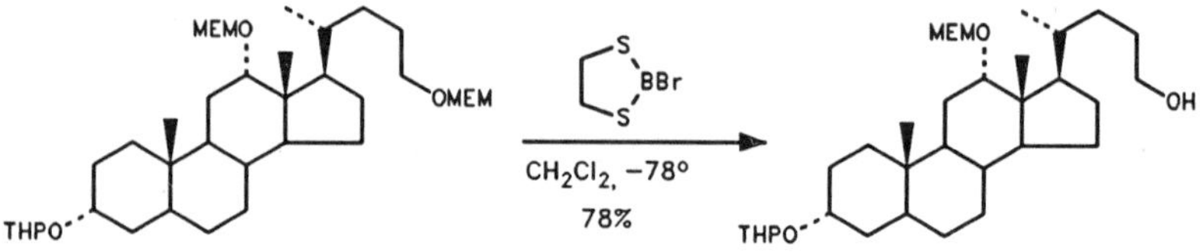

8. HBF_4, CH_2Cl_2, 0°, 3 h, 50–60% yield.[8]
9. In a study of the deprotection of the MEM ethers of hydroxyproline and serine derivatives it was found that the MEM group was stable to conditions that normally cleave the *t*-butyl and BOC groups (CF_3COOH, CH_2Cl_2, 1:1 (v:v)). The MEM group was also stable to 0.2 *N* HCl but it is not stable to 2.0 *N* HCl or HBr–AcOH.[9]

Removal time in TFA/CH_2Cl_2 (v/v)

	1:4	1:1	1:0
Z-Hyp(*t*-Bu)–ONb	45 min	15 min	5 min
Z-Hyp(MEM)OMe	10 h	6 h	2 h

Hyp = hydroxyproline, Nb = 4-nitrobenzoate.

10. (a) *n*-BuLi, THF; (b) $Hg(OAc)_2$, H_2O, THF, 81% yield.[10] In this case conventional methods to remove the MEM group were unsuccessful.

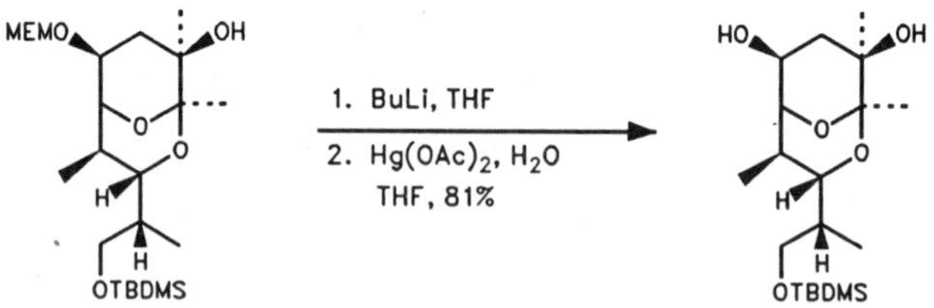

11.

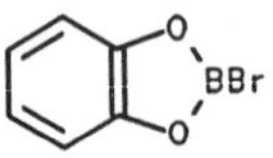

For a further discussion of this reagent, refer to the section on MOM ethers.[11]

12. Ph_2BBr, CH_2Cl_2, −78°, 71% yield.[12]

Ph$_2$BBr, CH$_2$Cl$_2$ −78°, 71%

1. E. J. Corey, J.-L. Gras, and P. Ulrich, *Tetrahedron Lett.*, 809 (1976).
2. G. H. Posner, A. Haces, W. Harrison, and C. M. Kinter, *J. Org. Chem.*, **52,** 4836 (1987).
3. Y. Quindon, H. E. Morton, and C. Yoakim, *Tetrahedron Lett.*, **24,** 3969 (1983).
4. E. J. Corey, D. H. Hua, and S. P. Seitz, *Tetrahedron Lett.*, **25,** 3 (1984).
5. H. Monti, G. Léandri, M. Klos-Ringuet, and C. Corriol, *Synth. Commun.*, **13,** 1021 (1983).
6. J. H. Rigby and J. Z. Wilson, *Tetrahedron Lett.*, **25,** 1429 (1984).
7. D. R. Williams and S. Sakdarat, *Tetrahedron Lett.*, **24,** 3965 (1983).
8. N. Ikota and B. Ganem, *J. Chem. Soc., Chem. Commun.*, 869 (1978).
9. D. Vadolas, H. P. Germann, S. Thakur, W. Keller, and E. Heidemann, *Int. J. Pept. Protein Res.*, **25,** 554 (1985).
10. R. E. Ireland, P. G. M. Wuts, and B. Ernst, *J. Am. Chem. Soc.*, **103,** 3205 (1981).
11. R. K. Boeckman, Jr., and J. C. Potenza, *Tetrahedron Lett.*, **26,** 1411 (1985).
12. M. Shibasaki, Y. Ishida, and N. Okabe, *Tetrahedron Lett.*, **26,** 2217 (1985).

14. 2,2,2-Trichloroethoxymethyl Ether: $Cl_3CCH_2OCH_2OR$

Formation

1. $Cl_3CCH_2OCH_2Cl$, NaH or KH, LiI, THF, 5 h, 70–90% yield.[1]
2. $Cl_3CCH_2OCH_2Cl$, $(i\text{-Pr})_2NEt$, CH_2Cl_2, 30–60% yield.[1]

Cleavage

1. Zn–Cu or Zn–Ag, MeOH, reflux, 97%.[1]
2. Zn, MeOH, Et_3N, AcOH, reflux 4 h, 90–100%.[1]
3. Li, NH_3.[1]

1. R. M. Jacobson and J. W. Clader, *Synth. Commun.*, **9,** 57 (1979).

15. Bis(2-chloroethoxy)methyl Ether: $ROCH(OCH_2CH_2Cl)_2$ (Chart 1)

The mixed ortho ester formed from tri(2-chloroethyl) orthoformate (100°, 10 min–2 h, 76% yield) is more stable to acid than is the unsubstituted derivative, but can be cleaved with 80% AcOH (20°, 1 h).[1]

1. T. Hata and J. Azizian, *Tetrahedron Lett.*, 4443 (1969).

16. 2-(Trimethylsilyl)ethoxymethyl Ether (SEMOR): $Me_3SiCH_2CH_2OCH_2OR$

Formation

1. $Me_3SiCH_2CH_2OCH_2Cl$, (*i*-Pr)$_2$NEt (DIPEA), CH_2Cl_2, 35–40°, 1–5 h, 86–100% yield.[1]
2. The above conditions failed in this example unless $Bu_4N^+I^-$ was added to prepare SEMI *in situ*.[2]

H OTBDMS S S H CH_2OBn OR O O R = H — SEMCl, CH_2Cl_2 / 1.1 eq $Bu_4N^+I^-$ DIPEA → R = SEM

Cleavage

SEM ethers are stable to the acidic conditions (AcOH, H_2O, THF, 45°, 7 h) that are used to cleave tetrahydropyranyl and *t*-butyldimethylsilyl ethers.

1. $Bu_4N^+F^-$, THF or HMPA, 45°, 8–12 h, 85–95% yield.[1,3] The cleavage of 2-(trimethylsilyl)ethyl glycosides (an example is shown in 2 below) is included here because they are functionally equivalent to the SEM group. They can be prepared by oxymercuration of a glycal with $Hg(OAc)_2$ and $TMSCH_2CH_2OH$, the reaction of a glycosyl halide using Koenig–Knorr conditions, by a Fisher glycosidation and by a glycal rearrangement.[4]
2. TFA, CH_2Cl_2 (2 : 1, v : v), 0°, 30 min, 93% yield.[4]

OBn BnO OTBDMS $TMSCH_2CH_2O$ O OBn — TFA, CH_2Cl_2 / 0°, 30 min / 93% → OBn BnO OTBDMS HO O OBn

The 4,6-*O*-benzylidene group is also cleaved under these conditions, but the anomeric linkage between sugars is not affected. Anomeric trimethylsilylethyl groups are also cleaved with $BF_3 \cdot Et_2O$[5] or $Ac_2O/FeCl_3$ (this reagent also cleaves the BOM group).[6]

3. $LiBF_4$, CH_3CN, 70°, 3–8 h, 81–90% yield.[7] This system of reagents also cleaves benzylidene acetals. Conventional reagents failed to cleave these glycosides. It is interesting to note that the β-anomers are cleaved more rapidly than the α-anomers and that the furanoside derivatives are not cleaved.
4. CsF, DMF, 130°, >89% yield.[8]
5. $Bu_4N^+F^-$, DMPU, 4 Å molecular sieves, 45–80°, 80–95% yield.[9] These conditions were especially effective in cleaving tertiary SEM derivatives and avoid the use of the toxic HMPA.

1. B. H. Lipshutz and J. J. Tegram, *Tetrahedron Lett.*, **21,** 3343 (1980).
2. B. H. Lipshutz, R. Moretti, and R. Crow, *Tetrahedron Lett.*, **30,** 15 (1989).
3. T. Kan, M. Hashimoto, M. Yanagiya, and H. Shirahama, *Tetrahedron Lett.*, **29,** 5417 (1988).
4. K. Jansson, T. Frejd, J. Kihlberg, and G. Magnusson, *Tetrahedron Lett.*, **29,** 361 (1988). For another case, see: R. H. Schlessinger, M. A. Poss, and S. Richardson, *J. Am. Chem. Soc.*, **108,** 3112 (1986).
5. A. Hasagawa, Y. Ito, H. Ishida, and M. Kiso, *J. Carbohydr. Chem.*, **8,** 125 (1989); K. Jansson, T. Frejd, J. Kihlberg, and G. Magnusson, *Tetrahedron Lett.*, **27,** 753 (1986).
6. K. P. R. Kartha, M. Kiso, and A. Hasegawa, *J. Carbohydr. Chem.*, **8,** 675 (1989).
7. B. H. Lipshutz, J. J. Pegram, and M. C. Morey, *Tetrahedron Lett.*, **22,** 4603 (1981).
8. K. Suzuki, T. Matsumoto, K. Tomooka, K. Matsumoto, and G.-I. Tsuchihashi, *Chem. Lett.*, 113 (1987).
9. B. H. Lipshutz and T. A. Miller, *Tetrahedron Lett.*, **30,** 7149 (1989).

17. Tetrahydropyranyl Ether (THP–OR) (Chart 1):

The introduction of a THP ether onto a chiral molecule results in the formation of diastereomers because of the additional stereogenic center present in the tetrahydropyran ring (which can make the interpretation of NMR spectra somewhat troublesome at times). Even so, this is one of the most widely used protective groups employed in chemical synthesis because of its low cost, the ease of its installation, its general stability to most nonacidic reagents, and the ease with which it can be removed.

Formation

1. Dihydropyran, TsOH, CH_2Cl_2, 20°, 1.5 h, 100% yield.[1]
2. Pyridinium *p*-toluenesulfonate (PPTS), dihydropyran, CH_2Cl_2, 20°, 4 h, 94–

100% yield.[2] The lower acidity of PPTS makes this a very mild method that is highly compatible with most functional groups.

3. Reillex 425·HCl, dihydropyran, 86°, 1.5 h, 84–98% yield.[3] The Reillex resin is a macroreticular polyvinylpyridine resin and is thus an insoluble form of the PPTS catalyst.
4. Amberlyst H-15 (SO_3H ion-exchange resin), dihydropyran, hexane, 1–2 h, 95% yield.[4]
5. Dihydropyran, K-10 clay, CH_2Cl_2, rt, 63–95% yield.[5]
6. Dihydropyran, $(TMSO)_2SO_2$, CH_2Cl_2, 92–100% yield.[6]
7. Dihydropyran, TMSI, CH_2Cl_2, rt, 80–96% yield.[7]
8. Dihydropyran, $Ph_3P·HBr$, 24 h, CH_2Cl_2, 88% yield.[8]

Cleavage

1. AcOH, THF, H_2O, (4:2:1), 45°, 3.5 h.[1] MEM ethers are stable to these conditions.[9]
2. PPTS, EtOH, (pH 3.0), 55°, 3 h, 95–100% yield.[2]
3. Amberlyst H-15, MeOH, 45°, 1 h, 95% yield.[3] Dowex-50W-X8, 25°, 1 h, MeOH, 99% yield.[10]
4. Boric acid, $EtOCH_2CH_2OH$, 90°, 2 h, 80–95% yield.[11]
5. TsOH, MeOH, 25°, 1 h, 94% yield.[12] TBDPS ethers are not affected by these conditions.[13]
6. $MgBr_2$, Et_2O, rt, 66–95% yield.[14] *t*-Butyldimethylsilyl and MEM ethers are not affected by these conditions, but the MOM ether is slowly cleaved. The THP derivatives of benzylic and tertiary alcohols give bromides.
7. Me_2AlCl, CH_2Cl_2, −25° → rt, 1 h, 89–100% yield.[15]

CO_2Me / OTBDMS / OTHP — Me_2AlCl, CH_2Cl_2, −65° → rt, 1 h, 89% → CO_2Me / OTBDMS / OH

8. MeOH, $(TMSO)_2SO_2$, 10–90 min, 93–100% yield.[6]
9. $(NCSBu_2Sn)_2O$ 1%, THF, H_2O.[16] Acetonides and TMS ethers are also cleaved under these conditions, but TBDMS, MTM, and MOM groups are stable. This catalyst has also been used to effect transesterifications.[17]
10. The THP ether can be converted directly to an acetate by refluxing in AcOH/AcCl (91% yield).[18] These conditions would probably convert other related acetals to acetates as well. The THP group can also be converted through the O–$SnBu_3$ to benzyl, MEM, benzoate, and tosylate groups.[19]
11. MeOH, reagent prepared by heating Bu_2SnO and Bu_3SnPO_4, heat 2 h, 90% yield.[20] This method is effective for primary, secondary, tertiary, benzylic,

and allylic THP derivatives. The MEM group and ketals are inert to this reagent, but TMS and TBDMS ethers are cleaved.

12. $Ph_3P \cdot Br_2$, CH_2Cl_2, $-50° \rightarrow 35°$, 85–94% yield.[21] Ethyl acetals and MOM groups are also cleaved with this reagent, but a THP ether can be selectively cleaved in the presence of a MOM ether.

Explosions have been reported on distillation of compounds containing a tetrahydropyranyl ether after a reaction with B_2H_6/H_2O_2–OH^- and with 40% CH_3CO_3H:

OTHP 1. B_2H_6 2. H_2O_2, NaOH → OTHP O

OTHP 40% CH_3CO_3H → OTHP O

It was thought that the acetal might have reacted with peroxy reagents, forming explosive peroxides. It was suggested that this could also occur with compounds such as tetrahydrofuranyl acetals, 1,3-dioxolanes, and methoxymethyl ethers.[22]

1. K. F. Bernady, M. B. Floyd, J. F. Poletto, and M. J. Weiss, *J. Org. Chem.*, **44,** 1438 (1979).
2. M. Miyashita, A. Yoshikoshi, and P. A. Grieco, *J. Org. Chem.*, **42,** 3772 (1977).
3. R. D. Johnston, C. R. Marston, P. E. Krieger, and G. L. Goe, *Synthesis*, 393 (1988).
4. A. Bongini, G. Cardillo, M. Orena, and S. Sandri, *Synthesis* 618 (1979).
5. S. Hoyer, P. Laszlo, M. Orlović, and E. Polla, *Synthesis*, 655 (1986).
6. Y. Morizawa, I. Mori, T. Hiyama, and H. Nozaki, *Synthesis*, 899 (1981).
7. G. A. Olah, A. Husain, and B. P. Singh, *Synthesis* 703 (1985).
8. V. Bolitt, C. Mioskowski, D.-S. Shin, and J. R. Falck, *Tetrahedron Lett.*, **29,** 4583 (1988).
9. E. J. Corey, R. L. Danheiser, S. Chandrasekaran, P. Siret, G. E. Keck, and J.-L. Gras, *J. Am. Chem. Soc.*, **100,** 8031 (1978).
10. R. Beier and B. P. Mundy, *Synth. Commun.*, **9,** 271 (1979).
11. J. Gigg and R. Gigg, *J. Chem. Soc. C*, 431 (1967).
12. E. J. Corey, H. Niwa, and J. Knolle, *J. Am. Chem. Soc.*, **100,** 1942 (1978).
13. A. B. Shenvi and H. Gerlach, *Helv. Chim. Acta*, **63,** 2426 (1980).
14. S. Kim and J. H. Park, *Tetrahedron Lett*, **28,** 439 (1987).
15. Y. Ogawa and M. Shibasaki, *Tetrahedron Lett.*, **25,** 663 (1984).
16. J. Otera and H. Nozaki, *Tetrahedron Lett.*, **27,** 5743 (1986).
17. J. Otera, T. Yano, A. Kawabata, and H. Nozaki, *Tetrahedron Lett.*, **27,** 2383 (1986).

18. M. Jacobson, R. E. Redfern, W. A. Jones, and M. H. Aldridge, *Science*, **170,** 543 (1970); T. Bakos and I. Vincze, *Synth. Commun.*, **19,** 523 (1989).
19. T. Sato, T. Tada, J. Otera, and H. Nozaki, *Tetrahedron Lett.*, **30,** 1665 (1989).
20. J. Otera, Y. Niibo, S. Chikada, and H. Nozaki, *Synthesis*, 328 (1988).
21. A. Wagner, M.-P. Heitz, and C. Mioskowski, *J. Chem. Soc., Chem. Commun.*, 1619 (1989).
22. A. I. Meyers, S. Schwartzman, G. L. Olson, and H.-C. Cheung, *Tetrahedron Lett.*, 2417 (1976).

18. 3-Bromotetrahydropyranyl Ether: 3-BrTHP–OR

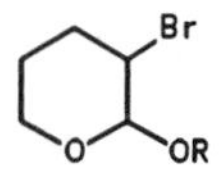

The 3-bromotetrahydropyranyl ether was prepared from a 17-hydroxy steroid and 2,3-dibromopyran (pyridine, benzene, 20°, 24 h); it was cleaved by zinc/ethanol.[1]

1. A. D. Cross and I. T. Harrison, *Steroids*, **6,** 397 (1965).

19. Tetrahydrothiopyranyl Ether (Chart 1):

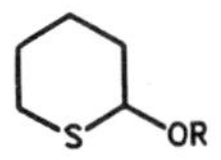

The tetrahydrothiopyranyl ether was prepared from a 3-hydroxy steroid and dihydrothiopyran (CF_3COOH, $CHCl_3$, 35% yield); it can be cleaved under neutral conditions ($AgNO_3$, aq. acetone, 85% yield).[1]

1. L. A. Cohen and J. A. Steele, *J. Org. Chem.*, **31,** 2333 (1966).

20. 1-Methoxycyclohexyl Ether:[1] A

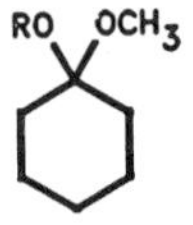

21. 4-Methoxytetrahydropyranyl Ether (MTHP–OR):[1] B (Chart 1)

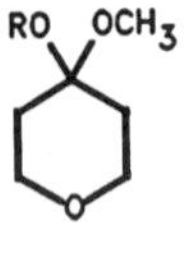

22. 4-Methoxytetrahydrothiopyranyl Ether:[2] C (Chart 1)

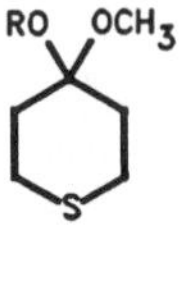

23. 4-Methoxytetrahydrothiopyranyl Ether *S,S*-Dioxide:[2] D

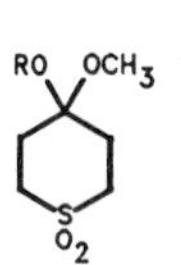

The preceding ethers have been examined as possible protective groups for the 2′-hydroxyl of ribonucleotides. The following rates of hydrolysis were found: A : B : C : D = 1 : 0.025 : 0.005 : 0.002.[3] These acetals can be prepared by the same methods used for the preparation of the THP derivative. Compounds B and C have been prepared from the vinyl ether and TMSCl as a catalyst.[4] Sulfoxide D was prepared from sulfide C by oxidation with m-$ClC_6H_4CO_3H$. An advantage of these ethers is that they do not introduce an additional stereogenic center into the molecules as does the THP group. The 4-methoxytetrahydropyranyl group has been used extensively in nucleoside synthesis but still suffers from excessive acid lability when the 9-phenylxanthen-9-yl group is used to protect 5′-hydroxy functions in ribonucleotides.[5] The recommended conditions for removal of this group are 0.01 *M* HCl at room temperature. Little, if any, use of these groups has been made by the general synthetic community, but the wide range of selectivities observed in their acidic hydrolysis should render them useful for the selective protection of polyfunctional molecules.

24. 1-[(2-Chloro-4-methyl)phenyl]-4-methoxypiperidin-4-yl Ether (CTMP Ether):[6]

This group was designed to have nearly constant acid stability with decreasing pH ($t_{1/2}$ = 80 min at pH = 3.0, $t_{1/2}$ = 33.5 min at pH = 0.5), which is in contrast to the MTHP group that is hydrolyzed faster as the pH is decreased ($t_{1/2}$ = 125 min at pH = 3, $t_{1/2}$ = 0.9 min at pH = 1.0). This group was reported to have excellent compatibility with the conditions used to remove the 9-(9-phenyl)xanthenyl ether (5.5 eq. CF_3COOH, 16.5 eq. pyrrole, CH_2Cl_2, rt, 30 s, 95.5% yield).[3,6,7]

1. C. B. Reese, R. Saffhill, and J. E. Sulston, *J. Am. Chem. Soc.*, **89**, 3366 (1967); *Idem.*, *Tetrahedron*, **26**, 1023 (1970).
2. J. H. van Boom, P. van Deursen, J. Meeuwse, and C. B. Reese, *J. Chem. Soc., Chem. Commun.*, 766 (1972).
3. C. B. Reese, H. T. Serafinowska, and G. Zappia, *Tetrahedron Lett.*, **27**, 2291 (1986).
4. H. C. P. F. Roelen, G. J. Ligtvoet, G. A. Van der Morel, and J. H. Van Boom, *Recl. Trav. Chim. Pays-Bas*, **106**, 545 (1987).

5. C. B. Reese and P. A. Skone, *Nucleic Acids Res.*, **13**, 5215 (1985).
6. For an improved preparation of the reagent, see: C. B. Reese and E. A. Thompson, *J. Chem. Soc., Perkin Trans. I*, 2881 (1988).
7. O. Sakatsume, M. Ohtsuki, H. Takaku, and C. B. Reese, *Nucleic Acid Symp. Ser.*, **20**, 77 (1988).

25. 1,4-Dioxan-2-yl Ether: 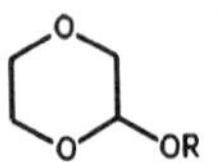

Formation

1. 1,4-Dihydrodioxin, $CuBr_2$, THF, rt, 50–88% yield.[1]

Cleavage

1. 6 *N* HCl, EtOH, reflux, 90% yield for cholesterol.[1] Although a direct stability comparison was not made, this group should be more stable than the THP group for the same reasons that the anomeric ethers of carbohydrates are more stable than their 2-deoxy counterparts.

1. M. Fetizon and I. Hanna, *Synthesis*, 806 (1985).

26. Tetrahydrofuranyl Ether (Chart 1): 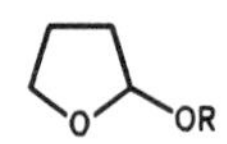

Formation

1. 2-Chlorotetrahydrofuran, Et_3N, 30 min, 82–98% yield.[1] 2-Chlorotetrahydrofuran is readily prepared from THF with SO_2Cl_2 (25°, 0.5 h, 85%).
2. Ph_2CHCO_2-2-tetrahydrofuranyl, 1% TsOH, CCl_4, 20°, 30 min, 90–99% yield.[1,2] The authors report that formation of the THF ether by reaction with 2-chlorotetrahydrofuran avoids a laborious procedure[3] that is required when dihydrofuran is used. In addition, the use of dihydrofuran to protect the 2′-OH of a nucleotide gives low yields (24–42%).[4] The tetrahydrofuranyl ester is reported to be a readily available, stable solid. A tetrahydrofuranyl ether can be cleaved in the presence of a THP ether.[1]
3. THF, $Et_3NHCe(III)(NO_3)_6$, 50–100°, 8 h, 30–98% yield.[5] Hindered alcohols give the lower yields. The method was also used to introduce the THP group with tetrahydropyran.

Cleavage

1. AcOH, H_2O, THF, (3 : 1 : 1), 25°, 30 min, 90% yield.[1]
2. 0.01 *N* HCl, THF (1 : 1), 25°, 10 min, 50% yield.[1]
3. pH 5, 25°, 3 h, 90% yield.[1]

1. C. G. Kruse, F. L. Jonkers, V. Dert, and A. van der Gen, *Recl. Trav. Chim. Pays-Bas*, **98,** 371 (1979).
2. C. G. Kruse, E. K. Poels, F. L. Jonkers, and A. van der Gen, *J. Org. Chem.*, **43,** 3548 (1978).
3. E. L. Eliel, B. E. Nowak, R. A. Daignault, and V. G. Badding, *J. Org. Chem.*, **30,** 2441 (1965).
4. E. Ohtsuka, A. Yamane, and M. Ikehara, *Chem. Pharm. Bull.*, **31,** 1534 (1983).
5. A. M. Maione and A. Romeo, *Synthesis*, 250 (1987).

27. Tetrahydrothiofuranyl Ether (Chart 1):

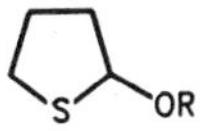

Formation

1. Dihydrothiofuran, $CHCl_3$, CF_3COOH, reflux, 6 days, 75% yield.[1]
2. 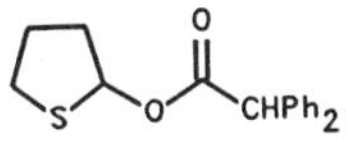 cat. TsOH, $CHCl_3$, 20°, 5 h, 85–95% yield.[2]

Cleavage

1. $AgNO_3$, acetone, H_2O, reflux, 90% yield.[1]
2. $HgCl_2$, CH_3CN, H_2O, 25°, 10 min, quant.[2]

Some of the methods used to cleave methylthiomethyl (MTM) ethers should also be applicable to the cleavage of tetrahydrothiofuranyl ethers.

1. L. A. Cohen and J. A. Steele, *J. Org. Chem.*, **31,** 2333 (1966).
2. C. G. Kruse, E. K. Poels, F. L. Jonkers, and A. van der Gen, *J. Org. Chem.*, **43,** 3548 (1978).

28. 2,3,3a,4,5,6,7,7a-Octahydro-7,8,8-trimethyl-4,7-methanobenzofuran-2-yl Ether: RO–MBF

Formation[1]

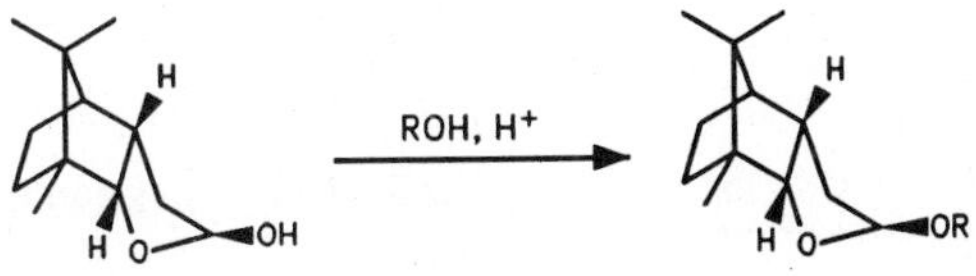

ROH, H^+

The advantage of this ketal is that unlike the THP group, only a single isomer is produced in the derivatization. Conditions used to hydrolyze the THP group can be used to hydrolyze this acetal. This group may also find applications in the resolution of racemic alcohols.

1. C. R. Noe, *Chem. Ber.*, **115,** 1576, 1591 (1982); C. R. Noe, M. Knollmüller, G. Steinbauer, E. Jangg, and H. Völlenkle, *Chem. Ber.*, **121,** 1231 (1988).

Substituted Ethyl Ethers

29. 1-Ethoxyethyl Ether (EE–OR): $ROCH(OC_2H_5)CH_3$ (Chart 1)

Formation

1. Ethyl vinyl ether, HCl (anhyd.).[1]
2. Ethyl vinyl ether, TsOH, 25°, 1 h.[2]
3. Ethyl vinyl ether, pyridinium tosylate (PPTS), CH_2Cl_2, rt, 0.5 h.[3]
4. The ethoxyethyl ether was selectively introduced on a primary alcohol in the presence of a secondary alcohol.[4]

OH O ... OH OTBDMS —(CH_2=CHOEt; TsOH, >81%)→ EEO O ... OH OTBDMS

5. $CH_3CH(Cl)OEt$, $PhNMe_2$, CH_2Cl_2, 0°, 10–60 min.[5] These conditions are effective for extremely acid-sensitive substrates or where conditions 1 and 2 fail.

Cleavage

1. 5% AcOH, 20°, 2 h, 100% yield.[1]
2. 0.5 *N* HCl, THF, 0°, 100% yield.[2] The ethoxyethyl ether is more readily cleaved by acidic hydrolysis than the THP ether, but it is more stable than the 1-methyl-1-methoxyethyl ether.
3. Pyridinium tosylate, *n*-PrOH, 80–85% yield.[6] An acetonide was not affected by these conditions.

1. S. Chládek and J. Smrt, *Chem. Ind. (London)*, 1719 (1964).
2. A. I. Meyers, D. L. Comins, D. M. Roland, R. Henning, and K. Shimizu, *J. Am. Chem. Soc.*, **101,** 7104 (1979).
3. A. Fukuzawa, H. Sato, and T. Masamune, *Tetrahedron Lett.*, **28,** 4303 (1987).
4. M. F. Semmelhack and S. Tomoda, *J. Am. Chem. Soc.*, **103,** 2427 (1981).
5. W. C. Still, *J. Am. Chem. Soc.*, **100,** 1481 (1978).
6. M. A. Tius and A. H. Faug, *J. Am. Chem. Soc.*, **108,** 1035 (1986).

30. 1-(2-Chloroethoxy)ethyl Ether (Cee–OR): $ROCH(CH_3)OCH_2CH_2Cl$

The Cee group was developed for the protection of the 2′-hydroxyl group of ribonucleosides.

Formation

1. $CH_2{=}CHOCH_2CH_2Cl$, PPTS, CH_2Cl_2, 80–83% yield.[1]

Cleavage

The relative rates of cleavage for a variety of uridine-protected acetals are given in the table below.

	1.5% Cl_2CHCO_2H in CH_2Cl_2		0.01 *N* HCl (pH 2)	
	$T_{1/2}$ (min)	T_∞ (min)	$T_{1/2}$ (min)	T_∞ (min)
$ROCH(CH_3)OCH_2CH_2Cl$	420	960	96	360
$ROCH(CH_3)O$–*i*-Pr	—	30 s	1	4
$ROCH(CH_3)OBu$	2	5	12	34
$ROCH(CH_3)OEt$	20 s	3	5	18
ROTHP	90	273	32	150

1. S.-i. Yamakage, O. Sakatsume, E. Furuyama, and H. Takaku, *Tetrahedron Lett.*, **30**, 6361 (1989).

31. 1-Methyl-1-methoxyethyl Ether: $ROC(OCH_3)(CH_3)_2$ (Chart 1)

Formation

1. $CH_2{=}C(CH_3)OMe$, cat. $POCl_3$, 20°, 30 min, 100% yield.[1]

Cleavage

1. 20% AcOH, 20°, 10 min.[1]
2. Pyridinium *p*-toluenesulfonate, 5°, 1 h.[2]

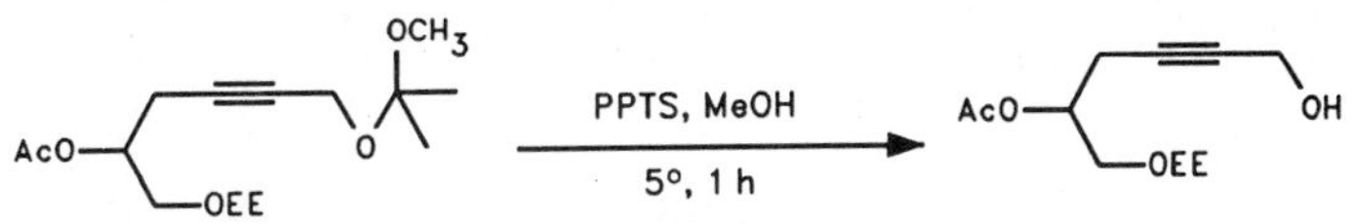

OCH3
AcO
O
OEE
PPTS, MeOH
5°, 1 h
AcO
OH
OEE

EE = Ethoxyethyl

1. A. F. Klug, K. G. Untch, and J. H. Fried, *J. Am. Chem. Soc.*, **94,** 7827 (1972).
2. G. Just, C. Luthe, and M. T. P. Viet, *Can. J. Chem.*, **61,** 712 (1983).

32. 1-Methyl-1-benzyloxyethyl Ether (MBE–OR): $ROC(OBn)(CH_3)_2$

Formation

1. $CH_2{=}C(OBn)(CH_3)$, $PdCl_2$(1,5-cyclooctadiene) [$PdCl_2$(COD)], 85–95% yield.[1]
2. $CH_2{=}C(OBn)(CH_3)$, $POCl_3$ or TsOH, 61–98% yield.[1] It should be noted that these conditions do not afford a cyclic acetal with a 1,3-diol. This ketal is stable to $LiAlH_4$, diisobutylaluminum hydride, NaOH, alkyllithiums, and Grignard reagents.

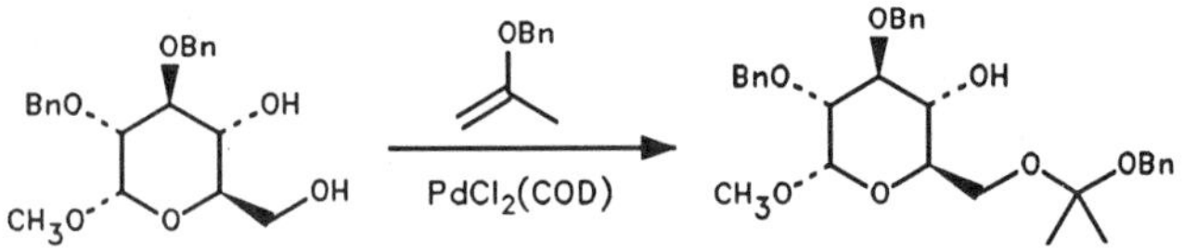

OBn BnO OH OH CH_3O O — OBn / $PdCl_2$(COD) → OBn BnO OH CH_3O O O OBn

Cleavage

1. H_2, 5% Pd–C, EtOH, rt, 92–99% yield.[1]
2. 3 *M* AcOH, H_2O, THF.[2]

33. 1-Methyl-1-benzyloxy-2-fluoroethyl Ether: $ROC(OBn)(CH_2F)(CH_3)$

Formation

1. $CH_2{=}C(OBn)CH_2F$, $PdCl_2$(COD), CH_3CN, rt, 24 h, 89–100% yield.[2] Protic acids can also be used to introduce this group, but the yields are sometimes lower. A primary alcohol can be protected in the presence of a secondary alcohol. This reagent also does not give cyclic acetals of 1,3-diols with palladium catalysis.

Cleavage

1. H_2, Pd–C, EtOH, 1 atm, 98–100% yield.[2] This group is stable to 3 *M* aqueous acetic acid at room temperature, conditions that cleave the TBDMS group and the 1-methyl-1-benzyloxyethyl ether.

1. T. Mukaiyama, M. Ohshima, and M. Murakami, *Chem. Lett.*, 265 (1984).
2. T. Mukaiyama, M. Ohshima, H. Nagaoka, and M. Murakami, *Chem. Lett.*, 615 (1984).

34. 2,2,2-Trichloroethyl Ether: Cl_3CCH_2OR

The anomeric position of a carbohydrate was protected as its trichloroethyl ether. Cleavage is effected with Zn, AcOH, AcONa (3 h, 92%).[1]

1. R. U. Lemieux and H. Driguez, *J. Am. Chem. Soc.*, **97,** 4069 (1975).

35. 2-Trimethylsilylethyl Ether: $Me_3SiCH_2CH_2OR$

Cleavage

1. $BF_3 \cdot Et_2O$, CH_2Cl_2, 0–25°, 79% yield.[1]

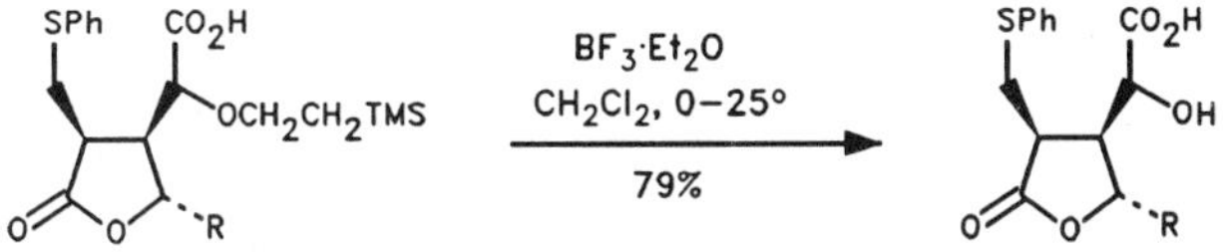

$BF_3 \cdot Et_2O$, CH_2Cl_2, 0–25°, 79%

1. S. D. Burke, G. J. Pacofsky, and A. D. Piscopio, *Tetrahedron Lett.*, **27,** 3345 (1986).

36. 2-(Phenylselenyl)ethyl Ether: $ROCH_2CH_2SePh$ (Chart 1)

This ether was prepared from an alcohol and 2-(phenylselenyl)ethyl bromide ($AgNO_3$, CH_3CN, 20°, 10–15 min, 80–90% yield); it is cleaved by oxidation (H_2O_2, 1 h; ozone; or $NaIO_4$), followed by acidic hydrolysis of the intermediate vinyl ether (dil. HCl, 65–70% yield).[1]

1. T.-L. Ho and T. W. Hall, *Synth. Commun.*, **5,** 367 (1975).

37. *t*-Butyl Ether: *t*-BuOR (Chart 1)

Formation

t-Butyl ethers can be prepared from a variety of alcohols, including allylic alcohols. They are stable to most reagents except strong acids. The *t*-butyl ether is probably one of the more underused alcohol protective groups considering its stability, the ease and efficiency of introduction, and the ease of cleavage.

1. Isobutylene, $BF_3 \cdot Et_2O$, H_3PO_4, 100% yield.[1,2]

OH

isobutylene
$BF_3 \cdot Et_2O$, H_3PO_4
100%

O–t–Bu

2. Isobutylene, Amberlyst H-15, hexane.[3]
3. Isobutylene, H_2SO_4.[4]
4. *t*-BuOC(=NH)CCl_3, $BF_3 \cdot Et_2O$, CH_2Cl_2, cyclohexane, 59–91% yield.[5]

Cleavage

1. Anhydrous CF_3COOH, 0–20°, 1–16 h, 80–90% yield.[2,4]
2. HBr, AcOH, 20°, 30 min.[6]
3. 4 *N* HCl, dioxane, reflux, 3 h.[7] In this case the *t*-butyl ether was stable to 10 *N* HCl, MeOH, 0–5°, 30 h.
4. Me_3SiI, CCl_4 or $CHCl_3$, 25°, <0.1 h, 100% yield.[8] Under suitable conditions this reagent also cleaves many other ethers, esters, ketals, and carbamates.[9]
5. Ac_2O, $FeCl_3$, Et_2O, 76–93% yield.[3,10] These conditions give the acetate of the alcohol, which can then be cleaved by simple basic hydrolysis.
6. $TiCl_4$, CH_2Cl_2, 0°, 1 min, 85% yield.[11]

1. R. A. Micheli, Z. G. Hajos, N. Cohen, D. R. Parrish, L. A. Portland, W. Sciamanna, M. A. Scott, and P. A. Wehrli, *J. Org. Chem.*, **40,** 675 (1975).
2. H. C. Beyerman and G. L. Heiszwolf, *J. Chem. Soc.*, 755 (1963).
3. A. Alexakis and J. M. Duffault, *Tetrahedron Lett.*, **29,** 6243 (1988); A. Alexakis, M. Gardette, and S. Colin, *Tetrahedron Lett.*, **29,** 2951 (1988).
4. H. C. Beyerman and J. S. Bontekoe, *Proc. Chem. Soc.*, 249 (1961).
5. A. Armstrong, I. Brackenridge, R. F. W. Jackson, and J. M. Kirk, *Tetrahedron Lett.*, **29,** 2483 (1988).
6. F. M. Callahan, G. W. Anderson, R. Paul, and J. E. Zimmerman, *J. Am. Chem. Soc.*, **85,** 201 (1963).
7. U. Eder, G. Haffer, G. Neef, G. Sauer, A. Seeger, and R. Wiechert, *Chem. Ber.*, **110,** 3161 (1977).
8. M. E. Jung and M. A. Lyster, *J. Org. Chem.*, **42,** 3761 (1977).
9. A. H. Schmidt, *Aldrichimica Acta*, **14,** 31 (1981).
10. B. Ganem and V. R. Small, Jr., *J. Org. Chem.*, **39,** 3728 (1974).
11. R. H. Schlessinger and R. A. Nugent, *J. Am. Chem. Soc.*, **104,** 1116 (1982).

38. Allyl Ether (Allyl–OR): CH_2=$CHCH_2$–OR (Chart 1)

The use of allyl ethers for the protection of alcohols is common in carbohydrate literature because allyl ethers are generally compatible with the various methods for glycoside formation.[1] Obviously the allyl ether is not compatible with powerful electrophiles such as bromine and catalytic hydrogenation. Allyl ethers are stable to moderately acidic conditions (1 *N* HCl, reflux, 10 h).[2] The synthesis of perdeuteroallyl bromide and its use as a protective group in carbohydrates has been reported. The perdeutero derivative has the advantage that the allyl resonances in the NMR no longer obscure other more diagnostic resonances such as those of the anomeric carbon in glycosides.[3]

Formation

1. $CH_2{=}CHCH_2Br$, NaOH, benzene, reflux, 1.5 h,[4] or NaH, benzene, 90–100% yield.[5]
2. $CH_2{=}CHCH_2OC({=}NH)CCl_3$, H^+.[6]
3. Allyl carbonates have been converted to allyl ethers.[7] In the case below acid- and base-catalyzed procedures failed because of the sensitivity of the $[(i\text{-}Pr)_2Si]_2O$ group.

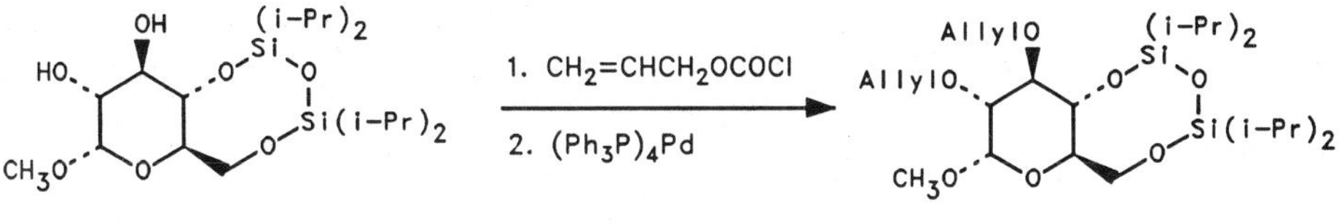

1. $CH_2{=}CHCH_2OCOCl$
2. $(Ph_3P)_4Pd$

4. Bu_2SnO, toluene, THF; $CH_2{=}CHCH_2Br$, $Bu_4N^+Br^-$, 96% yield.[8]

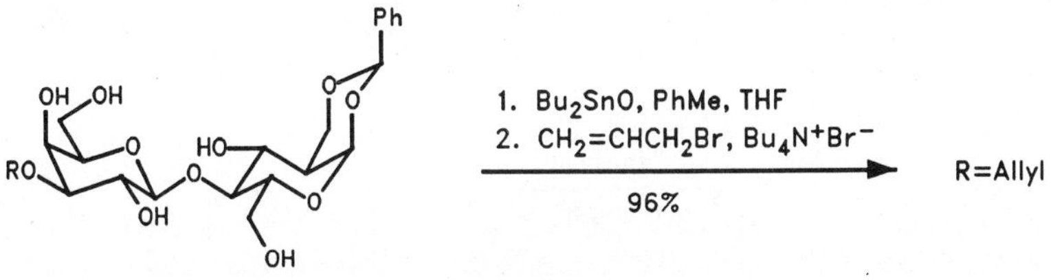

1. Bu_2SnO, PhMe, THF
2. $CH_2{=}CHCH_2Br$, $Bu_4N^+Br^-$

96%

R=Allyl

The crotyl ether has been introduced using similar methodology.[9]

5. $CH_2{=}CHCH_2OCO_2Et$, $Pd_2(dba)_3$, THF, 65°, 4 h, 70–97% yield.[10]

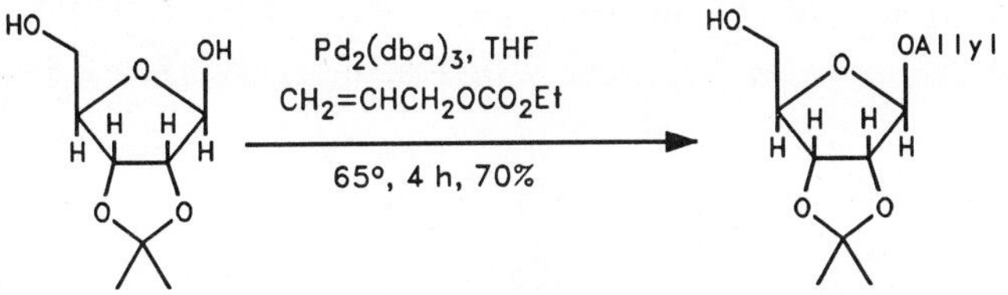

$Pd_2(dba)_3$, THF
$CH_2{=}CHCH_2OCO_2Et$

65°, 4 h, 70%

$Pd_2(dba)_3$ = tris(dibenzylideneacetone)dipalladium

This method is also effective for the protection of primary and secondary alcohols.

Cleavage

1. $ROCH_2CH{=}CH_2 \xrightarrow[100°,\ 15\ min]{t\text{-}BuOK,\ DMSO^{11}} ROCH{=}CHCH_3 \xrightarrow{i,\ ii,\ iii,\ or\ iv} ROH$

   i. 0.1 *N* HCl, acetone–water, reflux, 30 min.[8]

   ii. $KMnO_4$, NaOH–H_2O, 10°, 100% yield. These basic conditions avoid acid-catalyzed acetonide cleavage.[2]

   iii. $HgCl_2/HgO$, H_2O, acetone–H_2O, 5 min, 100% yield.[12]

   iv. Ozonolysis.[11]

2. $ROCH_2CH{=}CH_2 \xrightarrow[reflux,\ 3\ h]{(Ph_3P)_3RhCl,\ DABCO,\ EtOH} \xrightarrow{Hg(II),\ pH\ 2} ROH, > 90\%$[5]

   Allyl ethers are isomerized by $(Ph_3P)_3RhCl$, and *t*-BuOK/DMSO in the following order:[13]
   $(Ph_3P)_3RhCl$: allyl > 2-methylallyl > but-2-enyl
   t-BuOK: but-2-enyl > allyl > 2-methylallyl

3. It is possible to remove the allyl group in the presence of an allyloxycarbonyl (AOC) group using an $[Ir(COD)(Ph_2MeP)_2]PF_6$-catalyzed isomerization, but the selectivity is not complete. The allyloxycarbonyl group can be removed selectively in the presence of an allyl group using a palladium or rhodium catalyst.[14] Hydrogen-activated $[Ir(COD)(Ph_2MeP)_2]PF_6$ is a better catalyst for allyl isomerization (91–100% yield) because there is no reduction of the alkene, as is sometimes the case with $(Ph_3P)_3RhCl$.[15,16]

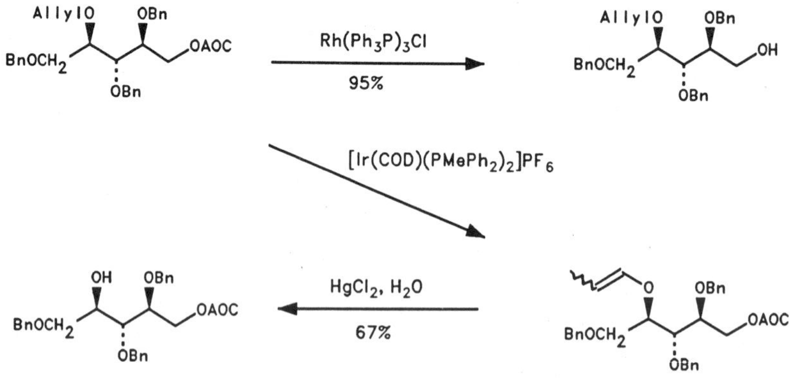

AllylO OBn BnOCH$_2$ OAOC OBn → Rh(Ph$_3$P)$_3$Cl, 95% → AllylO OBn BnOCH$_2$ OH OBn

→ [Ir(COD)(PMePh$_2$)$_2$]PF$_6$ → O OBn BnOCH$_2$ OAOC OBn

→ HgCl$_2$, H$_2$O, 67% → OH OBn BnOCH$_2$ OAOC OBn

4. Useful selectivity between allyl and 3-methylbut-2-enyl (prenyl) ethers has been achieved.[13]

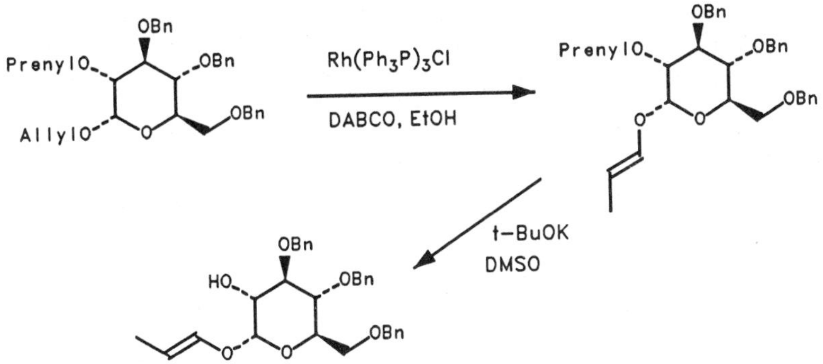

5. *trans*-$Pd(NH_3)_2Cl_2$/*t*-BuOH isomerizes allyl ethers to vinyl ethers that can then be hydrolyzed in 90% yield, but in the presence of an α-hydroxy group the intermediate vinyl ether cyclizes to an acetal.[17] Benzylidene acetals are not affected by this reagent.

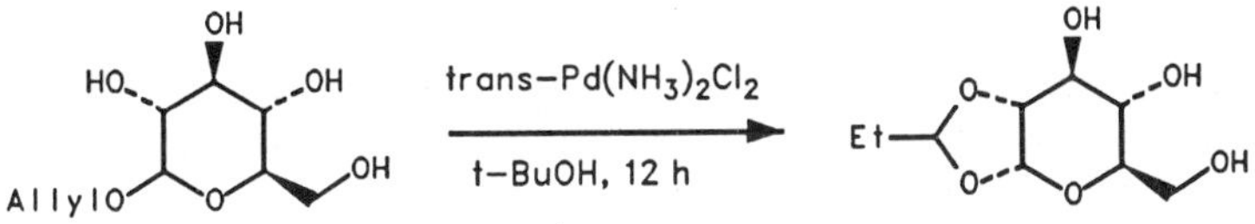

6. $ROCH_2CH{=}CH_2 \xrightarrow[\text{60–80°, 24 h, 80–95\%}]{\text{Pd/C, H}_2\text{O, MeOH, cat. TsOH or HClO}_4} ROH$[18]
7. $ROCH_2CH{=}CH_2 \xrightarrow[\text{reflux, 1 h, 50\%}]{\text{SeO}_2\text{, AcOH, dioxane}} [ROCH(OH)CH{=}CH_2] \rightarrow ROH$[19]
8. $ROCH_2CH{=}CH_2 \xrightarrow[\text{reflux, 2 h, 85–99\%}]{\text{NaTeH, EtOH, AcOH}} ROH$[20]
9. Ac_2O, $BF_3 \cdot Et_2O$, then MeONa/MeOH to hydrolyze the acetate.[21]
10. $RhCl_3$, DABCO, EtOH, H_2O; H_3O^+, EtOH.[22]

1. R. Gigg, *Am. Chem. Soc. Symp. Ser.*, **39,** 253 (1977); *ibid.*, **77,** 44 (1978); R. Gigg and R. Conant, *Carbohydr. Res.*, **100,** C5 (1982).
2. J. Cunningham, R. Gigg, and C. D. Warren, *Tetrahedron Lett.*, 1191 (1964).
3. J. Thiem, H. Mohn, and A. Heesing, *Synthesis*, 775 (1985).
4. R. Gigg and C. D. Warren, *J. Chem. Soc. C*, 2367 (1969).
5. E. J. Corey and W. J. Suggs, *J. Org. Chem.*, **38,** 3224 (1973).
6. T. Iversen and D. R. Bundle, *J. Chem. Soc., Chem. Commun.*, 1240 (1981); H.-P. Wessel, T. Iversen, and D. R. Bundle, *J. Chem. Soc., Perkin Trans. I*, 2247 (1985).
7. J. J. Oltvoort, M. Kloosterman, and J. H. Van Boom, *Recl: J. R. Neth. Chem. Soc.*, **102,** 501 (1983); F. Guibe and Y. Saint M'Leux, *Tetrahedron Lett.*, **22,** 3591 (1981).
8. S. Sato, S. Nunomura, T. Nakano, Y. Ito, and T. Ogawa, *Tetrahedron Lett.*, **29,** 4097 (1988).

9. A. K. M. Anisuzzaman, L. Anderson, and J. L. Navia, *Carbohydr. Res.*, **174,** 265 (1988).
10. R. Lakhmiri, P. Lhoste, and D. Sinou, *Tetrahedron Lett.*, **30,** 4669 (1989).
11. J. Gigg and R. Gigg, *J. Chem. Soc. C*, 82 (1966).
12. R. Gigg and C. D. Warren, *J. Chem. Soc. C*, 1903 (1968).
13. P. A. Gent and R. Gigg, *J. Chem. Soc., Chem. Commun.*, 277 (1974); R. Gigg, *J. Chem. Soc., Perkin Trans. I*, 738 (1980).
14. P. Boullanger, P. Chatelard, G. Descotes, M. Kloosterman, and J. H. Van Boom, *J. Carbohydr. Chem.*, **5,** 541 (1986).
15. J. J. Oltvoort, C. A. A. van Boeckel, J. H. de Koning, and J. H. van Boom, *Synthesis*, 305 (1981).
16. For hydrogenation during isomerization, see: C. D. Warren and R. W. Jeanloz, *Carbohydr. Res.*, **53,** 67 (1977); T. Nishiguchi, K. Tachi, and K. Fukuzumi, *J. Org. Chem.*, **40,** 237 (1975); C. A. A. van Boeckel and J. H. van Boom, *Tetrahedron Lett.*, 3561 (1979).
17. T. Bieg and W. Szeja, *J. Carbohydr. Chem.*, **4,** 441 (1985).
18. R. Boss and R. Scheffold, *Angew. Chem., Int. Ed. Engl.*, 15, 558 (1976).
19. K. Kariyone and H. Yazawa, *Tetrahedron Lett.*, 2885 (1970).
20. N. Shobana and P. Shanmugam, *Indian J. Chem., Sect. B*, **25B,** 658 (1986).
21. C. F. Garbers, J. A. Steenkamp, and H. E. Visagie, *Tetrahedron Lett.*, 3753 (1975).
22. M. Dufour, J.-C. Gramain, H.-P. Husson, M.-E. Sinibaldi, and Y. Troin, *Tetrahedron Lett.*, **30,** 3429 (1989).

39. *p*-Chlorophenyl Ether: *p*-ClC_6H_4–OR

Formation/Cleavage[1]

1. MsCl, Pyr
2. *p*-ClC_6H_4ONa

ROH → *p*-ClC_6H_4-OR

1. Li/NH_3
2. H_3O^+

(reverse arrow: *p*-ClC_6H_4-OR → ROH)

The *p*-chlorophenyl ether was used in this synthesis to minimize ring sulfonation during cyclization of a diketo ester with concentrated H_2SO_4/AcOH.[1]

1. J. A. Marshall and J. J. Partridge, *J. Am. Chem. Soc.*, **90,** 1090 (1968).

40. *p*-Methoxyphenyl Ether: *p*-$MeOC_6H_4OR$

Formation

1. *p*-$MeOC_6H_4OH$, DEAD, Ph_3P, THF, 82–99% yield.[1,2]

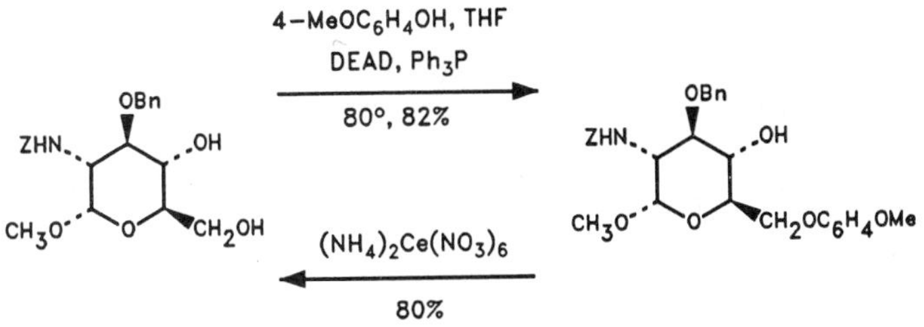

4-$MeOC_6H_4OH$, THF
DEAD, Ph_3P
80°, 82%

$(NH_4)_2Ce(NO_3)_6$
80%

Z = benzyloxycarbonyl, DEAD = diethyl azodicarboxylate

Cleavage

1. Ceric ammonium nitrate, CH_3CN, H_2O (4 : 1), 0°, 10 min, 80–85% yield.[1,2] This group is stable to 3*N* HCl, 100°; 3 *N* NaOH, 100°; H_2, 1200 psi; O_3, MeOH, −78°; Raney Ni, 100°; $LiAlH_4$; Jones reagent and pyridinium chlorochromate (PCC).

1. T. Fukuyama, A. A. Laud, and L. M. Hotchkiss, *Tetrahedron Lett.*, **26**, 6291 (1985).
2. M. Petitou, P. Duchaussoy, and J. Choay, *Tetrahedron Lett.*, **29**, 1389 (1988).

41. 2,4-Dinitrophenyl Ether: RODNP: 2,4-$(NO_2)_2$–C_6H_3OR

Formation

1. 2,4-Dinitrofluorobenzene, DABCO, DMF, 85% yield.[1] When this group was used to protect an anomeric center of a carbohydrate, only the β-isomer was formed, but this could be equilibrated to the α-isomer in 90% yield with K_2CO_3 in DMF.

1. H. J. Koeners, A. J. De Kok, C. Romers, and J. H. Van Boom, *Recl. Trav. Chim. Pays-Bas*, **99**, 355 (1980).

42. Benzyl Ether (BnOR): $PhCH_2OR$ (Chart 1)

Formation

1. BnCl, powdered KOH, 130–140°, 86% yield.[1]
2. BnCl, $Bu_4N^+HSO_4^-$, 50% KOH, benzene.[2]
3. NaH, THF, BnBr, $Bu_4N^+I^-$, 20°, 3 h, 100%.[3] This method was used to protect a hindered hydroxyl group. Increased reactivity is achieved by the *in situ* generation of benzyl iodide.

4. BnX (X = Cl, Br), Ag_2O, DMF, 25°, good yields.[4]
5. BnCl, $Ni(acac)_2$, reflux, 3 h, 80–90%.[5]
6. BnO—C(=NH)CCl_3, CF_3SO_3H.[6–8]
7. The primary alcohol below was selectively benzylated using NaH and BnBr at −70°.[9]

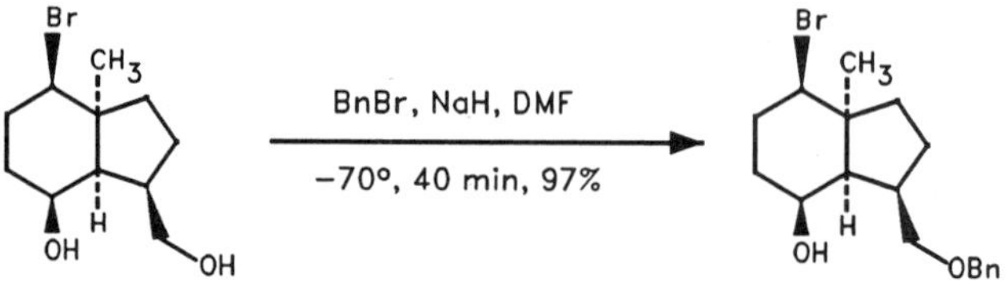

8. Ag_2O, BnBr, DMF, rt, 48 h, 76% yield.[10]

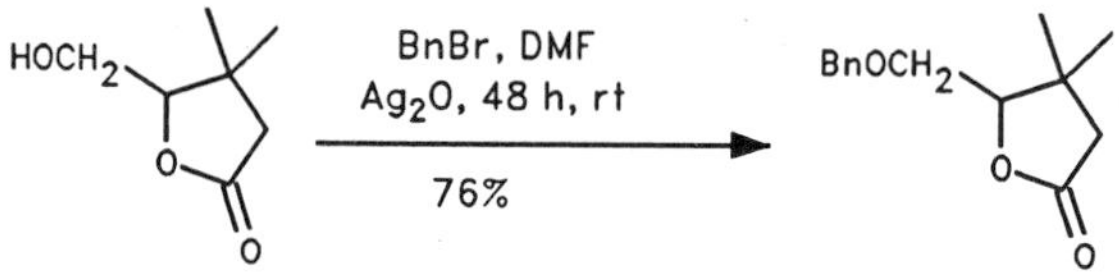

9. $(Bu_3Sn)_2O$, toluene, reflux; BnBr, *N*-methylimidazole, 95% yield.[11] Equatorial alcohols are benzylated in preference to axial alcohols in diol-containing substrates.

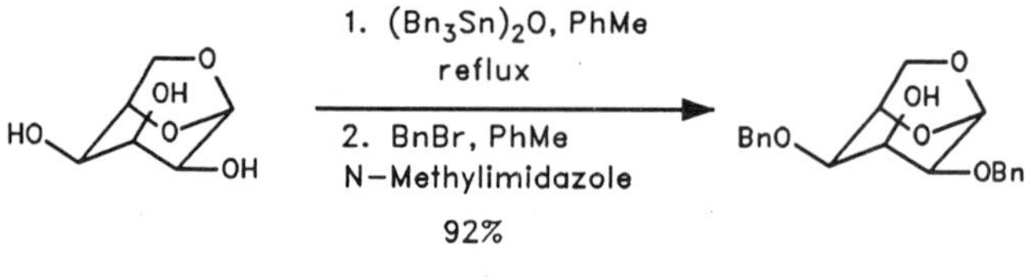

10. Bu_2SnO, benzene; BnBr, DMF, heat, 80% yield.[12] This method has also been used to protect selectively the anomeric hydroxyl in a carbohydrate derivative.[13]
11. BnI, NaH, rt, 90% yield.[14] Note that in this case the reaction proceeds without complication of the Payne rearrangement.

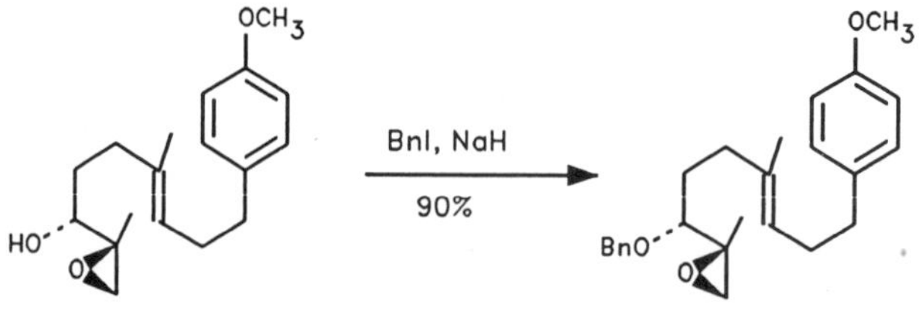

12. $PhCHN_2$, HBF_4, −40°, CH_2Cl_2, 66–92% yield.[15]

1. H. G. Fletcher, *Methods Carbohydr. Chem.*, **II,** 166 (1963).
2. H. H. Freedman and R. A. Dubois, *Tetrahedron Lett.*, 3251 (1975).
3. S. Czernecki, C. Georgoulis, and C. Provelenghiou, *Tetrahedron Lett.*, 3535 (1976); K. Kanai, I. Sakamoto, S. Ogawa, and T. Suami, *Bull. Chem. Soc. Jpn.*, **60,** 1529 (1987).
4. R. Kuhn, I. Löw, and H. Trishmann, *Chem. Ber.*, **90,** 203 (1957).
5. M. Yamashita and Y. Takegami, *Synthesis*, 803 (1977).
6. T. Iversen and K. R. Bundle, *J. Chem. Soc., Chem. Commun.*, 1240 (1981).
7. J. D. White, G. N. Reddy, and G. O. Spessard, *J. Am. Chem. Soc.* **110,** 1624 (1988).
8. U. Widmer, *Synthesis*, 568 (1987).
9. A. Fukuzawa, H. Sato, and T. Masamune, *Tetrahedron Lett.*, **28,** 4303 (1987).
10 L. Van Hijfte and R. D. Little, *J. Org. Chem.*, **50,** 3940 (1985).
11. C. Cruzado, M. Bernabe, and M. Martin-Lomas, *J. Org. Chem.*, **54,** 465 (1989).
12. W. R. Roush, M. R. Michaelides, D. F. Tai, B. M. Lesur, W. K. M. Chong, and D. J. Harris, *J. Am. Chem. Soc.*, **111,** 2984 (1989).
13. C. Bliard, P. Herczegh, A. Olesker, and G. Lukacs, *J. Carbohydr. Res.*, **8,** 103 (1989).
14. E. E. van Tamelen, S. R. Zawacky, R. K. Russell, and J. G. Carlson, *J. Am. Chem. Soc.*, **105,** 142 (1983).
15. L. J. Liotta and B. Ganem, *Tetrahedron Lett.*, **30,** 4759 (1989).

Cleavage

1. H_2/Pd–C, EtOH, 95% yield.[1,2] Palladium is the preferred catalyst since the use of platinum results in ring hydrogenation.[1] Hydrogenolysis of the benzyl group of threonine in peptides containing tryptophan often results in reduction of tryptophan to the 2,3-dihydro derivative.[3] The presence of nonaromatic amines can retard *O*-debenzylation.[4] Although it is possible to effect benzyl ether cleavage in the presence of an isolated olefin (H_2/5% Pd–C, 97% yield),[5] in general the degree of selectivity is dependent on the substitution pattern and the degree of steric hindrance. Good selectivity was achieved for hydrogenolysis of a benzyl group in the presence of a trisubstituted olefin conjugated to an ester.[6]
2. Pd–C using transfer hydrogenation. A number of methods have been developed where hydrogen is generated *in situ*. These include the use of cyclohexene (1–8 h, 80–90% yield),[7] cyclohexadiene (25°, 2 h, good yields),[8] HCO_2H,[9] ammonium formate (MeOH, reflux, 91% yield),[10] and isopropyl alcohol.[11] A benzylidene acetal is not cleaved when ammonium formate is used as the hydrogen source[10] and a trisubstituted olefin is not affected when formic acid is used as a hydrogen source.[12] In α-methyl 2,3-di-*O*-benzyl-4,6-*O*-benzylideneglucose the cleavage can be controlled to cleave the 2-benzyl group selectively (83%) when cyclohexene is used as the hydrogen source.[13]
3. Raney nickel W2 or W4, EtOH, 85–100% yield.[14] Mono- and dimethoxy-

substituted benzyl ethers and benzaldehyde acetals are not cleaved under these conditions.

4. Na/NH_3[15,16] or EtOH.[17]
5. Electrolytic reduction: -3.1 V, $R_4N^+F^-$, DMF.[18]
6. Me_3SiI, CH_2Cl_2, 25°, 15 min, 100% yield.[19] This reagent also cleaves most other ethers and esters, but selectivity can be achieved with the proper choice of conditions.
7. Lithium aluminum hydride will also cleave benzyl ethers, but this is seldom practical because of its high reactivity to other functional groups.[20]
8. Me_2BBr, $ClCH_2CH_2Cl$, 0°–rt, 70–93% yield.[21] This reagent also cleaves phenolic methyl ethers. Tertiary ethers give the bromide rather than the alcohol.
9. $FeCl_3$, Ac_2O, 55–75% yield.[22] The relative rates of cleavage for the 6-, 3-, and 2-*O*-benzyl groups of a glucose derivative are 125 : 24 : 1. Sulfuric acid has also been used as a catalyst.[23]
10. CrO_3/AcOH, 25°, 50% yield, [→ ROCOPh (→ ROH + $PhCO_2H$)].[24] This method was used to remove benzyl ethers from carbohydrates that contain functional groups sensitive to catalytic hydrogenation or dissolving metals. Esters are stable, but glycosides or acetals are cleaved.
11. RuO_2, $NaIO_4$, CCl_4, CH_3CN, H_2O, 54–96% yield.[25] In this case the benzyl group is oxidized to a benzoate that can be hydrolyzed under basic conditions.
12. Ozone, 50 min; then NaOMe, 60–88% yield.[26]
13. Electrolytic oxidation, 1.4–1.7 V, Ar_3N, CH_3CN, CH_2Cl_2, $LiClO_4$, lutidine.[27]
14. Ca/NH_3, ether or THF, 2 h; NH_4Cl, H_2O, 90% yield.[28] Acetylenes are not reduced under these conditions.
15. $PhSSiMe_3$, $Bu_4N^+I^-$, ZnI_2, $ClCH_2CH_2Cl$, 60°, 2 h, 75% yield.[29]

EtO_2C … OAc, OBn —[PhSTMS, ZnI_2, $Bu_4N^+I^-$; $ClCH_2CH_2Cl$, 60°, 2 h; 75%]→ EtO_2C … OAc, OH

16. Rh/Al_2O_3, H_2, 100%.[30]

i-PrO, BnO —[Rh/Al_2O_3, H_2; 100%]→ i-PrO, HO

17. $Ph_3C^+BF_4^-$, CH_2Cl_2.[31]

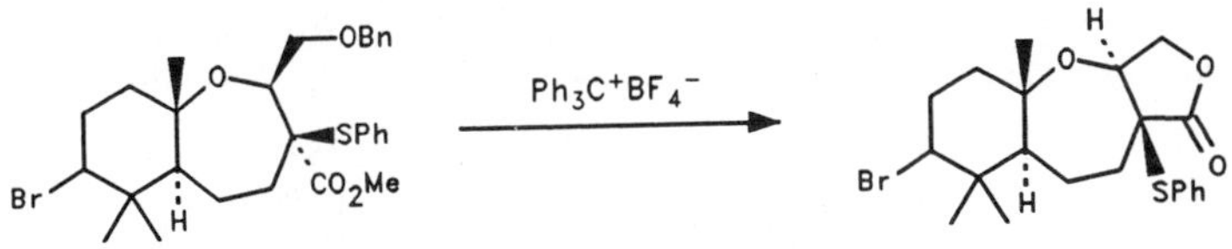

18. *t*-BuMgBr, benzene, 80°, 69%.[32] MeMgI fails in this reaction. In general, benzyl ethers are quite stable to Grignard reagents because these reactions are rarely run at such high temperatures.

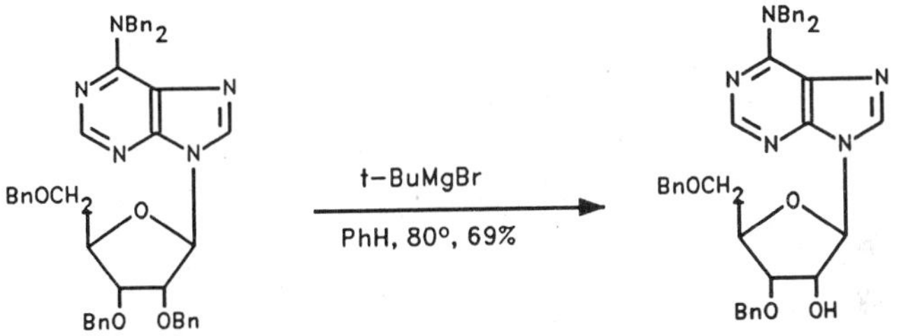

19. EtSH, $BF_3 \cdot Et_2O$, 63% yield.[33]
20. The fungus *Mortierella isabellina* NRRL 1757, 0–100% yield.[34]
21. $BF_3 \cdot Et_2O$, NaI, CH_3CN, 0°, 1 h; rt, 7 h, 80% yield.[35]
22. BCl_3, CH_2Cl_2, −78° → 0°; MeOH at −78°, 77% yield.[36]

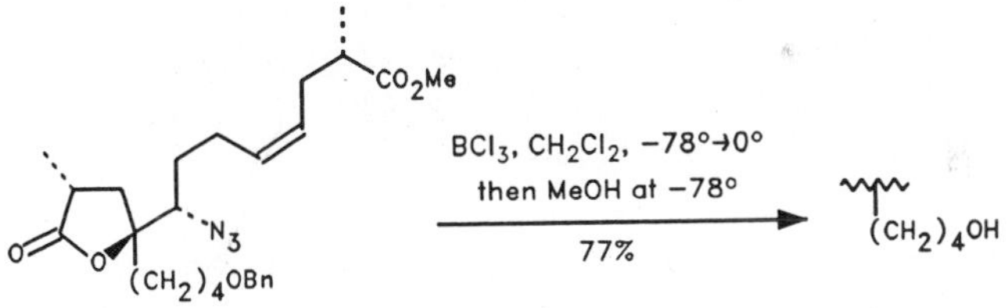

23. Me_3SiBr, thioanisole.[37] This reagent combination also cleaves a carbobenzoxy (Z) group, a $4\text{-}MeOC_6H_4CH_2SR$ group, and reduces sulfoxides to sulfides.
24. $SnCl_4$, CH_2Cl_2, rt, 30 min.[38]

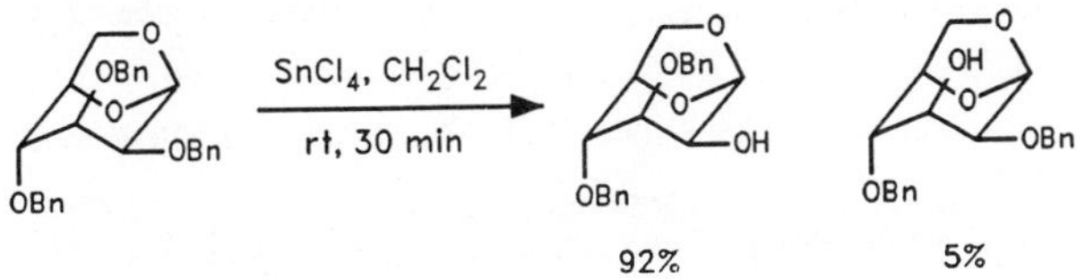

1. C. H. Heathcock and R. Ratcliffe, *J. Am. Chem. Soc.*, **93,** 1746 (1971).
2. W. H. Hartung and C. Simonoff, *Org. React.*, **7,** 263 (1953).

3. L. Kisfaludy, F. Korenczki, T. Mohacsi, M. Sajgo, and S. Fermandjian, *Int. J. Pept. Protein Res.*, **27,** 440 (1986).
4. B. P. Czech and R. A. Bartsch, *J. Org. Chem.*, **49,** 4076 (1984).
5. J. S. Bindra and A. Grodski, *J. Org. Chem.*, **43,** 3240 (1978).
6. D. Cain and T. L. Smith, Jr., *J. Am. Chem. Soc.*, **102,** 7568 (1980).
7. G. M. Anantharamaiah and K. M. Sivanandaiah, *J. Chem. Soc., Perkin Trans. 1*, 490 (1977); S. Hanessian, T. J. Liak, and B. Vanasse, *Synthesis*, 396 (1981).
8. A. M. Felix, E. P. Heimer, T. J. Lambros, C. Tzougraki, and J. Meienhofer, *J. Org. Chem.*, **43,** 4194 (1978).
9. B. ElAmin, G. M. Anantharamaiah, G. P. Royer, and G. E. Means, *J. Org. Chem.*, **44,** 3442 (1979).
10. T. Bieg and W. Szeja, *Synthesis*, 76 (1985).
11. M. Del Carmen Cruzado and M. Martin-Lomias, *Tetrahedron Lett.*, **27,** 2497 (1986).
12. M. E. Jung, Y. Usui, and C. T. Vu, *Tetrahedron Lett.*, **28,** 5977 (1987).
13. D. Beaupere, I. Boutbaiba, G. Demailly, and R. Uzan, *Carbohydr. Res.*, **180,** 152 (1988).
14. Y. Oikawa, T. Tanaka, K. Horita, and O. Yonemitsu, *Tetrahedron Lett.*, **25,** 5397 (1984); K. Horita, T. Yoshioka, T. Tanaka, Y. Oikawa, and O. Yonemitsu, *Tetrahedron*, **42,** 3021 (1986).
15. C. M. McCloskey, *Adv. Carbohydr. Chem.*, **12,** 137 (1957); I. Schön, *Chem. Rev.*, **84,** 287 (1984).
16. K. D. Philips, J. Zemlicka, and J. P. Horowitz, *Carbohydr. Res.*, **30,** 281 (1973).
17. E. J. Reist, V. J. Bartuska, and L. Goodman, *J. Org. Chem.*, **29,** 3725 (1964).
18. V. G. Mairanovsky, *Angew. Chem., Int. Ed. Engl.*, **15,** 281 (1976).
19. M. E. Jung and M. A. Lyster, *J. Org. Chem.*, **42,** 3761 (1977).
20. J. P. Kutney, N. Abdurahman, C. Gletsos, P. LeQuesne, E. Piers, and I. Vlattas, *J. Am. Chem. Soc.*, **92,** 1727 (1970).
21. Y. Guindon, C. Yoakim, and H. E. Morton, *Tetrahedron Lett.*, **24,** 2969 (1983).
22. K. P. R. Kartha, F. Dasgupta, P. P. Singh, and H. C. Srivastava, *J. Carbohydr. Chem.*, **5,** 437 (1986).
23. J. Sakai, T. Takeda, and Y. Ogihara, *Carbohydr. Res.*, **95,** 125 (1981).
24. S. J. Angyal and K. James, *Carbohydr. Res.*, **12,** 147 (1970).
25. P. F. Schuda, M. B. Cichowicz, and M. R. Heimann, *Tetrahedron Lett.*, **24,** 3829 (1983); P. F. Schuda and M. R. Heimann, *Tetrahedron Lett.*, **24,** 4267 (1983).
26. P. Angibeaud, J. Defaye, A. Gadelle, and J.-P. Utille, *Synthesis*, 1123 (1985).
27. W. Schmidt and E. Steckhan, *Angew. Chem. Int. Ed. Engl.*, **18,** 801 (1979); E. A. Mayeda, L. L. Miller, and J. F. Wolf, *J. Am. Chem. Soc.*, **94,** 6812 (1972).
28. J. R. Hwu, V. Chua, J. E. Schroeder, R. E. Barrans, Jr., K. P. Khoudary, N. Wang, and J. M. Wetzel, *J. Org. Chem.*, **51,** 4731 (1986).
29. K. C. Nicolaou, M. R. Pavia, and S. P. Seitz, *J. Am. Chem. Soc.*, **104,** 2027 (1982).
30. Y. Oikawa, T. Tanaka, and O. Yonemitsu, *Tetrahedron Lett.*, **27,** 3647 (1986).
31. T. R. Hoye, A. J. Caruso, J. F. Dellaria, Jr., and M. J. Kurth, *J. Am. Chem. Soc.*, **104,** 6704 (1982).
32. M. Kawana, *Chem. Lett.*, 1541 (1981).

33. S. M. Daly and R. W. Armstrong, *Tetrahedron Lett.*, **30**, 5713 (1989).
34. H. L. Holland, M. Conn, P. C. Chenchaiah, and F. M. Brown, *Tetrahedron Lett.*, **29**, 6393 (1988).
35. Y. D. Vankar and C. T. Rao, *J. Chem. Res., Synop.*, 232 (1985).
36. D. R. Williams, D. L. Brown, and J. W. Benbow, *J. Am. Chem. Soc.*, **111**, 1923 (1989).
37. N. Fujii, A. Otaka, N. Sugiyama, M. Hatano, and H. Yajima, *Chem. Pharm. Bull.*, **35**, 3880 (1987).
38. H. Hori, Y. Nishida, H. Ohrui, and H. Meguro, *J. Org. Chem.*, **54**, 1346 (1989).

Substituted Benzyl Ethers

Several methoxy-substituted benzyl ethers have been prepared and used as protective groups. Their utility lies in the fact that they are more readily cleaved oxidatively than the unsubstituted benzyl ethers. The table below gives the relative rates of cleavage with dichlorodicyanoquinone (DDQ).[1]

$(OMe)_n$ — DDQ (1.2 eq), CH_2Cl_2, H_2O (10:1), 20° → $PhCH_2CH_2OH$ + OHC—$(OMe)_n$

i ii iii

Cleavage of MPM, DMPM, and TMPM ethers with DDQ in CH_2Cl_2/H_2O at 20°

Protective Group	Time (h)	Yield ii	(%) iii	Protective Group	Time (h)	Yield ii	(%) iii
3,4-DMPM	<0.33	86	84	2-MPM	3.5	93	70
4-MPM	0.33	89	86	3,5-DMPM	8	73	92
2,3,4-TMPM	0.5	60	75	2,3-DMPM	12.5	75	73
3,4,5-TMPM	1	89	89	3-MPM	24	80	94
2,5-DMPM	2.5	95	16	2,6-DMPM	27.5	80	95

From the table it is clear that there are considerable differences in the cleavage rates of the various ethers that should prove quite useful.

43. *p*-Methoxybenzyl Ether: *p*-$MeOC_6H_4CH_2OR$, MPM–OR

Formation

1. The section on the formation of benzyl ethers should also be consulted.
2. *p*-$MeOC_6H_4CH_2OC(=NH)CCl_3$, H^+, 52–84% yield.[2]

3. p-$MeOC_6H_4CHN_2$, $SnCl_2$, ≈50% yield.[3] This method was used to introduce the MPM group at the 2′- and 3′-positions of ribonucleotides without selectivity for either the 2′- or 3′-isomer. The primary 5-hydroxyl was not affected.
4. NaH, p-$MeOC_6H_4CH_2Br$, DMF, −5°, 1 h, 65%.[4,5] Selectivity is probably achieved because of the increased acidity of the 2′-hydroxyl group.

4−$MeOC_6H_4CH_2Br$
DMF, NaH, 1 h, −5°
65%

5. NaH, p-$MeOC_6H_4CH_2Cl$, THF, 81% yield.[6]

Cleavage

1. The section on the cleavage of benzyl ethers should also be consulted.
2. Electrolytic oxidation: Ar_3N, CH_3CN, $LiClO_4$, 20°, 1.4–1.7 V, 80–90% yield.[7] Benzyl ethers are not affected by these conditions.
3. Dichlorodicyanoquinone (DDQ), CH_2Cl_2, H_2O, 40 min, rt, 84–93% yield.[8–10]. This method does not cleave simple benzyl ethers. This method was found effective in the presence of a boronate.[11] The following groups are stable to these conditions: ketones, epoxides, alkenes, acetonides, tosylates, MOM ethers, THP ethers, acetates, benzyloxymethyl (BOM) ethers, and TBDMS ethers.
4. Ozone, acetone, −78°, 42–82% yield.[12]
5. Ceric ammonium nitrite (CAN) or NBS, CH_2Cl_2, H_2O, 90% yield.[13]

CAN or Br_2 or NBS
CH_2Cl_2, H_2O

Phth = phthalimido

6. $Ph_3C^+ BF_4^-$, CH_2Cl_2 or CH_3CN, H_2O.[1,4] In this case the reaction with DDQ failed to go to completion. This was attributed to the reduced electron density on the aromatic ring because of its attachment at the more electron-poor anomeric center.
7. $h\nu$ >280 nm, H_2O, 1,4-dicyanonaphthalene, 70–81% yield.[14]
8. This example shows that overoxidation of allylic alcohols may occur with DDQ.[15]

DDQ, CH_2Cl_2

rt, 83%

+ the alcohol

44. 3,4-Dimethoxybenzyl Ether: 3,4-$(MeO)_2C_6H_3CH_2OR$, DMPM–OR

Formation

1. NaH, 3,4-$(MeO)_2C_6H_3CH_2Br$, DMF.[16]
2. 3,4-$(MeO)_2C_6H_3CH_2OC(=NH)CCl_3$, TsOH.[15]

This ether has properties similar to the *p*-methoxybenzyl (MPM) ether except that it can be removed from an alcohol with DDQ in the presence of an MPM group with 98% selectivity.[9–11] The selectivity is attributed to the lower oxidation potential of the DMPM group; 1.45 V for the DMPM versus 1.78 V for the MPM.

DDQ, 81%

The dimethoxybenzyl ether has also been used for protection of the anomeric hydroxyl in carbohydrates.[17]

1. N. Nakajima, R. Abe, and O. Yonemitsu, *Chem. Pharm. Bull.*, **36,** 4244 (1988).
2. H. Takaku, S. Ueda, and T. Ito, *Tetrahedron Lett.*, **24,** 5363 (1983); N. Nakajima, K. Horita, R. Abe, and O. Yonemitsu, *Tetrahedron Lett.*, **29,** 4139 (1988).
3. K. Kamaike, H. Tsuchiya, K. Imai, and H. Takaku, *Tetrahedron*, **42,** 4701 (1986).
4. H. Takaku and K. Kamaike, *Chem. Lett.*, 189 (1982).
5. H. Takaku, K. Kamaike, and H. Tsuchiya, *J. Org. Chem.*, **49,** 51 (1984).
6. J. L. Marco and J. A. Hueso-Rodriquez, *Tetrahedron Lett.*, **29,** 2459 (1988).
7. W. Schmidt and E. Steckhan, *Angew. Chem., Int. Ed. Engl.*, **18,** 801 (1979). See also: E. A. Mayeda, L. L. Miller, and J. F. Wolf, *J. Am. Chem. Soc.*, **94,** 6812 (1972); S. M. Weinreb, G. A. Epling, R. Comi, and M. Reitano, *J. Org. Chem.*, **40,** 1356 (1975).
8. K. Horita, T. Yoshioka, T. Tanaka, Y. Oikawa, and O. Yonemitsu, *Tetrahedron*, **42,** 3021 (1986); T. Tanaka, Y. Oikawa, T. Hamada, and O. Yonemitsu, *Tetrahedron Lett.*, **27,** 3651 (1986).
9. Y. Oikawa, T. Tanaka, K. Horita, and O. Yonemitsu, *Tetrahedron Lett.*, **25,** 5397 (1984).

10. Y. Oikawa, T. Yoshioka, and O. Yonemitsu, *Tetrahedron Lett.*, **23,** 885 (1982).
11. D. S. Matteson and A. A. Kandil, *J. Org. Chem.*, **52,** 5121 (1987).
12. M. Hirama and M. Shimizu, *Synth. Commun.*, **13,** 781 (1983).
13. B. Classon, P. J. Garegg, and B. Samuelsson, *Acta Chem. Scand. Ser. B*, **B38,** 419 (1984); R. Johansson and B. Samuelsson, *J. Chem. Soc., Perkin Trans. I*, 2371 (1984).
14. G. Pandey and A. Krishna, *Synth. Commun.*, **18,** 2309 (1988).
15. B. M. Trost and J. Y. L. Chung, *J. Am. Chem. Soc.*, **107,** 4586 (1985).
16. H. Takaku, T. Ito, and K. Imai, *Chem. Lett.*, 1005 (1986).
17. S. J. Danishefsky, H. G. Selnick, R. E. Zelle, and M. P. DeNinno, *J. Am. Chem. Soc.*, **110,** 4368 (1988); A. De Mesmaeker, P. Hoffmann, and B. Ernst, *Tetrahedron Lett.*, **30,** 3773 (1989).

45. *o*-Nitrobenzyl Ether: *o*-$NO_2C_6H_4CH_2OR$ (Chart 1)

46. *p*-Nitrobenzyl Ether: *p*-$NO_2C_6H_4CH_2OR$

The *o*-nitrobenzyl and *p*-nitrobenzyl ethers can be prepared and cleaved by many of the methods described for benzyl ethers.[1] The *p*-nitrobenzyl ether is also prepared from an alcohol and *p*-nitrobenzyl alcohol (trifluoroacetic anhydride, 2,6-lutidine, CH_2Cl_2, 67% yield).[2] In addition, the *o*-nitrobenzyl ether can be cleaved by irradiation (320 nm, 10 min, quant. yield of carbohydrate[3,4]; 280 nm, 95% yield of nucleotide[5]). The *p*-nitrobenzyl ether has been cleaved by electrolytic reduction (−1.1 V, DMF, $R_4N^+X^-$, 60% yield)[6] and by reduction with $Na_2S_2O_4$ (pH 8–9, 80–95% yield).[7] These ethers can also be cleaved oxidatively (DDQ or electrolysis) after reduction to the aniline derivative.[2]

1. D. G. Bartholomew and A. D. Broom, *J. Chem. Soc., Chem. Commun.*, 38 (1975).
2. K. Fukase, H. Tanaka, S. Torii, and S. Kusumoto, *Tetrahedron Lett.*, **31,** 389 (1990).
3. U. Zehavi, B. Amit, and A. Patchornik, *J. Org. Chem.*, **37,** 2281 (1972); U. Zehavi and A. Patchornik, *J. Org. Chem.*, **37,** 2285 (1972).
4. For reviews of photoremovable protective groups, see: V. N. R. Pillai, *Synthesis*, 1 (1980); V. N. R. Pillai, *Org. Photochem.*, **9,** 225 (1987).
5. E. Ohtsuka, S. Tanaka, and M. Ikehara, *J. Am. Chem. Soc.*, **100,** 8210 (1978).
6. V. G. Mairanovsky, *Angew. Chem., Int. Ed. Engl.*, **15,** 281 (1976).
7. E. Guibe-Jampel and M. Wakselman, *Synth. Commun.*, **12,** 219 (1982).

47. *p*-Halobenzyl Ethers: *p*-X-$C_6H_4CH_2OR$, X = Br, Cl

p-Halobenzyl ethers have been prepared to protect side-chain hydroxyl groups in amino acids. They are more stable to the conditions of acidic hydrolysis (50% CF_3COOH) than the unsubstituted benzyl ether; they are cleaved by HF (0°, 10 min).[1] These ethers also impart greater crystallinity, which often aids purification.[2]

1. D. Yamashiro, *J. Org. Chem*, **42,** 523 (1977).
2. S. Koto, S. Inada, N. Morishima, and S. Zen, *Carbohydr. Res.*, **87,** 294 (1980).

48. 2,6-Dichlorobenzyl Ether: $2,6\text{-}Cl_2C_6H_3CH_2OR$

Formation[1]

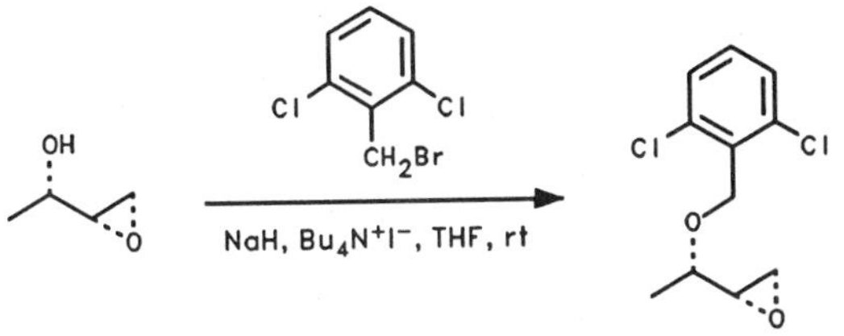

NaH, $Bu_4N^+I^-$, THF, rt

Cleavage[2]

This group is cleaved during an iodine-promoted tetrahydrofuran synthesis.

1. S. Hatakeyama, K. Sakurai, and S. Takano, *Heterocyles*, **24,** 633 (1986).
2. S. D. Rychnovsky and P. A. Bartlett, *J. Am. Chem. Soc.*, **103,** 3963 (1981).

49. *p*-Cyanobenzyl Ether: $p\text{-}CN\text{–}C_6H_4CH_2OR$

The *p*-cyanobenzyl ether, prepared from an alcohol and the benzyl bromide in the presence of sodium hydride (74% yield), can be cleaved by electrolytic reduction (−2.1 V, 71% yield). It is stable to electrolytic removal (−1.4 V) of a tritylone ether [i.e., 9-(9-phenyl-10-oxo)anthryl ether].[1]

1. C. van der Stouwe and H. J. Schäfer, *Tetrahedron Lett.*, 2643 (1979); *idem.*, *Chem. Ber.*, **114,** 946 (1981); J. P. Coleman, Naser-ud-din, H. G. Gilde, J. H. P. Utley, B. C. L. Weedon, and L. Eberson, *J. Chem. Soc., Perkin Trans. II*, 1903 (1973).

50. *p*-Phenylbenzyl Ether: $p\text{-}C_6H_5\text{–}C_6H_4CH_2OR$

Formation

The section on the formation of benzyl ethers should be consulted.

Cleavage

1. $FeCl_3$, CH_2Cl_2, 2–3 min, 68% yield.[1] Benzyl ethers are cleaved in 15–20 min under these conditions. Methyl glycosides, acetates and benzoates were not affected by this reagent.

1. M. H. Park, R. Takeda, and K. Nakanishi, *Tetrahedron Lett.*, **28,** 3823 (1987).

51. 2- and 4-Picolyl Ether: $C_5H_4NCH_2$-OR

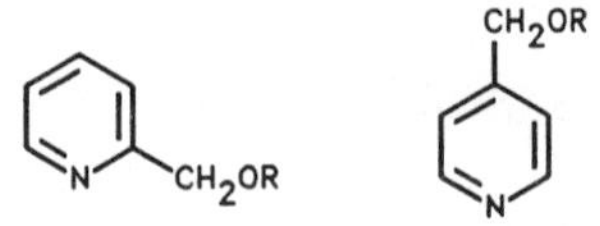

Picolyl ethers are prepared from their chlorides by a Williamson ether synthesis (68–83% yield). Some selectivity for primary versus secondary alcohols can be achieved (ratios = 4.3–4.6 : 1). They are cleaved electrolytically (−1.4 V, 0.5 *M* HBF_4, MeOH, 70% yield). Since picolyl chlorides are unstable as the free base, they must be generated from the hydrochloride prior to use.[1] These derivatives are relatively stable to acid (CF_3CO_2H, HF/anisole). Cleavage can also be effected by hydrogenolysis in acetic acid.[2]

1. S. Wieditz and H. J. Schaefer, *Acta Chem. Scand. Ser. B.*, **B37,** 475 (1983); A. Gosden, R. Macrae, and G. T. Young, *J. Chem. Res.* (*S*), 22 (1977).
2. J. Rizo, F. Albericio, G. Romero, C. G.-Esheverria, J. Claret, C. Muller, E. Giralt, and E. Pedroso, *J. Org. Chem.*, **53,** 5386 (1988).

52. 3-Methyl-2-picolyl *N*-Oxido Ether:

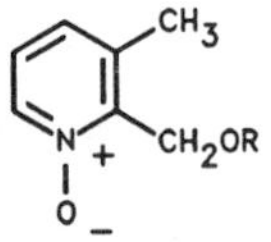

The authors prepared a number of substituted 2-diazomethylene derivatives of picolyl oxide to use for monoprotection of the *cis* glycol system in nucleosides. The 3-methyl derivative proved most satisfactory.[1]

Formation/Cleavage[1]

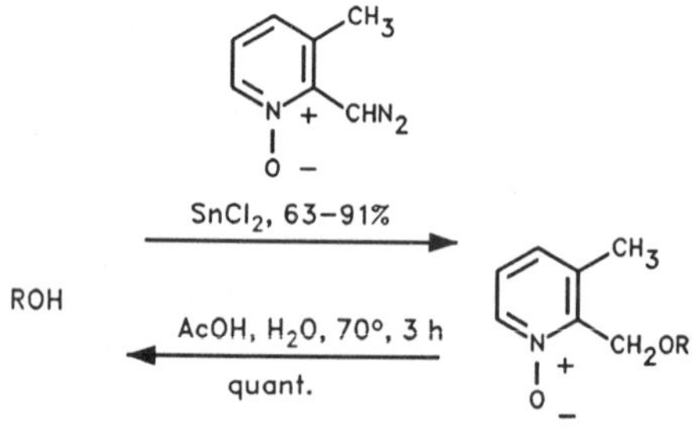

ROH → ($SnCl_2$, 63–91%) ; ← (AcOH, H_2O, 70°, 3 h, quant.)

1. Y. Mizuno, T. Endo, and K. Ikeda, *J. Org. Chem.*, **40,** 1385 (1975); Y. Mizuno, T. Endo, and T. Nakamura, *J. Org. Chem.*, **40,** 1391 (1975).

53. Diphenylmethyl Ether: Ph_2CHOR, DPM–OR

Formation

1. $(Ph_2CHO)_3PO$, cat. CF_3COOH, CH_2Cl_2, reflux, 4–9 h, 65–92% yield.[1] This methodology has been applied to the protection of amino acid alcohols.[2]
2. Ph_2CHOH, concd. H_2SO_4, 12 h, 70% yield.[3]
3. Ph_2CN_2, CH_3CN or benzene, 79–85% yield.[4]

Cleavage

1. Pd–C, $AlCl_3$, cyclohexene, reflux, 24 h, 91% yield.[5] Simple hydrogenation also cleaves this ether (71–100% yield).[5]
2. Electrolytic reduction: −3.0 V, DMF, $R_4N^+X^-$.[3]
3. 10% CF_3COOH, anisole, CH_2Cl_2.[2] Anisole is present to scavenge the diphenylmethyl cation liberated during the cleavage reaction.

1. L. Lapatsanis, *Tetrahedron Lett.*, 3943 (1978).
2. C. Froussios and M. Kolovos, *Synthesis* 1106 (1987); M. Kolovos and C. Froussios, *Tetrahedron Lett.*, **25,** 3909 (1984).
3. V. G. Mairanovsky, *Angew. Chem., Int. Ed. Engl.*, **15,** 281 (1976).
4. G. Jackson, H. F. Jones, S. Petursson, and J. M. Webber, *Carbohydr. Res.*, **102,** 147 (1982).
5. G. A. Olah, G. K. S. Prakash, and S. C. Narang, *Synthesis* 825 (1978).

54. *p,p'*-Dinitrobenzhydryl Ether: RO–DNB: $ROCH(C_6H_4\text{-}p\text{-}NO_2)_2$

Formation/Cleavage[1]

$$\text{ROH} \underset{\substack{\text{1. } PtO_2/H_2,\ Fe_3(CO)_{12} \text{ or } NaBH_4\text{–}Ni(OAc)_2 \\ \text{2. pH} < 5\text{, preferred is 3–4, 81–90\%}}}{\overset{(p\text{-}NO_2\text{–}C_6H_4)_2CN_2,\ BF_3\cdot Et_2O}{\rightleftarrows}} \text{RO—DNB}$$

The cleavage proceeds by initial reduction of the nitro groups followed by acid-catalyzed cleavage. The DNB group can be cleaved in the presence of allyl, benzyl, tetrahydropyranyl, methoxyethoxymethyl, methoxymethyl, silyl, trityl, and ketal protective groups.

1. G. Just, Z. Y. Wang, and L. Chan, *J. Org. Chem.*, **53,** 1030 (1988).

55. 5-Dibenzosuberyl Ether:

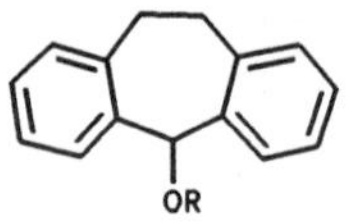

OR

The dibenzosuberyl ether is prepared from an alcohol and the suberyl chloride in the presence of triethylamine (CH_2Cl_2, 20°, 3 h, 75% yield). It is cleaved by acidic hydrolysis (1 *N* HCl/dioxane, 20°, 6 h, 80% yield). This group has also been used to protect amines, thiols, and carboxylic acids. The alcohol derivative can be cleaved in the presence of a dibenzosuberylamine.[1]

1. J. Pless, *Helv. Chim. Acta*, **59,** 499 (1976).

56. Triphenylmethyl Ether: Tr–OR, Ph_3C–OR (Chart 1)

Formation

1.

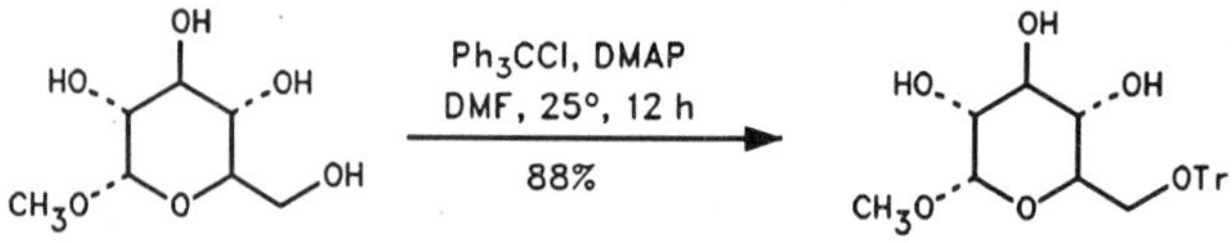

Ph_3CCl, DMAP
DMF, 25°, 12 h
88%

DMAP = 4-*N*, *N*-dimethylaminopyridine

A secondary alcohol reacts more slowly (40–45%, 18–24 h, 68–70% yield).[1]

2. $C_5H_5N^+CPh_3BF_4^-$, CH_3CN, Pyr, 60–70°, 75–90% yield.[2] Triphenylmethyl ethers can be prepared more readily with triphenylmethylpyridinium fluoroborate than with triphenylmethyl chloride–pyridine.
3. Ⓟ-*p*-$C_6H_4Ph_2CCl$, Pyr, 25°, 5 days, 90%, where Ⓟ = styrene–divinylbenzene polymer.[3] Triarylmethyl ethers of primary hydroxyl groups in glucopyranosides have been prepared using a polymeric form of triphenylmethyl chloride. Although the yields are not improved, the workup is simplified.
4. Ph_3CCl, 2,4,6-collidine, CH_2Cl_2, $Bu_4N^+ClO_4^-$, 15 min, 97% yield.[4] This is an improved procedure for installing the trityl group on polymer-supported nucleosides.
5. $Me_2NC_5H_5NCPh_3^+Cl^-$, CH_2Cl_2, 25°, 16 h, 95% yield.[5] In this case a primary alcohol is cleanly protected over a secondary alcohol. The reagent is a stable isolable salt. If the solvent is changed from CH_2Cl_2 to DMF, the amine of serine can be selectively protected.
6. The trityl group can migrate from one secondary center to another under acid catalysis.[6]

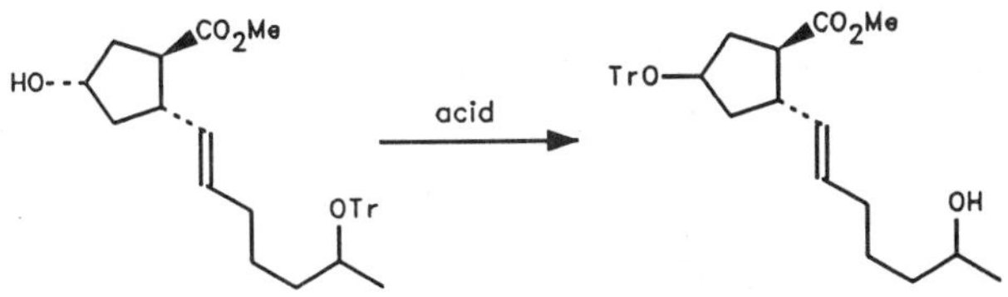

acid

7. $Ph_3COSiMe_3$, Me_3SiOTf, CH_2Cl_2, 0°, 0.5 h, 73–97% yield.[7] These conditions will also introduce the trityl group on a carboxyl group. The primary hydroxyl of persilylated ribose was selectively derivatized.

Cleavage

1. Formic acid, ether, 45 min, 88% yield.[8]

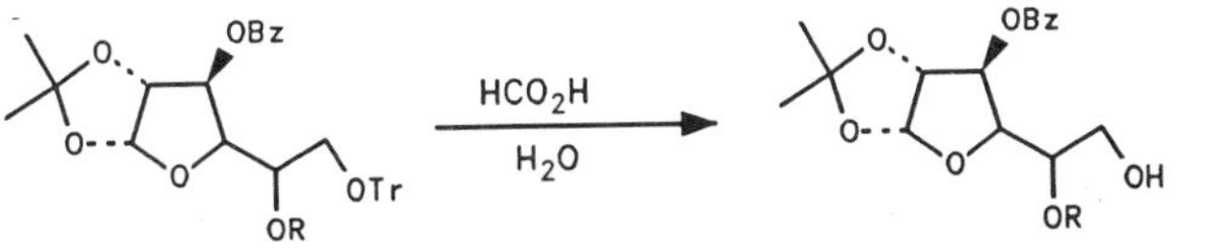

HCO_2H, H_2O

R = Ac 92%

R = TBDMS 88%

R = THP 60%/40% cleavage

2. $CuSO_4$ (anhydrous), benzene, heat, 89–100% yield.[9] In highly acylated carbohydrates, trityl removal proceeds without acyl migration.
3. Amberlyst 15-H, MeOH, rt, 5–10 min, 69–90%.[10]
4. AcOH, 56°, 7.5 h, 96%.[11]
5. 90% CF_3COOH, *t*-BuOH, 20°, 2–30 min, then Bio-Rad 1×2 (OH^-) resin.[12] These conditions were used to cleave the trityl group from the 5′-hydroxyl of a nucleoside. Bio-Rad resin neutralizes the hydrolysis and minimizes cleavage of glycosyl bonds.
6. SiO_2, benzene, 25°, 16 h, 81% yield.[13] This cleavage reaction is carried out on a column.
7. H_2/Pd, EtOH, 20°, 14 h, 80% yield.[14]
8. HCl(g), $CHCl_3$, 0°, 1 h, 91% yield.[15]
9. Electrolytic reduction: −2.9 V, $R_4N^+X^-$, DMF.[16]
10. $CH_3CH(OCPh_3)(CH_2)_4CH_2OCPh_3 \xrightarrow[20°,\ 15\ min,\ 91\%]{Ph_3C^+BF_4^-,\ CH_2Cl_2^{17}}$

    $CH_3CO(CH_2)_4CH_2OH$

    Since a secondary alcohol is oxidized in preference to a primary alcohol by $Ph_3C^+BF_4^-$, this reaction results in selective protection of a primary alcohol.
11. $SnCl_2$, Ac_2O, CH_3CN.[18] In this case a sulfoxide is also reduced.

12. Et_2AlCl, CH_2Cl_2, 3 min, 70–85% yield.[19] This method was used to remove the trityl group from various protected deoxyribonucleotides. The TBDPS group is stable to these conditions.
13. TsOH, MeOH, 25°, 5 h.[20]
14. $BF_3 \cdot Et_2O$, $HSCH_2CH_2SH$, 80% yield.[21]

$HSCH_2CH_2SH$, $BF_3 \cdot Et_2O$, 80%

15. Na, NH_3.[22] Benzyl groups are also removed under these conditions.
16. $ZnBr_2$, MeOH, 100% yield.[23]

57. α-Naphthyldiphenylmethyl Ether: RO–C(Ph)$_2$–α-$C_{10}H_7$ (Chart 1)

The α-naphthyldiphenylmethyl ether was prepared to protect, selectively, the 5′-OH group in nucleosides. It is prepared from α-naphthyldiphenylmethyl chloride in pyridine (65% yield), and cleaved selectively in the presence of a *p*-methoxyphenyldiphenylmethyl ether with sodium anthracenide, **a** (THF, 97% yield). The *p*-methoxyphenyldiphenylmethyl ether can be cleaved with acid in the presence of this group.[24]

Na^+

a

58. *p*-Methoxyphenyldiphenylmethyl Ether (MMTrOR): *p*-$MeOC_6H_4(Ph)_2C$–OR (Chart 1)

59. Di(*p*-methoxyphenyl)phenylmethyl Ether (DMTrOR): (*p*-$MeOC_6H_4)_2PhC$–OR

60. Tri(*p*-methoxyphenyl)methyl Ether (TMTrOR): (*p*-$MeOC_6H_4)_3C$–OR

These were originally prepared by Khorana[25] as selective protective groups for the 5′-OH of nucleosides and nucleotides. They were designed to be more acid-labile than the trityl group because depurination is often a problem in the acid-catalyzed removal of the trityl group. Introduction of *p*-methoxy groups increases the rate of hydrolysis by about one order of magnitude for each *p*-methoxy substituent. For 5′-protected uridine derivatives in 80% AcOH, 20°, the time for hydrolysis was as follows:

$$(p\text{-MeOC}_6\text{H}_4)_n(\text{Ph})_m\text{COR}$$

$n = 0, m = 3$, 48 h
$n = 1, m = 2$, 2 h
$n = 2, m = 1$, 15 min
$n = 3, m = 0$, 1 min

The trimethoxy derivative is too labile for most applications, but the mono and di-derivatives have been used extensively in the preparation of oligonucleotides and oligonucleosides. The monomethoxy derivative has been used for the selective protection of a primary allylic alcohol over a secondary allylic alcohol (MMTr, Pyr, −10°).[26]

Cleavage

In practice the various trityl derivatives are cleaved with acid, but the monomethoxy derivative can be cleaved with sodium naphthalenide in HMPA (90% yield).[27] It is not cleaved by sodium anthracenide, used to cleave α-naphthyldiphenylmethyl ethers.[24]

A solution of 3% CCl_3CO_2H in 95:5 CH_3NO_2/MeOH is recommended for removal of the DMTr group from the 5′-OH of deoxyribonucleotides because of reduced levels of depurination compared to Cl_3CO_2H/CH_2Cl_2, $PhSO_3H$/MeOH/CH_2Cl_2, and $ZnBr_2/CH_3NO_2$.[28]

61. 4-(4′-Bromophenacyloxyphenyl)diphenylmethyl Ether: *p*-(*p*-$BrC_6H_4C(O)CH_2O)C_6H_4(Ph)_2C$–OR

This group was developed for protection of the 5′-OH group in nucleosides. The derivative is prepared from the corresponding triarylmethyl chloride, and is cleaved by reductive cleavage (Zn/AcOH) of the phenacyl ether to the *p*-hydroxyphenyldiphenylmethyl ether followed by acidic hydrolysis with formic acid.[29]

62. 4,4′,4″-Tris(4,5-dichlorophthalimidophenyl)methyl Ether (CPTr–OR):

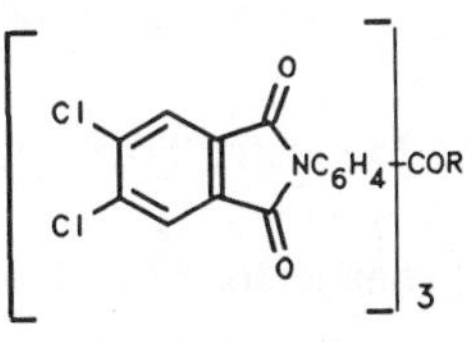

Cl Cl O O NC_6H_4–COR]$_3$

The CPTr group was developed for the protection of the 5′-OH of ribonucleosides. It is introduced with CPTrBr/$AgNO_3$/DMF (15 min) in 80–96% yield and can be removed by ammonia followed by 0.01 *M* HCl or 80% AcOH.[30] It can also be removed with hydrazine and acetic acid.[31]

63. 4,4′,4″-Tris(levulinoyloxyphenyl)methyl Ether (TLTr–OR):

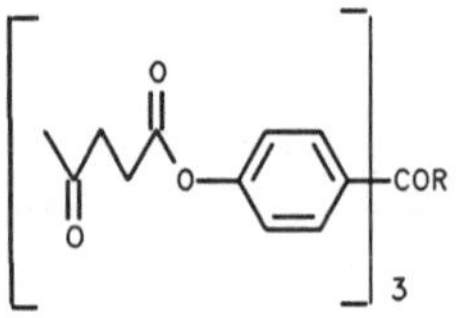

The TLTr group was developed for the protection of the 5′-OH of thymidine. It is introduced in 81% yield with TLTrBr/Pyr and is cleaved with hydrazine (3 min); Pyr–AcOH, 50°, 3 min, 81%. The $t_{1/2}$ in 80% AcOH is 24 h.[32]

64. 4,4′,4″-Tris(benzoyloxyphenyl)methyl Ether (TBTr–OR):

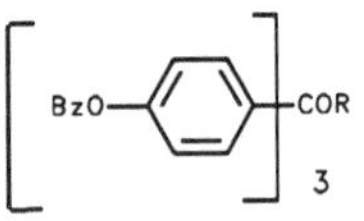

The TBTr group was prepared for 5′-OH protection in oligonucleotide synthesis. The group is introduced in >80% yield with TBTrBr/pyridine at 65°. It is 5 times more stable to 80% AcOH than the trityl group [$t_{1/2}$ (Tr) = 5 h; $t_{1/2}$ (TBTr) = 25 h]. The TBTr group is removed with 2 *M* NaOH. The di(4-methoxyphenyl)phenylmethyl (DMTr) group can be cleaved without affecting the TBTr derivative (80% AcOH, 95% yield).[33]

65. 3-(Imidazol-1-ylmethyl)bis(4′,4″-dimethoxyphenyl)methyl Ether (IDTr-OR):

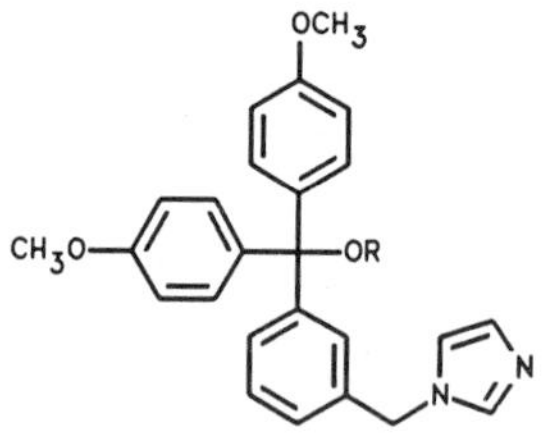

The IDTr group was developed to protect the 5′-OH of deoxyribonucleotides and to increase the rate of internucleotide bond formation through participation of the pendant imidazole group. Rate enhancements of ≈350 were observed except when $(i\text{-Pr})_2EtN$ was added to the reaction mixture, in which case reactions were complete in 30 s as opposed to the usual 5–6 h without the pendant imidazole group. The group is efficiently introduced with the bistetrafluoroborate salt, IDTr–BBF in DMF (70% yield). It is removed with 0.2 *M* Cl_2CHCO_2H or 1% CF_3COOH in CH_2Cl_2.[34]

66. 1,1-Bis(4-methoxyphenyl)-1′-pyrenylmethyl Ether (Bmpm–OR):

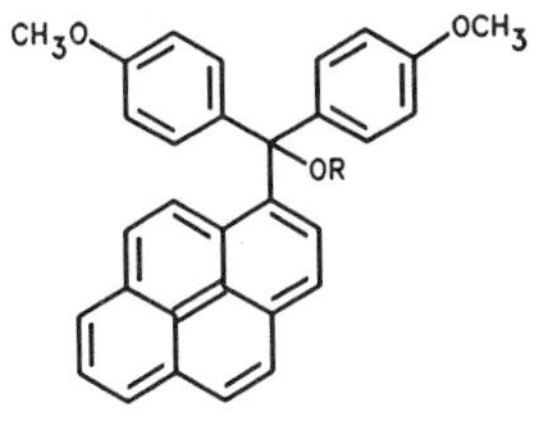

This substantial group was developed as a fluorescent, acid-labile protective group for oligonucleotide synthesis. It has properties very similar to those of the DMTr group except that it can be detected down to 10^{-10} *M* on TLC plates with 360-nm ultraviolet light.[35]

67. 9-Anthryl Ether: 9-anthryl-OR

This group is prepared by the reaction of the anion of 9-hydroxyanthracene and the tosylate of an alcohol. Since the formation of this group requires an S_N2 displacement on the alcohol to be protected, it is best suited for primary alcohols. It is cleaved by a novel singlet oxygen reaction followed by reduction of the endoperoxide with hydrogen and Raney nickel.[36]

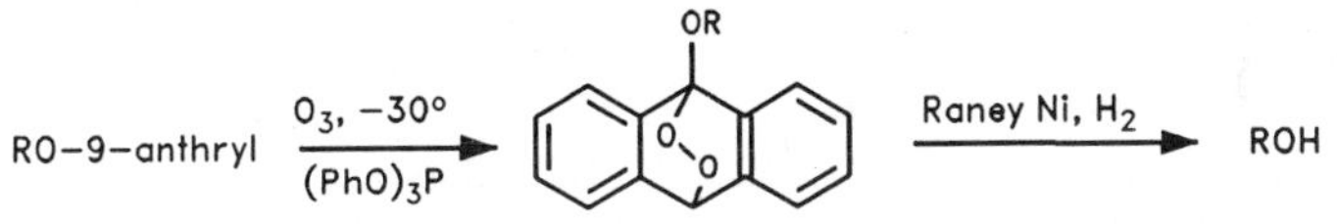

RO–9–anthryl $\xrightarrow[(PhO)_3P]{O_3,\ -30°}$ [endoperoxide, OR] $\xrightarrow{\text{Raney Ni, } H_2}$ ROH

68. 9-(9-Phenyl)xanthenyl Ether: pixyl-OR

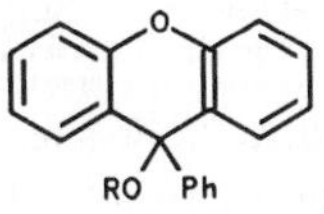

The pixyl ether is prepared from the xanthenyl chloride in 68–87% yield. This group has been used extensively in the protection of the 5′-OH of nucleosides; it is readily cleaved by acidic hydrolysis (80% AcOH, 20°, 8–15 min, 100% yield, or 3% trichloroacetic acid).[37] It can be cleaved under neutral conditions with $ZnBr_2$, thus reducing the extent of the often troublesome depurination of *N*-6-benzyloxyadenine residues during deprotection.[38] Conditions which remove the pixyl group also partially cleave the THP group ($t_{1/2}$ for THP at 2′-OH of ribonucleoside = 560 s in 3% Cl_2CHCO_2H/CH_2Cl_2).[39]

69. 9-(9-Phenyl-10-oxo)anthryl Ether (Tritylone Ether) (Chart 1):

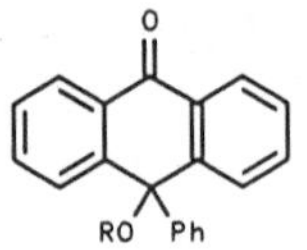

The tritylone ether is used to protect primary hydroxyl groups in the presence of secondary hydroxyl groups. It is prepared by the reaction of an alcohol with 9-phenyl-9-hydroxyanthrone under acid catalysis (cat. TsOH, benzene, reflux, 55–95% yield).[40,41] It can be cleaved under the harsh conditions of the Wolff-Kishner reduction (H_2NNH_2, NaOH, 200°, 88% yield),[24] and by electrolytic reduction (−1.4 V, LiBr, MeOH, 80–85% yield).[28] It is stable to 10% HCl, 55 h.[24]

1. S. K. Chaudhary and O. Hernandez, *Tetrahedron Lett.*, 95 (1979).
2. S. Hanessian and A. P. A. Staub, *Tetrahedron Lett.*, 3555 (1973).
3. J. M. J. Fréchet and K. E. Haque, *Tetrahedron Lett.*, 3055 (1975).
4. M. P. Reddy, J. B. Rampal, and S. L. Beaucage, *Tetrahedron Lett.*, **28,** 23 (1987).
5. O. Hernandez, S. K. Chaudhary, R. H. Cox, and J. Porter, *Tetrahedron Lett.*, **22,** 1491 (1981).
6. P. A. Bartlett and F. R. Green, III, *J. Am. Chem. Soc.*, **100,** 4858 (1978).
7. S. Murata and R. Noyori, *Tetrahedron Lett.*, **22,** 2107 (1981).
8. M. Bessodes, D. Komiotis, and K. Antonakis, *Tetrahedron Lett.*, **27,** 579 (1986).
9. G. Randazzo, R. Capasso, M. R. Cicala, and A. Evidente, *Carbohydr. Res.*, **85,** 298 (1980).
10. C. Malanga, *Chem. Ind. (London)*, 856 (1987).
11. R. T. Blickenstaff, *J. Am. Chem. Soc.*, **82,** 3673 (1960).
12. M. MacCoss and D. J. Cameron, *Carbohydr. Res.*, **60,** 206 (1978).
13. J. Lehrfeld, *J. Org. Chem.*, **32,** 2544 (1967).
14. R. N. Mirrington and K. J. Schmalzl, *J. Org. Chem.*, **37,** 2877 (1972); S. Hanessian and G. Rancourt, *Pure Appl. Chem*, **49,** 1201 (1977).
15. Y. M. Choy and A. M. Unrau, *Carbohydr. Res.*, **17,** 439 (1971).
16. V. G. Mairanovsky, *Angew. Chem., Int. Ed. Engl.*, **15,** 281 (1976).
17. M. E. Jung and L. M. Speltz, *J. Am. Chem. Soc.*, **98,** 7882 (1976).
18. B. M. Trost and L. H. Latimer, *J. Org. Chem.*, **43,** 1031 (1978).
19. H. Köster and N. D. Sinha, *Tetrahedron Lett.*, **23,** 2641 (1982).
20. A. Ichihara, M. Ubukata, and S. Sakamura, *Tetrahedron Lett.*, 3473 (1977).
21. P.-E. Sum and L. Weiler, *Can. J. Chem.*, **56,** 2700 (1978).
22. P. Kovác and S. Bauer, *Tetrahedron Lett.*, 2349 (1972).
23. V. Kohli, H. Bloecker, and H. Koester, *Tetrahedron Lett.*, **21,** 2683 (1983).
24. R. L. Letsinger and J. L. Finnan, *J. Am. Chem. Soc.*, **97,** 7197 (1975).

25. H. G. Khorana, *Pure Appl. Chem.*, **17,** 349 (1968); M. Smith, D. H. Rammler, I. H. Goldberg, and H. G. Khorana, *J. Am. Chem. Soc.*, **84,** 430 (1962).
26. J. Adams and J. Rokach, *Tetrahedron Lett.*, **25,** 35 (1984).
27. G. L. Greene and R. L. Letsinger, *Tetrahedron Lett.*, 2081 (1975).
28. H. Takaku, K. Morita, and T. Sumiuchi, *Chem. Lett.*, 1661 (1983).
29. A. T.-Rigby, Y.-H. Kim, C. J. Crosscup, and N. A. Starkovsky, *J. Org. Chem.*, **37,** 956 (1972).
30. M. Sekine and T. Hata, *J. Am. Chem. Soc.*, **108,** 4581 (1986).
31. M. D. Hagen, C. S.-Happ, E. Happ, and S. Chládek, *J. Org. Chem.*, **53,** 5040 (1988).
32. M. Sekine and T. Hata, *Bull. Chem. Soc. Jpn.*, **58,** 336 (1985).
33. M. Sekine and T. Hata, *J. Org. Chem.*, **48,** 3011 (1983).
34. M. Sekine and T. Hata, *J. Org. Chem.*, **52,** 946 (1987).
35. J. L. Fourrey, J. Varenne, C. Blonski, P. Dousset, and D. Shire, *Tetrahedron Lett*, **28,** 5157 (1987).
36. W. E. Barnett and L. L. Needham, *J. Chem. Soc., Chem. Commun.*, 1383 (1970); *idem, J. Org. Chem.*, **36,** 4134 (1971).
37. J. B. Chattopadhyaya and C. B. Reese, *J. Chem. Soc., Chem. Commun.*, 639 (1978).
38. M. D. Matteucci and M. H. Caruthers, *Tetrahedron Lett.*, **21,** 3243 (1980).
39. C. Christodoulou, S. Agrawal, and M. J. Gait, *Tetrahedron Lett.*, **27,** 1521 (1986).
40. W. E. Barnett, L. L. Needham, and R. W. Powell, *Tetrahedron*, **28,** 419 (1972).
41. C. van der Stouwe and H. J. Schäfer, *Tetrahedron Lett.*, 2643 (1979).

70. 1,3-Benzodithiolan-2-yl Ether (BDT–OR):

Formation

1. BDTO–*i*-Am, H^+, dioxane, rt, 81%.[1]
2. (1,3-benzodithiolylium BF_4^-) Pyr, CH_2Cl_2, 95%.[1]

The introduction of the BDT group proceeds under these rather neutral conditions, and this may prove advantageous for acid-sensitive substrates. The BDT group can also be reduced with Raney nickel to a methyl group or with Bu_3SnH to a 2-[(methylthio)phenylthio]methyl ether (MTPM ether).[2]

Cleavage

1. 80% AcOH, 100°, 30 min.[1]
2. 2% CF_3COOH, $CHCl_3$, 0°, 20 min, 97% yield.[1]

Half-lives for cleavage of 5′-protected thymidine in 80% AcOH at 15°

	DMTrT	mTHPT	BDTT	MMTrT	THPT	TBDT
$t_{1/2}$	3 min	23 min	38 min	48 min	3.5 h	2.5 h
$t_{complete}$	15 min	2.5 h	3 h	3 h	15 h	8 h

DMTrT = 5′-*O*-di-*p*-methoxytritylthymidine
mTHPT = 5′-*O*-(4-methoxytetrahydropyran-4-yl)thymidine
BDTT = 5′-*O*-(1,3-benzodithiol-2-yl)thymidine
MMTrT = 5′-*O*-mono-*p*-methoxytritylthymidine
THPT = 5′-tetrahydropyranylthymidine
TBDT = 3′-*O*-(1,3-benzodithiol-2-yl)thymidine

1. M. Sekine and T. Hata, *J. Am. Chem. Soc.*, **105,** 2044 (1983); *idem, J. Org. Chem.*, **48,** 3112 (1983).
2. M. Sekine and T. Nakanishi, *J. Org. Chem.*, **54,** 5998 (1989).

71. Benzisothiazolyl *S,S*-Dioxido Ether

Formation/Cleavage[1]

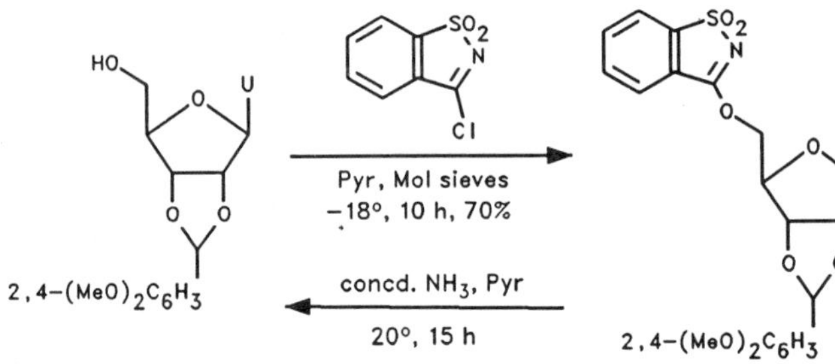

HO, U, O, O, O, 2,4-(MeO)$_2$C$_6$H$_3$

SO_2, N, Cl

Pyr, Mol sieves
−18°, 10 h, 70%

concd. NH_3, Pyr
20°, 15 h

1. H. Sommer and F. Cramer, *Chem. Ber.*, **107,** 24 (1974).

Silyl Ethers

72. Trimethylsilyl Ether (TMS Ether): $ROSi(CH_3)_3$ (Chart 1)

A large number of silylating agents exist for the introduction of the trimethylsilyl group onto a variety of alcohols. In general, the sterically least hindered alcohols are the most readily silylated, but these are also the most labile to hydrolysis with either acid or base. Trimethylsilylation is used extensively for derivatization of most functional groups to increase volatility for gas chromatography and mass spectrometry.

Formation

1. Me_3SiCl, Et_3N, THF, 25°, 8 h, 90% yield.[1]
2. Me_3SiCl, Li_2S, CH_3CN, 25°, 12 h, 75–95% yield.[2] Silylation occurs under neutral conditions with this combination of reagents.
3. $(Me_3Si)_2NH$, Me_3SiCl, Pyr, 20°, 5 min, 100% yield.[3] ROH is a carbohydrate. Hexamethyldisilazane (HMDS) is one of the most common silylating agents and readily silylates alcohols, acids, amines, thiols, phenols, hydroxamic acids, amides, thioamides, sulfonamides, phosphoric amides, phosphites, hydrazines, and enolizable ketones. It works best in the presence of a catalyst such as X–NH–Y, where at least one of the group X or Y is electron-withdrawing.[4]

HO~~~~OH —(0.5 eq HMDS, THF; 1 drop TMSCl; Reflux, 92%)→ HO~~~~OTMS   Ref. 5

4. $(Me_3Si)_2O$, PyH^+TsO^-, PhH, molecular sieves, reflux 4 days, 80–90% yield.[6] These mildly acidic conditions are suitable for acid-sensitive alcohols.
5. Me_3SiNEt_2.[7] Trimethylsilyldiethylamine selectively silylates equatorial hydroxyl groups in quantitative yield (4–10 h, 25°). The report indicated no reaction at axial hydroxyl groups. In the prostaglandin series the order of reactivity of trimethylsilyldiethylamine is $C_{11} > C_{15} >> C_9$ (no reaction). These trimethylsilyl ethers are readily hydrolyzed in aqueous methanol containing a trace of acetic acid.[8] The reagent is also useful for the silylation of amino acids.[9]

HO 9, CO_2Me, C_5H_{11}, 11 HO, 15 OH

6. $CH_3C(OSiMe_3){=}NSiMe_3$, DMF, 78°.[10] ROH is a C_{14}-hydroxy steroid. The sterically hindered silyl ether is stable to a Grignard reaction, but is hydrolyzed with 0.1 *N* HCl/10% aq. THF, 25°.[10] The reagent also silylates amides, amino acids, phenols, carboxylic acids, enols, ureas, and imides.[11] Most active hydrogen compounds can be silylated with this reagent.
7. $Me_3SiCH_2CO_2Et$, cat. $Bu_4N^+F^-$, 25°, 1–3 h, 90% yield. This reagent combination allows isolation of pure products under nonaqueous conditions. The reagent also converts aldehydes and ketones to trimethylsilyl

enol ethers.[12] The analogous methyl trimethylsilylacetate has also been used.[13]

8. $Me_3SiNHSO_2OSiMe_3$, CH_2Cl_2, 30°, 0.5 h, 92–98% yield. Higher yields of trimethylsilyl derivatives are realized by reaction of aliphatic, aromatic, and carboxylic hydroxyl groups with *N,O*-bis(trimethylsilyl)sulfamate than by reaction with *N,O*-bis(trimethylsilyl)acetamide.[14]
9. $Me_3SiNHCO_2SiMe_3$, THF, rapid, 80–95% yield. This reagent also silylates phenols and carboxyl groups.[15]
10. $MeCH{=}C(OMe)OSiMe_3$, CH_3CN or CH_2Cl_2, 50°, 30–50 min, 83–95% yield.[16] This reagent also silylates phenols, thiols, amides, and carboxyl groups.
11. $Me_3SiCH_2CH{=}CH_2$, TsOH, CH_3CN, 70–80°, 1–2 h, 90–95% yield.[17] This silylating reagent is stable to moisture. Allylsilanes can be used to protect alcohols, phenols, and carboxylic acids; there is no reaction with thiophenol except when CF_3SO_3H[18] is used as a catalyst. The method is also applicable to the formation of *t*-butyldimethylsilyl derivatives; the silyl ether of cyclohexanol was prepared in 95% yield from allyl-*t*-butyldimethylsilane. Iodine, bromine, trimethylsilyl bromide, and trimethylsilyl iodide have also been used as catalysts.[19] Nafion-H has been shown to be an effective catalyst.[20]
12. $(Me_3SiO)_2SO_2$.[21] This is a powerful silylating reagent, but has seen little application in organic chemistry.
13. *N,O*-Bis(trimethylsilyl)trifluoroacetamide.[22] The reagent is suitable for the silylation of carboxylic acids, alcohols, phenols, amides, and ureas. It has the advantage over bis(trimethylsilyl)acetamide in that the byproducts are more volatile.
14. *N,N'*-Bistrimethylsilylurea, CH_2Cl_2.[23] The reagent readily silylates carboxylic acids and alcohols. The byproduct urea is easily removed by filtration.
15. Me_3SiSEt.[24] Alcohols, thiols, amines and carboxylic acids are silylated.
16. Nafion–TMS, Et_3N, CH_2Cl_2, 100% yield.[25]
17. Isopropenyloxytrimethylsilane.[26] In the presence of an acid catalyst the reagent silylates alcohols and phenols. It also silylates carboxylic acids without added catalyst.
18. Methyl 3-trimethylsiloxy-2-butenoate.[27] This reagent silylates primary, secondary, and tertiary alcohols at room temperature without added catalyst.
19. *N*-Methyl-*N*-trimethylsilylacetamide.[28] This reagent has been used preparatively to silylate amino acids.[29]
20. Trimethylsilyl cyanide.[30] This reagent readily silylates alcohols, phenols, and carboxylic acids, and more slowly, thiols and amines. Amides and related compounds do not react with this reagent. The reagent has the advantage that a volatile gas (HCN is highly toxic) is the only byproduct.

21. $Me_3SiOC(O)NMe_2$.[31] This reagent produces only volatile byproducts and autocatalytically silylates alcohols, phenols, and carboxylic acids.
22. Trimethylsilylimidazole, CCl_4 or THF, rt.[32] This is a powerful silylating agent for hydroxyl groups. Basic amines are not silylated with this reagent, but as the acidity increases silylation can occur.
23. Trimethylsilyl trichloroacetate, K_2CO_3, 18-crown-6, 100–150°, 1–2 h, 80–90% yield.[33] This reagent silylates phenols, thiols, carboxylic acids, acetylenes, urethanes, and β-keto esters, producing CO_2 and chloroform as byproducts.
24. 3-Trimethylsilyloxazolidinone.[34] This reagent can be used to silylate most active hydrogen compounds.
25. Trimethylsilyl trifluoromethanesulfonate. This is an extremely powerful silylating agent, but probably is more useful for its many other applications in synthetic chemistry.[35]

Cleavage

Trimethylsilyl ethers are quite susceptible to acid hydrolysis, but acid stability is quite dependent on the local steric environment. For example, the 17α-TMS ether of a steroid is quite difficult to hydrolyze.

1. $Bu_4N^+F^-$, THF, aprotic conditions.[1]
2. K_2CO_3, anhydrous MeOH, 0°, 45 min, 100% yield.[36]

OTMS, TMSO, OTMS, OTMS, CH_3O, O → (K_2CO_3, MeOH, 0°, 45 min, 100%) → OTMS, TMSO, OTMS, OH, CH_3O, O

3. Citric acid, MeOH, 20°, 10 min, 100% yield.[37]
4. Rexyn 101 (polystyrene sulfonic acid), 80–91% yield.[38] This method does not cleave the *t*-butyldimethylsilyl ether.
5. HF, CH_3CN, H_2O.

1. E. J. Corey and B. B. Snider, *J. Am. Chem. Soc.*, **94,** 2549 (1972).
2. G. A. Olah, B. G. B. Gupta, S. C. Narang, and R. Malhotra, *J. Org. Chem.*, **44,** 4272 (1979).
3. C. C. Sweeley, R. Bentley, M. Makita, and W. W. Wells, *J. Am. Chem. Soc.*, **85,** 2497 (1963).
4. C. A. Bruynes and T. K. Jurriens, *J. Org. Chem.*, **47,** 3966 (1982).
5. R. K. Kanjolia and V. D. Gupta, *Z. Naturforsch. B*, **35B,** 767 (1980).

6. (a) H. W. Pinnick, B. S. Bal, and N. H. Lajis, *Tetrahedron Lett.*, 4261 (1978); (b) H. Matsumoto, Y. Hoshio, J. Nakabayashi, T. Nakano, and Y. Nagai, *Chem. Lett.*, 1475 (1980).
7. I. Weisz, K. Felföldi, and K. Kovács, *Acta Chim. Acad. Sci. Hung.*, **58,** 189 (1968).
8. (a) E. W. Yankee, U. Axen, and G. L. Bundy, *J. Am. Chem. Soc.*, **96,** 5865 (1974); (b) E. L. Cooper and E. W. Yankee, *J. Am. Chem. Soc.*, **96,** 5876 (1974).
9. K. Rühlmann, *J. Prakt. Chem.*, **9,** 315 (1959); K. Rühlmann, *Chem. Ber.*, **94,** 1876 (1961).
10. M. N. Galbraith, D. H. S. Horn, E. J. Middleton, and R. J. Hackney, *J. Chem. Soc., Chem. Commun.*, 466 (1968).
11. J. F. Klebe, H. Finkbeiner, and D. M. White, *J. Am. Chem. Soc.*, **88,** 3390 (1966).
12. E. Nakamura, T. Murofushi, M. Shimizu, and I. Kuwajima, *J. Am. Chem. Soc.*, **98,** 2346 (1976).
13. (a) L. A. Paquette and T. Sugimura, *J. Am. Chem. Soc.*, **108,** 3841 (1986); (b) T. Sugimura and L. A. Paquette, *J. Am. Chem. Soc.*, **109,** 3017 (1987).
14. B. E. Cooper and S. Westall, *J. Organomet. Chem.*, **118,** 135 (1976).
15. L. Birkofer and P. Sommer, *J. Organomet. Chem.*, **99,** Cl (1975).
16. Y. Kita, J. Haruta, J. Segawa, and Y. Tamura, *Tetrahedron Lett.*, 4311 (1979).
17. T. Morita, Y. Okamoto, and H. Sakurai, *Tetrahedron Lett.*, **21,** 835 (1980).
18. G. A. Olah, A. Husain, B. G. B. Gupta, G. F. Salem, and S. C. Narang, *J. Org. Chem.*, **46,** 5212 (1981).
19. H. Hosomi and H. Sakurai, *Chem. Lett.*, 85 (1981).
20. G. A. Olah, A. Husain, and B. P. Singh, *Synthesis*, 892 (1983).
21. L. H. Sommer, G. T. Kerr, and F. C. Whitmore, *J. Am. Chem. Soc.*, **70,** 445 (1948).
22. (a) D. L. Stalling C. W. Gehrke, and R. W. Zumalt, *Biochem. Biophys. Res. Commun.*, **31,** 616 (1968); (b) M. G. Horning, E. A. Boucher, and A. M. Moss, *J. Gas Chromatog.*, 297 (1967).
23. W. Verboom, G. W. Visser, and D. N. Reinhoudt, *Synthesis*, 807 (1981).
24. E. W. Abel, *J. Chem. Soc.*, 4406 (1960); *idem*, *ibid.*, 4933 (1961).
25. S. Murata and R. Noyori, *Tetrahedron Lett.*, **21,** 767 (1980).
26. M. Donike and L. Jaenicke, *Angew. Chem., Int. Ed. Engl.*, **8,** 974 (1969).
27. T. Veysoglu and L. A. Mitscher, *Tetrahedron Lett.*, **22,** 1303 (1981).
28. L. Birkofer and M. Donike, *J. Chromatogr.*, **26,** 270 (1967).
29. H. R. Kricheldorf, *Justus Liebigs Ann. Chem.*, **763,** 17 (1972).
30. K. Mai and G. Patil, *J. Org. Chem.*, **51,** 3545 (1986).
31. D. Knausz, A. Meszticzky, L. Szakacs, B. Csakvari, and K. D. Ujszaszy, *J. Organomet. Chem.*, **256,** 11 (1983); D. Knausz, A. Meszticzky, L. Szakacs, and B. Csakvari, *J. Organomet. Chem.*, **268,** 207 (1984).
32. S. Torkelson and C. Ainsworth, *Synthesis*, 722 (1976).
33. J. M. Renga and P.-C. Wang, *Tetrahedron Lett.*, **26,** 1175 (1985).
34. C. Palomo, *Synthesis*, 809 (1981); J. M. Aizpurua, C. Palomo, and A. L. Palomo, *Can. J. Chem.*, **62,** 336 (1984).
35. Review: H. Emde, D. Domsch, H. Feger, U. Frick, A. Götz, H. H. Hergott, K.

Hofmann, W. Kober, K. Krägeloh, T. Oesterle, W. Steppan, W. West, and G. Simchen, *Synthesis*, 1 (1982).
36. D. T. Hurst and A. G. McInnes, *Can. J. Chem.*, **43,** 2004 (1965).
37. G. L. Bundy and D. C. Peterson, *Tetrahedron Lett.*, 41 (1978).
38. R. A. Bunce and D. V. Hertzler, *J. Org. Chem.*, **51,** 3451 (1986).

73. Triethylsilyl Ether: TES–OR: Et_3SiOR

The triethylsilyl ether is approximately 10–100 times more stable[1] than the TMS ether and thus shows a greater stability to many reagents. Although TMS ethers can be cleaved in the presence of TES ethers, steric factors will play an important role in determining selectivity. The TES ether can be cleaved in the presence of a *t*-butyldimethylsilyl ether using 2% HF in acetonitrile.[2] In general, methods used to cleave the TBDMS ether are effective for cleavage of the TES ether.

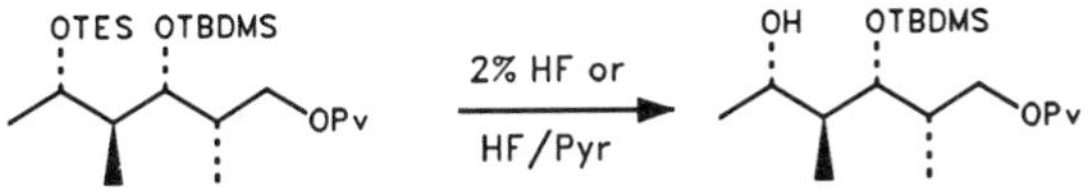

OTES OTBDMS ... OPv —(2% HF or HF/Pyr)→ OH OTBDMS ... OPv

Pv = Pivaloate

Formation

1. Et_3SiCl, Pyr. Triethylsilyl chloride is by far the most common reagent for the introduction of the TES group.[3] Silylation also occurs with imidazole and DMF[4] and with dimethylaminopyridine as a catalyst.[5] Phenols,[6] carboxylic acids,[1] and amines[7] have also been silylated with TESCl.

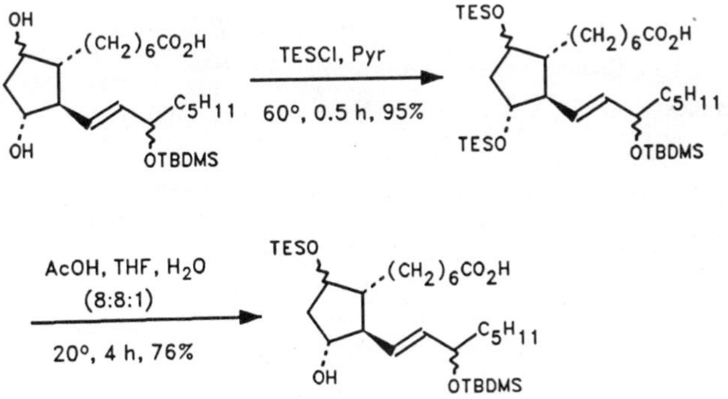

OH, $(CH_2)_6CO_2H$, C_5H_{11}, OH, OTBDMS —(TESCl, Pyr; 60°, 0.5 h, 95%)→ TESO, $(CH_2)_6CO_2H$, C_5H_{11}, TESO, OTBDMS

Ref. 3

—(AcOH, THF, H_2O (8:8:1); 20°, 4 h, 76%)→ TESO, $(CH_2)_6CO_2H$, C_5H_{11}, OH, OTBDMS

More acidic conditions [AcOH, THF, H_2O (6 : 1 : 3), 45°, 3 h] cleave all of the protective groups, 76% yield.

2. *N*-Methyl-*N*-triethylsilyltrifluoroacetamide.[8]
3. Allyltriethylsilane.[9]
4. Triethylsilane, CsF, imidazole.[10]
5. Triethylsilyl cyanide.[11]
6. *N*-Triethylsilylacetamide.[12]
7. Triethylsilyldiethylamine.[13]
8. 1-Methoxy-1-triethylsiloxypropene.[14]
9. 1-Methoxy-2-methyl-1-triethylsiloxypropene.[15]
10. Triethylsilyl triflate.[16]
11. Triethylsilyl perchlorate. This reagent represents an explosion hazard.[17]

1. C. E. Peishoff and W. L. Jorgensen, *J. Org. Chem.*, **48,** 1970 (1983).
2. D. Boschelli, T. Takemasa, Y. Nishitani, and S. Masamune, *Tetrahedron Lett.*, **26,** 5239 (1985).
3. T. W. Hart, D. A. Metcalfe, and F. Scheinmann, *J. Chem. Soc., Chem. Commun.*, 156 (1979).
4. W. Oppolzer, R. L. Snowden, and D. P. Simmons, *Helv. Chim. Acta*, **64,** 2002 (1981).
5. W. R. Roush and S. Russo-Rodriquez, *J. Org. Chem.*, **52,** 598 (1987).
6. T. L. McDonald, *J. Org. Chem.*, **43,** 3621 (1978).
7. R. West, P. Nowakowski, and P. Boudjouk, *J. Am. Chem. Soc.*, **98,** 5620 (1976).
8. M. Donike and J. Zimmermann, *J. Chromatogr.*, **202,** 483 (1980).
9. A. Hosomi and H. Sakurai, *Chem. Lett.*, 85 (1981).
10. L. Horner and J. Mathias, *J. Organomet. Chem.*, **282,** 175 (1985).
11. K. Mai and P. Patil, *J. Org. Chem.*, **51,** 3545 (1986).
12. J. Dieckman and C. Djerassi, *J. Org. Chem.*, **32,** 1005 (1967); J. Dieckman, J. B. Thompson, and C. Djerassi, *ibid.*, 3904 (1967).
13. A. R. Bassindale and D. R. M. Walton, *J. Organomet. Chem.*, **25,** 389 (1970).
14. Y. Kita, J. Haruta, J. Segawa, and Y. Tamura, *Tetrahedron Lett.*, 4311 (1979).
15. E. Yoshii and K. Takeda, *Chem. Pharm. Bull.*, **31,** 4586 (1983).
16. C. H. Heathcock, S. D. Young, J. P. Hagen, R. Pilli, and U. Badertscher, *J. Org. Chem.*, **50,** 2095 (1985).
17. (a) T. J. Barton and C. R. Tully, *J. Org. Chem.*, **43,** 3649 (1978); (b) D. B. Collum, J. H. McDonald, III, and W. C. Still, *J. Am. Chem. Soc.*, **102,** 2117 (1980); (c) For *O*-silylation of esters, see C. S. Wilcox and R. E. Babston, *Tetrahedron Lett.*, **25,** 699 (1984).

74. Triisopropylsilyl Ether (TIPS–OR): $(i\text{-Pr})_3SiOR$ (Chart 1)

The greater bulkiness of the TIPS group makes it more stable than the *t*-butyldimethylsilyl (TBDMS) group, but not as stable as the *t*-butyldiphenylsilyl (TBDPS) group to acidic hydrolysis. The TIPS group is more stable to basic hydrolysis than the TBDMS group and the TBDPS group.[1] Introduction of the TIPS group onto primary hydroxyls proceeds selectively over secondary hydroxyls.[2] The TIPS group has been used to prevent chelation with Grignard reagents during additions to carbonyls.[3]

Formation

1. TIPSCl, imidazole, DMF, 82% yield,[2]
2. TIPSCl, imidazole, DMAP, CH_2Cl_2.[4]
3. TIPSCl, pyridine, $AgNO_3$ or $Pb(NO_3)_2$, > 90% yield.[5] These conditions cleanly introduce the hindered TIPS group on the 3′-position of thymidine.
4. $TIPSOSO_2CF_3$, 2,6-lutidine, CH_2Cl_2.[6]
5. TIPSH, CsF, imidazole.[7]

Cleavage

1. 0.01 *N* HCl, EtOH, 90°, 15 min, 100% yield.[2]
2. HF, CH_3CN.[8]
3. $Bu_4N^+F^-$, THF.[9]
4. 80% AcOH, H_2O.[10]

1. R. F. Cunico and L. Bedell, *J. Org. Chem.*, **45,** 4797 (1980).
2. K. K. Ogilvie, E. A. Thompson, M. A. Quilliam, and J. B. Westmore, *Tetrahedron Lett.*, 2865 (1974).
3. S. V. Frye and E. L. Eliel, *Tetrahedron Lett.*, **27,** 3223 (1986).
4. M. Ohwa and E. L. Eliel, *Chem. Lett.*, 41 (1987).
5. S. Nishino, Y. Nagato, H. Yamamoto, and Y. Ishido, *J. Carbohydr. Chem.*, **5,** 199 (1986).
6. E. J. Corey, H. Cho, C. Rücker, and D. H. Hua, *Tetrahedron Lett.*, **22,** 3455 (1981); K. Tanaka, H. Yoda, Y. Isobe, and A. Kaji, *J. Org. Chem.*, **51,** 1856 (1986).
7. L. Horner and J. Mathias, *J. Organomet. Chem.*, **282,** 175 (1985).
8. J. L. Mascareñas, A. Mouriño, and L. Castedo, *J. Org. Chem.*, **51,** 1269 (1986).
9. J. C.-Y. Cheng, U. Hacksell, and G. P. Daves, Jr., *J. Org. Chem.*, **51,** 4941 (1986).
10. K. K. Ogilvie, S. L. Beaucage, D. W. Entwistle, E. A. Thompson, M. A. Quilliam, and J. B. Westmore, *J. Carbohyd., Nucleosides, Nucleotides*, **3,** 197 (1976).

75. Dimethylisopropylsilyl Ether (IPDMS–OR): $ROSiMe_2$–*i*-Pr (Chart 1)

Formation

1. $(i\text{-}PrMe_2Si)_2NH$, $i\text{-}PrMe_2SiCl$, 25°, 48 h, 98% yield.[1]
2. $i\text{-}PrMe_2SiCl$, imidazole, DMF, 26°, 2 h, 65% yield.[2]

Cleavage

1. AcOH/H_2O, (3 : 1), 35°, 10 min, 100% yield.[1] An IPDMS ether is more easily cleaved than a THP ether. It is not stable to Grignard or Wittig reactions, or to Jones oxidation.

1. E. J. Corey and R. K. Varma, *J. Am. Chem. Soc.*, **93,** 7319 (1971).
2. K. Toshima, K. Tatsuta, and M. Kinoshita, *Tetrahedron Lett.*, **27,** 4741 (1986).

76. Diethylisopropylsilyl Ether (DEIPS–OR): $ROSiEt_2$–*i*-Pr

This group is more labile to hydrolysis than the TBDMS group and has been used to protect an alcohol where the TBDMS group was too resistant to cleavage. The DEIPS group is approximately 90 times more stable than the TMS group to acid hydrolysis and 600 times more stable than the TMS group to base-catalyzed solvolysis.

Formation

1. Diethylisopropylsilyl chloride, imidazole, CH_2Cl_2, 25°, 1 h.[1]
2. Et_2-*i*-PrSiOTf, CH_2Cl_2, 2,6-lutidine, rt.[2]

Cleavage

1. 3 : 1 : 3 AcOH, H_2O, THF.[1] Any of the methods used to cleave the TBDMS ether also cleave the DEIPS ether.

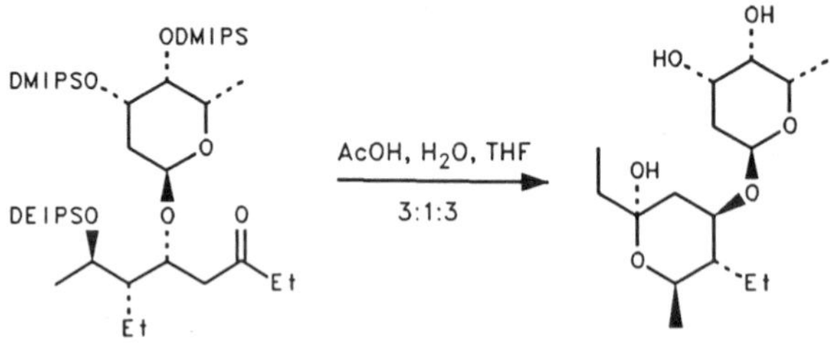

ODMIPS DMIPSO DEIPSO O O Et Et — AcOH, H_2O, THF 3:1:3 → OH HO OH O O O Et

DMIPS = Me_2–*i*-PrSi

2. AcOH, KF·HF, THF, H_2O, 30°, 46 h, 94% yield.[3] These conditions did not affect a secondary OTBDMS group.

1. K. Toshima, K. Tatsuta, and M. Kinoshita, *Tetrahedron Lett.*, **27,** 4741 (1986).
2. K. Toshima, S. Mukaiyama, M. Kinoshita, and K. Tatsuta, *Tetrahedron Lett.*, **30,** 6413 (1989).

3. K. Toshima, M. Misawa, K. Ohta, K. Tatsuta, and M. Kinoshita, *Tetrahedron Lett.*, **30,** 6417 (1989).

77. Dimethylthexylsilyl Ether: TDS–OR: $(CH_3)_2CHC(CH_3)_2Si(CH_3)_2$–OR

Both TDSCl and $TDSOSO_2CF_3$ are used to introduce the TDS group. In general, conditions similar to those used to introduce the TBDMS group are effective. This group is slightly more hindered than the TBDMS group, and the chloride has the advantage of being a liquid, which is useful for handling large quantities of material. Cleavage of this group can be accomplished with the same methods used to cleave the TBDMS group, but it is 2–3 times slower due to its increased steric bulk.[1]

1. H. Wetter and K. Oertle, *Tetrahedron Lett.*, **26,** 5515 (1985).

78. *t*-Butyldimethylsilyl Ether (TBDMS–OR): *t*-$BuMe_2SiOR$ (Chart 1)

The TBDMS ether has become one of the most popular silyl protective groups used in chemical synthesis. It is easily introduced with a variety of reagents, has the virtue of being quite stable to a variety of organic reactions, and is readily removed under conditions that do not attack other functional groups. It is approximately 10^4 times more stable to hydrolysis than is the trimethylsilyl (TMS) group. It has excellent stability toward base but is relatively sensitive to acid. The ease of introduction and removal of the TBDMS ether are influenced by steric factors that often allow for its selective introduction in polyfunctionalized, sterically differentiated molecules. It is relatively easy to introduce a primary TBDMS group in the presence of a secondary alcohol. The TBDMS group tends to migrate to adjacent hydroxyl groups under basic conditions, which include deprotection with the basic fluoride ion.[1] The migratory aptitude in nucleosides was found to be solvent dependent, with migration proceeding fastest in protic solvents.[2]

Formation

1. TBDMSCl, imidazole, DMF, 25°, 10 h, high yields.[3] This is the most common method for the introduction of the TBDMS group on alcohols with low steric demand. The method works best when the reactions are run in very concentrated solutions. This combination of reagents also silylates phenols,[4] hydroperoxides,[5] and hydroxyl amines.[6] Thiols, amines, and carboxylic acids are not effectively silylated under these conditions.[7]
2. $TBDMSOClO_3$, CH_3CN, Pyr, 20 min, 100% yield.[8] This reagent works well but has the disadvantage of being **explosive** and has been supplanted by $TBDMSOSO_2CF_3$.
3. $TBDMSOSO_2CF_3$, CH_2Cl_2, 2,6-lutidine, 0–25°.[9] This is one of the most powerful methods for introducing the TBDMS group. Other bases such as

triethylamine,[10] ethyldiisopropylamine,[11] and pyridine[12] have also been used successfully. In the presence of an ester or ketone it is possible simultaneously to form a silyl enol ether while silylating a hydroxyl group.[8]

4. TBDMSCl, Li_2S, CH_3CN, 25°, 5–8 h, 75–95% yield.[13] This reaction occurs under nearly neutral conditions.
5. TBDMSCl, DMAP, Et_3N, DMF, 25°, 12 h.[14] These conditions were used to silylate selectively a primary over a secondary alcohol.[15] Besides DMAP, other catalysts such as 1,1,3,3-tetramethylguanidine,[16] 1,8-diazabicyclo[5.4.0]undec-7-ene (83–99%),[7] 1,5-diazabicyclo[4.3.0]non-5-ene,[17] and ethyldiisopropylamine have been used.[18]
6. $TBDMSCH_2CH{=}CH_2$, TsOH, CH_3CN, 70–80°, 2.5 h, 95% yield.[19]
7. 4-*t*-Butyldimethylsiloxy-3-penten-2-one, DMF, TsOH, rt, 83–92% yield.[20]
8. 1-(*t*-Butyldimethylsilyl)imidazole.[21,22]
9. *N*-*t*-Butyldimethylsilyl-*N*-methyltrifluoroacetamide, CH_3CN, 5 min, 97–100% yield.[23] This reagent also silylates thiols, amines, amides, carboxylic acids, and enolizable carbonyl groups.
10. TBDMSCl, KH, 18-crown-6, THF, 0°–rt, 78% yield.[24] This combination of reagents is very effective in silylating extremely hindered alcohols.
11. 1-(*t*-Butyldimethylsiloxy)-1-methoxyethene, CH_3CN, 91–100% yield.[25] This reagent also silylates thiols and carboxylic acids.
12. TBDMSCN, 80°, 5 min, 95% yield.[26]
13. A secondary alcohol was selectively protected in the presence of a secondary allylic alcohol with TBDMSOTf, 2,6-lutidine at −78°.[27]

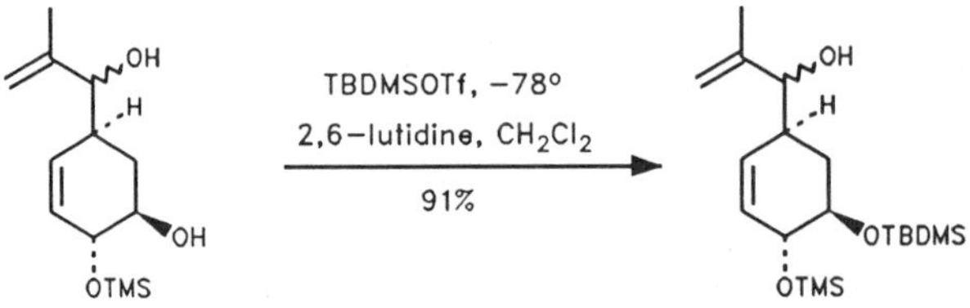

OH, H, OH, OTMS — TBDMSOTf, −78°, 2,6−lutidine, CH_2Cl_2, 91% → OH, H, OTBDMS, OTMS

The following schemes represent some interesting examples where the TBDMS group is introduced selectively on compounds with more than one alcohol.

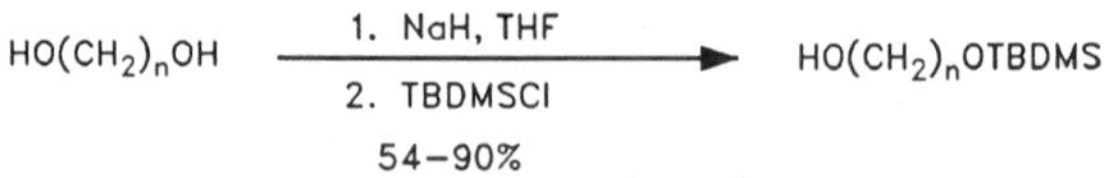

$HO(CH_2)_nOH$ — 1. NaH, THF; 2. TBDMSCl; 54−90% → $HO(CH_2)_nOTBDMS$ Ref. 28

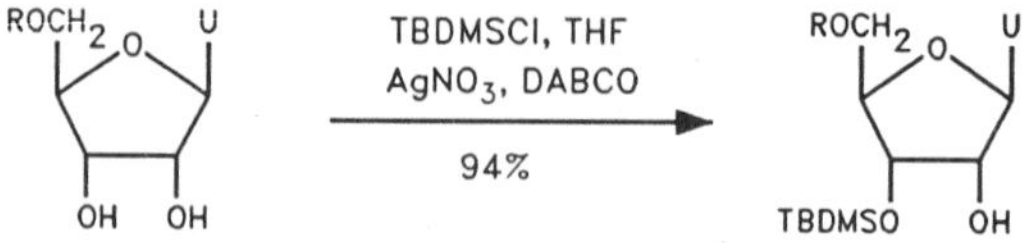

$ROCH_2$, O, U, OH, OH — TBDMSCl, THF, $AgNO_3$, DABCO, 94% → $ROCH_2$, O, U, TBDMSO, OH Ref. 29

R = DMTr or TBDMS

DMTr = $(4\text{-}MeOC_6H_4)_2PhC\text{-}$

HO, U, OH OH → TBDMSCl, Pyr / $AgNO_3$, 90% → TBDMSO, U, OH OTBDMS Ref. 30

HO, A, HO, OH → TBDMSCl, Pyr / $AgNO_3$ → TBDMSO, A, RO, OH + 3′,5′-isomer

TBDMSCl, $AgNO_3$ / 3-methylpyridine N-oxide (CH_3, $\overset{+}{N}-O^-$)

94% 3% Ref. 31

5% 90%

R = TBDMS

HO, SO_2Ph, HO → TBDMSCl, Im / DMF, >79% → HO, SO_2Ph, TBDMSO Ref. 32

1. C. A. A. Van Boeckel, S. F. Van Aelst, and T. Beetz, *Recl: J. R. Neth. Chem. Soc.*, **102,** 415 (1983); P. G. M. Wuts and S. S. Bigelow, *J. Org. Chem.*, **53,** 5023 (1988); F. Franke and R. D. Guthrie, *Aust. J. Chem.*, **31,** 1285 (1978); Y. Torisawa, M. Shibasaki, and S. Ikegami, *Tetrahedron Lett.*, 1865 (1979); K. K. Ogilvie, S. L. Beaucage, A. L. Schifman, N. Y. Theriault, and K. L. Sadana, *Can. J. Chem.*, **56,** 2768 (1978); S. S. Jones and C. B. Reese, *J. Chem. Soc., Perkin Trans. I*, 2762 (1979).
2. K. K. Ogilvie and D. W. Entwistle, *Carbohydr. Res.*, **89,** 203 (1981).
3. E. J. Corey and A. Venkateswarlu, *J. Am. Chem. Soc.*, **94,** 6190 (1972).
4. D. W. Hansen, Jr., and D. Pilipauskas, *J. Org. Chem.*, **50,** 945 (1985).
5. G. R. Clark, M. M. Nikaido, C. K. Fair, and J. Lin, *J. Org. Chem.*, **50,** 1994 (1985).
6. J. F. W. Keana, G. S. Heo, and G. T. Gaughan, *J. Org. Chem.*, **50,** 2346 (1985).
7. J. M. Aizpurua and C. Palomo, *Tetrahedron Lett.*, **26,** 475 (1985).
8. T. J. Barton and C. R. Tully, *J. Org. Chem.*, **43,** 3649 (1978).
9. E. J. Corey, H. Cho, C. Rücker, and D. H. Hua, *Tetrahedron Lett.*, **22,** 3455 (1981).
10. L. N. Mander and S. P. Sethi, *Tetrahedron Lett.*, **25,** 5953 (1984).
11. D. Boschelli, T. Takemasa, Y. Nishitani, and S. Masamune, *Tetrahedron Lett.*, **26,** 5239 (1985).
12. P. G. Gassman and L. M. Haberman, *J. Org. Chem.*, **51,** 5010 (1986).
13. G. A. Olah, B. G. B. Gupta, S. C. Narang, and R. Malhotra, *J. Org. Chem.*, **44,** 4272 (1979).
14. S. K. Chaudhary and O. Hernandez, *Tetrahedron Lett.*, 99 (1979).

15. See also: K. K. Ogilvie, A. L. Shifman, and C. L. Penney, *Can. J. Chem.*, **57,** 2230 (1979); W. Kinzy and R. R. Schmidt, *Liebigs Ann. Chem.*, 407 (1987).
16. S. Kim and H. Chang, *Synth. Commun.*, **14,** 899 (1984).
17. S. Kim and H. Chang, *Bull. Chem. Soc. Jpn.*, **58,** 3669 (1985).
18. L. Lombardo, *Tetrahedron Lett.*, **25,** 227 (1984).
19. T. Morita, Y. Okamoto, and H. Sakurai, *Tetrahedron Lett.*, **21,** 835 (1980).
20. T. Veysoglu and L. A. Mitscher, *Tetrahedron Lett.*, **22,** 1299 (1981).
21. M. T. Reetz and G. Neumeier, *Liebigs Ann. Chem.*, 1234 (1981).
22. G. R. Martinez, P. A. Grieco, E. Williams, K.-i. Kanai, and C. V. Srinivasan, *J. Am. Chem. Soc.*, **104,** 1436 (1982).
23. T. P. Mawhinney and M. A. Madson, *J. Org. Chem.*, **47,** 3336 (1982).
24. T. F. Braish and P. L. Fuchs, *Synth. Commun.*, **16,** 111 (1986).
25. Y. Kita, J.-i. Haruta, T. Fujii, J. Segawa, and Y. Tamura, *Synthesis*, 451 (1981).
26. K. Kai and G. Patil, *J. Org. Chem.*, **51,** 3545 (1986).
27. D. Askin, D. Angst, and S. Danishefsky, *J. Org. Chem.*, **52,** 622 (1987).
28. P. G. McDougal, J. G. Rico, Y.-I. Oh, and B. D. Condon, *J. Org. Chem.*, **51,** 3388 (1986).
29. G. H. Hakimelahi, Z. A. Proba, and K. K. Ogilvie, *Tetrahedron Lett.*, **22,** 5243 (1981).
30. C. H. Hakimelahi, Z. A. Proba, and K. K. Ogilvie, *Tetrahedron Lett.*, **22,** 4775 (1981).
31. K. K. Ogilvie, G. H. Hakimelahi, Z. A. Proba, and D. P. C. McGee, *Tetrahedron Lett.*, **23,** 1997 (1982); K. K. Ogilvie, D. P. C. McGee, S. M. Boisvert, G. H. Hakimelahi, and Z. A. Proba, *Can. J. Chem.*, **61,** 1204 (1983).
32. R. E. Donaldson and P. L. Fuchs, *J. Am. Chem. Soc.*, **103,** 2108 (1981).

Cleavage

1. $Bu_4N^+F^-$, THF, 25°, 1 h, >90% yield.[1] Fluoride ion is very basic especially under anhydrous conditions and thus may cause side reactions with base-sensitive substrates.[2] ArOTBDMS ethers can be cleaved in the presence of alkyl OTBDMS ethers.[3]
2. KF, 18-crown-6.[4]
3. $Bu_4N^+Cl^-$, $KF \cdot H_2O$, CH_3CN, 25°, 4 h, 95% yield.[5] This method generates fluoride ion *in situ* and is reported to be suitable for reactions that normally require anhydrous conditions.
4. Aqueous HF, CH_3CN (5:95), 20°, 1–3 h, 90–100% yield.[6] This reagent will cleave ROTBDMS ethers in the presence of ArOTBDMS ethers.[3] This reagent can be used to remove TBDMS groups from prostaglandins.
5. AcOH, H_2O, THF (3:1:1), 25–80°, 15 min–5 h.[1]
6. Dowex 50W-X8, MeOH, 20°.[7] Dowex 50W-X8 is a carboxylic acid resin, H^+ form.
7. $BF_3 \cdot Et_2O$, $CHCl_3$, 0–25°, 15 min–3 h, 70–90% yield.[8]
8. Pyridine–HF, THF, 0–25°, 70% yield.[9] Cyclic acetals and THP derivatives were found to be stable to these conditions.[10]

9. 57% HF in urea.[11]
10. Et_3N–HF, cyclohexane, rt, 30 min.[12]
11. $Ph_3C^+BF_4^-$, CH_3CN, CH_2Cl_2, rt, 60 h.[13]
12. Trifluoroacetic acid, H_2O (9 : 1), CH_2Cl_2, rt, 96 h.[14]
13. TsOH (0.1 eq.), THF, H_2O (20 : 1), 65% yield.[15]
14. 1% concd. HCl in EtOH.[11,16]
15. H_2SO_4.[17]
16. KO_2, DMSO, DME, 18-crown-6, 50–85% yield.[18]
17. $Bu_4Sn_2O(NCS)_2$, MeOH, reflux, 16 h, 70% yield.[19] This reagent also cleaves ketals and acetals, 77–97% yield.
18. Me_2BBr.[20]
19 $LiBF_4$, CH_3CN, CH_2Cl_2, 40–86% yield.[21] In this case $Bu_4N^+F^-$ or acid failed to remove a primary TBDMS group from a steroid.
20. Selectivity in the cleavage of a primary allylic TBDMS group was achieved with HF/CH_3CN in the presence of a more hindered secondary TBDMS group.[22]

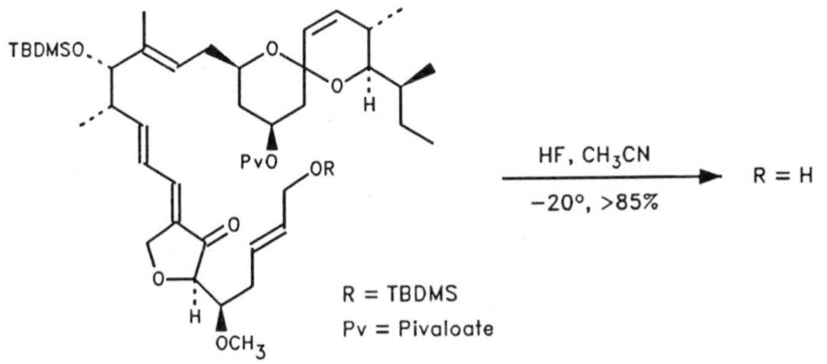

TBDMSO, PvO, OR, O, H, OCH$_3$

HF, CH_3CN / −20°, >85% → R = H

R = TBDMS
Pv = Pivaloate

21. Selective cleavage of one secondary TBDMS ether in the presence of a somewhat more hindered one was achieved with $Bu_4N^+F^-$ in THF.[23]

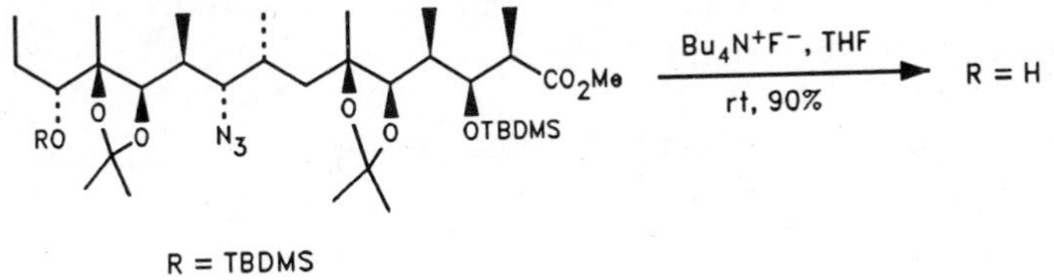

RO, O, N$_3$, OTBDMS, CO_2Me

$Bu_4N^+F^-$, THF / rt, 90% → R = H

R = TBDMS

22. NBS, DMSO, H_2O, rt, 17 h.[24] A trisubstituted steroidal alkene was not affected by these conditions.
23. 3 eq. *t*-BuOOH, 1.2 eq. $MoO_2(acac)_2$, CH_2Cl_2, 50–87% yield.[25]
24. 0.01 eq. $PdCl_2(CH_3CN)_2$, acetone, rt, 99% yield.[26] Acetals are also cleaved with this reagent, but the TBDPS, MEM, and THP groups are completely stable.

25. Pyridinium *p*-toluensulfonate, EtOH, 22–55°, 1.2–2 h, 80–92% yield.[27] These conditions were used to remove cleanly a TBDMS group in the presence of a TBDPS group.

26. Selective cleavage of a primary TBDMS group was achieved with acid in the presence of a secondary TBDMS group.[28]

BOCNH OTBDMS ... $CH_2OTBDMS$ — AcOH, H_2O, THF, 15 h, 30° → BOCNH OTBDMS ... CH_2OH

27. $^+NH_4F^-$, MeOH, H_2O, 60–65°, 65% yield.[29]

1. E. J. Corey and A. Venkateswarlu, *J. Am. Chem. Soc.*, **94,** 6190 (1972).
2. J. H. Clark, *Chem. Rev.*, **80,** 429 (1980).
3. E. W. Collington, H. Finch, and I. J. Smith, *Tetrahedron Lett.*, **26,** 681 (1985).
4. (a) G. Stork and P. F. Hudrlik *J. Am. Chem. Soc.*, **60,** 4462, 4464 (1968); (b) C. L. Liotta and H. P. Harris, *J. Am. Chem. Soc.*, **96,** 2250 (1974).
5. L. A. Carpino and A. C. Sau, *J. Chem. Soc., Chem. Commun.*, 514 (1979).
6. R. F. Newton, D. P. Reynolds, M. A. W. Finch, D. R. Kelly, and S. M. Roberts, *Tetrahedron Lett.*, 3981 (1979).
7. E. J. Corey, J. W. Ponder, and P. Ulrich, *Tetrahedron Lett.*, **21,** 137 (1980).
8. D. R. Kelly, S. M. Roberts, and R. F. Newton, *Synth. Commun.*, **9,** 295 (1979).
9. K. C. Nicolaou and S. E. Webber, *Synthesis*, 453 (1986).
10. S. Masamune, L. D.-L. Lu, W. P. Jackson, T. Kaiho, and T. Toyoda, *J. Am. Chem. Soc.*, **104,** 5523 (1982).
11. H. Wetter and K. Oertle, *Tetrahedron Lett.*, **26,** 5515 (1985).
12. J.-E. Nyström, T. D. McCanna, P. Helquist, and R. S. Iyer, *Tetrahedron Lett.*, **26,** 5393 (1985).
13. T. J. Barton and C. R. Tully, *J. Org. Chem.*, **43,** 3649 (1978).
14. R. Baker, W. J. Cummings, J. F. Hayes, and A. Kumar, *J. Chem. Soc., Chem. Commun.*, 1237 (1986).
15. E. J. Thomas and A. C. Williams, *J. Chem. Soc., Chem. Commun.*, 992 (1987).
16. R. F. Cunico and L. Bedell, *J. Org. Chem.*, **45,** 4797 (1980).
17. F. Franke and R. D. Guthrie, *Aust. J. Chem.*, **31,** 1285 (1978).
18. Y. Torisawa, M. Shibasaki, and S. Ikegami, *Chem. Pharm. Bull.*, **31,** 2607 (1983).
19. J. Otera and H. Nozaki, *Tetrahedron Lett.*, **27,** 5743 (1986).
20. Y. Guindon, C. Yoakim, and H. E. Morton, *J. Org. Chem.*, **49,** 3912 (1984).
21. B. W. Metcalf, J. P. Burkhart, and K. Jund, *Tetrahedron Lett.*, **21,** 35 (1980).
22. S. J. Danishefsky, D. M. Armistead, F. E. Wincott, H. G. Selnick, and R. Hungate, *J. Am. Chem. Soc.*, **109,** 8117 (1987).
23. T. Nakaba, M. Fukui, and T. Oishi, *Tetrahedron Lett.*, **29,** 2219, 2223 (1988).

24. R. J. Batten, A. J. Dixon, R. J. K. Taylor, and R. F. Newton, *Synthesis*, 234 (1980).
25. T. Hanamoto, T. Hayama, T. Katsuki, and M. Yamaguchi, *Tetrahedron Lett.*, **28,** 6329 (1987).
26. B. H. Lipshutz, D. Pollart, J. Monforte, and H. Kotsuki, *Tetrahedron Lett.*, **26,** 705 (1985).
27. C. Prakash, S. Saleh, and I. A. Blair, *Tetrahedron Lett.*, **30,** 19 (1989).
28. A. Kawai, O. Hara, Y. Hamada, and T. Shiari, *Tetrahedron Lett.*, **29,** 6331 (1988).
29. J. D. White, J. C. Amedio, Jr., S. Gut, and L. Jayasinghe, *J. Org. Chem.*, **54,** 4268 (1989).

79. *t*-Butyldiphenylsilyl Ether: TBDPS–OR: *t*-$BuPh_2SiOR$ (Chart 1)

The TBDPS group is considerably more stable ($\approx$100 times) than the TBDMS group toward acidic hydrolysis. The TBDPS group shows greater stability to many reagents that are incompatible with the TBDMS group, and it is less prone to undergo migration under basic conditions.[1] TBDPS ethers are stable to K_2CO_3/CH_3OH, to 9 *M* NH_4OH, 60°, 2 h, and to $NaOCH_3$ (cat.)/CH_3OH, 25°, 24 h. The ether is stable to 80% AcOH, used to cleave TBDMS, triphenylmethyl, and tetrahydropyranyl ethers. It is also stable to HBr/AcOH, 12°, 2 min, to 25–75% HCO_2H, 25°, 2–6 h, and to 50% aq. CF_3CO_2H, 25°, 15 min (conditions used to cleave acetals).[2]

Formation

1. TBDPSCl, imidazole, DMF, rt.[2] This is the original procedure used to introduce this group and is also the most widely employed method.
2. TBDPSCl, DMAP, Pyr.[3] Selective silylation of a primary hydroxyl was achieved under these conditions.
3. TBDPSCl, poly(vinylpyridine), HMPT, CH_2Cl_2.[4]
4. TBDPSCl, DMAP, triethylamine, CH_2Cl_2.[5] This combination of reagents was shown to be very selective for the silylation of a primary hydroxyl in the presence of a secondary hydroxyl.
5. Bu_2SnO, $PhCH_3$, heat, 10 h with H_2O removal, then $Bu_4N^+Br^-$, TBDPSCl, heat, 8 h. This method was used to introduce one TBDPS group on one of two primary alcohols in 92% yield.

Cleavage

1. $Bu_4N^+F^-$, THF, 25°, 1–5 h, >90% yield.[2]
2. 3% methanolic HCl, 25°, 3 h, 71% yield.[2] In benzoyl-protected carbohydrates this method gives clean deprotection without acyl migration.[6]
3. 5 *N* NaOH, EtOH, 25°, 7 h, 93% yield.[2]
4. 10% KOH, CH_3OH.[7]

5. Pyr·HF, THF.[8]
6. Amberlite 26 F^- form.[4]
7. HF, CH_3CN.[9]
8. KO_2, DMSO, 18-crown-6.[1]
9. $LiAlH_4$ has resulted in the cleavage of a TBDPS group, but this is not general.[10] TBDPS ethers are generally not affected by $LiAlH_4$.

OTBDPS / CONH2 / CO2Me —LiAlH4→ OH / NH2 / CH2OH

10. NaH, HMPA, 0°, 5 min; H_2O, 83–84% yield.[11] These conditions selectively cleave a TBDPS ether in the presence of a *t*-butyldimethylsilyl ether.

1. Y. Torisawa, M. Shibasaki, and S. Ikegami, *Chem. Pharm. Bull.*, **31,** 2607 (1983); W. W. Wood and A. Rashid, *Tetrahedron Lett.*, **28,** 1933 (1987).
2. S. Hanessian and P. Lavallee, *Can. J. Chem.*, **53,** 2975 (1975) *Idem., ibid.*, **55,** 562 (1977).
3. R. E. Ireland and D. M. Obrecht, *Helv. Chim. Acta.*, **69,** 1273 (1986); D. M. Clode, W. A. Laurie, D. McHale, and J. B. Sheridan, *Carbohydr. Res.*, **139,** 161 (1985).
4. G. Cardillo, M. Orena, S. Sandri, and C. Tomasihi, *Chem. Ind. (London)*, 643 (1983).
5. S. K. Chaudhary and O. Hernandez, *Tetrahedron Lett.*, 99 (1979); Y. Guindon, C. Yoakim, M. A. Bernstein, and H. E. Morton, *Tetrahedron Lett.*, **26,** 1185 (1985).
6. E. M. Nashed and C. P. J. Glaudemans, *J. Org. Chem.*, **52,** 5255 (1987).
7. A. A. Malik, R. J. Cormier, and C. M. Sharts, *Org. Prep. Proc. Int.*, **18,** 345 (1986).
8. K. C. Nicolaou, S. P. Seitz, M. R. Pavia, and N. A. Petasis, *J. Org. Chem.*, **44,** 4011 (1979); K. C. Nicolaou, S. P. Seitz, and M. R. Pavia, *J. Am. Chem. Soc.*, **103,** 1222 (1981).
9. Y. Ogawa, M. Nunomoto, and M. Shibasaki, *J. Org. Chem.*, **51,** 1625 (1986).
10. B. Rajashekhar and E. T. Kaiser, *J. Org. Chem.*, **50,** 5480 (1985).
11. M. S. Shekhani, K. M. Khan, K. Mahmood, P. M. Shah, and S. Malik, *Tetrahedron Lett.*, **31,** 1669 (1990).

80. Tribenzylsilyl Ether: $ROSi(CH_2C_6H_5)_3$ (Chart 1)

81. Tri-*p*-xylylsilyl Ether: $ROSi(CH_2C_6H_4\text{-}p\text{-}CH_3)_3$

To control the stereochemistry of epoxidation at the 10,11-double bond in intermediates in prostaglandin synthesis, a bulky protective group was used for the C_{15}–OH group. Epoxidation of the tribenzylsilyl ether yielded 88% α-oxide; epoxidation of the tri-*p*-xylylsilyl ether was less selective.[1]

Formation

1. $ClSi(CH_2C_6H_4\text{-}p\text{-}Y)_3$ (Y = H or CH_3), DMF, 2,6-lutidine, −20°, 24–36 h, 90–100% yield.[1]

Cleavage

1. AcOH, THF, H_2O (3:1:1), 26°, 6 h → 45°, 3 h, 85% yield.[1]

1. E. J. Corey and H. E. Ensley, *J. Org. Chem.*, **38,** 3187 (1973).

82. Triphenylsilyl Ether (TPS–OR): $ROSiPh_3$

The stability of the TPS group to basic hydrolysis is similar to that of the TMS group, but its stability to acid hydrolysis is about 400 times greater than the TMS group.[1]

Formation

1. Ph_3SiCl, Pyr.[2]
2. Ph_3SiBr, Pyr, −40°, 15 min.[3]
3. Ph_3SiH, cat.[4]

Cleavage

1. AcOH–H_2O–THF (3:1:1), 70°, 3 h, 70% yield.[3]
2. $Bu_4N^+F^-$.[5]
3. NaOH, EtOH.[2]
4. HCl.[6]

1. L. H. Sommer, *Stereochemistry, Mechanism and Silicon: An Introduction to the Dynamic Stereochemistry and Reaction Mechanisms of Silicon Centers*, McGraw-Hill, New York, 1965, p. 126.
2. S. A. Barker, J. S. Brimacombe, M. R. Harnden, and J. A. Jarvis, *J. Chem. Soc.*, 3403 (1963).
3. H. Nakai, N. Hamanaka, H. Miyake, and M. Hayashi, *Chem. Lett.*, 1499 (1979).
4. (a) E. Lukevics and M. Dzintara, *J. Organomet. Chem.*, **271,** 307 (1984); (b) L. Horner and J. Mathias, *J. Organomet. Chem.*, **282,** 175 (1985).
5. K. Maruoka, M. Hasegawa, H. Yamamoto, K. Suzuki, M. Shimazaki, and G.-i. Tsuchihashi, *J. Am. Chem. Soc.*, **108,** 3827 (1986).
6. R. G. Neville, *J. Org. Chem.*, **26,** 3031 (1961).

83. Diphenylmethylsilyl Ether: DPMS–OR: $Ph_2MeSiOR$

The DPMS group has stability intermediate between the TMS and TES (triethylsilyl) groups. It is incompatible with base, acid, BuLi, $LiAlH_4$, pyridinium chlorochromate, pyridinium dichromate, and CrO_3/pyridine. It is stable to Grignard reagents, Wittig reagents, *m*-chloroperoxybenzoic acid, and silica gel chromatography.[1]

Formation

1. $Ph_2MeSiCl$, DMF, imidazole, 83–92% yield.[1]

Cleavage

1. It can be cleaved with mild acid, fluoride ion, or base.[1]
2. NaN_3, DMF, 40°, 80–93% yield.[2]

1. S. E. Denmark, R. P. Hammer, E. J. Weber, and K. L. Habermas, *J. Org. Chem.*, **52,** 165 (1987).
2. S. J. Monger, D. M. Parry, and S. M. Roberts, *J. Chem. Soc., Chem. Commun.*, 381 (1989).

84. *t*-Butylmethoxyphenylsilyl Ether: TBMPS–OR: *t*-Bu(CH_3O)PhSiOR

The TBMPS group has a greater sensitivity to fluoride ion than the TBDMS and TBDPS groups, which allows for the selective cleavage of the TBMPS group in the presence of the latter two. The TBMPS group is also 140 times more stable to 0.01 *N* $HClO_4$ than the TBDMS group, thus allowing selective hydrolysis of the TBDMS group. The group can be introduced on primary, secondary, and tertiary hydroxyls in excellent yield when DMF is used as the solvent and can be selectively introduced onto primary hydroxyls when CH_2Cl_2 is used as solvent. The main problem with this group is that when it is introduced onto chiral molecules, diastereomers result, which may complicate NMR interpretation.[1]

Formation/Cleavage

$$\text{ROH} \underset{\text{Bu}_4\text{N}^+\text{F}^-,\ \text{THF or acid}}{\overset{t\text{-BuPhMeOSiBr, Et}_3\text{N, CH}_2\text{Cl}_2\text{ or DMF, 71–100\%}}{\rightleftarrows}} \text{ROTBMPS}$$

1. Y. Guindon, R. Fortin, C. Yoakim, and J. W. Gillard, *Tetrahedron Lett.*, **25,** 4717 (1984); J. W. Gillard, R. Fortin, H. E. Morton, C. Yoakim, C. A. Quesnelle, S. Daignault, and Y. Guindon, *J. Org. Chem.*, **53,** 2602 (1988).

CONVERSION OF SILYL ETHERS TO OTHER FUNCTIONAL GROUPS

The ability to convert a protective group to another functional group directly without first performing a deprotection is a potentially valuable transformation. Silyl-protected alcohols have been converted directly to aldehydes,[1] ketones,[2] bromides,[3] acetates,[4] and ethers[5] without first liberating the alcohol in a prior deprotection step.

1. G. A. Tolstikov, M. S. Miftakhov, N. S. Vostrikov, N. G. Komissarova, M. E. Adler, and O. Kuznetsov, *Zh. Org. Khim.*, **24,** 224 (1988); *Chem. Abstr.*, **110,** 7162c (1989).
2. F. P. Cossio, J. M. Aizpurua, and C. Palomo, *Can. J. Chem.*, **64,** 225 (1986).
3. H. Mattes and C. Benezra, *Tetrahedron Lett.*, **28,** 1697 (1987); S. Kim and J. H. Park, *J. Org. Chem.*, **53,** 3111 (1988); J. M. Aizpurua, F. P. Cossio, and C. Palomo, *J. Org. Chem.*, **51,** 4941 (1986).
4. S. J. Danishefsky and N. Mantlo, *J. Am. Chem. Soc.*, **110,** 8129 (1988); B. Ganem and V. R. Small, Jr., *J. Org. Chem.*, **39,** 3728 (1974); S. Kim and W. J. Lee, *Synth. Commun.*, **16,** 659 (1986); E.-F. Fuchs and J. Lehmann, *Chem. Ber.*, **107,** 721 (1974).
5. D. G. Saunders, *Synthesis*, 377 (1988).

ESTERS

See also Chapter 5, on the preparation of esters as protective groups for carboxylic acids.

1. Formate Ester: ROCHO (Chart 2)

Formation

1. 85% HCOOH, 60°, 1 h, 93% yield.[1]
2. 70% HCOOH, cat. $HClO_4$, 50–55°, good yields.[2]
3. $CH_3COOCHO$, Pyr, −20°, 80–100% yield.[3,4]
4. $Me_2N^+{=}CHOBz\ Cl^-$, Et_2O, overnight; dil. H_2SO_4, 60–96% yield.[5]
5. HCO_2H, $BF_3\cdot 2MeOH$, 90% yield.[6]

Cleavage

1. $KHCO_3$, H_2O, MeOH, 20°, 3 days.[3]
2. Dilute NH_3, pH 11.2, 22°, 62% yield.[7]

A formate ester can be cleaved selectively in the presence of an acetate [(MeOH, reflux)[4] or dil. NH_3 (formate is 100 times faster than an acetate)[7]] or benzoate ester (dil. NH_3).[7]

1. H. J. Ringold, B. Löken, G. Rosenkranz, and F. Sondheimer, *J. Am. Chem. Soc.*, **78,** 816 (1956).
2. I. W. Hughes, F. Smith, and M. Webb, *J. Chem. Soc.*, 3437 (1949).
3. F. Reber, A. Lardon, and T. Reichstein, *Helv. Chim. Acta*, **37,** 45 (1954).
4. J. Zemlicka, J. Beránek, and J. Smrt, *Collect. Czech. Chem. Commun.*, **27,** 2784 (1962).
5. J. Barluenga, P. J. Campos, E. Gonzalez-Nunez, and G. Asensio, *Synthesis*, 426 (1985).
6. M. Dymicky, *Org. Prep. Proc. Int.*, **14,** 177 (1982).
7. C. B. Reese and J. C. M. Stewart, *Tetrahedron Lett.*, 4273 (1968).

2. Benzoylformate Ester: ROCOCOPh

The benzoylformate ester can be prepared from the 3′-hydroxy group in a deoxyribonucleotide by reaction with benzoyl chloroformate (anh. Pyr, 20°, 12 h, 86% yield); it is cleaved by aqueous pyridine (20°, 12 h, 31% yield), conditions that do not cleave an acetate ester.[1]

1. R. L. Letsinger and P. S. Miller, *J. Am. Chem. Soc.*, **91,** 3356 (1969).

3. Acetate Ester: ROAc, CH_3CO_2R (Chart 2)

Formation

1. Ac_2O, Pyr, 20°, 12 h, 100% yield.[1] This is one of the most common methods for acetate introduction. By running the reaction at lower temperatures good selectivity can be achieved for primary alcohols over secondary alcohols.[2] Tertiary alcohols are generally not acylated under these conditions.
2. CH_3COCl, 25°, 16 h, 67–79% yield.[3]
3. Ac_2O, AcCl, Pyr, DMAP, 24–80°, 1–40 h, 72–95% yield.[4] The use of DMAP increases the rate of acylation by a factor of 10^4. These conditions will acylate most alcohols, including tertiary alcohols. The use of DMAP (4-*N*,*N*-dimethylaminopyridine) as a catalyst to improve the rate of esterification is quite general and works for other esters as well.
4. $AcOC_6F_5$, Et_3N, DMF, 80°, 12–60 h, 72–95% yield.[5] This reagent reacts with amines (25°, no Et_3N) selectively in the presence of alcohols to form *N*-acetyl derivatives in 80–90% yield.
5. The direct conversion of a THP-protected alcohol to an acetate is possible, thus avoiding a deprotection step.[6]

$$\text{CH}_3\text{CH=CH-CH}_2\text{-CH=CH}(\text{CH}_2)_8\text{OTHP} \xrightarrow[\text{reflux, 91\%}]{\text{AcCl, AcOH}} \text{CH}_3\text{CH=CH-CH}_2\text{-CH=CH}(\text{CH}_2)_8\text{OAc}$$

6. Ac–Imidazole, $PtCl_2(C_2H_4)$, 23°, 0.5–144 h, 51–87% yield.[7] Platinum(II) acts as a template to catalyze the acetylation of the pyridinyl alcohol, $C_5H_4N(CH_2)_nCH_2OH$. Normally acylimidazoles are not very reactive acylating agents with alcohols.
7. Ac_2O, CH_2Cl_2, 15 kbar (1.5 GPa), 79–98% yield.[8] This high pressure technique also works to introduce benzoates and TBDMS ethers onto highly hindered tertiary alcohols.
8. AcOEt, Al_2O_3, 75–80°, 24 h, 45–69% yield.[9] This method is selective for primary alcohols. Phenols do not react under these conditions. The use of $SiO_2 \cdot NaHSO_4$ as a solid support was also found to be effective.[10]
9. The monoacetylation of alpha–omega diols can be accomplished in excellent yields.[11]

$$HOCH_2(CH_2)_nCH_2OH \xrightarrow[60-90\%]{AcOH,\ H_2SO_4,\ H_2O \\ 30\ h - 1\ wk} AcOCH_2(CH_2)_nCH_2OH$$

A monoacetate can be isolated by continuous extraction with organic solvents such as cyclohexane/CCl_4.

10. Ac_2O, $BF_3 \cdot Et_2O$, THF, 0°.[12] These conditions give good chemoselectivity for the most nucleophilic hydroxyl group. Alcohols are acetylated in the presence of phenols.
11. $AcOCH_2CF_3$, porcine pancreatic lipase, THF, 60 h, 77% yield.[13] This enzymatic method was used to acetylate selectively the primary hydroxyl group of a variety of carbohydrates.
12. $CH_2{=}C{=}O$, *t*-BuOK, THF.[14] The 17α-hydroxy group of a steroid was acetylated by this method.
13. Bu_2SnO, $PhCH_3$, 110°, 2 h; AcCl, CH_2Cl_2, 0°, 30 min, 84% yield.[15]

$$HOCH_2CH(OH)CH_2OBn \xrightarrow[2.\ AcCl,\ CH_2Cl_2,\ 0°,\ 30\ min \quad 84\%]{1.\ Bu_2SnO,\ PhCH_3,\ 2\ h,\ 110°} AcOCH_2CH(OH)CH_2OBn$$

14. $AcOCH_2CCl_3$, pyridine, porcine pancreatic lipase, 85% yield.[16] These studies examined the selective acylation of carbohydrates. Mannose is acylated at the 6-position in 85% yield in one example.
15. AcOH, TMSCl, 81% yield.[17]

Cleavage

1. K_2CO_3, MeOH, H_2O, 20°, 1 h, 100% yield.[18]
2. KCN, 95% EtOH, 20° to reflux, 12 h, 93% yield.[19,20] Potassium cyanide is a mild transesterification catalyst, suitable for acid- or base-sensitive compounds. When used with 1,2-diol acetates hydrolysis proceeds slowly until the first acetate is removed.[21]
3. Guanidine, EtOH, CH_2Cl_2, rt, 85–100% yield.[22] Acetamides, benzoates, and pivaloates are stable under these conditions. Phenolic acetates can be removed in the presence of primary and secondary acetates with excellent selectivity.
4. *Candida cylindracea*, phosphate buffer pH 7, Bu_2O.[23] The 6-*O*-acetyl of α-methyl peracetylglucose was selectively removed. Porcine pancreatic lipase will also hydrolyze acetyl groups from carbohydrates. These lipases are not specific for acetate since they hydrolyze other esters as well. In general selectivity is dependent on the ester and the substrate.[13]
5. 50% NH_3, MeOH, 20°, 2.5 h, 85% yield.[24] The 3′-acetate is removed from cytosine in the presence of a 5′-benzoate. If the reaction time is extended to 2 days, the benzoate is removed as well as the benzoyl protection on nitrogen.
6. Bu_3SnOMe, $ClCH_2CH_2Cl$, 1 h, 77% yield.[25] These conditions selectively cleave the anomeric acetate of a glucose derivative in the presence of other acetates.
7. $BF_3 \cdot Et_2O$, wet CH_3CN, 96% yield.[26]

OAc NC H OAc H O OAc — $BF_3 \cdot Et_2O$, wet CH_3CN, 96% → OAc NC H OAc H O OH

8. Deprotection using enzymes can be quite useful. An added benefit is that a racemic or meso substrate can often by resolved with excellent enantioselectivity.[27]

OAc OAc — Porcine Pancreatic Lipase, 70%, 96% ee → OAc OH

9. 1,8-Diazabicyclo[5.4.0]undec-7-ene (DBU), benzene, 60°, 45 h, 47–97% yield.[28] Benzoates are not cleaved under these conditions.

1. H. Weber and H. G. Khorana, *J. Mol. Biol.*, **72,** 219 (1972); R. I. Zhdanov and S. M. Zhenodarova, *Synthesis*, 222 (1975).
2. G. Stork, T. Takahashi, I. Kawamoto, and T. Suzuki, *J. Am. Chem. Soc.*, **100,** 8272 (1978).
3. D. Horton, *Org. Synth., Collect. Vol. V*, 1 (1973).
4. G. Höfle, W. Steglich, and H. Vorbrüggen, *Angew. Chem., Int. Ed. Engl.*, **17,** 569 (1978).
5. L. Kisfaludy, T. Mohacsi, M. Low, and F. Drexler, *J. Org. Chem.*, **44,** 654 (1979).
6. M Jacobson, R. E. Redfern, W. A. Jones, and M. H. Aldridge, *Science*, **170,** 542 (1970).
7. J. C. Chottard, E. Mulliez, and D. Mansuy, *J. Am. Chem. Soc.*, **99,** 3531 (1977).
8. W. G. Dauben, R. A. Bunce, J. M. Gerdes, K. E. Henegar, A. F. Cunningham, Jr., and T. B. Ottoboni, *Tetrahedron Lett.*, **24,** 5709 (1983).
9. G. H. Posner and M. Oda, *Tetrahedron Lett.*, **22,** 5003 (1981); S. S. Rana, J. J. Barlow, and K. L. Matta, *Tetrahedron Lett.*, **22,** 5007 (1981).
10. T. Nishiguchi and H. Taya, *J. Am. Chem. Soc.*, **111,** 9102 (1989).
11. J. H. Babler and M. J. Coghlan, *Tetrahedron Lett.*, 1971 (1979).
12. Y. Nagao, E. Fujita, T. Kohno, and M. Yagi, *Chem. Pharm. Bull.*, **29,** 3202 (1981).
13. W. J. Hennen, H. M. Sweers, Y.-F. Wang, and C.-H. Wong, *J. Org. Chem.*, **53,** 4939 (1988). See also: E. W. Holla, *Angew. Chem., Int. Ed. Engl.*, **28,** 220 (1989).
14. J. N. Cardner, T. L. Popper, F. E. Carlon, O. Gnoj, and H. L. Herzog, *J. Org. Chem.*, **33,** 3695 (1968).
15. F. Aragozzini, E. Maconi, D. Potenza, and C. Scolastico, *Synthesis*, 225 (1989). For a review on the use of Sn–O derivatives to direct regioselective acylation and alkylation, see: S. David and S. Hanessian, *Tetrahedron*, **41,** 643 (1985).
16. M. Therisod and A. M. Klibanov, *J. Am. Chem. Soc.*, **108,** 5638 (1986); H. M. Sweers and C.-H. Wong, *J. Am. Chem. Soc.*, **108,** 6421 (1986).
17. R. Nakao, K. Oka, and T. Fukomoto, *Bull. Chem. Soc. Jpn.*, **54,** 1267 (1981).
18. J. J. Plattner, R. D. Gless, and H. Rapoport, *J. Am. Chem. Soc.*, **94,** 8613 (1972).
19. K. Mori, M. Tominaga, T. Takigawa, and M. Matsui, *Synthesis*, 790 (1973).
20. K. Mori and M. Sasaki, *Tetrahedron Lett.*, 1329 (1979).
21. J. Herzig, A. Nudelman, H. E. Gottlieb, and B. Fischer, *J. Org. Chem.*, **51,** 727 (1986).
22. N. Kunesch, C. Meit, and J. Poisson, *Tetrahedron Lett.*, **28,** 3569 (1987).
23. M. Kloosterman, E. W. J. Mosuller, H. E. Schoemaker, and E. M. Meijer, *Tetrahedron Lett.*, **28,** 2989 (1987).
24. T. Neilson and E. S. Werstiuk, *Can. J. Chem.*, **49,** 493 (1971).
25. A. Nudelman, J. Herzig, H. E. Gottlieb, E. Keinan, and J. Sterling, *Carbohydr. Res.*, **162,** 145 (1987).
26. D. Askin, C. Angst, and S. Danishefsky, *J. Org. Chem.*, **52,** 622 (1987).
27. Y.-F. Wang, C.-S. Chen, G. Girdaukas, and C. J. Sih, in *Enzymes in Organic Synthesis* (*Ciba Foundation Symposium, Vol. 111*), 128 (1985); K. Tsuji, Y. Terao, and K. Achiwa, *Tetrahedron Lett.*, **30,** 6189 (1989); R. Csuk and B. I. Glaenzer, *Z. Naturforsch. B, Chem. Sci.*, **43,** 1355 (1988). For examples in a cyclic series, see: K.

Laumen and M. Schneider, *Tetrahedron Lett.*, **26,** 2073 (1985); K. Naemura, N. Takahashi, and H. Chikamatsu, *Chem. Lett.*, 1717 (1988); C. R. Johnson and C. H. Senanayake, *J. Org. Chem.*, **54,** 735 (1989); D. R. Deardorff, A. J. Matthews, D. S. McMeekin, and C. L. Craney, *Tetrahedron Lett.*, **27,** 1255 (1986); N. W. Boaz, *Tetrahedron Lett.*, **30,** 2061 (1989).

28. L. H. B. Baptistella, J. F. dos Santos, K. C. Ballabio, and A. J. Marsaioli, *Synthesis*, 436 (1989).

4. Chloroacetate Ester: $ClCH_2CO_2R$

Formation

1. $(ClCH_2CO)_2O$, Pyr, 0°, 70–90% yield.[1]
2. $ClCH_2COCl$, Pyr, ether, 87% yield.[2]

Cleavage

In general, cleavage of chloroacetates can be accomplished in the presence of other esters such as acetates and benzoates because of the large difference in the hydrolysis rates for esters bearing electron-withdrawing groups. A study comparing the half-lives for hydrolysis of a variety of esters of 5′-*O*-acyluridines gave the following results.[3]

	$t_{1/2}$ (min)	
Acyl Group	Reagent I	Reagent II
Ac	191	59
MeOAc	10.4	2.5
PhOAc	3.9	<1[a]
Formyl	0.4	(0.22)[b]
$ClCH_2CO-$	0.28	(0.17)[b]

Reagent I = 155 m*M* NH_3/H_2O; reagent II = NH_3/MeOH.
[a]Reaction is too fast to measure.
[b]Time for complete solvolysis of the substrate.

The relative rates of hydrolysis of acetate, chloro-, dichloro-, and trichloroacetates have been compared and give the following relative rates: $1:760:1.6 \times 10^4:10^5$.[4]

1. $HSCH_2CH_2NH_2$ or $H_2NCH_2CH_2NH_2$ or *o*-phenylenediamine, Pyr, Et_3N, 1 h, rt.[1]
2. Thiourea, $NaHCO_3$, EtOH, 70°, 5 h, 70% yield.[2]
3. H_2O, Pyr, pH 6.7, 20 h, 100% yield.[5]
4. $NH_2NHC(S)SH$, lutidine, AcOH, 2–20 min, rt, 88–99% yield.[6] This method is superior to the use of thiourea in that it proceeds at lower temperatures and affords much higher yields. This reagent also serves to remove the re-

lated bromoacetyl esters that under these conditions are 5–10 times more labile. Cleavage occurs cleanly in the presence of an acetate.[7]

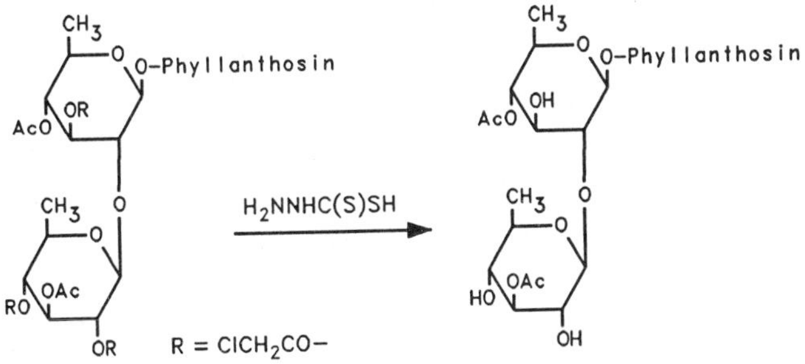

5. The lipase from *Pseudomonas sp.* K10 has also been used to cleave the chloroacetate, resulting in resolution of a racemic mixture since only one enantiomer was cleaved.[8]

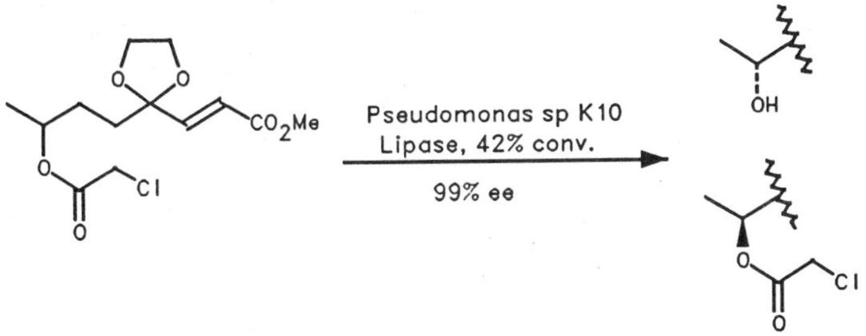

1. A. F. Cook and D. T. Maichuk, *J. Org. Chem.*, **35,** 1940 (1970).
2. M. Naruto, K. Ohno, N. Naruse, and H. Takeuchi, *Tetrahedron Lett.*, 251 (1979).
3. C. B. Reese, J. C. M. Stewart, J. H. van Boom, H. P. M. de Leeuw, J. Nagel, and J. F. M. de Rooy, *J. Chem. Soc., Perkin Trans. I*, 934 (1975).
4. N. S. Issacs, *Physical Organic Chemistry*, Halsted Press/Wiley, New York, 1987, p. 470.
5. F. Johnson, N. A. Starkovsky, A. C. Paton, and A. A. Carlson, *J. Am. Chem. Soc.*, **86,** 118 (1964).
6. C. A. A. van Boeckel and T. Beetz, *Tetrahedron Lett.*, **24,** 3775 (1983).
7. A. B. Smith, III, K. J. Hale, and H. A. Vaccaro, *J. Chem. Soc., Chem. Commun.*, 1026 (1987).
8. T. K. Ngooi, A. Scilimati, Z.-W. Guo, and C. J. Sih, *J. Org. Chem.*, **54,** 911 (1989).

5. Dichloroacetate Ester: Cl_2CHCO_2R

Formation

1. $Cl_2CHCOCl$.[1]
2. $(Cl_2CHCO)_2O$, Pyr, CH_2Cl_2.[2]

3. $Cl_2CHCOCCl_3$, DMF, 56% yield.[3] This reagent was used to acylate selectively the 6-position of an α-methylglucoside.

Cleavage

1. pH 9–9.5, 20°, 30 min.[1]
2. NH_3, MeOH.[3,4]
3. KOH, *t*-BuOH, H_2O, THF.[2]

1. J. R. E. Hoover, G. L. Dunn, D. R. Jakas, L. L. Lam, J. J. Taggart, J. R. Guarini, and L. Phillips, *J. Med. Chem.*, **17,** 34 (1974).
2. S. Masamune, W. Choy, F. A. J. Kerdesky, and B. Imperiali, *J. Am. Chem. Soc.*, **103,** 1566 (1981).
3. A. H. Haines and E. J. Sutcliffe, *Carbohydr. Res.*, **138,** 143 (1985).
4. C. B. Reese, J. C. M. Stewart, J. H. van Boom, H. P. M. de Leeuw, J. Nagel, and J. F. M. de Rooy, *J. Chem. Soc., Perkin Trans. I*, 934 (1975).

6. Trichloroacetate Ester: RO_2CCCl_3 (Chart 2)

Formation

1. Cl_3CCOCl, Pyr, DMF, 20°, 2 days, 60–90% yields.[1]

Cleavage

1. NH_3, EtOH, $CHCl_3$, 20°, 6 h, 81% yield.[1] Cleavage of the trichloroacetate occurs selectively in the presence of an acetate.
2. KOH, MeOH, 72% yield.[1] A formate ester was not hydrolyzed under these conditions.

1. V. Schwarz, *Collect. Czech. Chem. Commun.*, **27,** 2567 (1962).

7. Trifluoroacetate Ester: CF_3CO_2R, ROTFA

Formation

1. $(CF_3CO)_2O$, Pyr.[1]
2. CF_3CO_3H, 20°, 4 h, 83% yield.[2] In this case a hindered alcohol was converted to the TFA derivative. The use of TFA failed to give a trifluoroacetate. This method is probably not general.
3. Even with this highly reactive reagent excellent selectivity was achieved for one of two similar alcohols.[3]

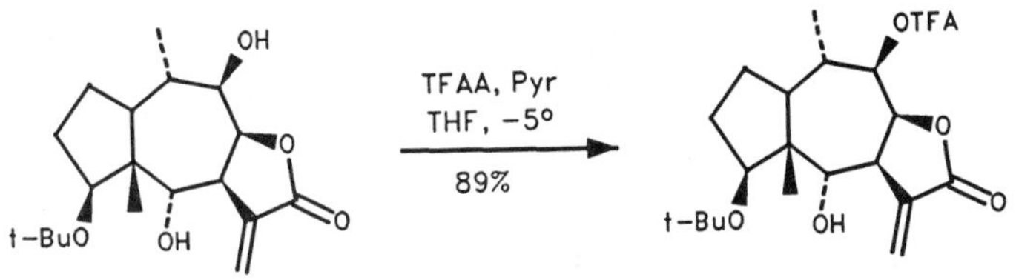

TFAA, Pyr
THF, −5°
89%

Cleavage

A series of nucleoside trifluoroacetates was hydrolyzed rapidly in 100% yield at 20°, pH 7.[4]

1. A. Lardon and T. Reichstein, *Helv. Chim. Acta*, **37,** 443 (1954).
2. G. W. Holbert and B. Ganem, *J. Chem. Soc., Chem. Commun.*, 248 (1978).
3. P. T. Lansbury, T. E. Nickson, J. P. Vacca, R. D. Sindelar, and J. M. Messinger, II, *Tetrahedron*, **43,** 5583 (1987).
4. F. Cramer, H. P. Bär, H. J. Rhaese, W. Sänger, K. H. Scheit, G. Schneider, and J. Tennigkeit, *Tetrahedron Lett.*, 1039 (1963).

8. Methoxyacetate Ester: $MeOCH_2CO_2R$

Formation

1. $MeOCH_2COCl$, Pyr.[1]

Cleavage

2. $NH_3/MeOH$ or NH_3/H_2O, 78% yield.[1] In nucleoside derivatives the methoxyacetate is cleaved 20 times faster than an acetate. It can be cleaved in the presence of a benzoate.

1. C. B. Reese and J. C. M. Stewart, *Tetrahedron Lett.*, 4273 (1968).

9. Triphenylmethoxyacetate Ester: $ROCOCH_2OCPh_3$

The triphenylmethoxyacetate was prepared in 53% yield from a nucleoside and the sodium acetate ($Ph_3COCH_2CO_2Na$, i-$Pr_3C_6H_2SO_2Cl$, Pyr) as a derivative that could be easily detected on TLC (i.e., it has a distinct orange–yellow color after it is sprayed with ceric sulfate). It is readily cleaved by $NH_3/MeOH$ (100% yield).[1]

1. E. S. Werstiuk and T. Neilson, *Can. J. Chem.*, **50,** 1283 (1972).

10. Phenoxyacetate Ester: $PhOCH_2CO_2R$ (Chart 2)

Formation

1. $(PhOCH_2CO)_2O$, Pyr.[1,2]
2. $(PhOCH_2CO)_2O$, Pyr, DMAP, CH_2Cl_2, 0°.[3]

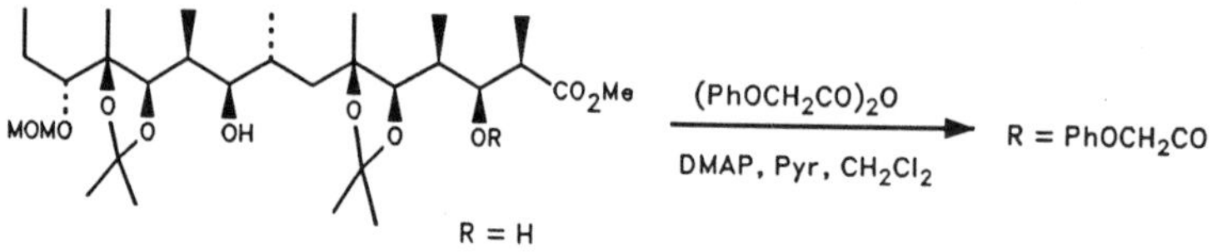

MOMO O O OH O O OR CO_2Me R = H $\xrightarrow[\text{DMAP, Pyr, } CH_2Cl_2]{(PhOCH_2CO)_2O}$ R = $PhOCH_2CO$

Cleavage

1. *t*-$BuNH_2$, MeOH.[2]
2. NH_3 in H_2O or MeOH.[1] The phenoxyacetate is 50 times more labile to aqueous ammonia than is an acetate.

1. C. B. Reese and J. C. M. Stewart, *Tetrahedron Lett.*, 4273 (1968).
2. T. Kamimura, T. Masegi, and T. Hata, *Chem. Lett.*, 965 (1982).
3. R. B. Woodward and 48 co-workers, *J. Am. Chem. Soc.*, **103,** 3210 (1981).

11. *p*-Chlorophenoxyacetate Ester: $ROCOCH_2C_6H_4$-*p*-Cl

The *p*-chlorophenoxyacetate, prepared to protect a nucleoside by reaction with the acetyl chloride, is cleaved by 0.2 *M* NaOH, dioxane–H_2O, 0°, 30 s.[1]

1. S. S. Jones and C. B. Reese, *J. Am. Chem. Soc.*. **101,** 7399 (1979).

12. *p*-Ⓟ-Phenylacetate Ester: $ROCOCH_2C_6H_4$–*p*-Ⓟ

Monoprotection of a symmetrical diol can be effected by reaction with a polymer-supported phenylacetyl chloride. The free hydroxyl group is then converted to an ether and the phenylacetate cleaved by aqueous ammonia–dioxane, 48 h.[1]

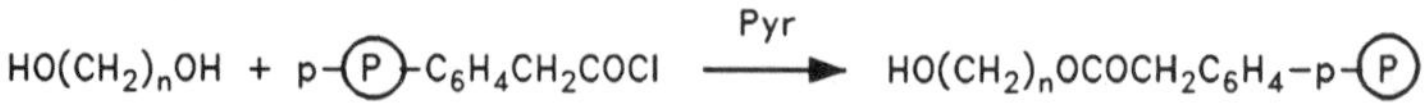

$HO(CH_2)_nOH$ + p-Ⓟ-$C_6H_4CH_2COCl$ $\xrightarrow{\text{Pyr}}$ $HO(CH_2)_nOCOCH_2C_6H_4$–p-Ⓟ

1. J. Y. Wong and C. C. Leznoff, *Can. J. Chem.*, **51,** 2452 (1973).

13. 3-Phenylpropionate Ester: $ROCOCH_2CH_2Ph$

The 3-phenylpropionate ester has been used in nucleoside synthesis.[1] It is cleaved by α-chymotrypsin (37°, 8–16 h, 70–90% yield).[2] It can be cleaved in the presence of an acetate.[3]

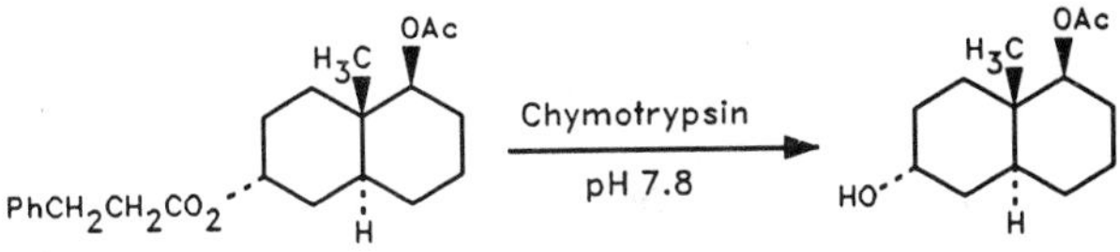

$PhCH_2CH_2CO_2$–(decalin, H_3C, OAc, H) —Chymotrypsin, pH 7.8→ HO–(decalin, H_3C, OAc, H)

1. H. S. Sachdev and N. A. Starkovsky, *Tetrahedron Lett.*, 733 (1969).
2. A. T.-Rigby, *J. Org. Chem.*, **38,** 977 (1973).
3. Y. Y. Lin and J. B. Jones, *J. Org. Chem.*, **38,** 3575 (1973).

14. 4-Oxopentanoate (Levulinate) Ester: $ROCOCH_2CH_2COCH_3$, LevOR

Formation

1. $(CH_3COCH_2CH_2CO)_2O$, Pyr, 25°, 24 h, 70–85% yield.[1]
2. $CH_3COCH_2CH_2CO_2H$, DCC, DMAP, 96% yield.[2]
3. 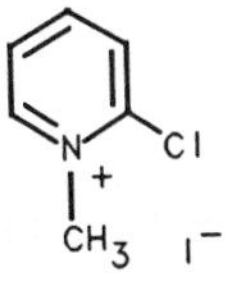 (2-chloro-1-methylpyridinium iodide structure: N⁺–CH_3, Cl, I^-) CMPI, $CH_3COCH_2CH_2CO_2H$, DABCO, 86% yield.[3]

   (CMPI = 2-chloro-1-methylpyridinium iodide)

Cleavage

1. $NaBH_4$, H_2O, pH 5–8, 20°, 20 min, 80–95% yield.[1] The byproduct 5-methyl-γ-butyrolactone is water-soluble and thus easily removed.
2. 0.5 *M* H_2NNH_2, H_2O, Pyr, AcOH, 2 min, 100% yield.[2] Normal esters are not cleaved under these conditions.[4]

15. 4,4-(Ethylenedithio)pentanoate Ester (Levulinoyldithioacetal Ester: ROLevS

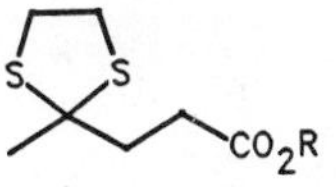

(1,3-dithiolane structure with CO_2R)

Formation

1. 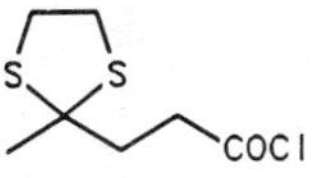 (1,3-dithiolane structure with COCl) 2,6-lutidine, 0°, 70% yield.[5]

2. 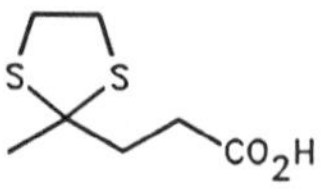 (2-methyl-1,3-dithiolan-2-yl)propionic acid (CO_2H) — CMPI, DABCO, dioxane, 2 h, 20°, 96% yield.[3] (CMPI = 2-chloro-1-methylpyridinium iodide)

Cleavage

The LevS group is converted to the Lev group with $HgCl_2/HgO$ (acetone/H_2O, 4 h, 20°, 74% yield). It can then be hydrolyzed using the conditions that remove the Lev group.[5]

The LevS group is stable to the conditions used for glycoside formation [$HgBr_2$, $Hg(CN)_2$].

1. A. Hassner, G. Strand, M. Rubinstein, and A. Patchornik, *J. Am. Chem. Soc.*, **97,** 1614 (1975).
2. J. H. van Boom and P. M. J. Burgers, *Tetrahedron Lett.*, 4875 (1976).
3. H. J. Koeners, J. Verhoeven, and J. H. van Boom, *Tetrahedron Lett.*, **21,** 381 (1980).
4. N. Jeker and C. Tamm, *Helv. Chim. Acta*, **71,** 1895, 1904 (1988).
5. H. J. Koeners, C. H. M. Verdegaal, and J. H. Van Boom, *Recl. Trav. Chim. Pays-Bas*, **100,** 118 (1981).

16. Pivaloate Ester: $(CH_3)_3CCO_2R$, PvOR (Chart 2)

Formation

1. PvCl, Pyr, 0–75°, 2.5 days, 99% yield.[1] In general, such extended reaction times are not required to obtain complete reaction. This is an excellent reagent for selective acylation of a primary alcohol over a secondary alcohol.[2-4]

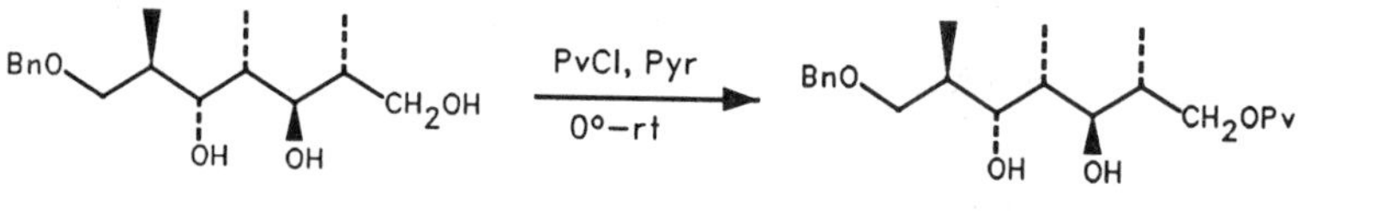

BnO–…(OH)(OH)–CH_2OH —PvCl, Pyr, 0°–rt→ BnO–…(OH)(OH)–CH_2OPv

2. Selective acylation can be obtained for one of two primary alcohols having slightly different steric environments.[5,6]

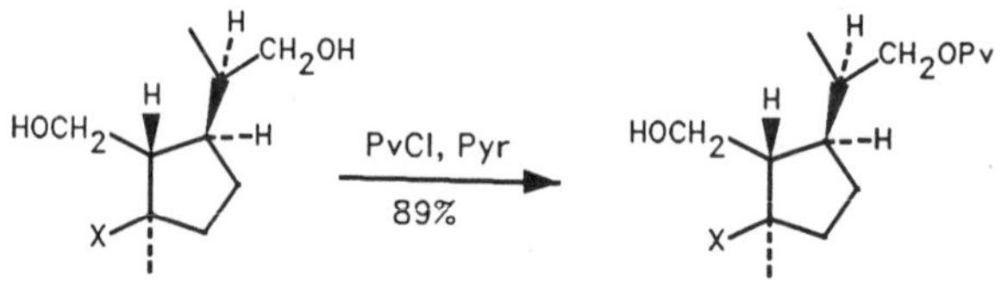

$HOCH_2$–(cyclopentane, X)–CH(CH_2OH) —PvCl, Pyr, 89%→ $HOCH_2$–(cyclopentane, X)–CH(CH_2OPv)

α-Methyl glucoside can be selectively acylated at positions 2- and 6- in 89% yield, and α-methyl 4,6-*O*-benzylidineglucoside can be selectively acylated at the 2-position in 77% yield.[7]

Cleavage

1. $Bu_4N^+OH^-$, 20°, 4 h.[8]
2. Aqueous $MeNH_2$, 20°, $t_{1/2}$ = 3 h.[9] In this case the 5′-position of uridine was deprotected. Acetates can be cleaved selectively in the presence of a pivaloate with NH_3/MeOH.
3. 0.5 *N* NaOH, EtOH, H_2O, 20°, 12 h, 58% yield.[10]
4. Li, NH_3, Et_2O; NH_4Cl, 70–85% yield.[11]
5. MeLi, Et_2O, 20°.[12]
6. *t*-BuOK, H_2O (8:2), 20°, 3 h, 94% yield.[13]
7. *i*-Bu_2AlH, CH_2Cl_2, −78°, 95% yield.[2]
8. Fungus, *Currulania lunata*, 6 h, 64% yield.[14] In this case a 21-pivaloate was removed from a steroid.
9. $K^+Et_3BH^-$, THF, −78°, 78% yield.[15]

PvO ... OPv → ($K^+Et_3BH^-$, −78°, THF, 78%) → HO ... OPv

1. M. J. Robins, S. D. Hawrelak, T. Kanai, J.-M. Siefert, and R. Mengel, *J. Org. Chem.*, **44,** 1317 (1979).
2. K. C. Nicolaou and S. E. Webber, *Synthesis*, 453 (1986).
3. D. Boschelli, T. Takemasa, Y. Nishitani, and S. Masamune, *Tetrahedron Lett.*, **26,** 5239 (1985).
4. H. Nagaoka, W. Rutsch, G. Schmid, H. Ilio, M. R. Johnson, and Y. Kishi, *J. Am. Chem. Soc.*, **102,** 7962 (1980).
5. N. Kato, H. Kataoka, S. Ohbuchi, S. Tanaka, and H. Takeshita, *J. Chem. Soc., Chem. Commun.*, 354 (1988).
6. P. F. Schuda and M. R. Heimann, *Tetrahedron Lett.*, **24,** 4267 (1983).
7. S. Tomic-Kulenovic and D. Keglevic, *Carbohydr. Res.*, **85,** 302 (1980).
8. C. A. A. van Boeckel and J. H. van Boom, *Tetrahedron Lett.*, 3561 (1979).
9. B. E. Griffin, M. Jarman, and C. B. Reese, *Tetrahedron*, **24,** 639 (1968).
10. K. K. Ogilvie and D. J. Iwacha, *Tetrahedron Lett.*, 317 (1973).
11. H. W. Pinnick and E. Fernandez, *J. Org. Chem.*, **44,** 2810 (1979).
12. B. M. Trost, S. A. Godleski, and J. L. Belletire, *J. Org. Chem.*, **44,** 2052 (1979).
13. P. G. Gassman and W. N. Schenk, *J. Org. Chem.*, **42,** 918 (1977).
14. H. Kosmol, F. Hill, U. Kerb, and K. Kieslich, *Tetrahedron Lett.*, 641 (1970).
15. S. J. Danishefsky, D. M. Armistead, F. E. Wincott, H. G. Selnick, and R. Hungate, *J. Am. Chem. Soc.*, **111,** 2967 (1989).

17. Adamantoate Ester: ROCO-1-adamantyl (Chart 2)

The adamantoate ester is formed selectively from a primary hydroxyl group (e.g., from the 5′-OH in a ribonucleoside) by reaction with adamantoyl chloride, Pyr (20°, 16 h). It is cleaved by alkaline hydrolysis (0.25 *N* NaOH, 20 min), but is stable to milder alkaline hydrolysis (e.g., NH_3, MeOH), conditions that cleave an acetate ester.[1]

1. K. Gerzon and D. Kau, *J. Med. Chem.*, **10,** 189 (1967).

18. Crotonate Ester: $ROCOCH{=}CHCH_3$

19. 4-Methoxycrotonate Ester: $ROCOCH{=}CHCH_2OCH_3$

The crotonate esters, prepared to protect a primary hydroxyl group in nucleosides, are cleaved by hydrazine (MeOH, Pyr, 2 h). The methoxycrotonate is 100-fold more reactive to hydrazinolysis and 2-fold less reactive to alkaline hydrolysis than the corresponding acetate.[1]

1. R. Arentzen and C. B. Reese, *J. Chem. Soc., Chem. Commun.*, 270 (1977).

20. Benzoate Ester: $PhCO_2R$, BzOR (Chart 2)

The benzoate ester is one of the more common esters used to protect alcohols. Benzoates are less readily hydrolyzed than acetates, and the tendency for benzoate migration to adjacent hydroxyls, in contrast to acetates, is not nearly as strong.[1] Benzoates can be forced to migrate to a thermodynamically more stable position.[2]

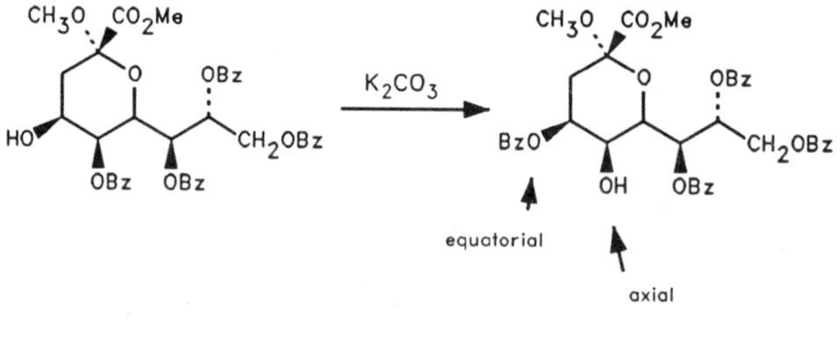

Formation

1. BzCl or Bz_2O, Pyr, 0°. Benzoyl chloride is the most common reagent for the introduction of the benzoate group. Reaction conditions vary depending on the nature of the alcohol to be protected. Cosolvents such as CH_2Cl_2 are often used with pyridine. Benzoylation in a polyhydroxylated system is much more selective than acetylation.[1] A primary alcohol is selectively

protected over a secondary allylic alcohol,[3] and an equatorial alcohol can be selectively protected in preference to an axial alcohol.[4]

2. Regioselective benzoylation of methyl 4,6-*O*-benzylidene-α-galactopyranoside can be effected by phase-transfer catalysis (BzCl, $Bu_4N^+Cl^-$, 40% NaOH, PhH, 69% yield of 2-benzoate; BzCl, $Bu_4N^+Cl^-$, 40% NaOH, HMPA, 62% yield of 3-benzoate).[5]

3. 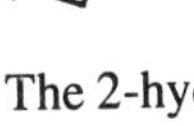 (1-benzoyl-1,2,4-triazole, NBz) Et_3N, DMF, 20°, 15 min, 90% yield.[6]

   The 2-hydroxyl of methyl 4,6-*O*-benzylidine-α-glucopyranoside was selectively protected.[7]

4. BzCN, Et_3N, CH_3CN, 5 min–2 h, >80% yield.[8,9] This reagent selectively acylates a primary hydroxyl group in the presence of a secondary hydroxyl group.[10]

5. $BzOCF(CF_3)_2$, TMEDA, 20°, 30 min, 90% yield.[11] This reagent also reacts with amines to form benzamides in high yields.

6. $BzOSO_2CF_3$, −78°, CH_2Cl_2, a few minutes.[12] With acid-sensitive substrates pyridine is used as a cosolvent. This reagent also reacts with ketals, epoxides,[12] and aldehydes.[13]

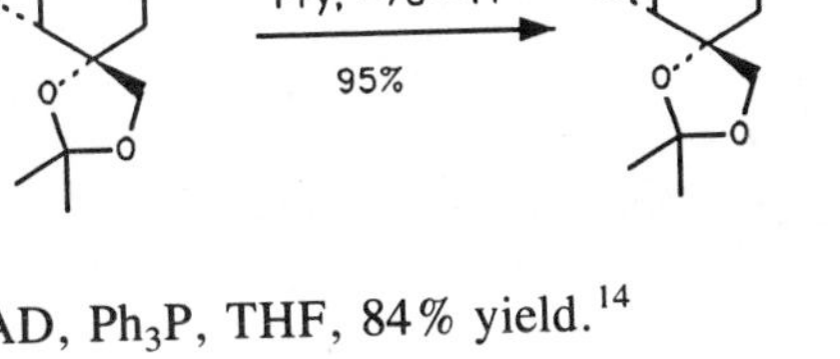

BzOTf, CH_2Cl_2
Pry, −78°–rt
95%

7. $PhCO_2H$, DIAD, Ph_3P, THF, 84% yield.[14]

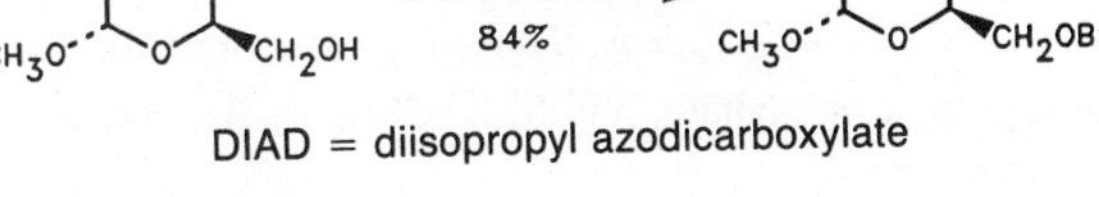

Ph_3P, DIAD
$PhCO_2H$, THF
84%

DIAD = diisopropyl azodicarboxylate

8. 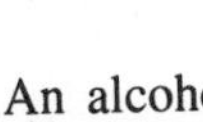 (1-benzoylimidazole, NBz) Ref. 15

9. An alcohol can be selectively benzoylated in the presence of a primary amine.[16]

BzCl, TEA

10. BuLi, BzCl, 10% Na_2CO_3, H_2O, 82% yield.[17] These conditions were used to monoprotect 1,4-butanediol.
11. BzOOBz, Ph_3P, CH_2Cl_2, 1 h, rt, ≈80% yield.[18] When these conditions are applied to unsymmetrical 1,2-diols, the benzoate of the kinetically and thermodynamically less stable isomer is formed.
12. $(Bu_3Sn)_2O$, BzCl.[19,20]

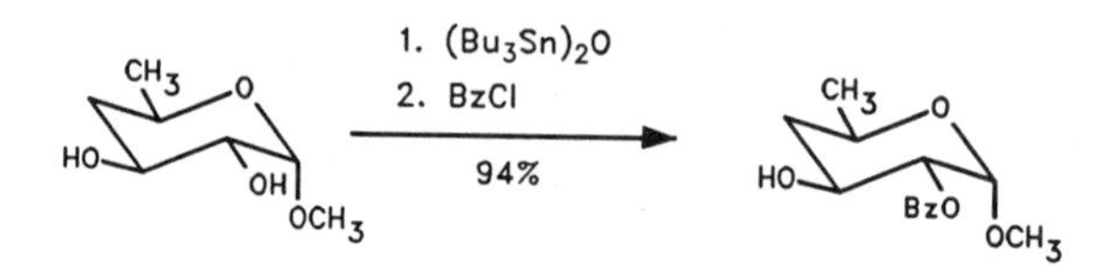

Cleavage

1. 1% NaOH, MeOH, 20°, 50 min, 90% yield.[21]
2. Et_3N, MeOH, H_2O (1:5:1), reflux, 20 h, 86% yield.[22]
3. MeOH, KCN.[23]
4. A benzoate ester can be cleaved in 60–90% yield by electrolytic reduction at −2.3 V.[24]
5. The following example illustrates the selective cleavage of a 2′-benzoate in a nucleotide derivative.[25]

BzO, B, O, OBz OBz → (H_2NNH_2, AcOH, Pyr (1:4); 20°, 7 days or 80°, 12 h; 80%) → BzO, B, O, OBz OH

6. $BF_3 \cdot Et_2O$, Me_2S.[26]

1. A. H. Haines, *Adv. Carbohydr. Chem. Biochem.*, **33,** 11 (1976).
2. S. J. Danishefsky, M. P. DeNinno, and S.-h. Chen, *J. Am. Chem. Soc.*, **110,** 3929 (1988).
3. R. H. Schlessinger and A. Lopes, *J. Org. Chem.*, **46,** 5252 (1981).
4. A. P. Kozikowski, X. Yan, and J. M. Rusnak, *J. Chem. Soc., Chem. Commun.*, 1301 (1988).
5. W. Szeja, *Synthesis*, 821 (1979).
6. J. Stawinski, T. Hozumi, and S. A. Narang, *J. Chem. Soc., Chem. Commun.*, 243 (1976).
7. S. Kim, H. Chang and W. J. Kim, *J. Org. Chem.*, **50,** 1751 (1985).
8. M. Havel, J. Velek, J. Pospišek, and M. Soucek, *Collect. Czech. Chem. Commun.*, **44,** 2443 (1979).
9. A. Holý and M. Soucek, *Tetrahedron Lett.*, 185 (1971).

10. R. M. Soll and S. P. Seitz, *Tetrahedron Lett.*, **28,** 5457 (1987).
11. N. Ishikawa and S. Shin-ya, *Chem. Lett.*, 673 (1976).
12. L. Brown and M. Koreeda, *J. Org. Chem.*, **49,** 3875 (1984).
13. K. Takeuchi, K. Ikai, M. Yoshida, and A. Tsugeno, *Tetrahedron*, **44,** 5681 (1988).
14. A. B. Smith, III, and K. J. Hale, *Tetrahedron Lett.*, **30,** 1037 (1989).
15. C. L. Brewer, S. David, and A. Veyrièrs, *Carbohydr. Res.*, **36,** 188 (1974).
16. Y. Ito, M. Sawamura, E. Shirakawa, K. Hayashizaki, and T. Hayashi, *Tetrahedron*, **44,** 5253 (1988). See also: T.-Y. Luh and Y. H. Chong, *Synth. Commun.*, **8,** 327 (1978).
17. A. J. Castellino and H. Rapoport, *J. Org. Chem.*, **51,** 1006 (1986).
18. A. M. Pautard and S. A. Evans, Jr., *J. Org. Chem.*, **53,** 2300 (1988).
19. S. Hanessian and R. Roy, *Can. J. Chem.*, **63,** 163 (1985).
20. For a mechanistic study of the tin-directed acylation, see: S. Roelens, *J. Chem. Soc., Perkin Trans. II*, 2105 (1988).
21. K. Mashimo and Y. Sato, *Tetrahedron*, **26,** 803 (1970).
22. K. Tsuzuki, Y. Nakajima, T. Watanabe, M. Yanagiya, and T. Matsumoto, *Tetrahedron Lett.*, 989 (1978).
23. J. Herzig, A. Nudelman, H. E. Gottlieb, and B. Fischer, *J. Org. Chem.*, **51,** 727 (1986).
24. V. G. Mairanovsky, *Angew. Chem., Int. Ed. Engl.*, **15,** 281 (1976).
25. Y. Ishido, N. Nakazaki, and N. Sakairi, *J. Chem. Soc., Perkin Trans. I*, 2088 (1979).
26. K. Fuji, T. Kawabata, and E. Fujita, *Chem. Pharm. Bull.*, **28,** 3662 (1980).

21. *p*-Phenylbenzoate Ester: $ROCOC_6H_4$-*p*-C_6H_5

The *p*-phenylbenzoate ester was prepared to protect the hydroxyl group of a prostaglandin intermediate by reaction with the benzoyl chloride (Pyr, 25°, 1 h, 97% yield). It was a more crystalline, more readily separated derivative than 15 other esters that were investigated.[1] It can be cleaved with K_2CO_3 in MeOH in the presence of a lactone.[2]

1. E. J. Corey, S. M. Albonico, U. Koelliker, T. K. Schaaf, and R. K. Varma, *J. Am. Chem. Soc.*, **93,** 1491 (1971).
2. T. V. RaganBabu, *J. Org. Chem.*, **53,** 4522 (1988).

22. 2,4,6-Trimethylbenzoate (Mesitoate) Ester: 2,4,6-$Me_3C_6H_2CO_2R$ (Chart 2)

Formation

1. $Me_3C_6H_2COCl$, Pyr, $CHCl_3$, 0°, 14 h → 23°, 1 h, 95% yield.[1]
2. $Me_3C_6H_2CO_2H$, $(CF_3CO)_2O$, PhH, 20°, 15 min.[2]

Cleavage

1. $LiAlH_4$, Et_2O, 20°, 2 h.[2]
2. *t*-BuOK, H_2O (8 : 1) "anhydrous hydroxide," 20°, 24–72 h, 50–72% yield.[3] A mesitoate ester is exceptionally stable to base: 2 *N* NaOH, 20°, 20 h; 12 *N* NaOH, EtOH, 50°, 15 min.

1. E. J. Corey, K. Achiwa, and J. A. Katzenellenbogen, *J. Am. Chem. Soc.*, **91,** 4318 (1969).
2. I. J. Bolton, R. G. Harrison, B. Lythgoe, and R. S. Manwaring, *J. Chem. Soc. C*, 2944 (1971).
3. P. G. Gassman and W. N. Schenk, *J. Org. Chem.*, **42,** 918 (1977).

Carbonates

Carbonates, like esters, can be cleaved by basic hydrolysis, but in general are much less susceptible to hydrolysis because of the resonance effect of the second oxygen. In general, carbonates are cleaved by taking advantage of the properties of the second alkyl substituent (e.g., zinc reduction of the 2,2,2-trichloroethyl carbonate). The reagents used to introduce the carbonate onto alcohols react readily with amines as well. As expected, basic hydrolysis of the resulting carbamate is considerably more difficult than basic hydrolysis of a carbonate.

23. Alkyl Methyl Carbonate: $ROCO_2CH_3$ (Chart 2)

Formation[1]

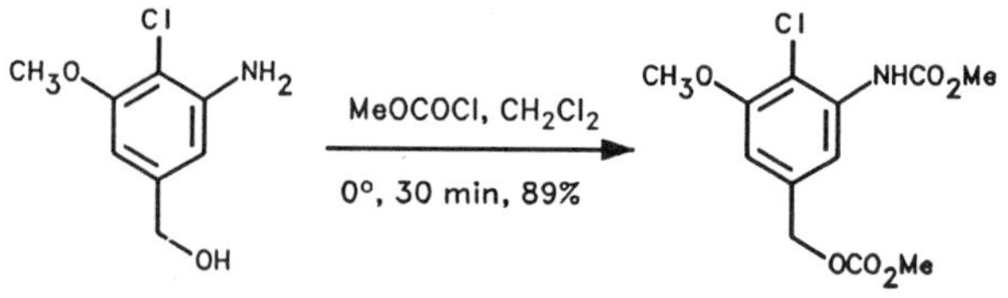

Cl, CH_3O, NH_2, OH — MeOCOCl, CH_2Cl_2 / 0°, 30 min, 89% → Cl, CH_3O, $NHCO_2Me$, OCO_2Me

Cleavage[1]

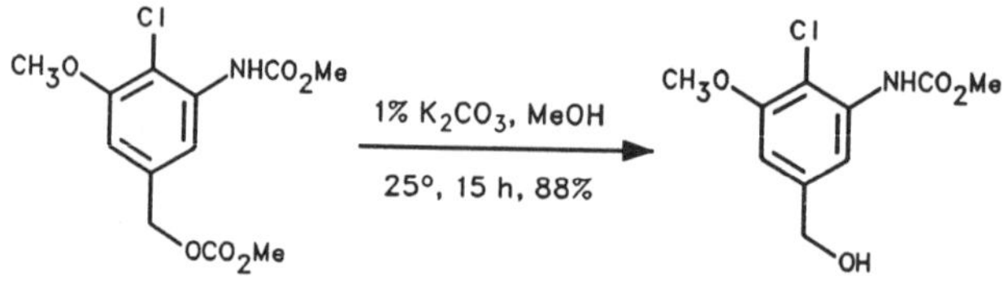

Cl, CH_3O, $NHCO_2Me$, OCO_2Me — 1% K_2CO_3, MeOH / 25°, 15 h, 88% → Cl, CH_3O, $NHCO_2Me$, OH

1. A. I. Meyers, K. Tomioka, D. M. Roland, and D. Comins, *Tetrahedron Lett.*, 1375 (1978).

24. 9-Fluorenylmethyl Carbonate: FmocOR

OCO_2R

Formation

1. FmocCl, Pyr, 20°, 40 min, 81–96% yield.[1]
2. ROH → ROFmoc (reagent: CF_3-benzotriazolyl-N-OCO_2Fm; DMAP, THF, CH_3CN; 53–95%) Ref. 2

Cleavage

1. Et_3N, Pyr, 2 h, 83–96% yield (half-life = 20 min).[1]

1. C. Gioeli and J. B. Chattopadhyaya, *J. Chem. Soc., Chem. Commun.*, 672 (1982).
2. K. Takeda, K. Tsuboyama, M. Hoshino, M. Kishino, and H. Ogura, *Synthesis*, 557 (1987).

25. Alkyl Ethyl Carbonate: $ROCO_2Et$

An ethyl carbonate, prepared and cleaved by conditions similar to those described for a methyl carbonate, was used to protect a hydroxyl group in glucose.[1]

1. F. Reber and T. Reichstein, *Helv. Chim. Acta*, **28,** 1164 (1945).

26. Alkyl 2,2,2-Trichloroethyl Carbonate (ROTroc): $ROCO_2CH_2CCl_3$ (Chart 2)

Formation

1. Cl_3CCH_2OCOCl, Pyr, 20°, 12 h.[1] The trichloroethyl carbonate can be introduced selectively onto a primary alcohol in the presence of a secondary alcohol.[2]

OR, TrocNH, OH, AllylO, O, CH_2OH → (Cl_3CH_2OCOCl; Pyr, >71%) → OR, TrocNH, OH, AllylO, O, CH_2OTroc

Cleavage

1. Zn, AcOH, 20°, 1–3 h, 80% yield.[1]
2. Zn, MeOH, reflux, short time.[1]
3. Zn–Cu, AcOH, 20°, 3.5 h, 100% yield.[3] A 2,2,2-tribromoethyl carbonate is cleaved by Zn–Cu/AcOH 10 times faster than trichloroethyl carbonate.
4. Electrolysis, −1.65 V, MeOH, $LiClO_4$, 80% yield.[4]

1. T. B. Windholz and D. B. R. Johnston, *Tetrahedron Lett.*, 2555 (1967).
2. M. Imoto, N. Kusunose, S. Kusumoto, and T. Shiba, *Tetrahedron Lett.*, **29,** 2227 (1988).
3. A. F. Cook, *J. Org. Chem.*, **33,** 3589 (1968).
4. M. F. Semmelhack and G. E. Heinsohn, *J. Am. Chem. Soc.*, **94,** 5139 (1972).

27. 2-(Trimethylsilyl)ethyl Carbonate (TMSEC): $Me_3SiCH_2CH_2OCO_2R$

Formation

1. $TMSCH_2CH_2OCOCl$, Pyr, 65–97% yield.[1]
2. $TMSCH_2CH_2OCO$–imidazole, DBU, benzene, 54% yield.[2]

Cleavage

1. 0.2 *M* $Bu_4N^+F^-$, THF, 20°, 10 min, 87–94% yield.[1]
2. $ZnCl_2$, CH_2Cl_2 or CH_3NO_2, 20°, 81–90% yield.[1]
3. $ZnBr_2$, CH_2Cl_2 or CH_3NO_2, 20°, 65–92% yield.[1]

1. C. Gioeli, N. Balgobin, S. Josephson, and J. B. Chattopadhyaya, *Tetrahedron Lett.*, **22,** 969 (1981).
2. W. R. Roush and T. A. Blizzard, *J. Org. Chem.*, **49,** 4332 (1984).

28. 2-(Phenylsulfonyl)ethyl Carbonate (Psec): $PhSO_2CH_2CH_2OCO_2R$

Formation

1. $PhSO_2CH_2CH_2OCOCl$, Pyr, 20°, 74–99% yield.[1] 4-Substituted phenylsulfonyl analogs of this protective group have also been prepared and their relative rates of cleavage studied: $t_{1/2}$ (min) (TEA, Pyr, 20°) 4-H, 180; 4-Me, 1140; 4-Cl, 60; 4-NO_2, 10.[2]

Cleavage

1. Et_3N, Pyr, 20 h, rt, 85–99% yield.[1]
2. NH_3, dioxane, H_2O (9:1), 7 min.[1]
3. K_2CO_3 (0.04 *M*), 1 min.[1]

1. N. Balgobin, S. Josephson, and J. B. Chattopadhyaya, *Tetrahedron Lett.*, **22,** 3667 (1981).
2. S. Josephson, N. Balgobin, and J. Chattopadhyaya, *Tetrahedron Lett.*, **22,** 4537 (1981).

29. 2-(Triphenylphosphonio)ethyl Carbonate (Peoc): $Ph_3P^+CH_2CH_2OCO_2R\ Cl^-$

Formation

1. $Ph_3P^+CH_2CH_2OCOCl\ Cl^-$, Pyr, CH_2Cl_2, 4 h, 0°, 65–94% yield.[1]

Cleavage

1. Me_2NH, MeOH, 0°, 75% yield.[1] *t*-Butyl esters could be cleaved with HCl without affecting the Peoc group.

1. H. Kunz and H.-H. Bechtolsheimer, *Synthesis*, 303 (1982).

30. Alkyl Isobutyl Carbonate: $ROCO_2CH_2CH(CH_3)_2$

An isobutyl carbonate was prepared by reaction with isobutyl chloroformate (Pyr, 20°, 3 days, 73% yield), to protect the 5′-OH group in thymidine. It was cleaved by acidic hydrolysis (80% AcOH, reflux, 15 min, 88% yield).[1]

1. K. K. Ogilvie and R. L. Letsinger, *J. Org. Chem.*, **32,** 2365 (1967).

31. Alkyl Vinyl Carbonate: $ROCO_2CH{=}CH_2$

Formation

1. $CH_2{=}CHOCOCl$, Pyr, CH_2Cl_2, 93% yield.[1]

Cleavage

1. Na_2CO_3, H_2O, dioxane, warm, 97% yield.[1] Phenols can be protected under similar conditions. Amines are converted by these conditions to carbamates

that are stable to alkaline hydrolysis with sodium carbonate. Carbamates are cleaved by acidic hydrolysis (HBr, MeOH, CH_2Cl_2, 8 h), conditions that do not cleave alkyl or aryl vinyl carbonates.

1. R. A. Olofson and R. C. Schnur, *Tetrahedron Lett.*, 1571 (1977).

32. Alkyl Allyl Carbonate: $ROCO_2CH_2CH{=}CH_2$ (Chart 2)

Formation

1. $CH_2{=}CHCH_2OCOCl$, Pyr, THF, 0° → 20°, 2 h, 90% yield.[1]

Cleavage

1. $Ni(CO)_4$, TMEDA, DMF, 55°, 4 h, 87–95% yield.[1] Because of the toxicity associated with nickel carbonyl, this method is rarely used and has largely been supplanted by palladium-based reagents.
2. $Pd(Ph_3P)_4$, HCO_2NH_4.[2]
3. $Pd(Ph_3P)_4$, Bu_3SnH, 90–100% yield.[3]
4. $PdCl_2(Ph_3P)_2$, dimedone, 91% yield.[4]

1. E. J. Corey and J. W. Suggs, *J. Org. Chem.*, **38,** 3223 (1973).
2. Y. Hayakawa, H. Kato, M. Uchiyama, H. Kajino, and R. Noyori, *J. Org. Chem.*, **51,** 2400 (1986).
3. F. Guibe and Y. Saint M'Leux, *Tetrahedron Lett.*, **22,** 3591 (1981).
4. H. X. Zhang, F. Guibe, and G. Balavoine, *Tetrahedron Lett.*, **29,** 623 (1988).

33. Alkyl *p*-Nitrophenyl Carbonate: $ROCOOC_6H_4$-*p*-NO_2 (Chart 2)

Formation/Cleavage[1]

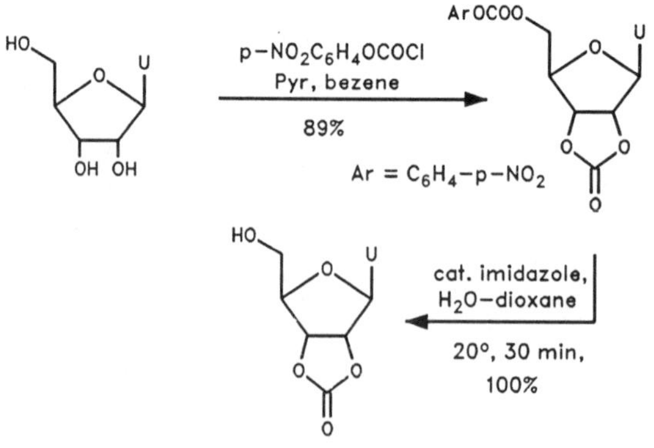

HO, U, O, OH OH
p-$NO_2C_6H_4OCOCl$, Pyr, bezene, 89%
ArOCOO, U, O, O, O, O
Ar = C_6H_4-p-NO_2
cat. imidazole, H_2O–dioxane, 20°, 30 min, 100%
HO, U, O, O, O, O

Acetates, benzoates, and cyclic carbonates are stable to these hydrolysis conditions. [Cyclic carbonates are cleaved by more alkaline conditions (e.g., dil. NaOH, 20°, 5 min, or aq. Pyr, warm, 15 min, 100% yield).][1]

1. R. L. Letsinger and K. K. Ogilvie, *J. Org. Chem.*, **32,** 296 (1967).

34. Alkyl Benzyl Carbonate: $ROCO_2Bn$ (Chart 2)

A benzyl carbonate was prepared in 83% yield from the sodium alkoxide of glycerol and benzyl chloroformate (20°, 24 h).[1] It is cleaved by hydrogenolysis (H_2/Pd–C, EtOH, 20°, 2 h, 2 atm, 76% yield)[1] and electrolytic reduction (−2.7 V, $R_4N^+X^-$, DMF, 70% yield).[2] A benzyl carbonate was used to protect the hydroxyl group in lactic acid during a peptide synthesis.[3]

1. B. F. Daubert and C. G. King, *J. Am. Chem. Soc.*, **61,** 3328 (1939).
2. V. G. Mairanovsky, *Angew. Chem., Int. Ed. Engl.*, **15,** 281 (1976).
3. G. Losse and G. Bachmann, *Chem. Ber.*, **97,** 2671 (1964).

35. Alkyl *p*-Methoxybenzyl Carbonate: *p*-$MeOC_6H_4CH_2OCO_2R$

36. Alkyl 3,4-Dimethoxybenzyl Carbonate: 3,4-$(MeO)_2C_6H_3CH_2OCO_2R$

These groups are readily cleaved with $Ph_3C^+BF_4^-$, 0°, 6 min, 90% yield; 0°, 15 min, 90% yield.[1] It should also be possible to cleave these carbonates with DDQ like the corresponding methoxy- and dimethoxyphenylmethyl ethers.

1. D. H. R. Barton, P. D. Magnus, G. Smith, G. Streckert, and D. Zurr, *J. Chem. Soc., Perkin Trans. I,* 542 (1972).

37. Alkyl *o*-Nitrobenzyl Carbonate

38. Alkyl *p*-Nitrobenzyl Carbonate (Chart 2)

The nitrobenzyl carbonates were prepared to protect a secondary hydroxyl group in a thienamycin precursor. The *o*-nitrobenzyl carbonate was prepared from the chloroformate (DMAP, CH_2Cl_2, 0° → 20°, 3 h) and cleaved by irradiation, pH 7.[1] The *p*-nitrobenzyl carbonate was prepared from the chloroformate (−78°, *n*-BuLi, THF, 85% yield) and cleaved by hydrogenolysis (H_2/Pd–C, dioxane, H_2O, EtOH, K_2HPO_4).[2] It is also cleaved by electrolytic reduction.[3]

1. L. D. Cama and B. G. Christensen, *J. Am. Chem. Soc.*, **100,** 8006 (1978).

2. D. B. R. Johnston, S. M. Schmitt, F. A. Bouffard, and B. G. Christensen, *J. Am. Chem. Soc.*, **100,** 313 (1978).
3. V. G. Mairanovsky, *Angew. Chem., Int. Ed., Engl.*, **15,** 281 (1976).

39. Alkyl *S*-Benzyl Thiocarbonate: $ROCOSCH_2Ph$ (Chart 2)

Formation

1. $PhCH_2SCOCl$, Pyr, 65–70% yield.[1]

Cleavage

1. H_2O_2, AcOH, AcOK, $CHCl_3$, 20°, 4 days, 50–55% yield.[1]

1. J. J. Willard, *Can. J. Chem.*, **40,** 2035 (1962).

40. 4-Ethoxy-1-naphthyl Carbonate

Formation/Cleavage[1]

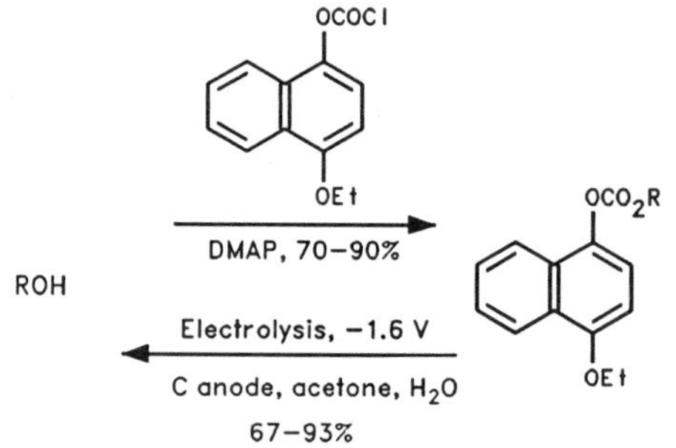

OCOCl
OEt
ROH
DMAP, 70–90%
Electrolysis, −1.6 V
C anode, acetone, H_2O
67–93%
OCO_2R
OEt

Amines can also be protected by this reagent; cleavage must be carried out in acidic media to avoid amine oxidation. The byproduct naphthoquinone can be removed by extraction with basic hydrosulfite. Ceric ammonium nitrate also serves as an oxidant for deprotection, but the yields are much lower.

1. R. W. Johnson, E. R. Grover, and L. J. MacPherson, *Tetrahedron Lett.*, **22,** 3719 (1981).

41. Methyl Dithiocarbonate: CH_3SCSOR

Formation[1]

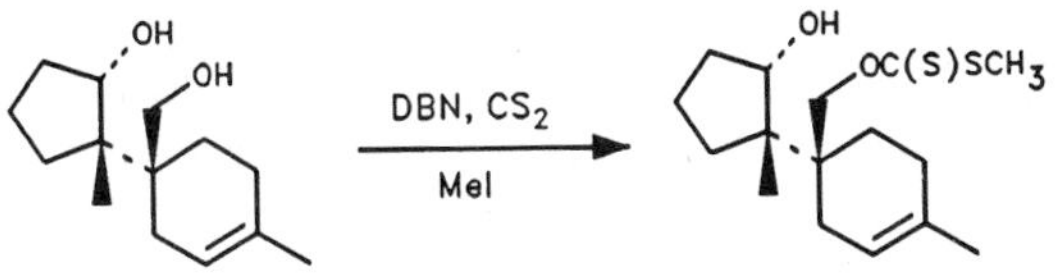

DBN, CS_2 / MeI

Most attempts to differentiate these hydroxyl groups with conventional derivatives resulted in the formation of a tetrahydrofuran. The dithiocarbonate can also be prepared by phase-transfer catalysis ($Bu_4N^+HSO_4^-$, 50% NaOH/H_2O, CS_2, MeI, rt, 1.5 h).[2]

Cleavage

1. These esters can be deoxygenated with Bu_3SnH[3] or as in the above example with $LiAlH_4$.[1]

1. R. H. Schlessinger and J. A. Schultz, *J. Org. Chem.*, **48,** 407 (1983).
2. A. W. M. Lee, W. H. Chan, H. C. Wong, and M. S. Wong, *Synth. Commun.*, **19,** 547 (1989).
3. D. H. R. Barton and S. W. McCombie, *J. Chem. Soc., Perkin Trans. I*, 1574 (1975).

Assisted Cleavage

The following derivatives represent protective groups that contain an auxiliary functionality, which when chemically modified, results in intramolecular, assisted cleavage, thus increasing the rate of cleavage over simple basic hydrolysis.

42. 2-Iodobenzoate Ester: 2-I–$C_6H_4CO_2R$

The 2-iodobenzoate is introduced by acylation of the alcohol with the acid (DCC, DMAP, CH_2Cl_2, 25°, 96% yield); it is removed by oxidation with Cl_2 (MeOH, H_2O, Na_2CO_3, pH > 7.5).[1]

43. 4-Azidobutyrate Ester: $N_3(CH_2)_3CO_2R$

The 4-azidobutyrate ester is introduced via the acid chloride. Cleavage occurs by pyrrolidone formation after the azide is reduced by hydrogenation, H_2S or Ph_3P.[2]

44. 4-Nitro-4-methylpentanoate Ester

Formation/Cleavage[3]

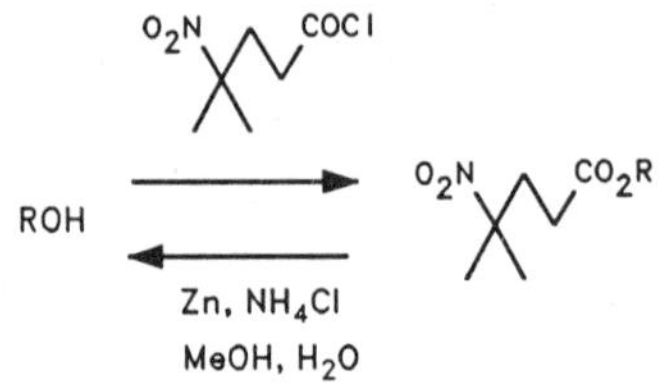

ROH → (O₂N-C(CH₃)₂CH₂CH₂COCl) → O₂N-C(CH₃)₂CH₂CH₂CO₂R; reverse: Zn, NH_4Cl, MeOH, H_2O

45. *o*-(Dibromomethyl)benzoate Ester: *o*-$(Br_2CH)C_6H_4CO_2R$

The *o*-(dibromomethyl)benzoate, prepared to protect nucleosides by reaction with the benzoyl chloride (CH_3CN, 65–90% yield), can be cleaved under nearly neutral conditions. The cleavage involves conversion of the $-CHBr_2$ group to $-CHO$ by silver ion-assisted hydrolysis. The benzoate group, *ortho* to the $-CHO$ group, now is rapidly hydrolyzed by neighboring-group participation (the morpholine and hydroxide ion-catalyzed hydrolyses of methyl 2-formylbenzoate are particularly rapid).[4]

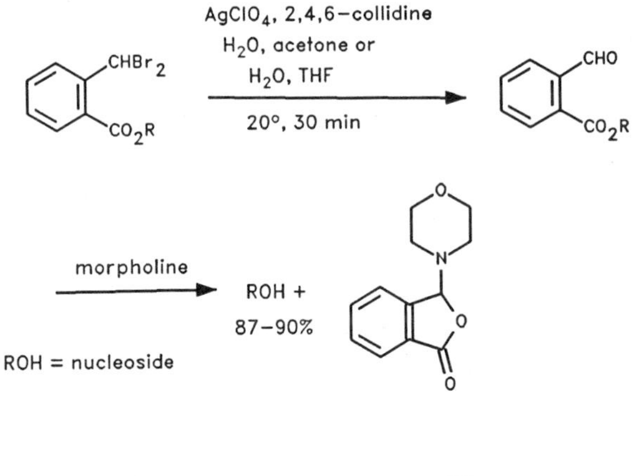

$AgClO_4$, 2,4,6-collidine, H_2O, acetone or H_2O, THF, 20°, 30 min

morpholine → ROH + (morpholino-phthalide) 87–90%

ROH = nucleoside

46. 2-Formylbenzenesulfonate Ester: o-$(OHC)C_6H_4SO_3R$

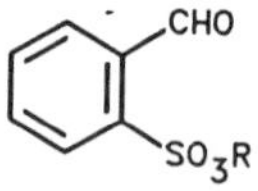

This sulfonate is prepared by reaction with the sulfonyl chloride. Cleavage occurs with 0.05 *M* NaOH (acetone, H_2O, 25°, 5 min, 83–93% yield). Here also cleavage is facilitated by intramolecular participation through the hydrate of the aldehyde.[5]

47. 2-(Methylthiomethoxy)ethyl Carbonate: MTMECOR: $CH_3SCH_2OCH_2CH_2OCO_2R$

Formation

1. $CH_3SCH_2OCH_2CH_2OCOCl$, 1-methylimidazole, CH_3CN, 1 h, >72% yield.[6]

Cleavage

1. $Hg(ClO_4)_2$, 2,4,6-collidine, acetone, H_2O (9 : 1), 5 h; NH_3, dioxane, H_2O (1 : 1).[6] In this case Hg(II) is used to cleave the MTM group, liberating a hydroxyl group, which assists in the cleavage of the carbonate on treatment with ammonia. Cleavage by ammonia is 500 times faster for this hydroxy derivative than for the initial MTM derivative.

48. 4-(Methylthiomethoxy)butyrate Ester: MTMBOR: $CH_3SCH_2O(CH_2)_3CO_2R$

Formation

1. 4-$(CH_3SCH_2O)(CH_2)_3CO_2H$, 2,6-dichlorobenzoyl chloride, Pyr, CH_3CN, 70% yield.[7] The MTMB group was selectively introduced onto the 5′-OH of thymidine.

Cleavage

1. $Hg(ClO_4)_2$, THF, H_2O, collidine, rt, 5 min; 1 *M* K_2CO_3, 10 min or TEA, 30 min.[7] Hg(II) cleaves the MTM group, liberating a hydroxyl group that assists in the cleavage of the ester.

49. 2-(Methylthiomethoxymethyl)benzoate Ester: MTMTOR: 2-$(CH_3SCH_2OCH_2)C_6H_4CO_2R$

This group was introduced and removed using the same conditions as for the MTMB group. The half-lives for ammonolysis of acetate, MTMB, and MTMT are 5 min, 15 min, and 6 h, respectively.[7]

1. R. A. Moss, P. Scrimin, S. Bhattacharya, and S. Chatterjee, *Tetrahedron Lett.*, **28,** 5005 (1987).
2. S. Kusumoto, K. Sakai, and T. Shiba, *Bull. Chem. Soc. Jpn.*, **59,** 1296 (1986).
3. T.-L. Ho, *Synth. Commun.*, **10,** 469 (1980).
4. J. B. Chattopadhyaya, C. B. Reese, and A. H. Todd, *J. Chem. Soc., Chem. Commun.*, 987 (1979); J. B. Chattopadhyaya and C. B. Reese, *Nucleic Acids Res.*, **8,** 2039 (1980).
5. M. S. Shashidhar and M. V. Bhatt, *J. Chem. Soc., Chem. Commun.*, 654 (1987).

6. S. S. Jones, C. B. Reese, and S. Sibanda, *Tetrahedron Lett.*, **22,** 1933 (1981).
7. J. M. Brown, C. Christodoulou, C. B. Reese, and G. Sindona, *J. Chem. Soc., Perkin Trans. I*, 1785 (1984).

Miscellaneous Esters

The following miscellaneous esters have been prepared as protective groups, but they have seen little use since publication of the first edition. Therefore they are simply listed for completeness, rather than described in detail.

50. 2,6-Dichloro-4-methylphenoxyacetate Ester[1]

51. 2,6-Dichloro-4-(1,1,3,3-tetramethylbutyl)phenoxyacetate Ester[1]

52. 2,4-Bis(1,1-dimethylpropyl)phenoxyacetate Ester[1]

53. Chlorodiphenylacetate Ester[2]

54. Isobutyrate Ester[3] (Chart 2)

55. Monosuccinoate Ester[4]

56. (*E*)-2-Methyl-2-butenoate (Tigloate) Ester[5]

57. *o*-(Methoxycarbonyl)benzoate Ester[6]

58. *p*-Ⓟ-Benzoate Ester[7]

59. α-Naphthoate Ester[8]

60. Nitrate[9] (Chart 2)

61. Alkyl *N,N,N',N'*-Tetramethylphosphorodiamidate: $[(CH_3)_2N]_2P(O)OR$[10]

1. C. B. Reese, *Tetrahedron*, **34,** 3143 (1978).
2. A. F. Cook and D. T. Maichuk, *J. Org. Chem.*, **35,** 1940 (1970).
3. H. Büchi and H. G. Khorana, *J. Mol. Biol.*, **72,** 251 (1972).
4. P. L. Julian, C. C. Cochrane, A. Magnani, and W. J. Karpel, *J. Am. Chem. Soc.*, **78,** 3153 (1956).
5. S. M. Kupchan, A. D. J. Balon, and E. Fujita, *J. Org. Chem.*, **27,** 3103 (1962).
6. G. Losse and H. Raue, *Chem. Ber.*, **98,** 1522 (1965).
7. R. D. Guthrie, A. D. Jenkins, and J. Stehlicek, *J. Chem. Soc. C*, 2690 (1971).
8. I. Watanabe, T. Tsuchiya, T. Takase, S. Umezawa, and H. Umezawa, *Bull. Chem. Soc. Jpn.*, **50,** 2369 (1977).
9. J. Honeyman and J. W. W. Morgan, *Adv. Carbohydr. Chem.*, **12,** 117 (1957); J. F. W. Keana, in *Steroid Reactions*, C. Djerassi, Ed., Holden-Day, San Francisco, 1963,

pp. 75–76; R. Boschan, R. T. Merrow, and R. W. Van Dolah, *Chem. Rev.*, **55,** 485 (1955); R. W. Binkley and D. J. Koholic, *J. Org. Chem.*, **44,** 2047 (1979); R. W. Binkley and D. J. Koholic, *J. Carbohydr. Chem.*, **3,** 85 (1984).
10. R. E. Ireland, D. C. Muchmore, and U. Hengartner, *J. Am. Chem. Soc.*, **94,** 5098 (1972).

62. Alkyl *N*-Phenylcarbamate: ROCONHPh (Chart 2)

Phenyl isocyanates are generally more reactive than alkyl isocyanates in their reactions with alcohols, but with CuCl catalysis even alkyl isocyanates will react readily with primary, secondary, or tertiary alcohols (45–95% yield).[1]

Formation

1. PhN=C=O, Pyr, 20°, 2–3 h, 100% yield.[2] This method was used to protect selectively the primary hydroxyl group in several pyranosides.[3]

Cleavage

1. MeONa, MeOH, reflux, 1.5 h, good yield.[4]
2. $LiAlH_4$, THF, or dioxane, reflux, 3–4 h, 90% yield.[3]
3. Cl_3SiH, Et_3N, CH_2Cl_2, 4–48 h, 25–80°, 80–95° yield.[5] Primary, secondary, tertiary, allylic, propargylic, or benzylic derivatives are cleaved by this method.

1. M. E. Duggan and J. S. Imagire, *Synthesis*, 131 (1989).
2. K. L. Agarwal and H. G. Khorana, *J. Am. Chem. Soc.*, **94,** 3578 (1972).
3. D. Plusquellec and M. Lefeuvre, *Tetrahedron Lett.*, **28,** 4165 (1987).
4. H. O. Bouveng, *Acta Chem. Scand.*, **15,** 87, 96 (1961).
5. W. H. Pirkle and J. R. Hauske, *J. Org. Chem.*, **42,** 2781 (1977).

63. Borate Ester: $(RO)_3B$

Formation

1. $BH_3 \cdot Me_2S$, 25°, 1 h, 80–90% yield.[1]
2. $B(OH)_3$, benzene, $-H_2O$, 100% yield.[2,3]

Cleavage

Borate esters are hydrolyzed with aqueous acid or base. More sterically hindered borates such as pinanediol derivatives are quite stable to hydrolysis.[4] Borates are stable to anhydrous acid and base, HBr/BzOOBz, NaH, and Wittig reactions.[3]

1. C. A. Brown and S. Krishnamurthy, *J. Org. Chem.*, **43,** 2731 (1978).
2. W. I. Fanta and W. F. Erman, *J. Org. Chem*, **37,** 1624 (1972).
3. W. I. Fanta and W. F. Erman, *Tetrahedron Lett.*, 4155 (1969).
4. D. S. Matteson and R. Ray, *J. Am. Chem. Soc.*, **102,** 7590 (1980).

64. Dimethylphosphinothioyl Ester: $(CH_3)_2P(S)OR$

The dimethylphosphinothioyl group has been used to protect hydroxyl groups in carbohydrates. It is prepared from the alcohol and $Me_2P(S)Cl$ (cat. DMAP, DBU). It is not prone to undergo acyl migration as are carboxylate esters. It is stable to the acidic conditions used to cleave acetonides and trityl ethers, to DBU/MeOH, $Bu_4N^+F^-$, Bu_3SnH, Grignard reagents, and cat. NaOMe/MeOH. The dimethylphosphinothioyl group is cleaved with $BnMe_3N^+OH^-$. It can also be cleaved by $Bu_4N^+F^-$ after conversion to the dimethylphosphonyl group with *m*-chloroperoxybenzoic acid.[1]

1. T. Inazu and T. Yamanoi, *Noguchi Kenkyusho Jiho*, 43–47 (1988); *Chem. Abstr.*, **111,** 7685w (1989).

65. Alkyl 2,4-Dinitrophenylsulfenate: $ROSC_6H_3$-2,4-$(NO_2)_2$ (Chart 2)

A nitrophenylsulfenate, cleaved by nucleophiles under very mild conditions, was developed as protection for a hydroxyl group during solid-phase nucleotide synthesis.[1] The sulfenate ester is stable to the acidic hydrolysis of acetonides.[2]

Formation

1. 2,4-$(NO_2)_2C_6H_3SCl$, Pyr, DMF or CH_2Cl_2, 20°, 1 h, 70–85% yield.[1]

Cleavage

1. Nu^-, MeOH, H_2O, 25°, 4 h, 63–80% yield:[1] Nu^- = $Na_2S_2O_3$, pH 8.9; NaCN, pH 8.9; Na_2S, pH 6.6; PhSH, pH 11.8.
2. H_2, Raney Ni, 54% yield.[1]
3. Al, $Hg(OAc)_2$, MeOH, 5 h, 67% yield.[2]
4. An *o*-nitrophenylsulfenate is cleaved by electrolytic reduction (−1.0 V, DMF, $R_4N^+X^-$).[3]

1. R. L. Letsinger, J. Fontaine, V. Mahadevan, D. A. Schexnayder, and R. E. Leone, *J. Org. Chem.*, **29,** 2615 (1964).
2. K. Takiura, S. Honda, and T. Endo, *Carbohydr. Res.*, **21,** 301 (1972).
3. V. G. Mairanovsky, *Angew. Chem., Int. Ed., Engl.*, **15,** 281 (1976).

Sulfonates

Sulfonate protective groups have been restricted largely to carbohydrates where they serve to protect the 2-OH with a nonparticipating group so that coupling gives predominantly 1,2-*cis* glycosides.

66. Sulfate: $ROSO_3^-$

Formation[1]/Cleavage[2]

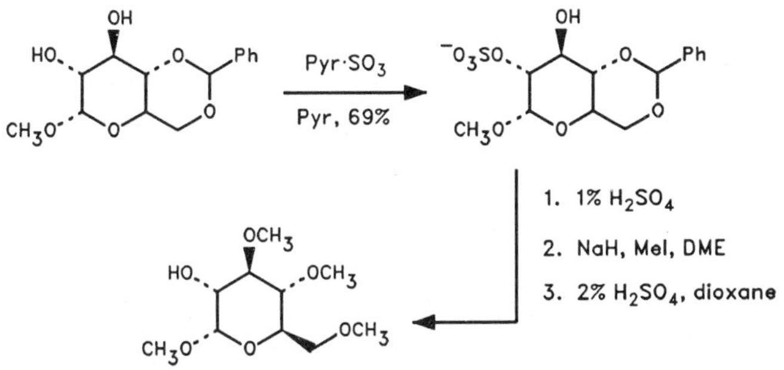

OH HO CH$_3$O O O Ph

$Pyr \cdot SO_3$

Pyr, 69%

OH ^-O_3SO CH$_3$O O O Ph

1. 1% H_2SO_4
2. NaH, MeI, DME
3. 2% H_2SO_4, dioxane

OCH$_3$ HO OCH$_3$ CH$_3$O O OCH$_3$

The α-anomer gives better selectivity for the 2-OH than does the β-anomer (3:2). Note that the conditions used to remove the 4,6-*O*-benzylidene group are sufficiently mild to retain the sulfate.[2]

67. Methanesulfonate (Mesylate): ROMs, $MeSO_3R$

Formation

1. MsCl, Et_3N, CH_2Cl_2, 0°, generally >90% yield.[3]

Cleavage

1. Na(Hg), 2-propanol, 84–98% yield.[4] The use of methanol or ethanol gives very slow reactions. Benzyl ethers are not affected by these conditions.

68. Benzylsulfonate: $ROSO_2Bn$

Formation

1. $BnSO_2Cl$, 2,6-lutidine, CH_2Cl_2, >72% yield.[5]

Cleavage

1. $NaNH_2$, DMF, 67–95% yield.[5]

69. Tosylate (TsOR): $CH_3C_6H_4SO_3R$

Formation

1. TsCl, Pyr.[6]

Cleavage

1. *hν*, 90% CH_3CN/H_2O, 1,5-dimethoxynaphthalene, NH_2NH_2 or $NaBH_4$ or Pyr·BH_3, 59–97% yield.[7]
2. *hν*, Et_3N, MeOH, 12 h, 91% yield.[8]
3. The tosyl group has also been removed by reductive cleavage with Na/NH_3 (65–73% yield),[9] Na/naphthalene (50–87% yield),[10] and Na(Hg)/MeOH (96.7% yield).[11]

1. A. Liav and M. B. Goren, *Carbohydr. Res.*, **131,** C8 (1984).
2. M. B. Goren and M. E. Kochansky, *J. Org. Chem.*, **38,** 3510 (1973); A. Liav and M. B. Goren, *Carbohydr. Res.*, **127,** 211 (1984).
3. A. Fürst and F. Koller, *Helv. Chim. Acta*, **30,** 1454 (1947).
4. K. T. Webster, R. Eby, and C. Schuerch, *Carbohydr. Res.*, **123,** 335 (1983).
5. L. F. Awad, El S. H. Ashry, and C. Schuerch, *Bull. Chem. Soc. Jpn.*, **59,** 1587 (1986).
6. L. F. Fieser and M. Fieser, *Reagents for Organic Synthesis*, Vol. 1, Wiley, New York, 1967, p. 1179.
7. A. Nishida, T. Hamada, and O. Yonemitsu, *J. Org. Chem.*, **53,** 3386 (1988).
8. R. W. Binkley and D. J. Koholic, *J. Org. Chem.*, **54,** 3577 (1989).
9. M. A. Miljkovic, M. Pesic, A. Jokic, and E. A. Davidson, *Carbohydr. Res.*, **15,** 162 (1970); J. Kovar, *Can. J. Chem.*, **48,** 2383 (1970).
10. H. C. Jarrell, R. G. S. Ritchie, W. A. Szarek, and J. K. N. Jones, *Can. J. Chem.*, **51,** 1767 (1973).
11. R. S. Tipson, *Methods Carbohydr. Chem.*, **II,** 250 (1963).

PROTECTION FOR 1,2- AND 1,3-DIOLS

The prevalence of diols in synthetic planning and in natural sources (e.g., in carbohydrates and nucleosides) has led to the development of a number of protective groups of varying stability to a substantial array of reagents. Dioxolanes and dioxanes are the most common protective groups for diols. The ease of formation follows the order:

$$HOCH_2C(CH_3)_2CH_2OH > HO(CH_2)_2OH > HO(CH_2)_3OH$$

In some cases the formation of a dioxolane or dioxane can result in the generation of a new stereogenic center, either with complete selectivity or as a mixture of the two possible isomers. Since the new center is removed on deprotection, it should not seriously complicate the use of those protective groups that generate a new stereogenic center.

Cyclic carbonates and cyclic boronates have also found considerable use as protective groups. In contrast to most acetals and ketals the carbonates are cleaved with strong base and sterically unencumbered boronates are readily cleaved by water.

Some of the protective groups for diols are listed in Reactivity Chart 3.

Cyclic Acetals and Ketals

1. Methylene Acetal (Chart 3)

Methylene acetals are the most stable acetals to acid hydrolysis. Difficulty in their removal is probably the reason that these compounds have not been used much.

Formation

1. 40% CH_2O, concd. HCl, 50°, 4 days, 68% yield.[1] The trismethylenedioxy derivative of a carbohydrate was formed.
2. Paraformaldehyde, H_2SO_4, AcOH, 90°, 1 h, good yield.[2]
3. DMSO, NBS, 50°, 12 h, 62% yield.[3]
4. CH_2Br_2, NaH, DMF, 0–30°, 40 h, 46% yield.[4]
5. $(MeO)_2CH_2$, 2,6-lutidine, TMSOTf, 0°, 15 min.[5] Similar conditions have been used to introduce MOM ethers on alcohols.

TMSOTf, 15 min, 2,6-lutidine, $(MeO)_2CH_2$, 0°

6. $(MeO)_2CH_2$, LiBr, TsOH, CH_2Cl_2, 23°, 83% yield.[6] In this case a 1,3-methylene acetal is formed in preference to a 1,2-methylene acetal from a 1,2,3-triol. These conditions also protect simple alcohols as their MOM derivatives.
7. CH_2Br_2, NaOH, CH_2Cl_2, cetyl$N^+Me_3Br^-$, heat, 81% yield.[7] This method is effective for both *cis*- and *trans*-1,2-diols.
8. DMSO, TMSCl.[8]
9. CH_2Br_2, powdered KOH, DMSO, rt, 49% yield.[9]
10. HCHO, cat. SO_2.[10]

Cleavage

1. BCl_3, CH_2Cl_2, −80°, 30 min, warm to 20°, 61% yield; isolated as the acetate derivative.[1]
2. 2 *N* HCl, 100°, 3 h.[2]
3. AcOH, Ac_2O, H_2SO_4, 2 h, 0°, 91.5% yield.[11]

AcOH, Ac_2O, H_2SO_4, 2 h, 0°, 91.5%

4. $Ph_3C^+BF_4^-$, CH_2Cl_2, reflux, 48 h; HCl, rt, 17.5 h, 86% yield.[12]
5. $(CF_3CO)_2O$, AcOH, CH_2Cl_2, 21°; MeOH, K_2CO_3, 92% yield.[13]

1. T. G. Bonner, *Methods Carbohydr. Chem.*, **II,** 314 (1963).
2. L. Hough, J. K. N. Jones, and M. S. Magson, *J. Chem Soc.*, 1525 (1952).
3. S. Hanessian, G. Y.-Chung, P. Lavallee, and A. G. Pernet, *J. Am. Chem. Soc.*, **94,** 8929 (1972).
4. J. S. Brimacombe, A. B. Foster, B. D. Jones, and J. J. Willard, *J. Chem. Soc. C*, 2404 (1967).
5. F. Matsuda, M. Kawasaki, and S. Terashima, *Tetrahedron Lett.*, **26,** 4639 (1985).
6. J. L. Gras, R. Nouguier, and M. Mchich, *Tetrahedron Lett.*, **28,** 6601 (1987).
7. D. G. Norman, C. B. Reese, and H. T. Serafinowska, *Synthesis*, 751 (1985). For a similar method, see: K. S. Kim and W. A. Szarek, *Synthesis*, 48 (1978).
8. B. S. Bal and H. W. Pinnick, *J. Org. Chem.*, **44,** 3727 (1979).
9. A. Liptak, V. A. Oláh, and J. Kerékgyártó, *Synthesis*, 421 (1982).
10. B. Burczyk, *J. Prakt. Chem.*, **322,** 173 (1980).
11. M. J. Wanner, N. P. Willard, G. J. Kooman, and U. K. Pandet, *Tetrahedron*, **43,** 2549 (1987).
12. H. Niwa, O. Okamoto, and K. Yamada, *Tetrahedron Lett.*, **29,** 5139 (1988).
13. J.-L. Gras, H. Pellissier, and R. Nouguier, *J. Org. Chem.*, **54,** 5675 (1989).

2. **Ethylidene Acetal** (Chart 3)

Formation

1. CH_3CHO, $CH_3CH(OMe)_2$, or paraldehyde, concd. H_2SO_4, 2–3 h, 60% yield.[1]

2. In the following example the ethylidene acetal was used because attempts to make the acetonide led to formation of a 1 : 1 mixture of the 1,3- and 1,4-acetonide.[2]

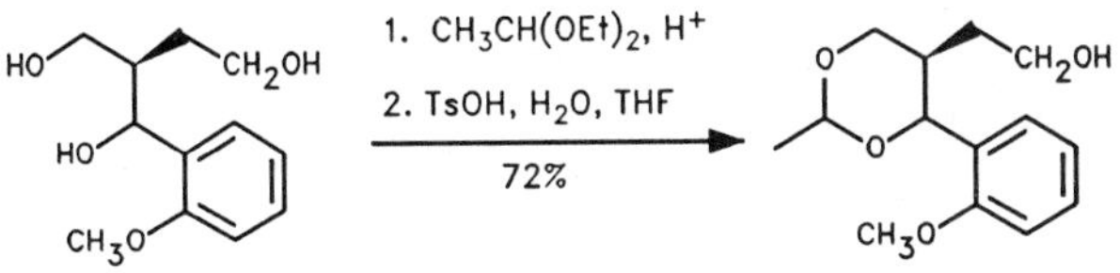

1. $CH_3CH(OEt)_2$, H^+
2. TsOH, H_2O, THF
72%

3. Diborane reduction of an ortho ester that is prepared from a triol with $CH_3C(OEt)_3$, PPTS.[3]

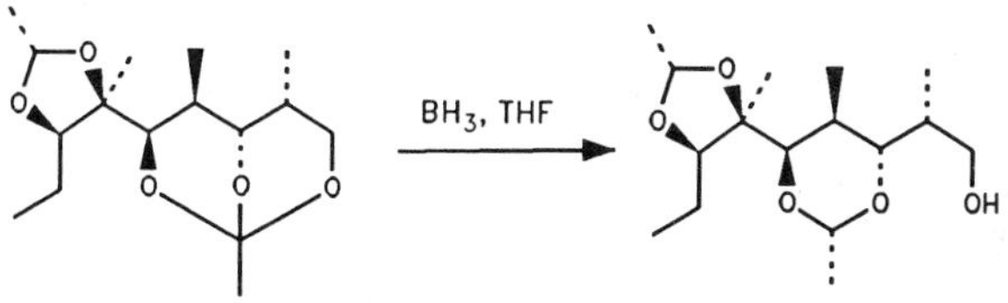

BH_3, THF

Cleavage

1. 0.67 N H_2SO_4, aq. acetone, reflux, 7 h.[1]
2. Ac_2O, cat. H_2SO_4, 20°, 5 min, 60% yield.[1] The ethylidene acetal is cleaved to form an acetate that can be hydrolyzed with base.
3. 80% AcOH, reflux, 1.5 h.[4]
4. O_3, CH_2Cl_2, 75% yield.[3]

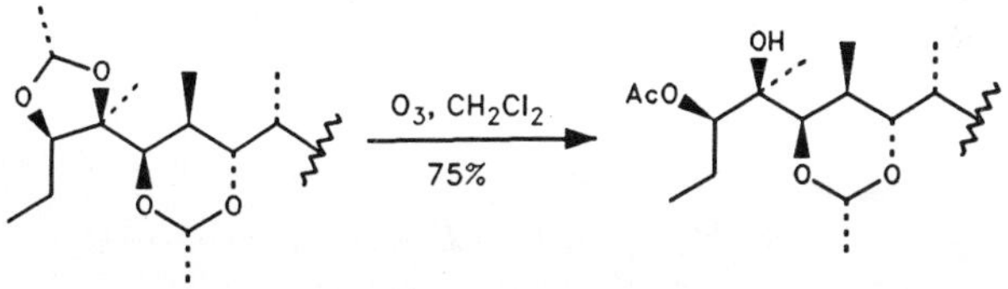

O_3, CH_2Cl_2
75%

1. T. G. Bonner, *Methods Carbohydr. Chem.*, **II,** 309 (1963); D. M. Hall, T. E. Lawler, and B. C. Childress, *Carbohydr. Res.*, **38,** 359 (1974).
2. A. G. Brewster and A. Leach, *Tetrahedron Lett.*, **27,** 2539 (1986).
3. G. Stork and S. D. Rychnovsky, *J. Am. Chem. Soc.*, **109,** 1565 (1987).
4. J. W. Van Cleve and C. E. Rist, *Carbohydr. Res.*, **4,** 82 (1967).

3. 1-*t*-Butylethylidene Ketal

4. 1-Phenylethylidene Ketal

1-*t*-Butyl- and 1-phenylethylidene ketals were prepared selectively from the C_4–C_6, 1,3-diol in glucose by an acid-catalyzed transketalization reaction [e.g.,

$Me_3CC(OMe)_2CH_3$, TsOH/DMF, 24 h, 79% yield; $PhC(OMe)_2Me$, TsOH, DMF, 24 h, 90% yield, respectively]. They are cleaved by acidic hydrolysis: AcOH, 20°, 90 min, 100% yield and AcOH, 20°, 3 days, 100% yield, respectively.[1]

1. M. E. Evans, F. W. Parrish, and L. Long, Jr., *Carbohydr. Res.*, **3,** 453 (1967).

5. (4-Methoxyphenyl)ethylidene Acetal

Formation/Cleavage[1]

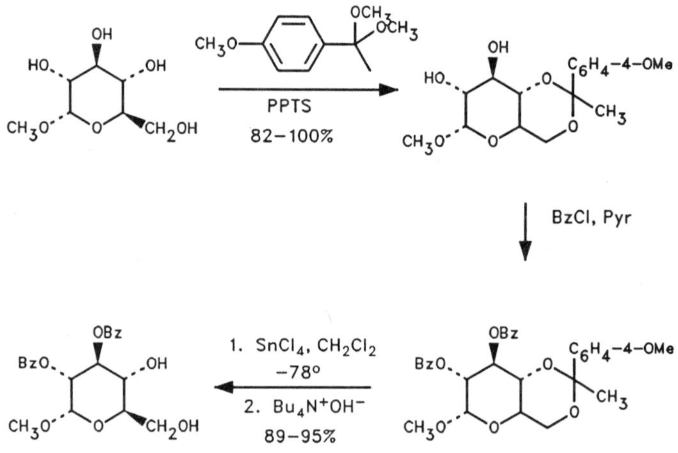

PPTS
82–100%

BzCl, Pyr

1. $SnCl_4$, CH_2Cl_2
−78°
2. $Bu_4N^+OH^-$
89–95%

PPTS = Pyridinium *p*-toluenesulfonate

1. B. H. Lipshutz and M. C. Morey, *J. Org. Chem.*, **46,** 2419 (1981).

6. 2,2,2-Trichloroethylidene Acetal

Trichloroacetaldehyde (chloral) reacts with glucose in the presence of sulfuric acid to form two monoacetals and four diacetals. The trichloro acetal is cleaved by reduction (H_2, Raney Ni, 50% NaOH, EtOH, 15 min).[1] The trichloro acetal can probably be cleaved with Zn/AcOH [cf. $ROCH(R')OCH_2CCl_3$ cleaved by Zn/AcOH, AcONa, 20°, 3 h, 90% yield[2]].

1. S. Forsén, B. Lindberg, and B.-G. Silvander, *Acta Chem. Scand.*, **19,** 359 (1965).
2. R. U. Lemieux and H. Driguez, *J. Am. Chem. Soc.*, **97,** 4069 (1975).

7. **Acetonide (Isopropylidene Ketal)** (Chart 3)

Acetonide formation is the most commonly used protection for 1,2- and 1,3-diols. The acetonide has been used extensively in carbohydrate chemistry to mask selectively the hydroxyls of the many different sugars.[1] In preparing acetonides of triols, the 1,2-derivative is generally favored over the 1,3-derivative, but the extent to which the 1,2-acetonide is favored is dependent on structure.[2-5] Note that the 1,2-selectivity for the ketal from 3-pentanone is better than that from acetone.[5]

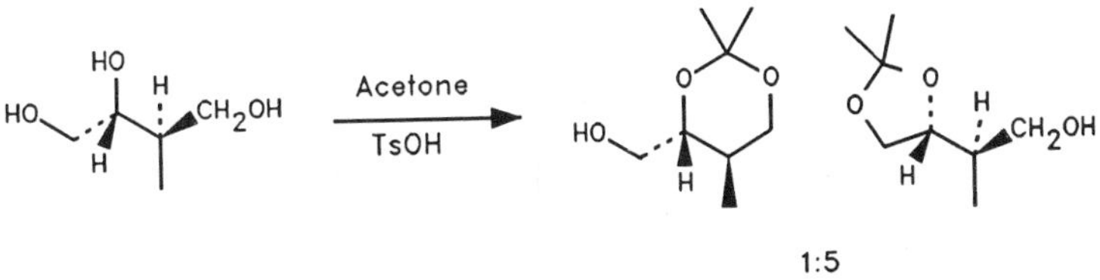

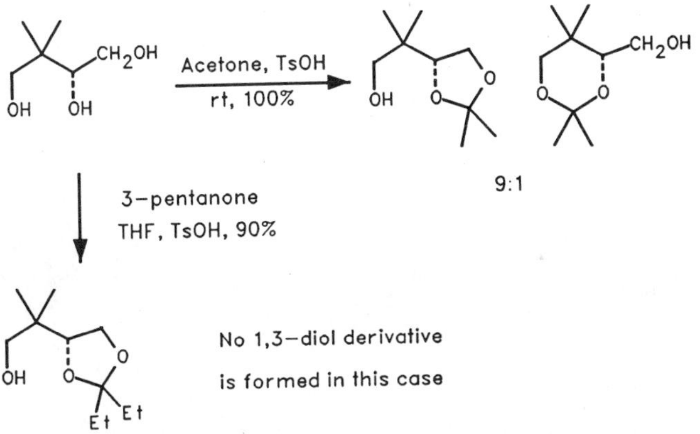

Acetone, TsOH

1:5

Acetone, TsOH, rt, 100%

9:1

3-pentanone, THF, TsOH, 90%

No 1,3-diol derivative is formed in this case

In cases where two 1,2-acetonides are possible, the thermodynamically favored one prevails. Secondary alcohols have a greater tendency to form cyclic acetals than do primary alcohols,[6,7] but an acetonide from a primary alcohol is preferred over an acetonide from two *trans*, secondary alcohols.

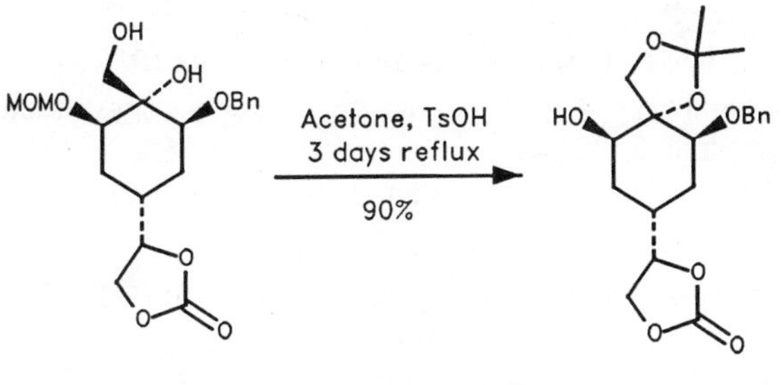

Acetone, TsOH, 3 days reflux, 90%

Ref. 8

In the case below, **i** is isomerized to **ii**, producing a *trans* derivative, but acetonide **iii** fails to isomerize to the internal derivative because the less favorable *cis* product would be formed.[9]

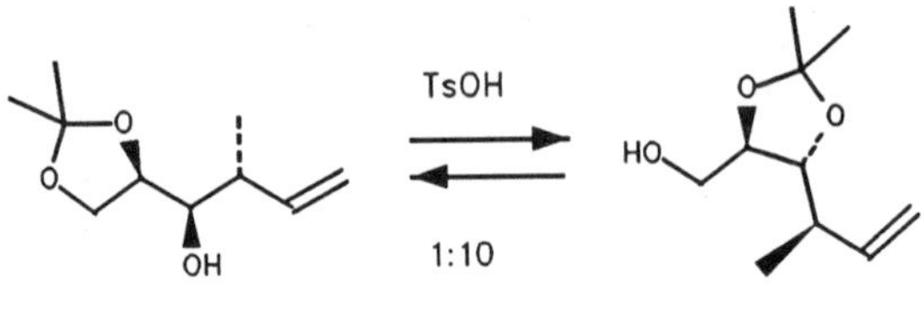

TsOH

1:10

OH

i

HO

ii

OH

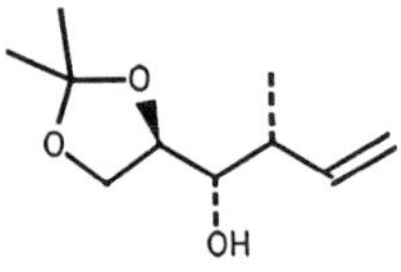

iii

Formation

1. $CH_3C(OCH_3){=}CH_2$, dry HBr, CH_2Cl_2, 0°, 16 h, 75% yield.[10]
2.

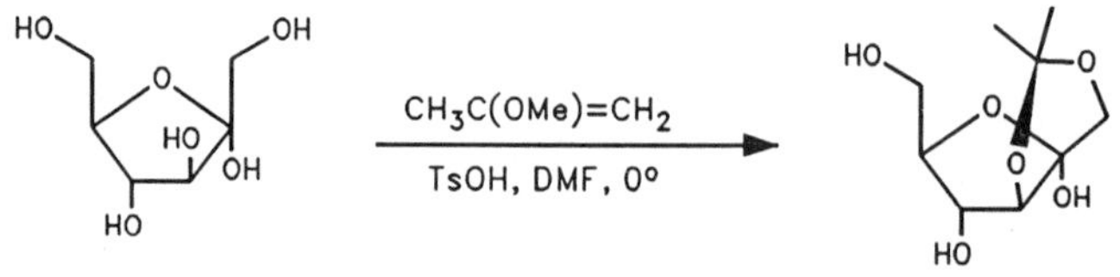

HO OH HO OH HO

$CH_3C(OMe){=}CH_2$

TsOH, DMF, 0°

HO O OH HO

Under these conditions 2-methoxypropene reacts to form the kinetically controlled 1,3-*O*-isopropylidene, instead of the thermodynamically more stable 1,2-*O*-isopropylidene.[11]

3. TsOH, DMF, $Me_2C(OMe)_2$, 24 h.[12,13] This method has become one of the most popular methods for the preparation of acetonides. It generally gives high yields and is compatible with acid-sensitive protective groups such as the TBDMS group.
4. $Me_2C(OMe)_2$, DMF, pyridinium *p*-toluenesulfonate (PPTS).[14] The use of PPTS for acid-catalyzed reactions has been quite successful and is particularly useful when TsOH acid is too strong an acid for the functionality in a given molecule. TBDMS groups are stable under these conditions.[15]
5. Anhydrous acetone, $FeCl_3$, 36°, 5 h, 60–70% yield.[16]
6. $Me_2C(OMe)_2$, di-*p*-nitrophenyl hydrogen phosphate, 3–5 h, 90–100% yield.[17]

7. $MeC(OEt){=}CH_2$, cat. HCl, DMF, 25°, 12 h, 90–100% yield.[18] These conditions are used to obtain the kinetic acetonide.[19]
8. $MeC(OTMS){=}CH_2$, concd. HCl or TMSCl, 10–30 min, 80–85% yield.[20] This method is effective for the formation of *cis*- or *trans*-acetonides of 1,2-cyclohexanediol.
9. The classical method for acetonide formation is by reaction of a diol with acetone and an acid catalyst.[21,22]
10. Acetone, I_2, 70–85% yield, rt or reflux.[23]
11. Acetone, $CuSO_4$, H_2SO_4, 90% yield.[24] If PPTS replaces H_2SO_4 as the acid, the acetonide can be formed in the presence of a trityl group.[25]

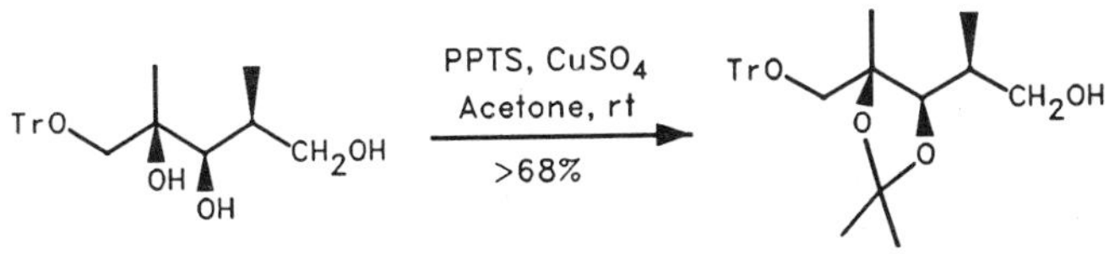

TrO ... OH, OH, CH_2OH → (PPTS, $CuSO_4$, Acetone, rt, >68%) → TrO ... O, O (acetonide), CH_2OH

12. Trimethylsilylated diols are converted to acetonides with acetone and TMSOTf, −78°, 3.5 h, >76% yield.[26]
13. Acetone, $AlCl_3$, Et_2O, rt, 3.5 h, 80% yield.[27] Other methods failed in this sterically demanding case.

Cleavage

1. Cat. I_2, MeOH, rt, 24 h, >80% yield.[28] Benzylidene ketals and thioketals are also cleaved under these conditions.
2. 1 *N* HCl, THF (1 : 1), 20°.[6]
3. 2 *N* HCl, 80°, 6 h.[29]
4. 60–80% AcOH, 25°, 2 h, 92% yield of *cis*-1,2-diol.[30] MOM groups are stable to these conditions.[31]
5. 80% AcOH, reflux, 30 min, 78% yield of *trans*-1,2-diol.[30]
6. TsOH, MeOH, 25°, 5 h.[32] These conditions failed to cleave the acetonide of a 2′,3′-ribonucleoside.[33]
7. Dowex 50-W (H^+), water, 70°, excellent yield.[34]
8. BCl_3, 25°, 2 min, 100% yield.[35]
9. Br_2, Et_2O.[22]
10. $PdCl_2(CH_3CN)_2$, CH_3CN, H_2O, rt.[36] When the solvent is changed to wet acetone the reagent cleaves an ethylene glycol ketal from ketones in 82–100% yield. TBDPS and MEM groups are stable, but TBDMS and THP groups are cleaved under these conditions.
11. CF_3CO_2H, THF, H_2O, 83% yield.[37]

TBDPSO / C5H11 acetonide —TFA, THF / H_2O, 83%→ TBDPSO / OH / OH / C5H11

12. In the following example the acetonide protective group is selectively converted to one of two *t*-butyl groups. The reaction appears to be general, but the alcohol bearing the *t*-butyl group varies with structure.[38] Benzylidene ketals are also cleaved.

CH_2OBn / H / H / CH_2OH acetonide —Me_3Al, rt / $PhCH_3$, 70%→ CH_2OBn / H—O–t–Bu / H—OH / CH_2OH

13. $(Bu_2SnNCS)_2O$, diglyme, H_2O, 100°, 82% yield.[39] The THP group is also cleaved by this reagent.
14. $FeCl_3$–SiO_2, $CHCl_3$, 74% yield.[40] When used in acetone, this reagent cleaves the trityl and TBDMS groups. These conditions also cleave THP and TMS groups, but TBDMS, MTM, and MOM groups are not affected when $CHCl_3$ is used as solvent.
15. MeOH, PPTS, heat, high yield.[41]

1. For a review, see: D. M. Clode, *Chem. Rev.*, **79,** 491 (1979).
2. D. R. Williams and S.-Y. Sit, *J. Am. Chem. Soc.*, **106,** 2949 (1984).
3. P. Lavallee, R. Ruel, L. Grenier, and M. Bissonnette, *Tetrahedron Lett.*, **27,** 679 (1986).
4. A. I. Meyers and J. P. Lawson, *Tetrahedron Lett.*, **23,** 4883 (1982).
5. S. Hanessian, *Aldrichimica Acta*, **22,** 3 (1989).
6. S. J. Angyal and R. J. Beveridge, *Carbohydr. Res.*, **65,** 229 (1978).
7. P. A. Grieco, Y. Yokoyama, G. P. Withers, F. J. Okuniewicz, and C.-L. J. Wang, *J. Org. Chem.*, **43,** 4178 (1978).
8. S. Nishiyama, Y. Ikeda, S. Yoshida, and S. Yamamura, *Tetrahedron Lett.*, **30,** 105 (1989).
9. J. W. Coe and W. R. Roush, *J. Org. Chem.*, **54,** 915 (1989).
10. E. J. Corey, S. Kim, S. Yoo, K. C. Nicolaou, L. S. Melvin, Jr., D. J. Brunelle, J. R. Falck, E. J. Trybulski, R. Lett, and P. W. Sheldrake, *J. Am. Chem. Soc.*, **100,** 4620 (1978).
11. E. Fanton, J. Gelas, and D. Horton, *J. Chem. Soc., Chem. Commun.*, 21 (1980).
12. M. E. Evans, F. W. Parrish, and L. Long, Jr., *Carbohydr. Res.*, **3,** 453 (1967).
13. B. H. Lipshutz and J. C. Barton, *J. Org. Chem.*, **53,** 4495 (1988).
14. M. Kitamura, M. Isobe, Y. Ichikawa, and T. Goto, *J. Am. Chem. Soc.*, **106,** 3252 (1984).

15. K. Mori and S. Maemoto, *Liebigs Ann. Chem.*, 863 (1987).
16. P. P. Singh, M. M. Gharia, F. Dasgupta, and H. C. Srivastava, *Tetrahedron Lett.*, 439 (1977).
17. A. Hampton, *J. Am. Chem. Soc.*, **83,** 3640 (1961).
18. S. Chládek and J. Smrt, *Collect. Czech. Chem. Commun.*, **28,** 1301 (1963).
19. J. Gelas and D. Horton, *Heterocycles*, **16,** 1587 (1981).
20. G. L. Larson and A. Hernandez, *J. Org. Chem.*, **38,** 3935 (1973).
21. O. Th. Schmidt, *Methods Carbohydr. Chem.*, **II,** 318 (1963).
22. A. N. de Belder, *Adv. Carbohydr. Chem.*, **20,** 219 (1965).
23. K. P. R. Kartha, *Tetrahedron Lett.*, **27,** 3415 (1986).
24. P. Rollin and J.-R. Pougny, *Tetrahedron*, **42,** 3479 (1986).
25. T. Nakata, M. Fukui, and T. Oishi, *Tetrahedron Lett.*, **29,** 2219 (1988).
26. S. D. Rychnovsky, *J. Org. Chem.*, **54,** 4982 (1989).
27. B. Lal, R. M. Gidwani, and R. H. Rupp, *Synthesis*, 711 (1989).
28. W. A. Szarek, A. Zamojski, K. N. Tiwari, and E. R. Isoni, *Tetrahedron Lett.*, **27,** 3827 (1986).
29. T. Ohgi, T. Kondo, and T. Goto, *Tetrahedron Lett.*, 4051 (1977).
30. M. L. Lewbart and J. J. Schneider, *J. Org. Chem.*, **34,** 3505 (1969).
31. S. Hanessian, D. Delorme, P. C. Tyler, G. Demailly, and Y. Chapleur, *Can. J. Chem.*, **61,** 634 (1983).
32. A. Ichihara, M. Ubukata, and S. Sakamura, *Tetrahedron Lett.*, 3473 (1977).
33. J. Kimura and O. Mitsunobu, *Bull. Chem. Soc. Jpn.*, **51,** 1903 (1978).
34. P.-T. Ho, *Tetrahedron Lett.*, 1623 (1978).
35. T. J. Tewson and M. J. Welch, *J. Org. Chem.*, **43,** 1090 (1978).
36. B. H. Lipshutz, D. Pollart, J. Monforte, and H. Kotsuki, *Tetrahedron Lett.*, **26,** 705 (1985).
37. Y. Leblanc, B. J. Fitzsimmons, J. Adams, F. Perez, and J. Rokach, *J. Org. Chem.*, **51,** 789 (1986).
38. S. Takano, T. Ohkawa, and K. Ogasawara, *Tetrahedron Lett.*, **29,** 1823 (1988).
39. J. Otera and H. Nozaki, *Tetrahedron Lett.*, **27,** 5743 (1986).
40. K. S. Kim, Y. H. Song, B. H. Lee, and C. S. Hahn, *J. Org. Chem.*, **51,** 404 (1986).
41. R. Van Rijsbergen, M. J. O. Anteunis, and A. De Bruyn, *J. Carbohydr. Chem.*, **2,** 395 (1983)

8. Cyclopentylidene Ketal, i

9. Cyclohexylidene Ketal, ii

10. Cycloheptylidene Ketal, iii

Compounds **i**, **ii**, and **iii** can be prepared by an acid-catalyzed reaction of a diol and the cycloalkanone in the presence of ethyl orthoformate and mesitylenesulfonic acid.[1] The relative ease of acid-catalyzed hydrolysis [0.53 *M* H_2SO_4, H_2O, PrOH (65 : 35), 20°] for compounds **i**, **iii**, acetonide, and **ii** is $C_5 \approx C_7 >$ acetonide >> C_6 (e.g., $t_{1/2}$ for 1,2-*O*-alkylidene-α-D-glucopyranoses of C_5, C_7, acetonide, and C_6 derivatives are 8, 10, 20, and 124 h, respectively).[1]

The cyclohexylidene ketal has also been prepared from dimethoxycyclohexane and TsOH;[2] $HC(OEt)_3$, cyclohexanone, TsOH, EtOAc, heat, 5 h, 78%; 1-(trimethylsiloxy)cyclohexene, concd. HCl, 20°, 10–30 min, 70–75% yield,[3] and cyclohexanone, TsOH, $CuSO_4$.[4] The cyclohexylidene derivative of a *trans*-1,2-diol has also been prepared.[5] Cyclohexylidene derivatives are cleaved by acidic hydrolysis: 10% HCl, Et_2O, 25°, 5 min;[3] TFA, H_2O, 20°, 6 min–2 h, 65–85% yield;[6] 0.1 *N* HCl, dioxane;[5] BCl_3, CH_2Cl_2, −80°, 15 h, 90% yield.[7] The cyclohexylidene derivative is also subject to cleavage with Grignard reagents, but under harsh reaction conditions (MeMgI, PhH, 85°, 58% yield.)[8] *trans*-Cyclohexylidene ketals are preferentially cleaved in the presence of *cis*-cyclohexylidene ketals.[9]

The cyclopentylidene ketal has also been prepared from dimethoxycyclopentane, TsOH, CH_3CN, and can be cleaved with 2:1 $AcOH/H_2O$, rt, 2 h.[10] Methoxycyclopentene (PPTS, CH_2Cl_2, rt, 100%) has also been used to introduce this group.[11]

1. W. A. R. van Heeswijk, J. B. Goedhart, and J. F. G. Vliegenthart, *Carbohydr. Res.*, **58,** 337 (1977).
2. C. Kuroda, P. Theramongkol, J. R. Engebrecht, and J. D. White, *J. Org. Chem.*, **51,** 956 (1986).
3. G. L. Larson and A. Hernandez, *J. Org. Chem.*, **38,** 3935 (1973).
4. W. R. Rousch, M. R. Michaelides, D. F. Tai, B. M. Lesur, W. K. M. Chong, and D. J. Harris, *J. Am. Chem. Soc.*, **111,** 2984 (1989).
5. D. Askin, C. Angst, and S. Danishefsky, *J. Org. Chem.*, **50,** 5005 (1985).
6. S. L. Cook and J. A. Secrist, *J. Am. Chem. Soc.*, **101,** 1554 (1979).
7. S. D. Géro, *Tetrahedron Lett.*, 591 (1966).
8. M. Kawana and S. Emoto, *Bull. Chem. Soc. Jpn.*, **53,** 230 (1980).
9. Y.-C. Liu and C.-S. Chen, *Tetrahedron Lett.*, **30,** 1617 (1989).
10. C. B. Reese and J. G. Ward, *Tetrahedron Lett.*, **28,** 2309 (1987).
11. R. M. Soll and S. P. Seitz, *Tetrahedron Lett.*, **28,** 5457 (1987).

11. **Benzylidene Acetal** (Chart 3)

A benzylidene acetal is a commonly used protective group for 1,2- and 1,3-diols. In the case of a 1,2,3-triol the 1,3-acetal is the preferred product. It has the advantage that it can be removed under neutral conditions by hydrogenolysis or by acid hydrolysis. Benzyl groups[1] and isolated olefins[2] have been hydrogenated in the presence of 1,3-benzylidene acetals. Benzylidene acetals of 1,2-diols are more susceptible to hydrogenolysis than are those of 1,3-diols. In fact, the former can be removed in the presence of the latter.[3] A polymer-bound benzylidene acetal has also been prepared.[4]

Formation

1. PhCHO, $ZnCl_2$, 28°, 4 h.[5]
2. PhCHO, DMSO, concd. H_2SO_4, 25°, 4 h.[6]
3. K_2CO_3 or Pyr, CH_2Cl_2, 25°, 16 h, 45–82% yield.[7] 2X⁻, X = FSO_3^- or BF_4^-

   This method is suitable for the protection of 1,2-, 1,3-, and 1,4-diols.
4. PhCHO, TsOH, reflux, $-H_2O$, 72% yield.[8]
5. $PhCH(OMe)_2$, HBF_4, Et_2O, DMF, 97% yield.[9,10]

Cleavage

1. H_2/Pd–C, AcOH, 25°, 30–45 min, 90% yield.[11]
2. Na, NH_3, 85% yield.[12]
3. The benzylidene acetal is cleaved by acidic hydrolysis (e.g., 0.01 *N* H_2SO_4, 100°, 3 h, 92% yield;[13] 80% AcOH, 25°, $t_{1/2}$ for uridine = 60 h[14]), conditions that do not cleave a methylenedioxy group.[13]
4. Electrolysis: −2.9 V, $R_4N^+X^-$, DMF.[15]
5. BCl_3, 100% yield. This reagent also cleaves a number of other ketal-type protective groups.[16]
6. I_2, MeOH, 85% yield.[17]
7. $FeCl_3$, CH_2Cl_2, 3–30 min, 68–85% yield.[18] Benzyl groups are also cleaved by this reagent.
9. $Pd(OH)_2$, cyclohexene, 98% yield.[1]
10. Pd–C, hydrazine, MeOH.[19] In this case a 1,2-benzylidene acetal was cleaved in the presence of a 1,3-benzylidene acetal.
11. Pd–C, HCO_2NH_4, 97% yield.[3]

Benzylidene acetals have the useful property that one of the two C—O bonds can be selectively cleaved. The direction of cleavage is dependent on steric and electronic factors as well as on the nature of the cleavage reagent.

12. $(i\text{-Bu})_2AlH$, CH_2Cl_2 or $PhCH_3$, 0°–rt, yields generally >80%.[20,21]

$(i\text{-Bu})_2AlH$, 0°, CH_2Cl_2

13. TMSCN, $BF_3 \cdot Et_2O$.[22]

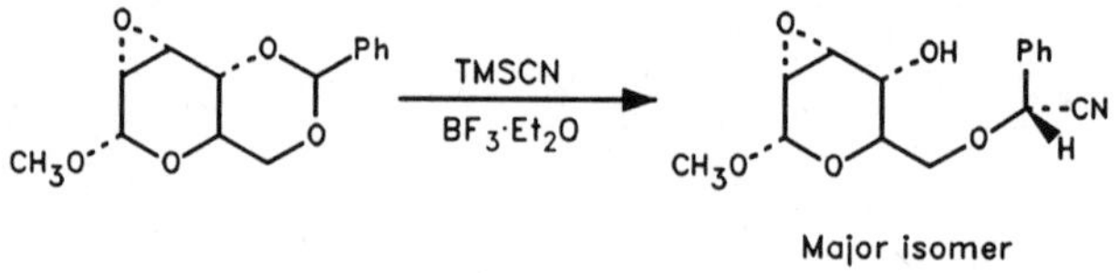

TMSCN, $BF_3 \cdot Et_2O$

Major isomer

14. $Zn(BH_4)_2$, TMSCl, Et_2O, 25°, 45 min, 77–97% yield.[23] In this case reduction takes place to form a monobenzyl derivative of a diol.
15. NBS, CCl_4, H_2O, 75% yield.[24]

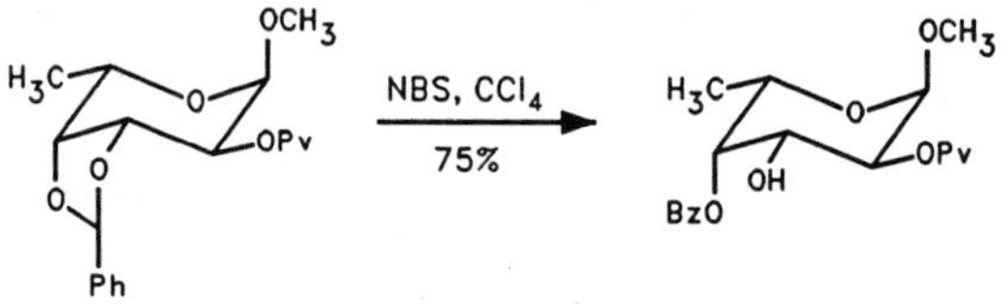

NBS, CCl_4, 75%

In this type of cleavage reaction it appears that the axial benzoate is the preferred product. If water is excluded from the reaction, a bromo benzoate is obtained.[25]

16. $BrCCl_3$, CCl_4, *hν*, 30 min, 100% yield.[26]

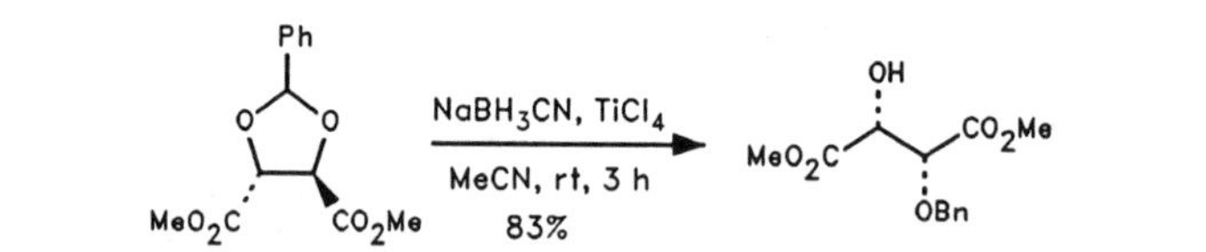

$BrCCl_3$, CCl_4, hv, 30 min, 100%

17. $NaBH_3CN$, $TiCl_4$, CH_3CN, rt, 3 h, 83% yield.[27]

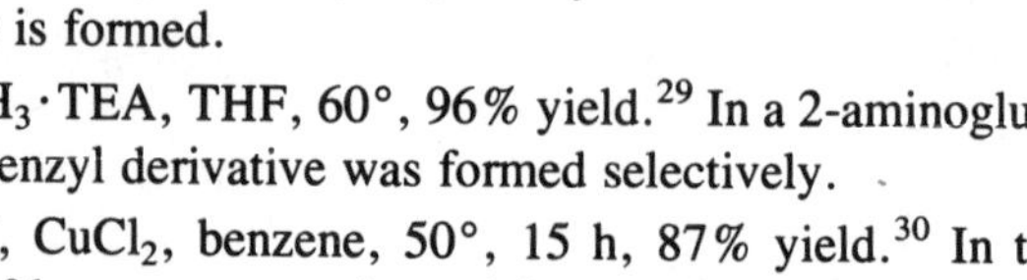

$NaBH_3CN$, $TiCl_4$, MeCN, rt, 3 h, 83%

18. $Ph_3C^+BF_4^-$, CH_3CN, 25°, 8 h, 80% yield.[28] A 1:1 mixture of diol monobenzoates is formed.
19. $AlCl_3$, $BH_3 \cdot TEA$, THF, 60°, 96% yield.[29] In a 2-aminoglucose derivative the 6-*O*-benzyl derivative was formed selectively.
20. *t*-BuOOH, $CuCl_2$, benzene, 50°, 15 h, 87% yield.[30] In this case a 1:1 mixture of benzoates was formed from a 4,6-benzylidene glucose derivative.

21. Ozonolysis, Ac_2O, NaOAc, −78°, 1 h, 95% yield.[31] In this case the benzylidene acetal is converted to a diester.

1. S. Hanessian, T. J. Liak, and B. Vanasse, *Synthesis*, 396 (1981).
2. A. B. Smith, III, and K. J. Hale, *Tetrahedron Lett.*, **30,** 1037 (1989).
3. T. Bieg and W. Szeja, *Carbohydr. Res.*, **140,** C7 (1985).
4. J. M. J. M. Fréchet and G. Pellé, *J. Chem. Soc., Chem. Commun.*, 225 (1975).
5. H. G. Fletcher, Jr., *Methods Carbohydr. Chem.*, **II,** 307 (1963).
6. R. M. Carman and J. J. Kibby, *Aust. J. Chem.*, **29,** 1761 (1976).
7. R. M. Munavu and H. H. Szmant, *Tetrahedron Lett.*, 4543 (1975).
8. D. A. McGowan and G. A. Berchtold, *J. Am. Chem. Soc.*, **104,** 7036 (1982).
9. R. Albert, K. Dax, R. Pleschko, and A. Stütz, *Carbohydr. Res.*, **137,** 282 (1985); T. Yamanoi, T. Akiyama, E. Ishida, H. Abe, M. Amemiya, and T. Inazu, *Chem. Lett.*, 335 (1989).
10. M. T. Crimmins, W. G. Hollis, Jr., and G. J. Lever, *Tetrahedron Lett.*, **28,** 3647 (1987).
11. W. H. Hartung and R. Simonoff, *Org. React.*, **7,** 263–326 (1953); see pp. 271, 284, 302.
12. M. Zaoral, J. Jezek, R. Straka, and K. Masek, *Collect. Czech. Chem. Commun.*, **43,** 1797 (1978).
13. R. M. Hann, N. K. Richtmyer, H. W. Diehl, and C. S. Hudson, *J. Am. Chem. Soc.*, **72,** 561 (1950).
14. M. Smith, D. H. Rammler, I. H. Goldberg, and H. G. Khorana, *J. Am. Chem. Soc.*, **84,** 430 (1962).
15. V. G. Mairanovsky, *Angew. Chem., Int. Ed. Engl.*, **15,** 281 (1976).
16. T. G. Bonner, E. J. Bourne, and S. McNally, *J. Chem. Soc.*, 2929 (1960).
17. W. A. Szarek, A. Zamojski, K. N. Tiwari, and E. R. Ison, *Tetrahedron Lett.*, **27,** 3827 (1986).
18. M. H. Park, R. Takeda, and K. Nakanishi, *Tetrahedron Lett.*, **28,** 3823 (1987).
19. T. Bieg and W. Szeja, *Synthesis*, 317 (1986).
20. S. Takano, M. Akiyama, S. Sato, and K. Ogasawara, *Chem. Lett.*, 1593 (1983); S. Hatakeyama, K. Sakurai, K. Saijo, and S. Takano, *Tetrahedron Lett.*, **26,** 1333 (1985).
21. S. L. Schreiber, Z. Wang, and G. Schulte, *Tetrahedron Lett.*, **29,** 4085 (1988).
22. F. G. De las Heras, A. San Felix, A. Calvo-Mateo, and P. Fernandez-Resa, *Tetrahedron*, **41,** 3867 (1985).
23. H. Kotsuki, Y. Ushio, N. Yoshimura, and M. Ochi, *J. Org. Chem.*, **52,** 2594 (1987).
24. R. W. Binkley, G. S. Goewey, and J. C. Johnston, *J. Org. Chem.*, **49,** 992 (1984).
25. O. Han and H.-w. Liu, *Tetrahedron Lett.*, **28,** 1073 (1987).
26. P. M. Collins, A. Manro, E. C. Opara-Mottah, and M. H. Ali, *J. Chem. Soc., Chem. Commun.*, 272 (1988).
27. G. Adam and D. Seebach, *Synthesis*, 373 (1988).
28. S. Hanessian and A. P. A. Staub, *Tetrahedron Lett.*, 3551 (1973).

29. B. Classon, P. J. Garegg, and A.-C. Helland, *J. Carbohydr. Chem.*, **8**, 543 (1989).
30. K. Sato, T. Igarashi, Y. Yanagisawa, N. Kawauchi, H. Hashimoto, and J. Yoshimura, *Chem. Lett.*, 1699 (1988).
31. P. Deslongchamps, C. Moreau, D. Fréhel, and R. Chênevert, *Can. J. Chem.*, **53**, 1204 (1975).

12. *p*-Methoxybenzylidene Acetal (Chart 3)

The *p*-methoxybenzylidene acetal is a versatile protective group for diols that undergoes acid hydrolysis 10 times faster than the benzylidene group.[1]

Formation

1. *p*-$MeOC_6H_4CHO$, acid, 70–95% yield.[1]
2. *p*-$MeOC_6H_4CH_2OMe$, DDQ, CH_2Cl_2, rt, 30 min, 49–82% yield.[2]
3. *p*-$MeOC_6H_4CHO$, $ZnCl_2$.[3]
4. *p*-$MeOC_6H_4CH(OMe)_2$, acid.[4] The related *o*-methoxybenzylidene acetal has been prepared by this method.[5]
5. The *p*-methoxybenzylidene ketal can be prepared by DDQ oxidation of a *p*-methoxybenzyl group that has a neighboring hydroxyl.[6] This methodology has been used to advantage in a number of syntheses.[7,8] The following example illustrates the reaction.[9]

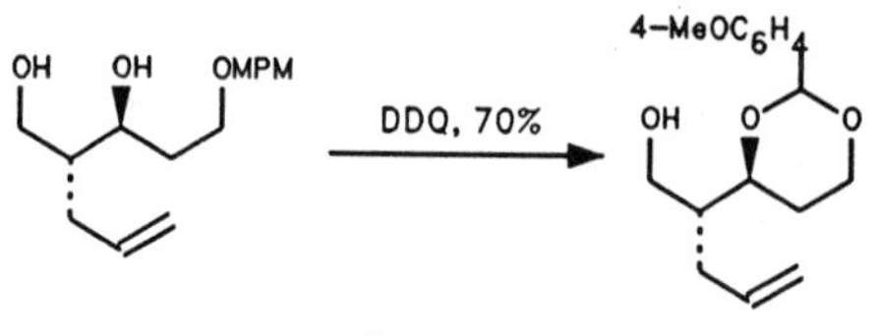

OH OH OMPM — DDQ, 70% → 4-$MeOC_6H_4$; OH O O

MPM = PMB[9] = *p*-methoxybenzyl

Cleavage

1. 80% AcOH, 25°, 10 h, 100% yield.[1]
2. $Ce(NH_4)_2(NO_3)_6$, CH_3CN, H_2O.[10]

As with the benzylidene group a variety of methods shown below have been developed to effect cleavage of one of the two C—O bonds in this acetal.

3. $(i\text{-}Bu)_2AlH$, $PhCH_3$, 75% yield.[10]
4. DDQ, water, 87% yield.[7] This method results in the formation of a mixture of the two possible monobenzoates.
5. Selective cleavage to give either the more or less substituted derivative is possible with the proper choice of reagents.[4,10–12]

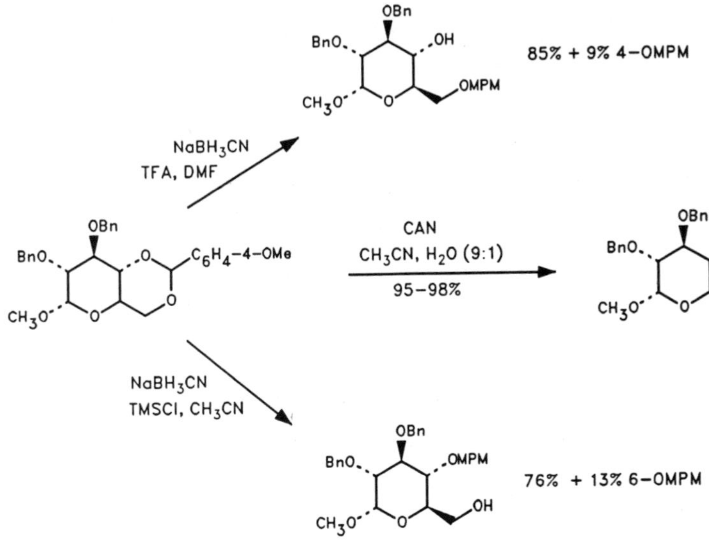

Treatment with $LiAlH_4/AlCl_3$,[12] $BH_3 \cdot NMe_3/AlCl_3$,[4] BH_3, THF, heat,[13] or $NaBH_3CN/TMSCl$, CH_3CN[11] results in cleavage at the least hindered side of the ketal, whereas treatment with $NaBH_3CN/HCl$[4] or $NaBH_3CN/TFA/DMF$[11] results in formation of an MPM ether at the least hindered alcohol.

1. M. Smith, D. H. Rammler, I. H. Goldberg, and H. G. Khorana, *J. Am. Chem. Soc.*, **84,** 430 (1962).
2. Y. Oikawa, T. Nishi, and O. Yonemitsu, *Tetrahedron Lett.*, **24,** 4037 (1983).
3. S. Hanessian, J. Kloss, and T. Sugawara, *Trends in Synthetic Carbohydrate Chemistry, ACS Symposium Series 386*, D. Horton, L. D. Hawkins and G. J. McGarvey, Eds., American Chemical Society, Washington, DC, 1989, p. 64.
4. M. Kloosterman, T. Slaghek, J. P. G. Hermans, and J. H. Van Boom, *Recl: J. R. Neth. Chem. Soc.*, **103,** 335 (1984).
5. V. Box, R. Hollingsworth, and E. Roberts, *Heterocycles*, **14,** 1713 (1980).
6. Y. Oikawa, T. Yoshioka, and O. Yonemitsu, *Tetrahedron Lett.*, **23,** 889 (1982).
7. A. F. Sviridov, M. S. Ermolenko, D. V. Yaskunsky, V. S. Borodkin, and N. K. Kochetkov, *Tetrahedron Lett.*, **28,** 3835 (1987).
8. J. S. Yadav, M. C. Chander, and B. V. Joshi, *Tetrahedron Lett.*, **29,** 2737 (1988).
9. A. B. Jones, M. Yamaguchi, A. Patten, S. J. Danishefsky, J. A. Ragan, D. B. Smith, and S. L. Schreiber, *J. Org. Chem.*, **54,** 17 (1989); A. B. Smith, III, K. J. Hale, L. M. Laakso, K. Chen, and A. Riera, *Tetrahedron Lett.*, **30,** 6963 (1989).
10. R. Johansson and B. Samuelsson, *J. Chem. Soc., Chem. Commun.*, 201 (1984).
11. R. Johansson and B. Samuelsson, *J. Chem. Soc., Perkin Trans. I*, 2371 (1984); P. J. Garegg, H. Hultberg, and S. Wallin, *Carbohydr. Res.*, **108,** 97 (1982).

12. D. Joniak, B. Kôsíková, and L. Kosáková, *Collect. Czech. Chem. Commun.*, **43,** 769 (1978).
13. T. Tsuri and S. Kamata, *Tetrahedron Lett.*, **26,** 5195 (1985).

13. 2,4-Dimethoxybenzylidene Ketal

Formation

1. 2,4-$(MeO)_2C_6H_3CHO$, benzene, TsOH, heat, >81% yield.[1]

Cleavage

This acetal is stable to hydrogenation with W4–Raney Ni, which was used to cleave a benzyl group in 99% yield.[2]

1. M. Smith, D. H. Rammler, I. H. Goldberg, and H. G. Khorana, *J. Am. Chem. Soc.*, **84,** 430 (1962).
2. K. Horita, T. Yoshioka, T. Tanaka, Y. Oikawa, and O. Yonemitsu, *Tetrahedron*, **42,** 3021 (1986).

14. 3,4-Dimethoxybenzylidene Acetal

Formation/Cleavage[1]

$AlEt_2Cl$, $PhCH_3$, CH_2Cl_2, Et_3SiH, 1.5 h, −55°, 65%

The acetal can also be cleaved with DDQ (CH_2Cl_2, H_2O, 66% yield) to afford the monobenzoate.[2] Treatment of a 3,4-dimethoxybenzyl ether containing a free hydroxyl with DDQ (benzene, 3 Å molecular sieves, rt) affords the 3,4-dimethoxybenzylidene acetal.[2]

1. M. J. Wanner, N. P. Willard, G. J. Koomen, and U. K. Pandet, *Tetrahedron*, **43,** 2549 (1987).
2. K. Nozaki and H. Shirahama, *Chem. Lett.*, 1847 (1988).

15. 2-Nitrobenzylidene Acetal

The 2-nitrobenzylidene acetal has been used to protect carbohydrates. It can be cleaved by photolysis (45 min, MeOH; CF_3CO_3H, CH_2Cl_2, 0°, 95% yield) to form primarily axial 2-nitrobenzoates from diols containing at least one axial alcohol.[1]

1. P. M. Collins and V. R. N. Munasinghe, *J. Chem. Soc., Perkin Trans. I*, 921 (1983).

Cyclic Ortho Esters

A variety of cyclic ortho esters,[1,2] including cyclic orthoformates, have been developed to protect *cis*-1,2-diols. Cyclic ortho esters are more readily cleaved by acidic hydrolysis (e.g., by a phosphate buffer, pH 4.5–7.5, or by 0.005–0.05 *M* HCl)[3] than are acetonides. Careful hydrolysis or reduction can be used to prepare selectively monoprotected diol derivatives.

16. Methoxymethylene Acetal (Chart 3)

17. Ethoxymethylene Acetal

Formation

1. $CH_3C(OMe)_3$ or $CH_3C(OEt)_3$, acid catalyst, 77% or 45–80% yields, respectively.[4–6]

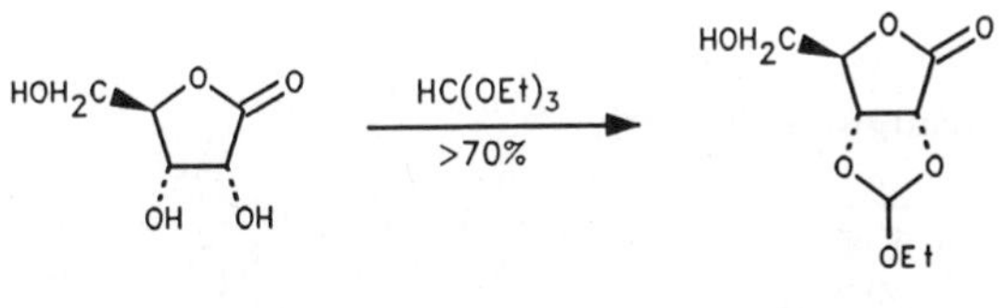

HOH$_2$C, OH, OH — HC(OEt)$_3$, >70% → HOH$_2$C, OEt

Cleavage

1. Acidic hydrolysis affords a monoacetate.[7]

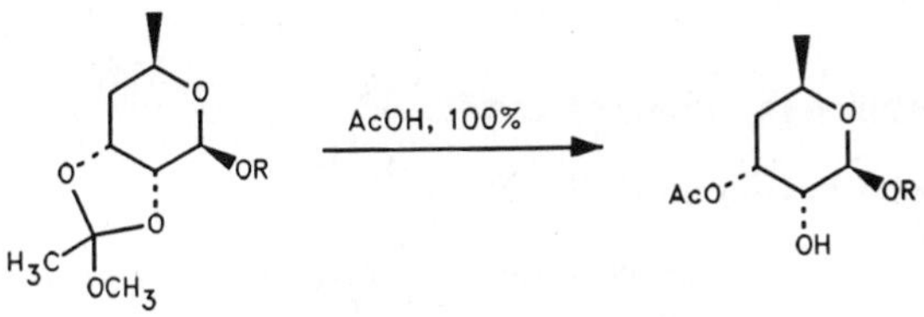

H$_3$C, OCH$_3$, OR — AcOH, 100% → AcO, OH, OR

2. Reduction with $(i\text{-Bu})_2AlH$ affords a diol with one hydroxyl group protected as a MOM group. The more substituted hydroxyl bears the MOM group.[8]

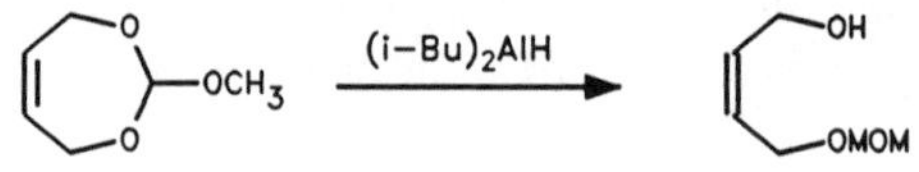

$(i\text{-Bu})_2AlH$

1. C. B. Reese, *Tetrahedron*, **34,** 3143 (1978).
2. V. Amarnath and A. D. Broom, *Chem. Rev.*, **77,** 183 (1977).
3. M. Ahmad, R. G. Bergstrom, M. J. Cashen, A. J. Kresge, R. A. McClelland, and M. F. Powell, *J. Am. Chem. Soc.*, **99,** 4827, (1977).
4. B. E. Griffin, M. Jarman, C. B. Reese, and J. E. Sulston, *Tetrahedron*, **23,** 2301 (1967).
5. J. Zemlicka, *Chem. Ind. (London)*, 581 (1964); F. Eckstein and F. Cramer, *Chem. Ber.*, **98,** 995 (1965).
6. R. M. Ortuño, R. Mercé, and J. Font, *Tetrahedron Lett.*, **27,** 2519 (1986).
7. S. Hanessian and R. Roy, *Can. J. Chem.*, **63,** 163 (1985).
8. M. Takasu, Y. Naruse, and H. Yamamoto, *Tetrahedron Lett.*, **29,** 1947 (1988).

The following ortho esters have been prepared to protect the diols of nucleosides. They are readily hydrolyzed with mild acid to afford monoester derivatives, generally as a mixture of positional isomers.

18. Dimethoxymethylene Ortho Ester[1] (Chart 3)

19. 1-Methoxyethylidene Ortho Ester[2]

20. 1-Ethoxyethylidine Ortho Ester[3]

With this ortho ester good selectivity for the axial alcohol is achieved in the acidic hydrolysis of a pyranoside derivative.[4]

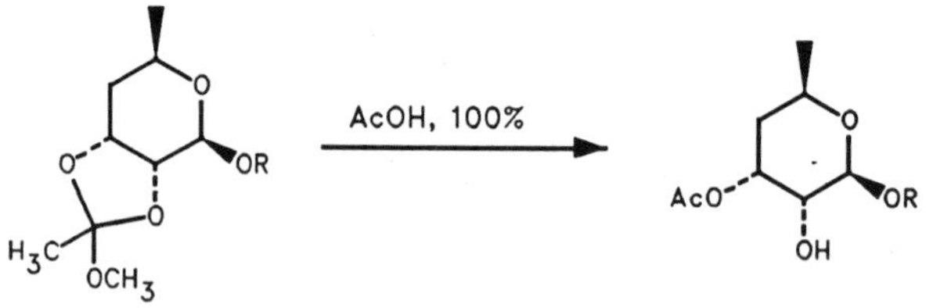

AcOH, 100%

21. 1,2-Dimethoxyethylidene Ortho Ester[5]

22. α-Methoxybenzylidene Ortho Ester[2]

23. 1-(*N*,*N*-Dimethylamino)ethylidene Derivative[6]

24. α-(*N*,*N*-Dimethylamino)benzylidene Derivative[6]

1. G. R. Niaz and C. B. Reese, *J. Chem. Soc., Chem. Commun.*, 552 (1969).
2. C. B. Reese and J. E. Sulston, *Proc. Chem. Soc.*, 214 (1964).
3. V. P. Miller, D.-y. Yang, T. M. Weigel, O. Han, and H.-w. Liu, *J. Org. Chem.*, **54,** 4175 (1989).
4. S. Hanessian and R. Roy, *Can. J. Chem.*, **63,** 163 (1985).
5. J. H. van Boom, G. R. Owen, J. Preston, T. Ravindranathan, and C. B. Reese, *J. Chem. Soc. C*, 3230 (1971).
6. S. Hanessian and E. Moralioglu, *Can. J. Chem.*, **50,** 233 (1972).

25. 2-Oxacyclopentylidene Ortho Ester

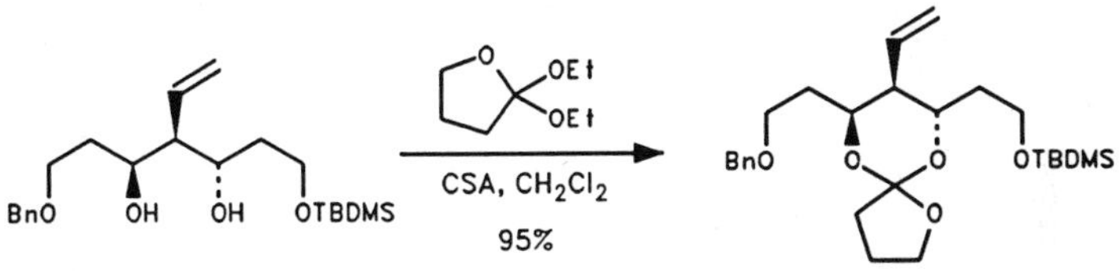

BnO OH OH OTBDMS → (2,2-diethoxytetrahydrofuran; CSA, CH_2Cl_2; 95%) → BnO, OTBDMS ortho ester

CSA = camphorsulfonic acid

This ortho ester does not form a monoester upon deprotection as do acyclic ortho esters, thus avoiding a hydrolysis step.[1]

1. R. M. Kennedy, A. Abiko, T. Takemasa, H. Okumoto, and S. Masamune, *Tetrahedron Lett.*, **29,** 451 (1988).

Silyl Derivatives

26. Di-*t*-butylsilylene Group: $DTBS(OR)_2$

The DTBS group is probably the most useful of the bifunctional silyl ethers. Dimethylsilyl and diisopropylsilyl derivatives of diols are very susceptible to hydrolysis even in water and therefore are of limited use.

Formation

1. $(t\text{-Bu})_2SiCl_2$, CH_3CN, TEA, HOBt, 65°.[1,2] Tertiary alcohols will not react under these conditions. The reagent is effective for both 1,2- and 1,3-diols.
2. $(t\text{-Bu})_2Si(OTf)_2$, 2,6-lutidine, 0–25°, $CHCl_3$.[3,4] This reagent readily silylates 1,2-, 1,3- and 1,4-diols even when one of the alcohols is tertiary.

Cleavage

Derivatives of 1,3- and 1,4-diols are stable to pH 4–10 at 22° for several hours, but derivatives of 1,2-diols undergo rapid hydrolysis under basic conditions (5:1 THF–pH 10 buffer, 22°, 5 min) to form monosilyl ethers of the parent diol.

1. 48% aq. HF, CH_3CN, 25°, 15 min, 95% yield.[3]
2. $Bu_3NH^+F^-$, THF.[5]
3. Pyr·HF, THF, 25°, 88% yield.[1]

OTBDMS H H SC_6H_4OMe OAc H H O O O Si t-Bu_2 —Pyr·HF / THF→ OTBDMS H H SC_6H_4OMe OAc H H OH OH O

1. B. M. Trost and C. G. Caldwell, *Tetrahedron Lett.*, **22,** 4999 (1981).
2. B. M. Trost, C. G. Caldwell, E. Murayama, and D. Heissler, *J. Org. Chem.*, **48,** 3252 (1983).
3. E. J. Corey and P. B. Hopkins, *Tetrahedron Lett.*, **23,** 4871 (1982).
4. K. Furusawa, K. Ueno and T. Katsura, *Chem. Lett.*, 97 (1990).
5. K. Furusawa, *Chem. Lett.*, 509 (1989).

27. 1,3-(1,1,3,3-Tetraisopropyldisiloxanylidene) Derivative: TIPDS(OR)$_2$

Formation

1. $TIPDSCl_2$, DMF, imidazole.[1-3]

OH OH —$TIPDSCl_2$ / DMF, Im→ O—Si(i-Pr)$_2$ O O—Si(i-Pr)$_2$

2. $TIPDSCl_2$, Pyr.[4-6]

In polyhydroxylated systems the regiochemical outcome is determined by initial reaction at the sterically less hindered alcohol.[7]

Cleavage

1. $Bu_4N^+F^-$, THF.[1,5,8] When $Bu_4N^+F^-$ is used to remove the TIPDS group, ester groups can migrate because of the basic nature of fluoride ion. Migration can be prevented by the addition of Pyr·HCl.[9]

2. TEA·HF.[10]
3. 0.2 *M* HCl, dioxane, H_2O or MeOH.[1]
4. 0.2 *M* NaOH, dioxane, H_2O.[1]
5. TMSI, CH_2Cl_2, 0°, 0.5 h, 83% yield.[11]
6. Ac_2O, AcOH, H_2SO_4.[2]
7. The TIPDS derivative can be induced to isomerize from the thermodynamically less stable eight-membered ring to the more stable seven-membered ring derivative.[4,12] The isomerization occurs only in DMF.

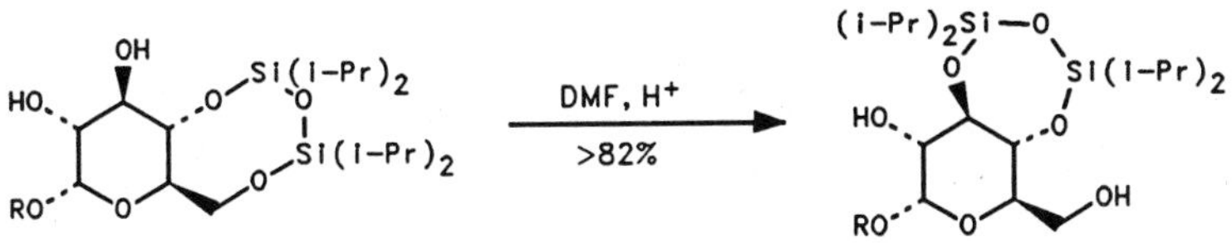

DMF, H^+
>82%

28. Tetra-*t*-butoxydisiloxane-1,3-diylidene Derivative: $TBDS(OR)_2$

Formation

1. $TBDSCl_2$, Pyr, rt, 50–87% yield.[13]

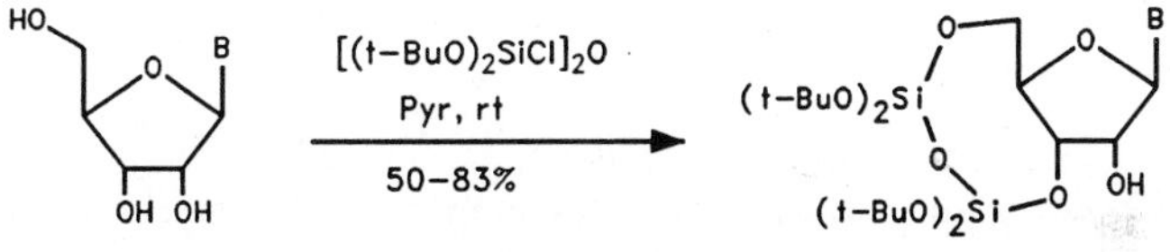

$[(t\text{-BuO})_2SiCl]_2O$
Pyr, rt
50–83%

B = pyrimidine or purine residue

Cleavage

1. $Bu_4N^+F^-$, THF, 2 min.[13] The TBDS group is less reactive toward triethylammonium fluoride than is the TIPDS group. It is stable to 2 *M* HCl, aq. dioxane, overnight. Treatment with 0.2 *M* NaOH, aq. dioxane leads to cleavage of only the Si—O bond at the 5′-position of the uridine derivative. The TBDS derivative is 25 times more stable than the TIPDS derivative to basic hydrolysis.

1. W. T. Markiewicz, *J. Chem. Res. Synop.*, 24 (1979).
2. J. P. Schaumberg, G. C. Hokanson, J. C. French, E. Smal, and D. C. Baker, *J. Org. Chem.*, **50,** 1651 (1985).
3. E. Ohtsuka, M. Ohkubo, A. Yamane, and M. Ikebara, *Chem. Pharm. Bull.*, **31,** 1910 (1983).
4. C. A. A. van Boeckel and J. H. van Boom, *Tetrahedron*, **41,** 4545, 4557 (1985).

5. J. Thiem, V. Duckstein, A. Prahst, and M. Matzke, *Liebigs Ann. Chem.*, 289 (1987).
6. J. S. Davies, E. J. Tremeer, and R. C. Treadgold, *J. Chem. Soc., Perkin Trans. I*, 1107 (1987).
7. W. T. Markiewicz, N. Sh. Padyukova, S. Samek, and J. Smrt, *Collect. Czech. Chem. Commun.*, **45,** 1860 (1980).
8. M. D. Hagen, C. S.-Happ, E. Happ, and S. Chládek, *J. Org. Chem.*, **53,** 5040 (1988).
9. J. J. Oltvoort, M. Kloosterman, and J. H. Van Boom, *Recl: J. R. Neth. Chem. Soc.*, **102,** 501 (1983).
10. W. T. Markiewicz, E. Biala, and R. Kierzek, *Bull. Pol. Acad. Sci., Chem.*, **32,** 433 (1984).
11. T. Tatsuoka, K. Imao, and K. Suzuki, *Heterocycles*, **24,** 617 (1986).
12. C. H. M. Verdegaal, P. L. Jansse, J. F. M. de Rooij, and J. H. Van Boom, *Tetrahedron Lett.*, **21,** 1571 (1980).
13. W. T. Markiewicz, B. Nowakowska, and K. Adrych, *Tetrahedron Lett.*, **29,** 1561 (1988).

29. Cyclic Carbonates (Chart 3)

Cyclic carbonates[1,2] are very stable to acidic hydrolysis (AcOH, HBr, and H_2SO_4/MeOH) and are more stable to basic hydrolysis than esters.

Formation

1. Cl_2CO, Pyr, 20°, 1 h.[3]
2. *p*-$NO_2C_6H_4OCOCl$, Pyr, 20°, 5 days, 72% yield.[4]
3. *N,N'*-Carbonyldiimidazole, PhH, heat, 12 h–4 days, 90% yield.[5,6]
4. Cl_3CCOCl, Pyr, 1 h, rt, >80% yield.[7]

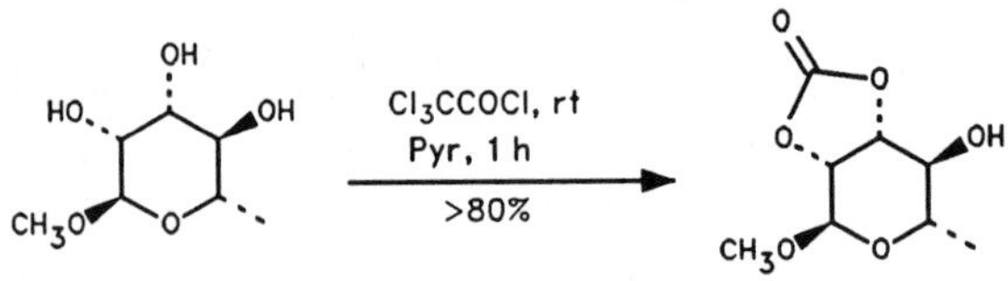

5. CO, S, Et_3N, 80°, 4 h; $CuCl_2$, rt, 18 h, 66–100% yield.[8]

Cleavage

1. $Ba(OH)_2$, H_2O, 70°.[9]
2. Pyr, H_2O, reflux, 15 min, 100% yield.[4] These conditions were used to remove the carbonate from uridine.

3. 0.5 *M* NaOH, 50% aq. dioxane, 25°, 5 min, 100% yield.[4]

1. L. Hough, J. E. Priddle, and R. S. Theobald, *Adv. Carbohydr. Chem.*, **15,** 91–158 (1960).
2. V. Amarnath and A. D. Broom, *Chem. Rev.*, **77,** 183 (1977).
3. W. N. Haworth and C. R. Porter, *J. Chem. Soc.*, 151 (1930).
4. R. L. Letsinger and K. K. Ogilvie, *J. Org. Chem.*, **32,** 296 (1967).
5. J. P. Kutney and A. H. Ratcliffe, *Synth. Commun.*, **5,** 47 (1975).
6. K. Narasaka, in *Trends in Synthetic Carbohydrate Chemistry, ACS Symposium Series 386*, D. Horton, L. D. Hawkins, and G. J. McGarvey, Eds., American Chemical Society, Washington, DC, 1989, p. 290.
7. K. Tatsuta, K. Akimoto, M. Annaka, Y. Ohno, and M. Kinoshita, *Bull. Chem. Soc. Jpn.*, **58,** 1699 (1985).
8. T. Mizuno, F. Nakamura, Y. Egashira, I. Nishiguchi, T. Hirashima, A. Ogawa, N. Kambe, and N. Sonoda, *Synthesis*, 636 (1989).
9. W. G. Overend, M. Stacey, and L. F. Wiggins, *J. Chem. Soc.*, 1358 (1949).

30. Cyclic Boronates

Although boronates are quite susceptible to hydrolysis, they have been found useful for the protection of carbohydrates.[1,2] It should be noted that as the steric demands of the diol increase, the rate of hydrolysis decreases. For example, pinacol boronates are rather difficult to hydrolyze; in fact, they can be isolated from aqueous systems with no hydrolysis.

31. Ethyl Boronate[3] (Chart 3)

Formation

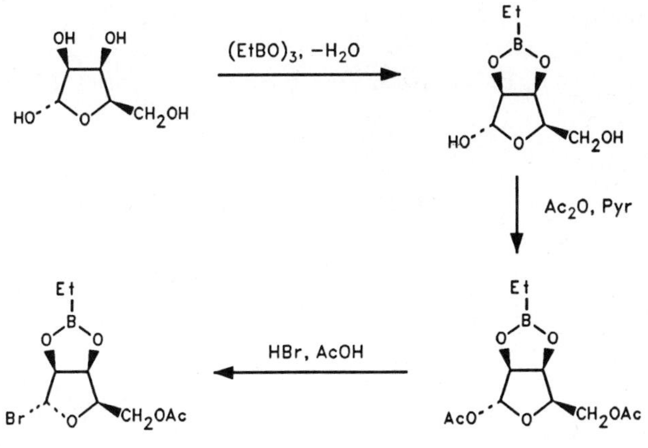

OH OH

HO O CH_2OH

$(EtBO)_3$, $-H_2O$

Et B O O

HO O CH_2OH

Ac_2O, Pyr

Refs. 4, 5

Et B O O

Br O CH_2OAc

HBr, AcOH

Et B O O

AcO O CH_2OAc

1. $(EtBO)_3$, PhH, azeotropic H_2O removal, 95% yield.[4,5]
2. $[t\text{-}C_4H_9CO_2B(C_2H_5)]_2O$, Pyr; then concentrate under reduced pressure.[5,6]
3. $LiEt_3BH$, THF, 0°–rt, 98% yield.[7]

32. Phenyl Boronate

Formation

1. $PhB(OH)_2$, PhH[8] or Pyr.[9]

A polymeric version of the phenyl boronate has been developed.[10]

Cleavage

1. 1,3-Propanediol, acetone.[1] This method removes the boronate by exchange.
2. Acetone, H_2O (4 : 1), 30 min, 83% yield.[10]
3. H_2O_2, EtOAc, >80% yield.[11]

1. R. J. Ferrier, *Adv. Carbohydr. Chem. Biochem.*, **35,** 31–80 (1978).
2. W. V. Dahlhoff and R. Köster, *Heterocycles*, **18,** 421 (1982).
3. W. V. Dahlhoff and R. Köster, *J. Org. Chem.*, **41,** 2316 (1976), and references cited therein.
4. W. V. Dahlhoff, A. Geisheimer and R. Köster, *Synthesis*, 935 (1980).
5. W. V. Dahlhoff and R. Köster, *Synthesis*, 936 (1980).
6. R. Köster, K. Taba, and W. V. Dahlhoff, *Liebigs Ann. Chem.*, 1422 (1983).
7. L. Garlaschelli, G. Mellerio, and G. Vidari, *Tetrahedron Lett.*, **30,** 597 (1989).
8. R. J. Ferrier, *Methods Carbohydr. Chem.*, **VI,** 419–426 (1972).
9. J. M. J. Fréchet, L. J. Nuyens, and E. Seymour, *J. Am. Chem. Soc.*, **101,** 432 (1979).
10. N. P. Bullen, P. Hodge, and F. G. Thorpe, *J. Chem. Soc., Perkin Trans. I*, 1863 (1981).
11. D. A. Evans and R. P. Polniaszek, *Tetrahedron Lett.*, **27,** 5683 (1986).

3

PROTECTION FOR PHENOLS AND CATECHOLS

The phenolic hydroxyl group occurs widely in plant and animal life, both land-based and aquatic, as demonstrated by the vast number of natural products that contain this group. In developing a synthesis of any phenol-containing product, protection is often mandatory to prevent reaction with oxidizing agents and electrophiles or reaction of the nucleophilic phenoxide ion with even mild alkylating and acylating agents. Many of the protective groups developed for alcohol protection are also applicable to phenol protection and thus the chapter on alcohol protection should also be consulted. Ethers are the most widely used protective groups for phenols and in general they are more easily cleaved than the analogous ethers of simple alcohols.[1] Esters are also important protective groups for phenols, but are not as stable to hydrolysis as the related alcohol derivatives. Simple esters are easily hydrolyzed with mild base (e.g., $NaHCO_3$/aq. MeOH, 25°), but more sterically demanding esters (e.g., pivaloate) require harsher conditions to effect hydrolysis. Catechols can be protected in the presence of phenols as cyclic acetals or ketals or cyclic esters.

Some of the more important phenol and catechol protective groups are included in Reactivity Chart 4.[2]

1. For a review on ether cleavage, see: M. V. Bhatt and S. U. Kulkarni, *Synthesis*, 249 (1983).
2. See also: E. Haslam, "Protection of Phenols and Catechols," in *Protective Groups in Organic Chemistry*, J. F. W. McOmie, Ed., Plenum Press, New York and London, 1973, pp. 145–182.

PROTECTION FOR PHENOLS

Ethers

Historically, simple *n*-alkyl ethers formed from a phenol and a halide or sulfate were cleaved under rather drastic conditions (e.g., refluxing HBr). New ether protective groups have been developed that are removed under much milder conditions (e.g., via nucleophilic displacement, hydrogenolysis of benzyl ethers, and mild acid hydrolysis of acetal-type ethers) that seldom affect other functional groups in a molecule.

1. **Methyl Ether:** $ArOCH_3$ (Chart 4)

Formation

1. MeI, K_2CO_3, acetone, reflux, 6 h.[1,2] This is a very common and often very efficient method for the preparation of phenolic methyl ethers; it is also applicable to the formation of phenolic benzyl ethers.
2. Me_2SO_4, NaOH, EtOH, reflux, 3 h, 71–74% yield.[1]

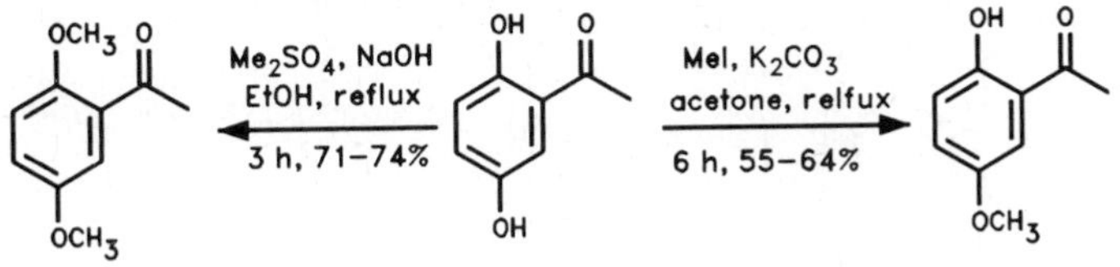

Me_2SO_4, NaOH, EtOH, reflux, 3 h, 71–74% ← (2,5-dihydroxyacetophenone) → MeI, K_2CO_3, acetone, relfux, 6 h, 55–64%

3. Li_2CO_3, MeI, DMF, 55°, 18 h, 54–90%.[3] This method selectively protects phenols with $pK_a < 8$ as a result of electron-withdrawing *ortho-* or *para-*substituents.

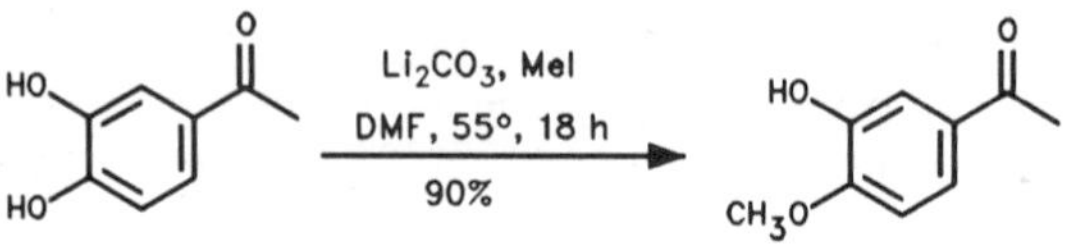

$$\text{3,4-(HO)}_2\text{C}_6\text{H}_3\text{COCH}_3 \xrightarrow[\text{90\%}]{\text{Li}_2\text{CO}_3,\ \text{MeI, DMF, 55°, 18 h}} \text{3-HO-4-CH}_3\text{O-C}_6\text{H}_3\text{COCH}_3$$

4. RX, or R'_2SO_4, NaOH, CH_2Cl_2, H_2O, $PhCH_2N^+Bu_3Br^-$, 25°, 2–13 h, 75–95% yield.

   Ar = simple; 2- or 2,6-disubstituted[4,5]
   R = Me, allyl, CH_2—CHCH$_2$— (epoxide, bridged by O), *n*-Bu, *c*-C_5H_{11}, $PhCH_2$, $-CH_2CO_2Et$
   R′ = Me, Et

5. Methyl, ethyl, and benzyl ethers have been prepared in the presence of tetraethylammonium fluoride as a Lewis base (alkyl halide, DME, 20°, 3 h, 60–85% yields).[5]
6. $p\text{-NO}_2\text{-C}_6\text{H}_4\text{ONa} + \text{MeN(NO)CONH}_2 \xrightarrow{\text{DME, 0° → 25°, 6 h}}$ $[p\text{-NO}_2\text{-C}_6\text{H}_4\text{O}^- + \text{CH}_2\text{N}_2] \rightarrow p\text{-NO}_2\text{-C}_6\text{H}_4\text{OMe}$, >90%[6]
7. Phenols protected as *t*-BuMe$_2$Si ethers can be converted directly to methyl or benzyl ethers (MeI or BnBr, KF, DMF, rt, >90% yield).[7]

Cleavage

1. Me_3SiI, $CHCl_3$, 25–50°, 12–140 h.[8] Iodotrimethylsilane in quinoline (180°, 70 min) selectively cleaves an aryl methyl group, in 72% yield, in the presence of a methylenedioxy group.[9] Me_3SiI cleaves esters more slowly than ethers and cleaves alkyl aryl ethers (48 h, 25°) more slowly than alkyl alkyl ethers (1.3–48 h, 25°), but benzyl, trityl, and *t*-butyl ethers are cleaved quite rapidly (0.1 h, 25°).[7]
2. EtSNa, DMF, reflux, 3 h, 94–98% yield.[10,11] Potassium thiophenoxide has been used to cleave an aryl methyl ether without causing migration of a double bond.[12] Sodium benzylselenide ($PhCH_2SeNa$) and sodium thiocresolate ($p\text{-}CH_3C_6H_4SNa$) cleave a dimethoxyaryl compound regioselectively, reportedly as a result of steric factors in the former case[13] and electronic factors in the latter case.[14]

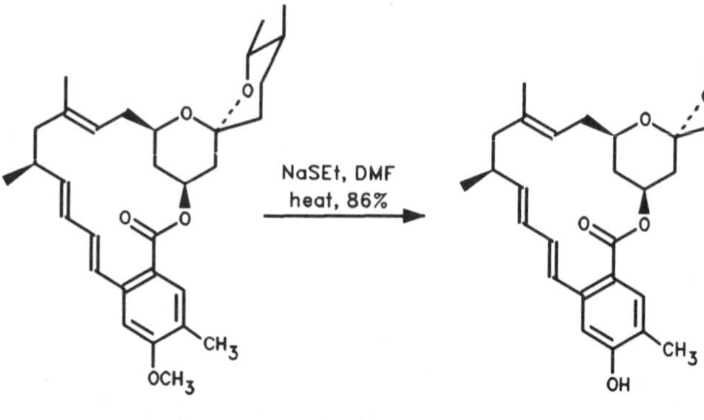

Macrolide aryl methyl ether (OCH_3, CH_3) → phenol (OH, CH_3); NaSEt, DMF, heat, 86%. Ref. 15

3. Sodium sulfide in *N*-methylpyrrolidone (140°, 2–4 h) cleaves aryl methyl ethers in 78–85% yield.[16]
4. Lithium diphenyphosphide (THF, 25°, 2 h; HCl, H_2O, 87% yield) selectively cleaves an aryl methyl ether in the presence of an ethyl ether.[17] It also cleaves a phenyl benzyl ether and a phenyl allyl ether to phenol in 88% and 78% yield, respectively.[18]
5. DMSO, NaCN, 125–180°, 5–48 h, 65–90% yield.[19] This cleavage reaction is successful for aromatic systems containing ketones, amides, and carboxylic acids; mixtures are obtained from nitro-substituted aromatic compounds; there is no reaction with 5-methoxyindole (180°, 48 h).
6. LiI, collidine, reflux, 10 h, quant.[20] Aryl ethyl ethers are cleaved more slowly; dialkyl ethers are stable to these conditions.
7. $AlBr_3$, EtSH, 25°, <1 h, 94% yield.[21]

Both methyl aryl and methyl alkyl ethers are cleaved under these conditions. A methylenedioxy group, used to protect a catechol, is cleaved under similar conditions in satisfactory yield; methyl and ethyl esters are stable (0–20°, 2 h).[20]

Regioselective cleavage of dimethoxyaryl derivatives with methanesulfonic acid/methionine has been reported.[22]

8. BBr_3, CH_2Cl_2, −80° → 20°, 12 h, 77–86% yield.[23] Methylenedioxy groups and diphenyl ethers are stable to these cleavage conditions. Benzyloxycarbonyl and *t*-butoxycarbonyl groups, benzyl esters,[24] and 1,3-dioxolanes are cleaved with this reagent.

Either an aryl methyl ether or a methylenedioxy group can be cleaved with boron trichloride under various conditions.[25]

Boron triiodide rapidly cleaves methyl ethers of *ortho*-, *meta*-, or *para*-substituted aromatic aldehydes (0°, 25°; 0.5–5 min; 40–86% yield).[26]

Boron tribromide is reported to be more effective than iodotrimethylsilane for cleaving aryl methyl ethers.[27]

9. $BBr_3 \cdot S(CH_3)_2$, $ClCH_2CH_2Cl$, 83°, 50–99% yield.[28] The advantage of this method is that the reagent is a stable, easily handled solid. Methylenedioxy groups are also cleaved by this reagent.
10. 9-Bromo-9-borabicyclo[3.3.0]nonane (9-Br-BBN), CH_2Cl_2, reflux, 87–100% yield.[29] 9-Br–BBN also cleaves dialkyl ethers, allyl aryl ethers, and methylenedioxy groups.
11. 2-Bromo-1,3,2-benzodioxaborole, CH_2Cl_2 (cat. $BF_3 \cdot Et_2O$), 25°, 0.5–36 h, 95–98% yield. Aryl benzyl ethers, methyl esters, and aromatic benzoates are also cleaved.[30]
12. Pyr·HCl, 220°, 6 min, 34% yield of morphine from codeine.[31]
13. Excess MeMgI, 155–165°, 15 min, 80% yield.[32]
14. 48% HBr, AcOH, reflux, 30 min, 85%.[33] The efficiency of this method is significantly improved if a phase-transfer catalyst (n-$C_{16}H_{33}P^+Bu_3\ Br^-$) is added to the mixture.[34]

15. BCl_3, CH_2Cl_2, −20°, 94% yield.[35]

1. BCl_3, CH_2Cl_2 −20°
2. Na_2CO_3, H_2O
94%

16. $AlCl_3$, 3 h, 0°, 75% yield.[36,37]

$AlCl_3$, 75%
0°, 3 h

Ref 35

17. LiCl, DMF, heat, 4–72 h.[38]

3 eq. LiCl, DMF
heat, 4–72 h, 98%

18. CF_3SO_3H, PhSMe, 0–25°.[39,40] In this case *O*-methyltyrosine was deprotected without evidence for O → C migration, which is often a problem when removing protective groups from tyrosine.

1. G. N. Vyas and N. M. Shah, *Org. Synth., Collect. Vol. IV*, 836 (1963).
2. A. R. MacKenzie, C. J. Moody, and C. W. Rees, *Tetrahedron*, **42,** 3259 (1986).
3. W. E. Wymann, R. Davis, J. W. Patterson, Jr., and J. R. Pfister, *Synth. Commun.*, **18,** 1379 (1988).
4. A. McKillop, J.-C. Fiaud, and R. P. Hug, *Tetrahedron*, **30,** 1379 (1974).
5. J. M. Miller, K. H. So, and J. H. Clark, *Can. J. Chem.*, **57,** 1887 (1979).
6. S. M. Hecht and J. W. Kozarich, *Tetrahedron Lett.*, 1307 (1973).
7. A. K. Sinhababu, M. Kawase, and R. T. Borchardt, *Tetrahedron Lett.*, **28,** 4139 (1987).
8. M. E. Jung and M. A. Lyster, *J. Org. Chem.*, **42,** 3761 (1977).
9. J. Minamikawa and A. Brossi, *Tetrahedron Lett.*, 3085 (1978).
10. G. I. Feutrill and R. N. Mirrington, *Tetrahedron Lett.*, 1327 (1970); *idem*, *Aust. J. Chem.*, **25,** 1719, 1731 (1972).
11. A. S. Kende and J. P. Rizzi, *Tetrahedron Lett.*, **22,** 1779 (1981).
12. J. W. Wildes, N. H. Martin, C. G. Pitt, and M. E. Wall, *J. Org. Chem.*, **36,** 721 (1971).

13. R. Ahmad, J. M. Saá, and M. P. Cava, *J. Org. Chem.*, **42,** 1228 (1977).
14. C. Hansson and B. Wickberg, *Synthesis*, 191 (1976).
15. A. B. Smith, III, S. R. Schow, J. D. Bloom, A. S. Thompson, and K. N. Winzenberg, *J. Am. Chem. Soc.*, **104,** 4015 (1982).
16. M. S. Newman, V. Sankaran, and D. R. Olson, *J. Am. Chem. Soc.*, **98,** 3237 (1976).
17. R. E. Ireland and D. M. Walba, *Org. Synth.*, **56,** 44 (1977); *Collect. Vol. VI*, 567 (1988).
18. F. G. Mann and M. J. Pragnell, *Chem. Ind. (London)*, 1386 (1964).
19. J. R. McCarthy, J. L. Moore, and R. J. Crege, *Tetrahedron Lett.*, 5183 (1978).
20. I. T. Harrison, *J. Chem. Soc., Chem. Commun.*, 616 (1969).
21. M. Node, K. Nishide, K. Fuji, and E. Fujita, *J. Org. Chem.*, **45,** 4275 (1980).
22. N. Fujii, H. Irie, and H. Yajima, *J. Chem. Soc., Perkin Trans. I*, 2288 (1977).
23. J. F. W. McOmie and D. E. West, *Org. Synth., Collect. Vol. V*, 412 (1973).
24. A. M. Felix, *J. Org. Chem.*, **39,** 1427 (1974).
25. M. Gerecke, R. Borer, and A. Brossi, *Helv. Chim. Acta*, **59,** 2551 (1976).
26. J. M. Lansinger and R. C. Ronald, *Synth. Commun.*, **9,** 341 (1979).
27. E. H. Vickery, L. F. Pahler, and E. J. Eisenbraun, *J. Org. Chem.*, **44,** 4444 (1979).
28. P. G. Williard and C. B. Fryhle, *Tetrahedron Lett.*, **21,** 3731 (1980).
29. M. V. Bhatt, *J. Organomet. Chem.*, **156,** 221 (1978).
30. P. F. King and S. G. Stroud, *Tetrahedron Lett.*, **26,** 1415 (1985).
31. M. Gates and G. Tschudi, *J. Am. Chem. Soc.*, **78,** 1380 (1956).
32. R. Mechoulam and Y. Gaoni, *J. Am. Chem. Soc.*, **87,** 3273 (1965).
33. I. Kawasaki, K. Matsuda, and T. Kaneko, *Bull. Chem. Soc. Jpn.*, **44,** 1986 (1971).
34. D. Landini, F. Montanari, and F. Rolla, *Synthesis*, 771 (1978).
35. H. Nagaoka, G. Schmid, H. Iio, and Y. Kishi, *Tetrahedron Lett.*, **22,** 899 (1981).
36. K. A. Parker and J. J. Petraitis, *Tetrahedron Lett.*, **22,** 397 (1981).
37. T.-t. Li and Y. L. Wu, *J. Am. Chem. Soc.*, **103,** 7007 (1981).
38. A. M. Bernard, M. R. Ghiani, P. P. Piras, and A. Rivoldini, *Synthesis*, 287 (1989).
39. Y. Kiso, S. Nakamura, K. Ito, K. Ukawa, K. Kitagawa, T. Akita, and H. Moritoki, *J. Chem. Soc., Chem. Commun.*, 971 (1979).
40. Y. Kiso, K. Ukawa, S. Nakamura, K. Ito, and T. Akita, *Chem. Pharm. Bull.*, **28,** 673 (1980).

2. Methoxymethyl Ether (MOM Ether): $ArOCH_2OCH_3$ (Chart 4)

Formation

1. $ClCH_2OMe$, CH_2Cl_2, NaOH–H_2O, Adogen (phase-transfer cat.), 20°, 20 min, 80–95% yield.[1] This method has been used to protect selectively a phenol in the presence of an alcohol.[2]
2. $ClCH_2OMe$, CH_3CN, 18-crown-6, 80% yield.[3]
3. $MeOCH_2OMe$, TsOH, CH_2Cl_2, molecular sieves, N_2, reflux, 12 h, 60–80%

yield.[4] This method of formation avoids the use of the carcinogen chloromethyl methyl ether.

Cleavage

1. HCl, *i*-PrOH, THF, 25°, 12 h, quant.[4]
2. 2 *N* HOAc, 90°, 40 h, high yield.[5] This group has been used in a synthesis of 13-desoxydelphonine from *o*-cresol, a synthesis that required the group to be stable to many reagents.[6]
3. NaI, acetone, cat. HCl, 50°, 85% yield.[7]
4. P_2I_4, CH_2Cl_2, 0° → rt, 30 min, 70–90% yield.[8] This method is also effective for removal of the SEM and MEM groups.

1. F. R. van Heerden, J. J. van Zyl, G. J. H. Rall, E. V. Brandt, and D. G. Roux, *Tetrahedron Lett.*, 661 (1978).
2. T. R. Kelly, C. T. Jagoe and Q. Li, *J. Am. Chem. Soc.*, **111,** 4522 (1989).
3. G. J. H. Rall, M. E. Oberholzer, D. Ferreira, and D. G. Roux, *Tetrahedron Lett.*, 1033 (1976).
4. J. P. Yardley and H. Fletcher, 3rd, *Synthesis*, 244 (1976).
5. M. A. A.-Rahman, H. W. Elliott, R. Binks, W. Küng, and H. Rapoport, *J. Med. Chem.*, **9,** 1 (1966).
6. K. Wiesner, *Pure Appl. Chem.*, **51,** 689 (1979).
7. D. R. Williams, B. A. Barner, K. Nishitani, and J. G. Phillips, *J. Am. Chem. Soc.*, **104,** 4708 (1982).
8. H. Saimoto, Y. Kusano, and T. Hiyama, *Tetrahedron Lett.*, **27,** 1607 (1986).

3. Benzyloxymethyl Ether (BOM Ether): $C_6H_5CH_2OCH_2OAr$

Formation

1. BOMCl, NaH, DMF, >81% yield.[1]

Cleavage

1. MeOH, Dowex 50W-X8 (H^+), 90% yield.[1]

1. W. R. Roush, M. R. Michaelides, D. F. Tai, B. M. Lesur, W. K. M. Chong, and D. J. Harris, *J. Am. Chem. Soc.*, **111,** 2984 (1989).

4. Methoxyethoxymethyl Ether (MEM Ether): $ArOCH_2OCH_2CH_2OCH_3$ (Chart 4)

In an attempt to metalate a MEM-protected phenol with BuLi, the methoxy group was eliminated, forming the vinyloxymethyl ether. This was attributed to intramolecular proton abstraction.[1]

A 2-methoxyethoxymethyl ether was used to protect one phenol group during a total synthesis of gibberellic acid.[2]

Formation

1. NaH, THF, 0°; $MeOCH_2CH_2OCH_2Cl$, 0 → 25°, 2 h, 75% yield.[2]

Cleavage

1. CF_3CO_2H, CH_2Cl_2, 23°, 1 h, 74% yield.[1]
2. For other methods of cleavage the chapter on alcohol protection should be consulted.

1. J. Mayrargue, M. Essamkaoui, and H. Moskowitz, *Tetrahedron Lett.*, **30,** 6867 (1989).
2. E. J. Corey, R. L. Danheiser, S. Chandrasekaran, P. Siret, G. E. Keck, and J.-L. Gras, *J. Am. Chem. Soc.*, **100,** 8031 (1978).

5. 2-(Trimethylsilyl)ethoxymethyl (SEM) Ether: $(CH_3)_3SiCH_2CH_2OCH_2OAr$

Formation

1. SEMCl, DMAP, Et_3N, benzene, reflux, 3 h, 98% yield.[1]
2. SEMCl, $(i\text{-Pr})_2NEt$, CH_2Cl_2, 97% yield.[2]

Cleavage

1. $Bu_4N^+F^-$, HMPA, 40°, 2 h, 23–51% yield.[3]
2. H_2SO_4, MeOH, THF, 90% yield.[1]
3. P_2I_4, CH_2Cl_2, 0° → rt, 30 min, 62–86% yield.[2] These conditions also cleave methoxymethyl and methoxyethoxymethyl ethers.

1. T. L. Shih, M. J. Wyvratt, and H. Mrozik, *J. Org. Chem.*, **52,** 2029 (1987).
2. H. Saimoto, Y. Kusano and T. Hiyama, *Tetrahedron Lett.*, **27,** 1607 (1986).
3. A. Leboff, A.-C. Carbonnelle, J.-P. Alazard, C. Thal, and A. S. Kende, *Tetrahedron Lett.*, **28,** 4163 (1987).

6. Methylthiomethyl Ether (MTM Ether): $ArOCH_2SCH_3$ (Chart 4)

Formation

1. NaOH, $ClCH_2SMe$, HMPA, 25°, 16 h, 91–94% yield.[1]

Cleavage

1. $HgCl_2$, $CH_3CN–H_2O$, reflux, 10 h, 90–95% yield.[1] Aryl methylthiomethyl ethers are stable to the conditions used to hydrolyze primary alkyl MTM ethers (e.g., $HgCl_2/CH_3CN–H_2O$, 25°, 6 h). They are moderately stable to acidic conditions (95% recovered from $HOAc/THF–H_2O$, 25°, 4 h).
2. Ac_2O, Me_3SiCl, 25 min, rt, 95% yield.[2]

1. R. A. Holton and R. G. Davis, *Tetrahedron Lett.*, 533 (1977).
2. N. C. Barua, R. P. Sharma, and J. N. Baruah, *Tetrahedron Lett.*, **24,** 1189 (1983).

7. Phenylthiomethyl Ether (PTM Ether): $C_6H_5SCH_2OAr$

Formation

1. NaI, $PhSCH_2Cl$, NaH, HMPA, 87–94% yield.[1]

Cleavage

1. $CH_3CN:H_2O$ (4 : 1), $HgCl_2$, 24 h, 90–94% yield. The methylthiomethyl ether group can be removed in the presence of the phenylthiomethyl ether.[1]

1. R. A. Holton and R. V. Nelson, *Synth. Commun.*, **10,** 911 (1980).

8. 2,2-Dichloro-1,1-difluoroethyl Ether: $CHCl_2CF_2OAr$

Formation/Cleavage

$$ArOH \xrightarrow[Bu_4N^+HSO_4^-,\ 92\%]{F_2C{=}CCl_2,\ 40\%\ KOH} ArOCF_2CHCl_2$$

$$ArOCF_2CHCl_2 \xrightarrow[rt,\ 85\%]{6\%\ KOH,\ H_2O,\ DMSO} ArOH$$

This group decreases the electron density on the aromatic ring and thus inhibits solvolysis of the tertiary alcohol **i** and the derived acetate **ii**.[1]

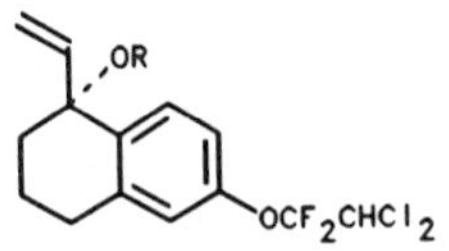

i R = H
ii R = Ac

1. S. G. Will, P. Magriotis, E. R. Marinelli, J. Dolan, and F. Johnson, *J. Org. Chem.*, **50,** 5432 (1985).

9. Tetrahydropyranyl Ether (THP Ether): ArO–2-tetrahydropyranyl

The tetrahydropyranyl ether, prepared from a phenol and dihydropyran (HCl/EtOAc, 25°, 24 h), is cleaved by aqueous oxalic acid (MeOH, 50–90°, 1–2 h).[1]

1. H. N. Grant, V. Prelog, and R. P. A. Sneeden, *Helv. Chim. Acta*, **46,** 415 (1963).

10. Phenacyl Ether: $ArOCH_2COC_6H_5$ (Chart 4)

11. *p*-Bromophenacyl Ether: $ArOCH_2COC_6H_4$-*p*-Br

Formation

1. $BrCH_2COPh$, K_2CO_3, acetone, reflux, 1–2 h, 85–95% yield.[1]

Cleavage

1. Zn, HOAc, 25°, 1 h, 88–96% yield.[1] Phenacyl and *p*-bromophenacyl ethers of phenols are stable to 1% ethanolic alkali (reflux, 2 h), and to 5 *N* sulfuric acid in ethanol–water. The phenacyl ether, prepared from β-naphthol, is cleaved in 82% yield by 5% ethanolic alkali (reflux, 2 h).

1. J. B. Hendrickson and C. Kandall, *Tetrahedron Lett.*, 343 (1970).

12. Cyclopropylmethyl Ether: $ArOCH_2$–*c*-C_3H_5

For a particular phenol, the authors required a protective group that would be stable to reduction (by complex metals, catalytic hydrogenation, and Birch conditions) and that could be easily and selectively removed.

Formation

1. KO-*t*-Bu, DMF, 0°, 30 min; *c*-$C_3H_5CH_2Br$, 20°, 20 min → 40°, 6 h, 80% yield.[1]

Cleavage

1. Aqueous HCl, MeOH, reflux, 2 h, 94% yield.[1]

1. W. Nagata, K. Okada, H. Itazaki, and S. Uyeo, *Chem. Pharm. Bull.*, **23,** 2878 (1975).

13. Allyl Ether: $ArOCH_2CH=CH_2$ (Chart 4)

Allyl ethers can be prepared by reaction of a phenol and the allyl bromide in the presence of base.[1] Several reagents have been used to effect their cleavage:

Cleavage

1. Pd–C, TsOH, H_2O or MeOH, 60–80°, 6 h, > 95% yield.[2]
2. SeO_2/HOAc, dioxane, reflux, 1 h, 40–75% yield.[3]
3. $Ph_3P/Pd(OAc)_2$, HCOOH, 90°, 1 h.[4]
4. $NaAlH_2(OCH_2CH_2OCH_3)_2$, $PhCH_3$, reflux, 10 h, 62% yield.[5] An aryl allyl ether is selectively cleaved by this reagent (which also cleaves aryl benzyl ethers) in the presence of an *N*-allylamide.
5. Pd(0) cat., Bu_3SnH, AcOH, *p*-NO_2–C_6H_4OH.[6]
6. EtOH, $RhCl_3$, reflux, 86% yield.[1]
7. $LiPPh_2$, THF, 4 h, reflux, 78% yield.[7]
8. $SiCl_4$, NaI, CH_2Cl_2, CH_3CN, 8 h, 84% yield.[8]
9. The section on the cleavage of allyl ethers of alcohols should also be consulted.

1. See, for example: S. F. Martin and P. J. Garrison, *J. Org. Chem.*, **47,** 1513 (1982).
2. R. Boss and R. Scheffold, *Angew. Chem., Int. Ed., Engl.*, **15,** 558 (1976).
3. K. Kariyone and H. Yazawa, *Tetrahedron Lett.*, 2885 (1970).
4. H. Hey and H.-J. Arpe, *Angew. Chem., Int. Ed., Engl.*, **12,** 928 (1973).
5. T. Kametani, S.-P. Huang, M. Ihara, and K. Fukumoto, *J. Org. Chem.*, **41,** 2545 (1976).
6. P. Four and F. Guibe, *Tetrahedron Lett.*, **23,** 1825 (1982).
7. F. G. Mann and M. J. Pragnell, *J. Chem. Soc.*, 4120 (1965).
8. M. V. Bhatt and S. S. El-Morey, *Synthesis*, 1048 (1982).

14. Isopropyl Ether: $ArOCH(CH_3)_2$

An isopropyl ether was developed as a phenol protective group that would be more stable to Lewis acids than an aryl benzyl ether.[1] The isopropyl group has also been tested for use in protection of the phenolic oxygen of tyrosine during peptide synthesis.[2]

Formation

1. Me_2CHBr, K_2CO_3, DMF, acetone, 20°, 19 h.[1]

Cleavage

1. BCl_3, CH_2Cl_2, 0°, rapid; or $TiCl_4$, CH_2Cl_2, 0°, slower.[1] There was no reaction with $SnCl_4$.[1]
2. $SiCl_4$, NaI, 14 h, CH_2Cl_2, CH_3CN, 80% yield.[3]

1. T. Sala and M. V. Sargent, *J. Chem. Soc., Perkin Trans. I*, 2593 (1979).
2. See cyclohexyl ether in this section: M. Engelhard and R. B. Merrifield, *J. Am. Chem. Soc.*, **100**, 3559 (1978).
3. M. V. Bhatt and S. S. El-Morey, *Synthesis*, 1048 (1982).

15. Cyclohexyl Ether: $ArO\text{-}c\text{-}C_6H_{11}$ (Chart 4)

Formation[1]

$$p\text{-}HOC_6H_4CH_2CH(NHCOCF_3)CO_2Me \xrightarrow[CH_2Cl_2,\ \text{reflux, 24 h, } 60\%]{\text{cyclohexene, } BF_3\cdot Et_2O} p\text{-}c\text{-}C_6H_{11}OC_6H_4CH_2CH(NHCOCF_3)CO_2Me$$

Cleavage[1]

1. HF, 0°, 30 min, 100% yield.
2. 5.3 *N* HBr/AcOH, 25°, 2 h, 99% yield.

An ether that would not undergo rearrangement to a 3-alkyl derivative during acid-catalyzed removal of —NH protective groups was required to protect the phenol group in tyrosine. Four compounds were investigated: *O*-cyclohexyl-, *O*-isobornyl-, *O*-[1-(5-pentamethylcyclopentadienyl)ethyl]-, and *O*-isopropyltyrosine.

The *O*-isobornyl- and *O*-[1-(5-pentamethylcyclopentadienyl)ethyl]-derivatives do not undergo rearrangement, but are very labile in trifluoroacetic acid (100% cleaved in 5 min). The cyclohexyl and isopropyl derivatives are more stable to acid, but undergo some rearrangement. The cyclohexyl group combines minimal rearrangement with ready removal.[1]

A comparison has been made with several other common protective groups for tyrosine, and the degree of side-chain alkylation decreases in the order: Bn >

$2\text{-ClC}_6H_4CH_2$ > $2,6\text{-Cl}_2C_6H_3CH_2$ > cyclohexyl > *t*-Bu ~ benzyloxycarbonyl ~ 2-Br-benzyloxycarbonyl.[2]

1. M. Engelhard and R. B. Merrifield, *J. Am. Chem. Soc.*, **100,** 3559 (1978).
2. J. P. Tam, W. F. Heath, and R. B. Merrifield, *Int. J. Pept. Protein Res.*, **21,** 57 (1983).

16. *t*-Butyl Ether: $ArOC(CH_3)_3$ (Chart 4)

Formation

1. Isobutylene, cat. concd. H_2SO_4, CH_2Cl_2, 25°, 6–10 h, 93% yield.[1] These conditions also convert carboxylic acids to *t*-butyl esters.
2. Isobutylene, CF_3SO_3H, CH_2Cl_2, −78°, 70–90% yield.[2] These conditions will protect a phenol in the presence of a primary alcohol.
3. *t*-Butyl halide, Pyr, 20–30°, a few hours, 65–95% yield.[3]

Cleavage

The section on *t*-butyl ethers of alcohols should also be consulted.

1. Anhydrous CF_3CO_2H, 25°, 16 h, 81% yield.[1]
2. CF_3CH_2OH, CF_3SO_3H, −5°, 60 s, 100% yield.[2]

1. H. C. Beyerman and J. S. Bontekoe, *Recl. Trav. Chim. Pays-Bas*, **81,** 691 (1962).
2. J. L. Holcombe and T. Livinghouse, *J. Org. Chem.*, **51,** 111 (1986).
3. H. Masada and Y. Oishi, *Chem. Lett.*, 57 (1978).

17. Benzyl Ether: $ArOCH_2C_6H_5$ (Chart 4)

Formation

1. In general, benzyl ethers are prepared from a phenol by treating an alkaline solution of the phenol with a benzyl halide.[1]
2. $CHCl_3$, MeOH, K_2CO_3, BnBr, 4 h, heat.[2] In this case some (5 : 1) selectivity was achieved for a less hindered phenol in the presence of a more hindered one.

Cleavage

1.[1] CH_3O ... OBn, CHO, OBn, OH

1. H_2, Pd–C, Ac_2O, AcONa, PhH, 1.5 h
2. H_2, Pd–C, EtOAc
3. Ac_2O, Pyr

CH_3O ... O, O, OAc, OAc, OH

Catalytic hydrogenation in acetic anhydride–benzene removes the aromatic benzyl ether and forms a monoacetate; hydrogenation in ethyl acetate removes the aliphatic benzyl ether to give, after acetylation, the diacetate.[3]

2. Pd–C, 1,4-cyclohexadiene, 25°, 1.5 h, 95–100% yield.[4] Palladium black, a more reactive catalyst than Pd–C, must be used to cleave the more stable aliphatic benzyl ethers.
3. Na, *t*-BuOH, 70–80°, 2 h, 78%.[5] In this example sodium in *t*-butyl alcohol cleaves two aryl benzyl ethers and reduces a double bond that is conjugated with an aromatic ring; nonconjugated double bonds are stable.

OBn, OBn, $CH{=}CH(CH_2)_5CH{=}CHC_6H_{13}$ —(Na, t–BuOH; 70–80°, 2 h, 78%)→ OH, OH, $(CH_2)_7CH{=}CHC_6H_{13}$

4. $BF_3 \cdot Et_2O$, EtSH, 25°, 40 min, 80–90% yield.[6] Addition of sodium sulfate prevents hydrolysis of a dithioacetal group present in the compound; replacement of ethanethiol with ethanedithiol prevents cleavage of a dithiolane group.
5. CF_3OSO_2F or CH_3OSO_2F, $PhSCH_3$, CF_3CO_2H, 0°, 30 min, 100% yield.[7] Thioanisole suppresses acid-catalyzed rearrangement of the benzyl group to form 3-benzyltyrosine. The more acid-stable 2,6-dichlorobenzyl ether is cleaved in a similar manner.
6. Me_3SiI, CH_3CN, 25–50°, 100% yield.[8] Selective removal of protective groups is possible with this reagent since a carbamate, $=NCOOCMe_3$, is cleaved in 6 min at 25°; an aryl benzyl ether is cleaved in 100% yield, with no formation of 3-benzyltyrosine, in 1 h at 50°, at which time a methyl ester begins to be cleaved.
7. 2-Bromo-1,3,2-benzodioxaborole, CH_2Cl_2, 95% yield.[9]
8. CF_3CO_2H, $PhSCH_3$, 25°, 3 h.[10] The use of dimethyl sulfide or anisole as a cation scavenger was not as effective because of side reactions. Benzyl ethers of serine and threonine were slowly cleaved (30% in 3 h; complete cleavage in 30 h). The use of pentamethylbenzene had been shown to increase the rate of deprotection of *O*-Bn-tyrosine.[11]

1. See, for example: M. C. Venuti, B. E. Loe, G. H. Jones, and J. M. Young, *J. Med. Chem.*, **31,** 2132 (1988).
2. H. Schmidhammer and A. Brossi, *J. Org. Chem.*, **48,** 1469 (1983).
3. G. Büchi and S. M. Weinreb, *J. Am. Chem. Soc.*, **93,** 746 (1971).
4. A. M. Felix, E. P. Heimer, T. J. Lambros, C. Tzougraki, and J. Meienhofer, *J. Org. Chem.*, **43,** 4194 (1978).
5. B. Loev and C. R. Dawson, *J. Am. Chem. Soc.* **78,** 6095 (1956).
6. K. Fuji, K. Ichikawa, M. Node, and E. Fujita, *J. Org. Chem.*, **44,** 1661 (1979).

7. Y. Kiso, H. Isawa, K. Kitagawa, and T. Akita, *Chem. Pharm. Bull*, **26,** 2562 (1978).
8. R. S. Lott, V. S. Chauhan, and C. H. Stammer, *J. Chem. Soc., Chem. Commun.*, 495 (1979).
9. P. F. King and S. G. Stroud, *Tetrahedron Lett.*, **26,** 1415 (1985).
10. Y. Kiso, K. Ukawa, S. Nakamura, K. Ito, and T. Akita, *Chem. Pharm. Bull*, **28,** 673 (1980).
11. H. Yoshino, Y. Tsuchiya, I. Saito, and M. Tsujii, *Chem. Pharm. Bull.*, **35,** 3438 (1987).

18. 2,6-Dimethylbenzyl Ether: $2,6\text{-}(CH_3)_2C_6H_3CH_2OAr$

The 2,6-dimethylbenzyl ether is considerably more stable to hydrogenolysis than is the benzyl ether. It has a half-life of 15 h at 1 atm of hydrogen in the presence of Pd–C whereas the benzyl ether has a half-life of ~45 min. This added stability allows hydrogenation of azides, nitro groups, and olefins in the presence of a dimethylbenzyl group.[1]

1. R. Davis and J. M. Muchowski, *Synthesis*, 987 (1982).

19. 4-Methoxybenzyl Ether: $4\text{-}CH_3OC_6H_4CH_2OAr$

Formation

1. $MeOC_6H_4CH_2Cl$, $Bu_4N^+I^-$, K_2CO_3, acetone, 55°, 96% yield.[1]
2. $MeOC_6H_4CH_2Br$, $(i\text{-}Pr)_2NEt$, CH_2Cl_2, rt, 80% yield.[2]

Cleavage

1. CF_3CO_2H, CH_2Cl_2, 85% yield.[1]
2. Camphorsulfonic acid, $(CH_3)_2C(OCH_3)_2$, rt.[2]

1. J. D. White and J. C. Amedio, Jr., *J. Org. Chem.*, **54,** 736 (1989).
2. H. Nagaoka, G. Schmid, H. Iio, and Y. Kishi, *Tetrahedron Lett.*, **22,** 899 (1981).

20. *o*-Nitrobenzyl Ether: $o\text{-}NO_2\text{—}C_6H_4CH_2OAr$ (Chart 4)

CO_2Me NHPG OH

1. NaOMe
2. $2\text{-}NO_2C_6H_4CH_2Cl$
25°, 15 min, 80%

350 nm, MeOH
25°, 12 h, 98%

CO_2Me NHPG $OCH_2C_6H_4\text{-}2\text{-}NO_2$

An *o*-nitrobenzyl ether can be cleaved by photolysis. In tyrosine this avoids the use of acid-catalyzed cleavage and the attendant conversion to 3-benzyltyrosine.[1] (Note that this unwanted conversion can also be suppressed by the addition of thioanisole; see section on benzyl ether cleavage.)

1. B. Amit, E. Hazum, M. Fridkin, and A. Patchornik, *Int. J. Pept. Protein Res.*, **9,** 91 (1977).

21. 2,6-Dichlorobenzyl Ether: $ArOCH_2C_6H_3$–2,6-Cl_2

This group is readily cleaved by a mixture of CF_3SO_3H, $PhSCH_3$, and CF_3CO_2H.[1] Of the common benzyl protecting groups used to protect the hydroxyl of tyrosine the 2,6-dichlorobenzyl shows a low incidence of alkylation at the 3-position of tyrosine during cleavage with HF/anisole. A comparative study on deprotection of X-Tyr in HF/anisole gives the following percentages of side reactions for various X groups: Bn, 24.5; 2-ClBn, 9.8; 2,6-Cl_2Bn, 6.5; cyclohexyl, 1.5; *t*-Bu, <0.2; Cbz, 0.5; 2-Br–Cbz, 0.2.[2]

1. Y. Kiso, M. Satomi, K. Ukawa, and T. Akita, *J. Chem. Soc., Chem. Commun.*, 1063 (1980).
2. J. P. Tam, W. F. Heath, and R. B. Merrifield, *Int. J. Pept. Protein Res.*, **21,** 57 (1983).

22. 4-(Dimethylaminocarbonyl)benzyl Ether: $(CH_3)_2NCOC_6H_4CH_2OAr$

The 4-(dimethylaminocarbonyl)benzyl ether has been used to protect the phenolic hydroxyl of tyrosine. It is stable to CF_3CO_2H (120 h), but not to HBr/AcOH (complete cleavage in 16 h). It can also be cleaved by hydrogenolysis (H_2/Pd–C).[1]

1. V. S. Chauhan, S. J. Ratcliffe, and G. T. Young, *Int. J. Pept. Protein Res.*, **15,** 96 (1980).

23. 9-Anthrylmethyl Ether: $ArOCH_2$–9-anthryl (Chart 4)

9-Anthrylmethyl ethers, formed from the sodium salt of a phenol and 9-anthrylmethyl chloride in DMF, can be cleaved with CH_3SNa (DMF, 25°, 20 min, 85–99% yield). They are also cleaved by CF_3CO_2H/CH_2Cl_2 (0°, 10 min, 100% yield); they are stable to CF_3CO_2H/dioxane (25°, 1 h).[1]

1. N. Kornblum and A. Scott, *J. Am. Chem. Soc.*, **96,** 590 (1974).

24. 4-Picolyl Ether: $ArOCH_2$–4-pyridyl (Chart 4)

Formation[2]*/Cleavage*[1,2]

1. 4–picolyl chloride, NaOH, EtOH; 2. EDTA → O-4-picolyl tyrosine; Electrolysis[1], 0.5 N H_2SO_4 → Tyr–OH

EDTA = ethylenediaminetetraacetic acid

An aryl 4-picolyl ether is stable to trifluoroacetic acid, used to cleave an *N-t*-butoxycarbonyl group.[2]

1. P. M. Scopes, K. B. Walshaw, M. Welford, and G. T. Young, *J. Chem. Soc.*, 782 (1965).
2. A. Gosden, D. Stevenson, and G. T. Young, *J. Chem. Soc., Chem. Commun.*, 1123 (1972).

25. Heptafluoro-*p*-tolyl Ethers: $ArOC_6F_4$–CF_3,

26. Tetrafluoro-4-pyridyl Ethers: $ArOC_5F_4N$

Formation/Cleavage[1,2]

Estradiol → 3-$CF_3C_6F_4$–estradiol: $C_6F_5CF_3$ (1 eq.), NaOH, CH_2Cl_2, $Bu_4N^+HSO_4^-$, 95%

3-$CF_3C_6F_4$–estradiol → Estradiol: NaOMe, DMF, 1 h, 10°, 87%

If 2 eq. of reagent are used, both hydroxyls can be protected and the phenolic hydroxyl can be selectively cleaved with NaOMe. The tetrafluoropyridyl derivative is introduced under similar conditions. The use of this methodology has been reviewed.[3]

1. M. Jarman and R. McCague, *J. Chem. Soc., Chem. Commun.*, 125 (1984).
2. M. Jarman and R. McCague, *J. Chem. Res., Synop*, 114 (1985).
3. M. Jarman, *J. Fluorine Chem.*, **42,** 3 (1989).

Silyl Ethers

Aryl and alkyl trimethylsilyl ethers can often be cleaved by refluxing in aqueous methanol, an advantage for acid- or base-sensitive substrates. The ethers are stable to Grignard and Wittig reagents, and to reduction with lithium aluminum hydride at −15°. Aryl *t*-butyldimethylsilyl ethers require acid- or fluoride ion-catalyzed hydrolysis for removal.

27. Trimethylsilyl Ether (TMS Ether): $ArOSi(CH_3)_3$

Formation

1. Me_3SiCl, Pyr, 30–35°, 12 h, satisfactory yield.[1]
2. $(Me_3Si)_2NH$, cat. concd. H_2SO_4, reflux, 2 h, 97% yield.[2]

Cleavage

Trimethylsilyl ethers are readily cleaved by fluoride ion, mild acids, and bases. If the TMS derivative is somewhat hindered, it also becomes less susceptible to cleavage. A phenolic TMS ether can be cleaved in the presence of an alkyl TMS ether [Dowex 1X8(HO^-), EtOH, rt, 6 h, 78% yield].[3]

1. Cl. Moreau, F. Roessac, and J. M. Conia, *Tetrahedron Lett.*, 3527 (1970).
2. S. A. Barker and R. L. Settine, *Org. Prep. Proced. Int.*, **11**, 87 (1979).
3. Y. Kawazoe, M. Nomura, Y. Kondo, and K. Kohda, *Tetrahedron Lett.*, **28**, 4307 (1987).

28. *t*-Butyldimethylsilyl Ether (TBDMS Ether): $ArOSi(CH_3)_2C(CH_3)_3$ (Chart 4)

Formation

1. *t*-$BuMe_2SiCl$, DMF, imidazole, 25°, 3 h, 96% yield.[1,2]

Cleavage

1. 0.1 *M* HF, 0.1 *M* NaF, pH 5, THF, 25°, 2 days, 77% yield.[1] In this substrate a mixture of products resulted from attempted cleavage of the *t*-butyldimethylsilyl ether with tetra-*n*-butylammonium fluoride, the reagent generally used.[3]
2. KF, 48% aq. HBr, DMF, rt, 91% yield.[4]

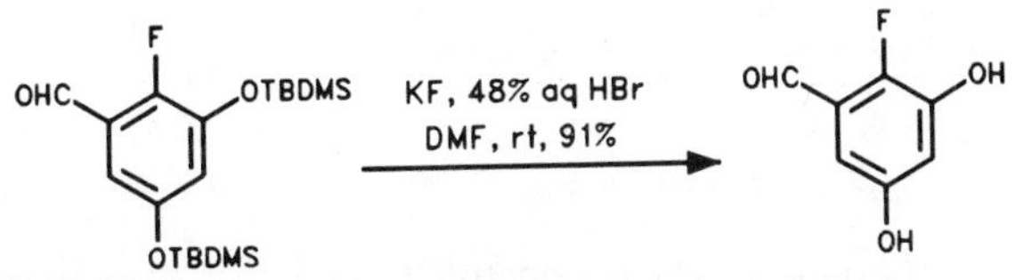

F
OHC OTBDMS
OTBDMS
KF, 48% aq HBr
DMF, rt, 91%
F
OHC OH
OH

The use of $Bu_4N^+F^-$ results in decomposition of this substrate.

1. P. M. Kendall, J. V. Johnson, and C. E. Cook, *J. Org. Chem.*, **44,** 1421 (1979).
2. R. C. Ronald, J. M. Lansinger, T. S. Lillie, and C. J. Wheeler, *J. Org. Chem.*, **47,** 2451 (1982).
3. E. J. Corey and A. Venkateswarlu, *J. Am. Chem. Soc.*, **94,** 6190 (1972).
4. A. K. Sinhababu, M. Kawase, and R. T. Borchardt, *Synthesis*, 710 (1988).

Esters

Aryl esters, prepared from the phenol and an acid chloride or anhydride in the presence of base, are readily cleaved by saponification. In general they are more readily cleaved than the related esters of alcohols, thus allowing selective removal of phenolic esters. 9-Fluorenecarboxylates and 9-xanthenecarboxylates are also cleaved by photolysis. To permit selective removal, a number of carbonate esters have been investigated: aryl benzyl carbonates can be cleaved by hydrogenolysis; aryl 2,2,2-trichloroethyl carbonates, by Zn/THF–H_2O.

29. Aryl Acetate: $ArOCOCH_3$ (Chart 4)

Formation

1. AcCl, NaOH, dioxane, $Bu_4N^+HSO_4^-$, 25°, 30 min, 90% yield.[1] Phase-transfer catalysis with tetra-*n*-butylammonium hydrogen sulfate effects acylation of sterically hindered phenols and selective acylation of a phenol in the presence of an aliphatic secondary alcohol.
2. 1-Acetyl-*v*-triazolo[4,5-*b*]pyridine, THF, 1 *N* NaOH, 30 min.[2]

COMe
N
N
N
HO
OH
THF, rt, 1 h, 89%
AcO
OH

This method is also effective in the selective introduction of a benzoate ester.

Cleavage

1. $NaHCO_3$/aq. MeOH, 25°, 0.75 h, 94% yield.[3]
2. 3 *N* HCl, acetone, reflux, 2 h.[4]
3. Aqueous NH_3, 0°, 48 h.[4]
4. $NaBH_4$, $HO(CH_2)_2OH$, 40°, 18 h, 87% yield.[5] Lithium aluminum hydride can be used to effect efficient ester cleavage if no other functional group is present that can be attacked by this strong reducing agent.[6]

The following set of conditions will selectively remove a phenolic acetate in the presence of an alcoholic acetate.

5. TsOH, SiO_2, toluene, 80°, 6–40 h, 79–100% yield.[7]
6. $(NH_2)_2C{=}NH$, MeOH, 50°, 95% yield.[8]
7. $Me_2NCH_2C(O)N(OH)Me$, MeOH or THF/H_2O, 84% yield.[9]
8. Zn, MeOH, 91–100% yield.[10]

1. V. O. Illi, *Tetrahedron Lett.*, 2431 (1979).
2. M. P. Paradisi, G. P. Zecchini, and I. Torrini, *Tetrahedron Lett.*, **27,** 5029 (1986).
3. For example, see G. Büchi and S. M. Weinreb, *J. Am. Chem. Soc.*, **93,** 746 (1971).
4. E. Haslam, G. K. Makinson, M. O. Naumann, and J. Cunningham, *J. Chem. Soc.*, 2137 (1964).
5. J. Quick and J. K. Crelling, *J. Org. Chem.*, **43,** 155 (1978).
6. H. Mayer, P. Schudel, R. Rüegg, and O. Isler, *Helv. Chim. Acta*, **46,** 650 (1963).
7. G. Blay, M. L. Cardona, M. B. Garcia, and J. P. Pedro, *Synthesis*, 438 (1989).
8. N. Kunesch, C. Miet, and J. Poisson, *Tetrahedron Lett.*, **28,** 3569 (1987).
9. M. Ono and I. Itoh, *Tetrahedron Lett.*, **30,** 207 (1989).
10. A. G. González, Z. D. Jorge, H. L. Dorta, and F. R. Luis, *Tetrahedron Lett.*, **22,** 335 (1981).

30. Aryl Levulinate: $CH_3COCH_2CH_2CO_2Ar$

Cleavage[1]

PhO_2C (aryl levulinate) → PhO_2C (phenol, OH)

Na_2SO_3, $Na_2S_2O_5$ (4:1), 30 min, rt; THF or CH_3CN or EtOH, H_2O; 85–95%

1. M. Ono and I. Itoh, *Chem. Lett.*, 585 (1988).

31. Aryl Pivaloate (ArOPv): $(CH_3)_3CCO_2Ar$ (Chart 4)

Formation/Cleavage[1]

Pivaloyl chloride reacts selectively with the less hindered phenol group.

OH, CMe_3, OH → OH, CMe_3, $OCOCMe_3$

Forward: Me_3CCOCl, Pyr, 5–10°, 4 days, 84%

Reverse: KOH, 50% aq EtOH, reflux, 64 h, N_2, 87%

1. L. K. T. Lam and K. Farhat, *Org. Prep. Proced. Int.*, **10,** 79 (1978).

32. Aryl Benzoate: $ArOCOC_6H_5$ (Chart 4)

Aryl benzoates, stable to alkylation conditions using K_2CO_3/Me_2SO_4, are cleaved by more basic hydrolysis (KOH).[1] They are stable to anhydrous hydrogen chloride,[2] but are cleaved by hydrochloric acid.[3]

Formation

1. $(ClCO)_2$, Me_2NCHO, PhCOOH; Pyr, 20°, 2 h, 90% yield.[4]
2. 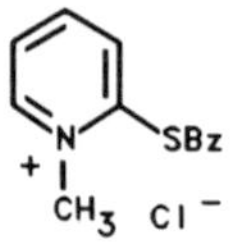 aq. $NaHCO_3$ or aq. NaOH, 80% yield.[5]

   This reagent forms aryl benzoates under aqueous conditions. (It also acylates amines and carboxylic acids.)
3. Monoesterification of a symmetrical dihydroxy aromatic compound can be effected by reaction with polymer-bound benzoyl chloride (Pyr, benzene, reflux, 15 h) to give a polymer-bound benzoate, which can be alkylated with diazomethane to form, after basic hydrolysis (0.5 *M* NaOH, dioxane, H_2O, 25°, 20 h, or 60°, 3 h), a monomethyl ether.[6]

Cleavage

1. Under anhydrous conditions, cesium carbonate or bicarbonate quantitatively cleaves an aryl dibenzoate or diacetate to the monoester; yields are considerably lower with potassium carbonate.[7]

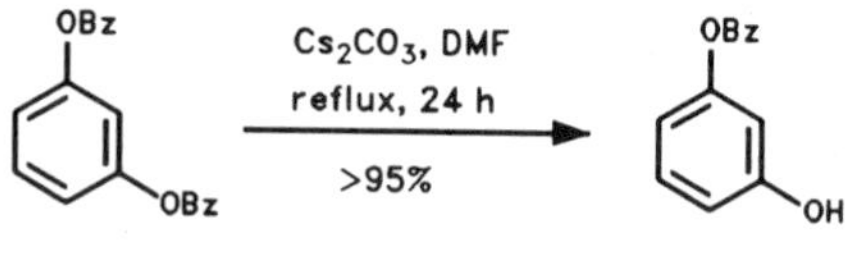

   Cs_2CO_3, DMF, reflux, 24 h, >95%

2. $BuNH_2$, benzene, rt, 1–24 h, >85% yield.[8]

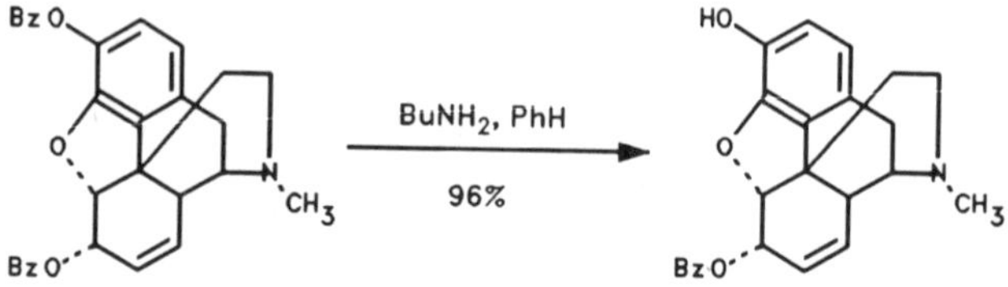

   $BuNH_2$, PhH, 96%

   This method is generally selective for phenolic esters.

3. 2-Bromo-1,3,2-benzodioxaborole, CH_2Cl_2 (cat. $BF_3 \cdot Et_2O$), 25°, 0.25 h, 71% yield.[9]

1. M. Gates, *J. Am. Chem. Soc.*, **72,** 228 (1950).
2. D. D. Pratt and R. Robinson, *J. Chem. Soc.*, 1577 (1922).
3. A. Robertson and R. Robinson, *J. Chem. Soc.*, 1710 (1927).
4. P. A. Stadler, *Helv. Chim. Acta*, **61,** 1675 (1978).
5. M. Yamada, Y. Watabe, T. Sakakibara, and R. Sudoh, *J. Chem. Soc., Chem. Commun.*, 179 (1979).
6. C. C. Leznoff and D. M. Dixit, *Can. J. Chem.*, **55,** 3351 (1977).
7. H. E. Zaugg, *J. Org. Chem.*, **41,** 3419 (1976).
8. K. H. Bell, *Tetrahedron Lett.*, **27,** 2263 (1986).
9. P. F. King and S. G. Stroud, *Tetrahedron Lett.*, **26,** 1415 (1985).

33. Aryl 9-Fluorenecarboxylate (Chart 4):

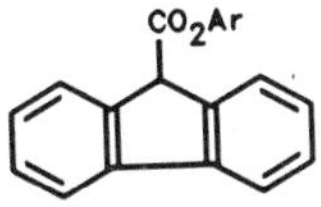

CO_2Ar

Aryl 9-fluorenecarboxylates prepared from the phenol and the acid chloride (9-fluorenecarbonyl chloride, Pyr, C_6H_6, 25°, 1 h, 65% yield) can be cleaved by photolysis ($h\nu$, Et_2O, reflux, 4 h, 60% yield). The related aryl xanthenecarboxylates (**i**) were prepared and cleaved in the same way.[1]

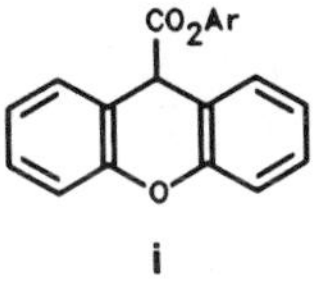

CO_2Ar

O

i

1. D. H. R. Barton, Y. L. Chow, A. Cox, and G. W. Kirby, *J. Chem. Soc.*, 3571 (1965).

Carbonates

34. Aryl Methyl Carbonate: $ArOCO_2CH_3$ (Chart 4)

In an early synthesis a methyl carbonate, prepared by reaction of a phenol with methyl chloroformate, was cleaved selectively in the presence of a phenyl ester.[1]

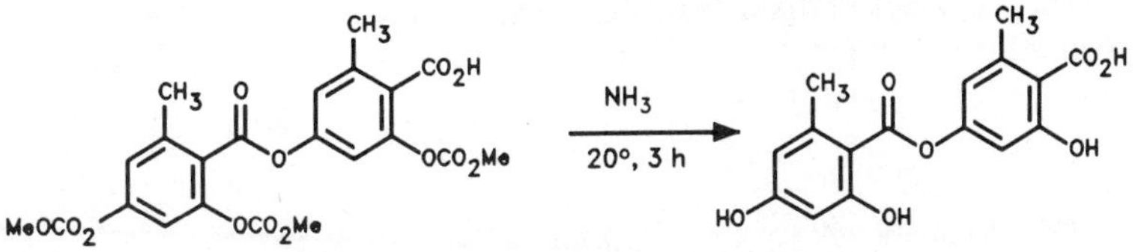

CH_3 CO_2H CH_3 O O OCO_2Me $MeOCO_2$ OCO_2Me

NH_3

20°, 3 h

CH_3 CO_2H CH_3 O O OH HO OH

An ethyl carbonate was cleaved by refluxing in acetic acid for 6 h.[2]

1. E. Fischer and H. O. L. Fischer, *Ber.*, **46,** 1138 (1913).
2. E. Haslam, R. D. Haworth, and G. K. Makinson, *J. Chem. Soc.*, 5153 (1961).

35. Aryl 2,2,2-Trichloroethyl Carbonate: $ArOCOOCH_2CCl_3$ (Chart 4)

Formation

1. Cl_3CCH_2OCOCl, Pyr or aq. NaOH, 25°, 12 h.[1]

Cleavage

1. Zn, HOAc, 25°, 1–3 h, or Zn, CH_3OH, heat, a few minutes.[1]
2. Zn, THF–H_2O, pH 4.2, 25°, 4 h.[2] The authors suggest that selective cleavage should be possible by this method since at pH 4.2, 25°, 2,2,2-trichloroethyl esters are cleaved in 10 min, 2,2,2-trichloroethyl carbamates are cleaved in 30 min, and the 2,2,2-trichloroethyl carbonate of estrone, formed in 87% yield from estrone and the acid chloride, is cleaved in 4 h (97% yield).

1. T. B. Windholz and D. B. R. Johnston, *Tetrahedron Lett.*, 2555 (1967).
2. G. Just and K. Grozinger, *Synthesis*, 457 (1976).

36. Aryl Vinyl Carbonate: $ArOCO_2CH{=}CH_2$ (Chart 4)

Formation

1. $CH_2{=}CHOCOCl$, Pyr, 95% yield.[1]

Cleavage

1. Na_2CO_3, warm aq. dioxane, 96% yield. Selective protection of an aryl —OH or an amine —NH group is possible by reaction of the compound with vinyl chloroformate. Vinyl carbamates ($RR'NCO_2CH{=}CH_2$) are stable to the basic conditions (Na_2CO_3) used to cleave vinyl carbonates. Conversely, vinyl carbonates are stable to the acidic conditions ($HBr/CH_3OH/CH_2Cl_2$) used to cleave vinyl carbamates. Vinyl carbonates are cleaved by more acidic conditions: 2 *N* anhyd. HCl/dioxane, 25°, 3 h, 10% yield; HBF_4, 25°, 12 h, 30% yield; 2 *N* $HCl/CH_3OH–H_2O$ (4 : 1), 60°, 8 h, 100% yield.[1]

1. R. A. Olofson and R. C. Schnur, *Tetrahedron Lett.*, 1571 (1977).

37. Aryl Benzyl Carbonate: $ArOCOOCH_2C_6H_5$ (Chart 4)

Formation

1. $PhCH_2OCOCl$, Pyr, CH_2Cl_2, THF.[1]

Cleavage

1. H_2/Pd–C, EtOH, 20°.[1] *o*-Bromobenzyl carbonates have been developed for use in solid-phase peptide synthesis. An aryl *o*-bromobenzyl carbonate is stable to acidic cleavage (CF_3CO_2H) of a *t*-butyl carbamate; a benzyl carbonate is cleaved.

   The *o*-bromo derivative is quantitatively cleaved with hydrogen fluoride (0°, 10 min).[2]

1. M. Kuhn and A. von Wartburg, *Helv. Chim. Acta*, **52,** 948 (1969).
2. D. Yamashiro and C. H. Li, *J. Org. Chem.*, **38,** 591 (1973).

38. Aryl Carbamates: ArOCONHR

Formation

1. RNCO (R = Ph, *i*-Bu), 60°, 2 h, 65–85% yield.[1]

Cleavage

1. 2 *N* NaOH, 20°, 2 h, 78% yield.[1]
2. $H_2NNH_2 \cdot H_2O$, DMF, 20°, 3 h, 59–87% yield.[1]

1. G. Jäger, R. Geiger, and W. Siedel, *Chem. Ber.*, **101,** 2762 (1968).

Phosphinates

39. Dimethylphosphinyl Ester (Dmp Ester): $(CH_3)_2P(O)OAr$

Formation

1. $Me_2P(O)Cl$, Et_3N, $CHCl_3$, 76% yield.[1] The Dmp group was used to protect tyrosine for use in peptide synthesis. It is stable to 1 *M* HCl/MeOH, 1 *M* HCl/AcOH, CF_3CO_2H, HBr/AcOH, and H_2/Pd–C.

Cleavage

The Dpm group can be cleaved by the following reagents: liquid HF (0°, 1 h); 1 *M* Et_3N/MeOH (rt, 7 h); 0.1 *M* NaOH (rt, < 5 min); 5% aq. $NaHCO_3$ (rt, 5 h); 20% hydrazine/MeOH (rt, < 5 min); 50% pyridine/DMF (rt, 6 h); $Bu_4N^+F^-$ (rt, < 5 min).[1]

40. Dimethylthiophosphinyl Ester (Mpt Ester): $(CH_3)_2P(S)OAr$

Formation

1. MptCl, CH_2Cl_2, Et_3N, 66% yield.[2]

Cleavage

The *O*-Mpt group is quite stable to acidic conditions (HBr/AcOH, CF_3CO_2H, 1 *M* HCl/AcOH), but is slowly cleaved under basic conditions (1 *M* NaOH/MeOH, 5 min; 1 *M* Et_3N/MeOH, reflux, 12 h). In contrast, the *N*-Mpt group is readily cleaved with acid (CF_3CO_2H, 60 min; 1 *M* HCl/AcOH, 15 min; HBr/AcOH, 5 min), but not with base. The Mpt group was used to protect tyrosine during peptide synthesis.[2] The Mpt group can be removed with aq. $AgNO_3$ or $Hg(OAc)_2$[3] or fluoride ion.[4]

1. M. Ueki, Y. Sano, I. Sori, K. Shinozaki, H. Oyamada, and S. Ikeda, *Tetrahedron Lett.*, **27,** 4181 (1986).
2. M. Ueki and T. Inazu, *Bull. Chem. Soc. Jpn.*, **55,** 204 (1982).
3. M. Ueki and K. Shinozaki, *Bull. Chem. Soc. Jpn.*, **56,** 1187 (1983).
4. M. Ueki and K. Shinozaki, *Bull. Chem. Soc. Jpn.*, **57,** 2156 (1984).

Sulfonates

An aryl methane- or toluenesulfonate ester is stable to reduction with lithium aluminum hydride, to the acidic conditions used for nitration of an aromatic ring (HNO_3/HOAc),[1] and to the high temperatures (200–250°) of an Ullman reaction. Aryl sulfonate esters, formed by reaction of a phenol with a sulfonyl chloride in pyridine or aqueous sodium hydroxide, are cleaved by warming in aqueous sodium hydroxide.[2]

1. E. M. Kampouris, *J. Chem. Soc.*, 2651 (1965).
2. F. G. Bordwell and P. J. Boutan, *J. Am. Chem. Soc.*, **79,** 717 (1957).

41. Aryl Methanesulfonate: $ArOSO_2CH_3$ (Chart 4)

In a synthesis of decinine a phenol was protected as a methanesulfonate that was stable during an Ullman coupling reaction and during condensation, catalyzed by calcium hydroxide, of an amine with an aldehyde. Aryl methanesulfonates are cleaved by warm sodium hydroxide solution.[1,2]

An aryl methanesulfonate was cleaved to a phenol by phenyllithium or phenylmagnesium bromide;[3] it was reduced to an aromatic hydrocarbon by sodium in liquid ammonia.[4]

1. I. Lantos and B. Loev, *Tetrahedron Lett.*, 2011 (1975).
2. J. E. Rice, N. Hussain, and E. J. LaVoie, *J. Labelled Compd. Radiopharm.*, **24,** 1043 (1987).
3. J. E. Baldwin, D. H. R. Barton, I. Dainis, and J. L. C. Pereira, *J. Chem. Soc. C*, 2283 (1968).
4. G. W. Kenner and N. R. Williams, *J. Chem. Soc.*, 522 (1955).

42. Aryl Toluenesulfonate: $ArOSO_2C_6H_4\text{-}p\text{-}CH_3$

Formation[1]

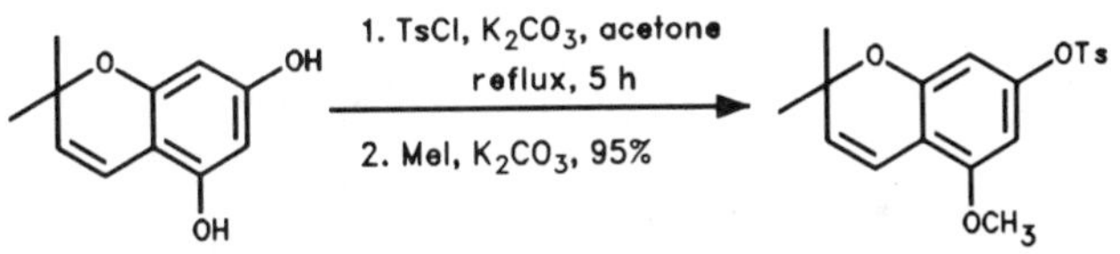

1. TsCl, K_2CO_3, acetone reflux, 5 h
2. MeI, K_2CO_3, 95%

Cleavage[1]

KOH, H_2O, EtOH
reflux, 1 h, 60 %

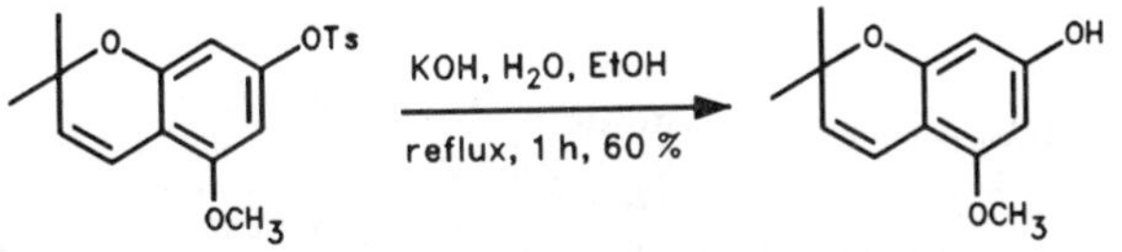

An aryl toluenesulfonate is stable to lithium aluminum hydride (Et_2O, reflux, 4 h) and to *p*-toluenesulfonic acid ($C_6H_5CH_3$, reflux, 15 min).

o-Aminophenol can be selectively protected as a sulfonate or a sulfonamide.[2]

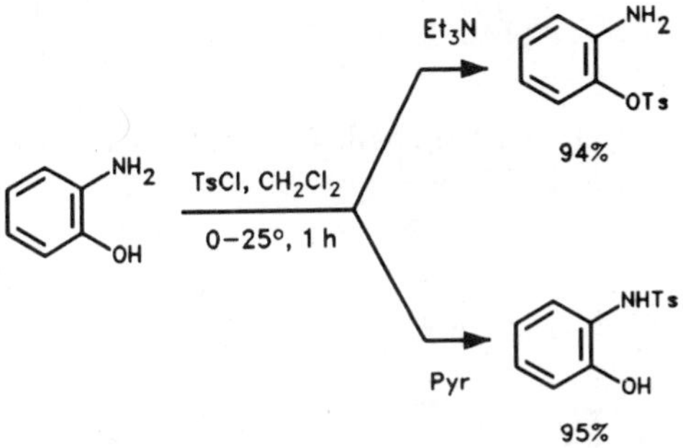

TsCl, CH_2Cl_2
0–25°, 1 h
Et_3N → OTs, NH_2 94%
Pyr → NHTs, OH 95%

1. M. L. Wolfrom, E. W. Koos, and H. B. Bhat, *J. Org. Chem.*, **32,** 1058 (1967).
2. K. Kurita, *Chem. Ind. (London)*, 345 (1974).

43. Aryl 2-Formylbenzenesulfonate:

The formylbenzenesulfonate prepared from a phenol (2-CHO–$C_6H_4SO_2Cl$, Et_3N) can be cleaved with NaOH (aq. acetone, rt, 5 min) in the presence of a hindered acetate.[1]

1. M. S. Shashidhar and M. V. Bhatt, *J. Chem. Soc., Chem. Commun.*, 654 (1987).

PROTECTION FOR CATECHOLS (1,2-Dihydroxybenzenes)

Catechols can be protected as diethers or diesters by methods that have been described to protect phenols. However, formation of cyclic acetals and ketals (e.g., methylenedioxy, acetonide, cyclohexylidenedioxy, diphenylmethylenedioxy derivatives) or cyclic esters (e.g., borates or carbonates) selectively protects the two adjacent hydroxyl groups in the presence of isolated phenol groups.

Cyclic Acetals and Ketals

44. Methylene Acetal (Chart 4):

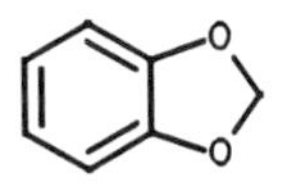

The methylenedioxy group, often present in natural products, is stable to many reagents. Efficient methods for both formation and removal of the group are available.

Formation

1. CH_2Br_2, NaOH, H_2O, Adogen, reflux, 3 h, 76–86% yield.[1] Adogen = $R_3N^+CH_3Cl^-$, phase-transfer catalyst (R = C_8–C_{10} straight-chain alkyl groups). Earlier methods required anhydrous conditions and aprotic solvents.
2. CH_2X_2 (X = Br, Cl), DMF, KF or CsF, 110°, 1.5 h, 70–98% yield.[2]

Cleavage

1. $AlBr_3$, EtSH, 0°, 0.5–1 h, 73–78% yield.[3] Aluminum bromide cleaves aryl and alkyl methyl ethers in high yield; methyl esters are stable.

Selective cleavage of an aryl methylenedioxy group, or an aryl methyl ether, by boron trichloride has been investigated.[4–6]

2. PCl_5, CH_2Cl_2, reflux; H_2O; reflux, 3 h, 61% yield.[7]

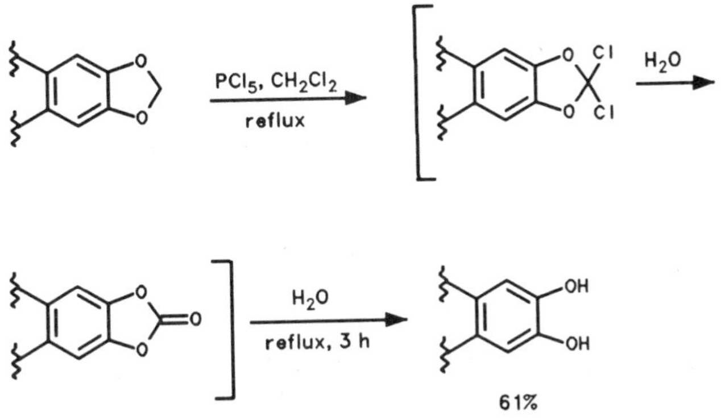

PCl_5, CH_2Cl_2
reflux
H_2O
H_2O
reflux, 3 h
61%

3. BCl_3, CH_3SCH_3, $ClCH_2CH_2Cl$, 83°, 98% yield.[8]
4. 9-Br–BBN, 24 h, 40°, CH_2Cl_2.[9]
5. A 4-nitro-1,2-methylenedioxybenzene has been cleaved to a catechol with 2 *N* NaOH, 90°, 30 min;[10] a similar compound substituted with a 4-nitro or 4-formyl group has been cleaved by $NaOCH_3$/DMSO, 150°, 2.5 min (13–74% catechol, 6–60% recovered starting material).[11]
6. $Pb(OAc)_4$, benzene, 50°, 8 h.[12]

1. A. P. Bashall and J. F. Collins, *Tetrahedron Lett.*, 3489 (1975).
2. J. H. Clark, H. L. Holland, and J. M. Miller, *Tetrahedron Lett.*, 3361 (1976).
3. M. Node, K. Nishide, M. Sai, K. Ichikawa, K. Fuji, and E. Fujita, *Chem. Lett.*, 97 (1979).
4. M. Gerecke, R. Borer, and A. Brossi, *Helv. Chim. Acta*, **59,** 2551 (1976).
5. S. Teitel, J. O'Brien, and A. Brossi, *J. Org. Chem.*, **37,** 3368 (1972).
6. F. M. Dean, J. Goodchild, L. E. Houghton, J. A. Martin, R. B. Morton, B. Parton, A. W. Price, and N. Somvichien, *Tetrahedron Lett.*, 4153 (1966).
7. G. L. Trammell, *Tetrahedron Lett.*, 1525 (1978).
8. P. G. Williard and C. B. Fryhle, *Tetrahedron Lett.*, **21,** 3731 (1980).
9. M. V. Bhatt, *J. Organomet. Chem.*, **156,** 221 (1978).

10. E. Haslam and R. D. Haworth, *J. Chem. Soc.*, 827 (1955).
11. S. Kobayashi, M. Kihara, and Y. Yamahara, *Chem. Pharm. Bull.*, **26,** 3113 (1978).
12. Y. Ikeya, H. Taguchi, and I. Yoshioka, *Chem. Pharm. Bull.*, **29,** 2893 (1981).

45. Acetonide Derivative (Chart 4):

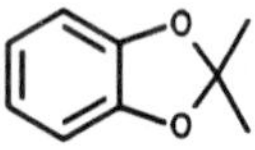

A catechol can be protected as an acetonide (acetone, 70% yield). It is cleaved with 6 *N* HCl (reflux, 2 h, high yield)[1] or by refluxing in acetic acid/H_2O (100°, 18 h, 90% yield).[2]

1. K. Ogura and G.-i. Tsuchihashi, *Tetrahedron Lett.*, 3151 (1971).
2. E. J. Corey and S. D. Hurt, *Tetrahedron Lett.*, 3923 (1977).

46. Cyclohexylidene Ketal:

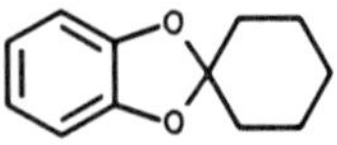

The cyclohexylidene ketal, prepared from a catechol and cyclohexanone (Al_2O_3/TsOH, CH_2Cl_2, reflux, 36 h),[1] is stable to metalation conditions (RX/BuLi) that cleave aryl methyl ethers.[2] The ketal is cleaved by acidic hydrolysis (concd. HCl/EtOH, reflux, 1.5 h, → 20°, 12 h); it is stable to milder acidic hydrolysis that cleaves tetrahydropyranyl ethers (1 *N* HCl/EtOH, reflux, 5 h, 91% yield).[3]

1. G. Schill and E. Logemann, *Chem. Ber.*, **106,** 2910 (1973).
2. G. Schill and K. Murjahn, *Chem. Ber.*, **104,** 3587 (1971).
3. J. Boeckmann and G. Schill, *Chem. Ber.*, **110,** 703 (1977).

47. Diphenylmethylene Ketal (Chart 4):

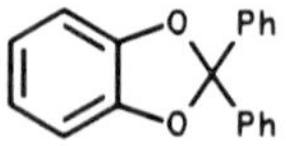

The diphenylmethylene ketal prepared from a catechol (Ph_2CCl_2, Pyr, acetone, 12 h)[1] or (Ph_2CCl_2, neat, 170°, 5 min, 59%)[2] can be cleaved by hydrogenolysis (H_2/Pd–C, THF).[3] It has also been prepared from a 1,2,3-trihydroxybenzene (Ph_2CCl_2, 160°, 5 min, 80% yield) and cleaved by acidic hydrolysis (HOAc, reflux, 7 h).[4]

1. W. Bradley, R. Robinson, and G. Schwarzenbach, *J. Chem. Soc.*, 793 (1930).
2. S. Bengtsson and T. Högberg, *J. Org. Chem.*, **54,** 4549 (1989).
3. E. Haslam, R. D. Haworth, S. D. Mills, H. J. Rogers, R. Armitage, and T. Searle, *J. Chem. Soc.*, 1836 (1961).
4. L. Jurd, *J. Am. Chem. Soc.*, **81,** 4606 (1959).

Cyclic Esters

48. Cyclic Borate (Chart 4): 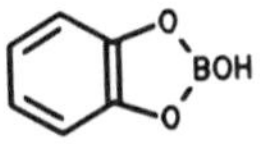

A cyclic borate can be used to protect a catechol group during base-catalyzed alkylation or acylation of an isolated phenol group; the borate ester is then readily hydrolyzed by dilute acid.[1]

Formation

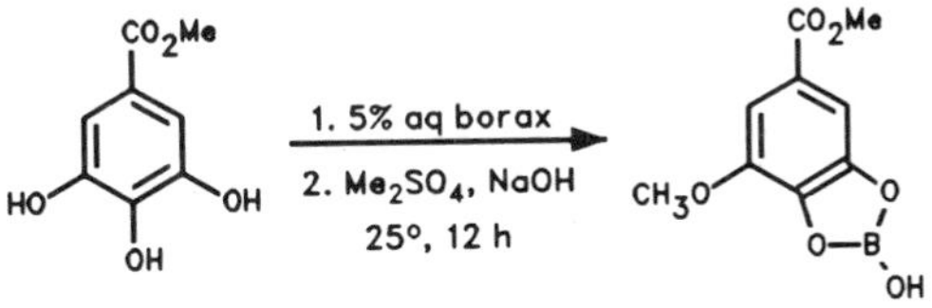

CO_2Me
HO
OH
OH
1. 5% aq borax
2. Me_2SO_4, NaOH
25°, 12 h
CO_2Me
CH_3O
O
O—B
OH

Cleavage

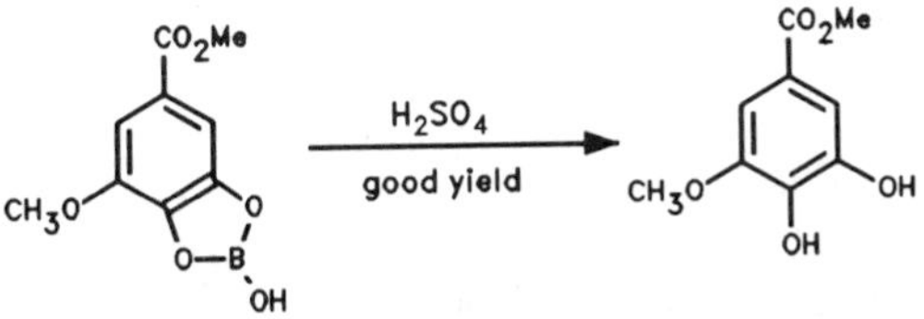

CO_2Me
CH_3O
O
O—B
OH
H_2SO_4
good yield
CO_2Me
CH_3O
OH
OH

1. R. R. Scheline, *Acta Chem. Scand.*, **20,** 1182 (1966).

49. Cyclic Carbonate (Chart 4): 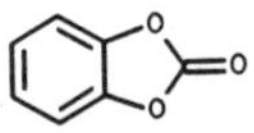

Cyclic carbonates have been used to a limited extent only (since they are readily hydrolyzed) to protect the catechol group in a polyhydroxy benzene.

Formation[1]

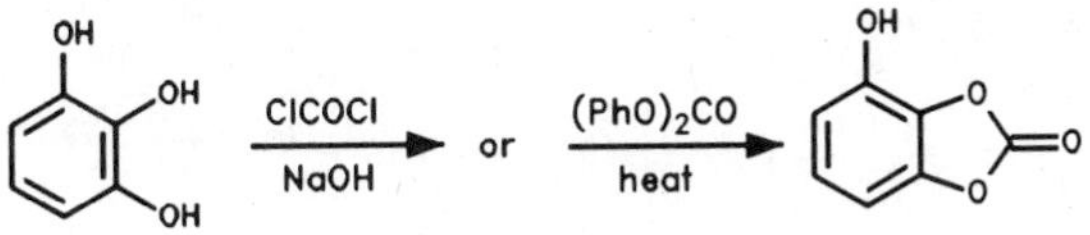

OH
OH
OH
ClCOCl
NaOH
or
$(PhO)_2CO$
heat
OH
O
O
O

Cleavage

The cyclic carbonate is easily cleaved by refluxing in water for 30 min.[2] It can be converted to the 1,2-dimethoxybenzene derivative (aq. NaOH, Me_2SO_4, reflux, 3 h).[3]

1. A. Einhorn, J. Cobliner, and H. Pfeiffer, *Ber.*, **37,** 100 (1904).
2. H. Hillemann, *Ber.*, **71,** 34 (1938).
3. W. Baker, J. A. Godsell, J. F. W. McOmie, and T. L. V. Ulbricht, *J. Chem. Soc.*, 4058 (1953).

PROTECTION FOR 2-HYDROXYBENZENETHIOLS

Two derivatives have been prepared that may prove useful as protective groups for 2-hydroxybenzenethiols.

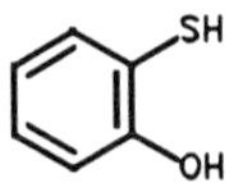

SH
OH

Formation[1,2]

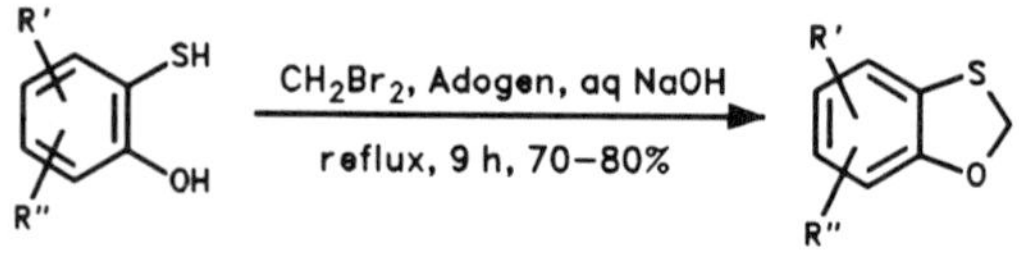

CH_2Br_2, Adogen, aq NaOH
reflux, 9 h, 70–80%

Ref. 1

R′, R″ = H, Me, Cl
Adogen = $MeR_3N^+Cl^-$, phase transfer catalyst
R = C_8–C_{10} straight chain alkyl groups

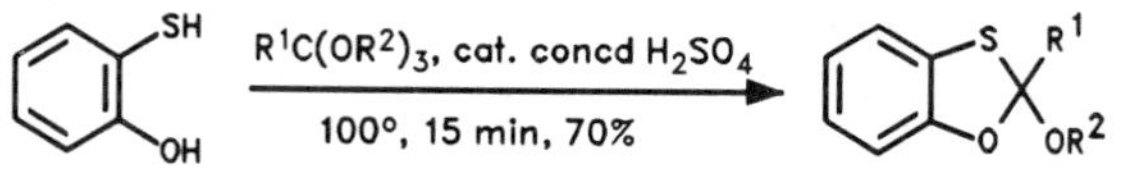

$R^1C(OR^2)_3$, cat. concd H_2SO_4
100°, 15 min, 70%

Ref. 2

R^1 = H, Me, Ph; R^2 = Me, Et

1. S. Cabiddu, A. Maccioni, and M. Secci, *Synthesis*, 797 (1976).
2. S. Cabiddu, S. Melis, L. Bonsignore, and M. T. Cocco, *Synthesis*, 660 (1975).

4

PROTECTION FOR THE CARBONYL GROUP

During a synthetic sequence a carbonyl group may have to be protected against attack by various reagents such as strong or moderately strong nucleophiles, including organometallic reagents; acidic, basic, catalytic, or hydride reducing agents; and some oxidants. Because of the order of reactivity of the carbonyl group [e.g., aldehydes (aliphatic > aromatic) > acyclic ketones and cyclohexanones > cyclopentanones > α,β-unsaturated ketones or α,α-disubstituted ketones >> aromatic ketones], it may be possible to protect a reactive carbonyl group selectively in the presence of a less reactive one. In keto steroids the order of reactivity to ketalization is C_3 or Δ^4-C_3 > C_{17} > C_{12} > C_{20} > $C_{17,21\text{-(OH)}_2}C_{20}$ > C_{11}.[1] A review discusses the relative rates of hydrolysis of acetals, ketals, and ortho esters.[2]

The most useful protective groups are the acyclic and cyclic acetals or ketals, and the acyclic or cyclic thioacetals or ketals. The protective group is introduced by treating the carbonyl compound in the presence of acid with an alcohol, diol, thiol, or dithiol. Cyclic and acyclic acetals and ketals are stable to aqueous and nonaqueous bases, to nucleophiles including organometallic reagents, and to hydride reduction. A 1,3-dithiane or 1,3-dithiolane, prepared to protect an aldehyde, is converted by strong base to an anion. The oxygen derivatives are stable to neutral and basic catalytic reduction, and to reduction by sodium in ammonia. Although the sulfur analogs poison hydrogenation catalysts, they can be cleaved by Raney Ni and by sodium/ammonia.

The oxygen derivatives are stable to most oxidants; the sulfur derivatives are cleaved by a wide range of oxidants. The oxygen, but not the sulfur, analogs are readily cleaved by acidic hydrolysis. Sulfur derivatives are cleaved under neutral conditions by mercury(II), silver(I), or copper(II) salts; oxygen analogs are stable to those conditions. The properties of oxygen and sulfur derivatives are combined in the cyclic 1,3-oxathianes and 1,3-oxathiolanes.

The carbonyl group forms a number of other very stable derivatives. They are less used as protective groups because of the greater difficulty involved in their removal. Such derivatives include cyanohydrins, hydrazones, imines, oximes, and semicarbazones. Enol ethers are used to protect one carbonyl group in a 1,2- or 1,3-dicarbonyl compound.

Although IUPAC no longer uses the term "ketal," we have retained it to indicate compounds formed from ketones.

Derivatives of carbonyl compounds that have been used as protective groups in synthetic schemes are described in this chapter; some of the more important protective groups are listed in Reactivity Chart 5.[3,4]

1. H. J. E. Loewenthal, *Tetrahedron*, **6,** 269 (1959).
2. E. H. Cordes and H. G. Bull *Chem. Rev.*, **74,** 581–603 (1974).
3. See also: H. J. E. Loewenthal, "Protection of Aldehydes and Ketones," in *Protective Groups in Organic Chemistry*, J. F. W. McOmie, Ed., Plenum Press, New York and London, 1973, pp. 323–402.
4. J. F. W. Keana, in *Steroid Reactions*, C. Djerassi, Ed., Holden-Day, San Francisco, 1963, pp. 1–66, 83–87.

ACETALS AND KETALS

Acyclic Acetals and Ketals

Methods similar to those used to form and cleave dimethyl acetal and ketal derivatives can be used for other dialkyl acetals and ketals.

1. Dimethyl Acetals and Ketals: $R_2C(OCH_3)_2$ (Chart 5)

Formation

1. MeOH, dry HCl, 2 min.[1]

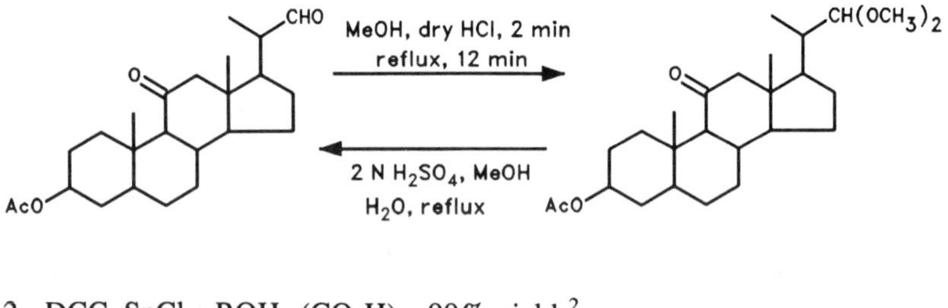

CHO
MeOH, dry HCl, 2 min
reflux, 12 min
$CH(OCH_3)_2$
O
O
2 N H_2SO_4, MeOH
H_2O, reflux
AcO
AcO

2. DCC–$SnCl_4$; ROH, $(CO_2H)_2$, 90% yield.[2]
3. $CH(OMe)_3$, $MeNO_2$, CF_3COOH, reflux, 4 h, 81–93% yield.[3] This procedure was reported to be particularly effective for the preparation of ketals of diaryl ketones.

4. MeOH, $LaCl_3$, $(MeO)_3CH$, 25°, 10 min, 80–100% yield.[4] Dimethyl acetals can be prepared efficiently under neutral conditions by catalysis with lanthanoid halides, but the results of the reaction with ketones are unpredictable.
5. Me_3SiOCH_3, Me_3SiOTf, CH_2Cl_2, −78°, 86% yield.[5] A norbornyl ketone was not ketalized under these conditions.
6. $(MeO)_3CH$, anhydrous MeOH, TsOH, reflux, 2 h.[6] Diethyl ketals have been prepared under similar conditions (EtOH, TsOH, 0–23°, 15 min–6 h, 80–95% yield) in the presence of molecular sieves to shift the equilibrium by adsorbing water.[7] Amberlyst-15[8] or graphite bisulfate[9] and $(EtO)_3CH$ have been used to prepare diethyl ketals.

OEt CHOH —— $(MeO)_3CH$, anhyd MeOH, TsOH, reflux, 2 h ——→ OEt $CH(OMe)_2$

In the following example a mixture of the *cis*- and *trans*-decalones is converted completely to the *cis*-isomer, in general the thermodynamically less favored isomer.[10]

$(MeO)_3CH$, TsOH, MeOH; TsOH, H_2O, acetone; $(MeO)_3CH$, TsOH, CH_2Cl_2

7. MeOH, $(MeO)_4Si$, dry HCl, 25°, 3 days.[11]
8. MeOH, acidic ion-exchange resin, 7–86% yield.[12]
9. $(MeO)_3CH$, Montmorillonite Clay K-10, 5 min–15 h, >90% yield.[13] Diethyl ketals have been prepared in satisfactory yield by reaction of the carbonyl compound and ethanol in the presence of montmorillonite clay.[14]
10. MeOH, NH_4Cl, reflux, 1.5 h, 66% yield.[15]

MeOH, NH_4Cl, reflux, 1.5 h, 66%

11. MeOH, $PhSO_2NHOH$, 25°, 15 min, 75–85% yield.[16]
12. Me_2SO_4, 2 *N* NaOH, MeOH, H_2O, reflux, 30 min, 85% yield.[17] In this case the hemiacetal of phthaldehyde is alkylated with methyl sulfate; this use is probably restricted to cases that are stable to the strongly basic conditions.

Cleavage

1. 50% CF_3COOH, $CHCl_3$, H_2O, 0°, 90 min, 96% yield.[18]

50% TFA, $CHCl_3$–H_2O, 0°, 90 min, 96%

2. TsOH, acetone.[19]
3. SiO_2 and oxalic or sulfuric acid, 0.5–24 h, 90–95% yield.[20]
4. Me_3SiI, CH_2Cl_2, 25°, 15 min, 85–95% yield.[21] Under these cleavage conditions 1,3-dithiolanes, alkyl and trimethylsilyl enol ethers, and enol acetates are stable. 1,3-Dioxolanes give complex mixtures. Alcohols, epoxides, trityl, *t*-butyl, and benzyl ethers and esters are reactive. Most other ethers and esters, amines, amides, ketones, olefins, acetylenes, and halides are expected to be stable.
5. $TiCl_4$, LiI, Et_2O, rt, 3 h, 75–90% yield.[22]
6. $LiBF_4$, wet CH_3CN, 96% yield.[23] Unsubstituted 1,3-dioxolanes are hydrolyzed only slowly, but substituted dioxolanes are completely stable.[23] This reagent proved excellent for hydrolysis of the dimethyl ketal in the presence of the acid-sensitive oxazolidine.[24]

$LiBF_4$, CH_3CN, 2% H_2O, 60°, 30 min, 95%

7. HCO_2H, pentane, 1 h, 20°.[25] Under these conditions a β,γ-olefin does not migrate into conjugation.
8. Amberlyst-15, acetone, H_2O, 20 h.[26] Aldehyde acetals conjugated with

electron-withdrawing groups tend to be slow to hydrolyze. The use of HCl/THF or PPTS/acetone in the case below was slow and caused considerable isomerization. A TBDMS group is stable under these conditions.[27]

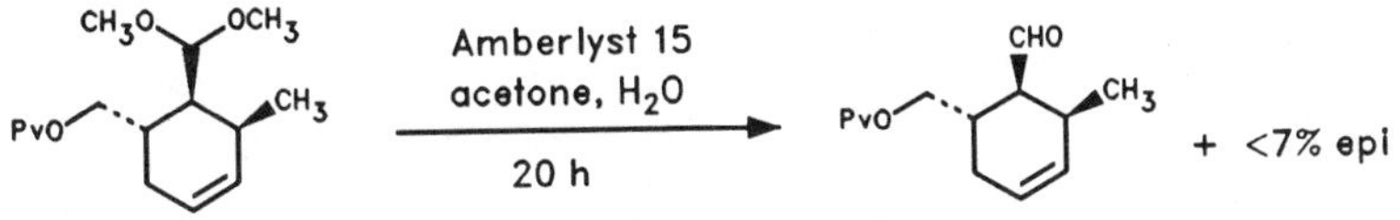

Amberlyst 15, acetone, H_2O, 20 h; + <7% epi

9. 70% H_2O_2, Cl_3CCO_2H, CH_2Cl_2, *t*-BuOH; dimethyl sufide, 80% yield.[28] Other methods cleaved the epoxide. This method also cleaves aliphatic THP and trityl ethers.

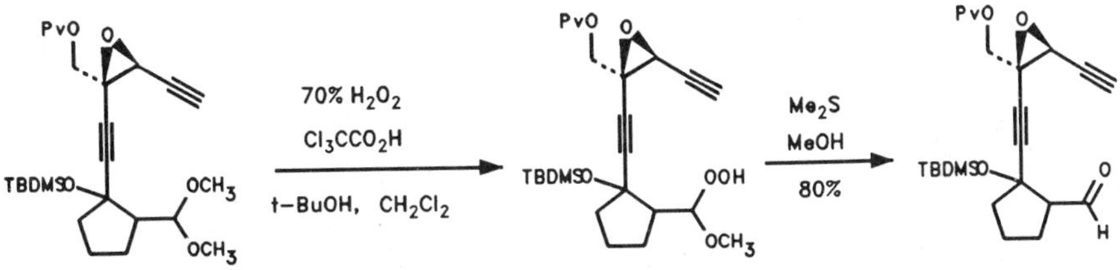

70% H_2O_2, Cl_3CCO_2H, t-BuOH, CH_2Cl_2; Me_2S, MeOH, 80%

10. CF_3COOH, rt; $NaHCO_3$, 98% yield.[29]

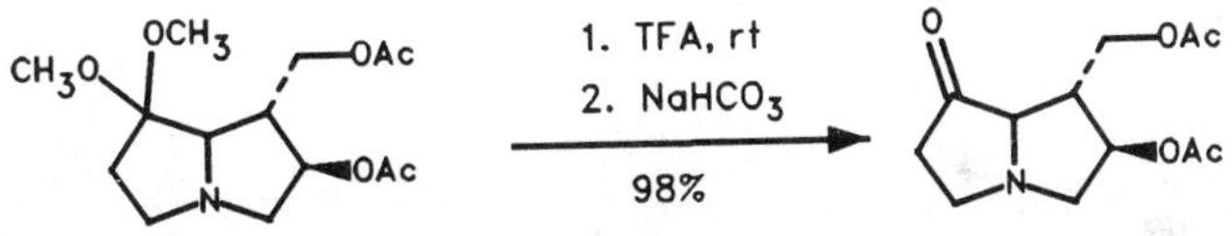

1. TFA, rt 2. $NaHCO_3$ 98%

11. AcOH, H_2O, 89% yield.[30] A factor of 400 in the relative rate of hydrolysis is attributed to a conformational effect where the lone pair on oxygen in the silyl ketals does not overlap with the incipient cation during hydrolysis.

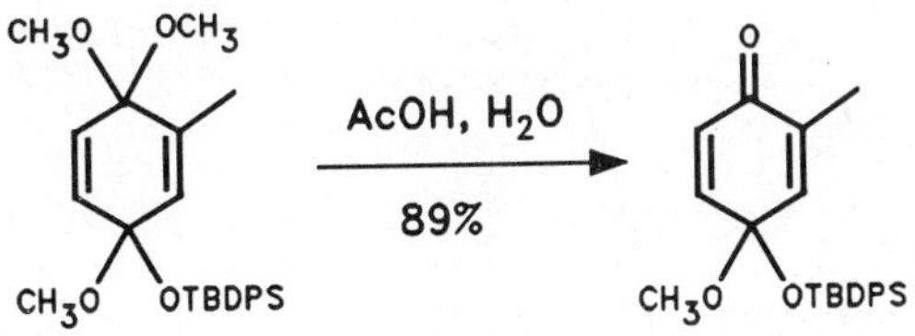

AcOH, H_2O 89%

12. Oxalic acid, THF, H_2O, rt, 12 min, 72% yield.[31]

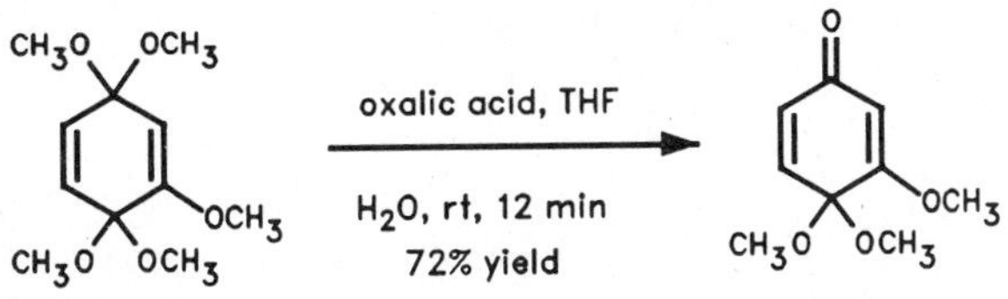

oxalic acid, THF, H_2O, rt, 12 min, 72% yield

13. $BF_3 \cdot Et_2O$, $Et_4N^+I^-$, $CHCl_3$, 69–82% yield.[32]
14. 10% H_2O, silica gel, CH_2Cl_2, 18 h, rt.[33] In this example attempts to use HCl resulted in THP cleavage followed by cyclization to form a furan.

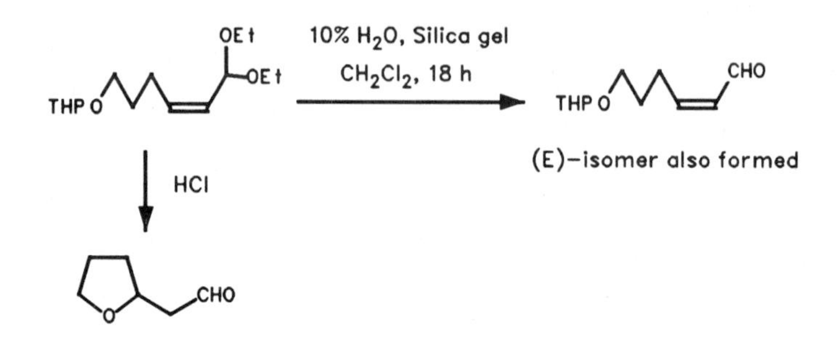

15. DMSO, H_2O, dioxane, reflux, 12 h, 65–99% yield.[34] These conditions cleave a dimethyl ketal in the presence of a *t*-butyldimethylsilyl ether.

1. A. F. B. Cameron, J. S. Hunt, J. F. Oughton, P. A. Wilkinson, and B. M. Wilson, *J. Chem. Soc.*, 3864 (1953).
2. N. H. Andersen and H.-S. Uh, *Synth. Commun.*, **3,** 125 (1973).
3. A. Thurkauf, A. E. Jacobson, and K. C. Rice, *Synthesis*, 233 (1988).
4. A. L. Gemal and J.-L. Luche, *J. Org. Chem.*, **44,** 4187 (1979).
5. M. Vandewalle, J. Van der Eycken, W. Oppolzer, and C. Vullioud, *Tetrahedron*, **42,** 4035 (1986).
6. E. Wenkert and T. E. Goodwin, *Synth. Commun.*, **7,** 409 (1977).
7. D. P. Roelofsen, E. R. J. Wils, and H. Van Bekkum, *Recl. Trav. Chim. Pays-Bas*, **90,** 1141 (1971).
8. S. A. Patwardhan and S. Dev, *Synthesis*, 348 (1974).
9. J. P. Alazard, H. B. Kagan, and R. Setton, *Bull. Soc. Chim. Fr.*, 499 (1977).
10. J. B. P. A. Wijnberg, R. P. W. Kesselmans, and A. de Groot, *Tetrahedron Lett.*, **27,** 2415 (1986).
11. W. W. Zajac and K. J. Byrne, *J. Org. Chem.*, **35,** 3375 (1970).
12. N. B. Lorette, W. L. Howard, and J. H. Brown, Jr., *J. Org. Chem.*, **24,** 1731 (1959).
13. E. C. Taylor and C.-S. Chiang, *Synthesis*, 467 (1977). Montmorillonite clay is activated $Al_2O_3/SiO_2/H_2O$.
14. V. M. Thuy and P. Maitte, *Bull. Soc. Chim. Fr.*, 2558 (1975).
15. J. I. DeGraw, L. Goodman, and B. R. Baker, *J. Org. Chem.*, **26,** 1156 (1961).
16. A. Hassner, R. Wiederkehr, and A. J. Kascheres, *J. Org. Chem.*, **35,** 1962 (1970).
17. E. Schmitz, *Chem. Ber.*, **91,** 410 (1958).
18. R. A. Ellison, E. R. Lukenbach, and C.-W. Chiu, *Tetrahedron Lett.*, 499 (1975).
19. E. W. Colvin, R. A. Raphael, and J. S. Roberts, *J. Chem. Soc., Chem. Commun.*, 858 (1971).
20. F. Huet, A. Lechevallier, M. Pellet, and J. M. Conia, *Synthesis*, 63 (1978).
21. M. E. Jung, W. A. Andrus, and P. L. Ornstein, *Tetrahedron Lett.*, 4175 (1977).

22. G. Balme and J. Goré, *J. Org. Chem.*, **48,** 3336 (1983).
23. B. H. Lipshutz and D. F. Harvey, *Synth. Commun.*, **12,** 267 (1982).
24. M. Bonin, J. Royer, D. S. Grierson, and H.-P. Husson, *Tetrahedron Lett.*, **27,** 1569 (1986).
25. F. Barbot and P. Miginiac, *Synthesis*, 651 (1983).
26. G. M. Cappola, *Synthesis*, 1021 (1984).
27. A. E. Greene, M. A. Teixeira, E. Barreiro, A. Cruz, and P. Crabbé, *J. Org. Chem.*, **47,** 2553 (1982).
28. A. G. Meyers, M. A. M. Fundy, and P. A. Linstrom, Jr., *Tetrahedron Lett.*, **29,** 5609 (1988).
29. J. J. Tufariello and K. Winzenberg, *Tetrahedron Lett.*, **27,** 1645 (1986).
30. A. J. Stern and J. S. Swenton, *J. Org. Chem.*, **54,** 2953 (1989).
31. D. A. Evans, S. P. Tanis, and D. J. Hart, *J. Am. Chem. Soc.*, **103,** 5813 (1981).
32. A. K. Mandal, P. Y. Shrotri, and A. D. Ghogare, *Synthesis*, 221 (1986).
33. L. Crombie and D. Fisher, *Tetrahedron Lett.*, **26,** 2477 (1985).
34. T. Kametani, H. Kondoh, T. Honda, H. Ishizone, Y. Suzuki, and W. Mori, *Chem. Lett.*, 901 (1989).

2. Bis(2,2,2-trichloroethyl) Acetals and Ketals: $R_2C(OCH_2CCl_3)_2$ (Chart 5)

Formation[1]

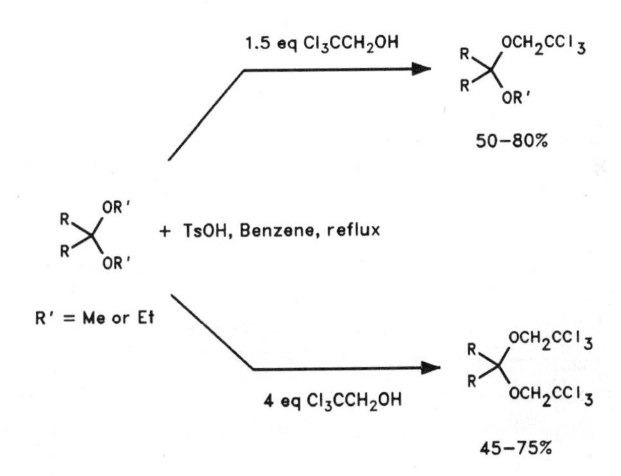

Cleavage

1. Zn/EtOAc or THF, reflux, 3–12 h, 40–100% yield.[1] It is more efficient to prepare this ketal by an exchange reaction with the dimethyl or diethyl ketal than directly from the carbonyl compound. Hydrolysis can also be effected by acid catalysis.

1. J. L. Isidor and R. M. Carlson, *J. Org. Chem.*, **38,** 554 (1973).

3. Dibenzyl Acetals and Ketals: $R_2C(OCH_2Ph)_2$

Formation/Cleavage[1]

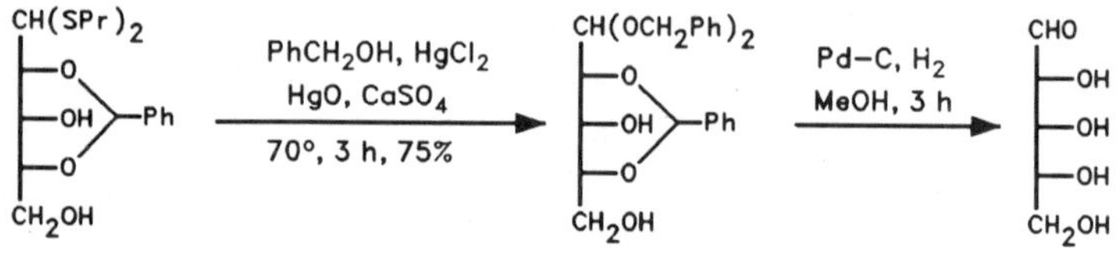

$CH(SPr)_2$ … CH_2OH (O,O-benzylidene, Ph) → [$PhCH_2OH$, $HgCl_2$, HgO, $CaSO_4$; 70°, 3 h, 75%] → $CH(OCH_2Ph)_2$ … CH_2OH (O,O-benzylidene, Ph) → [Pd–C, H_2; MeOH, 3 h] → CHO, OH, OH, OH, CH_2OH

1. J. H. Jordaan and W. J. Serfontein, *J. Org. Chem.*, **28,** 1395 (1963).

4. Bis(2-nitrobenzyl) Acetals and Ketals: $R_2C(OCH_2C_6H_4\text{-}2\text{-}NO_2)_2$

Formation

1. $2\text{-}NO_2C_6H_4CH_2OSiMe_3$, Me_3SiOTf, −78°, 78–95% yield.[1]

Cleavage

1. Photolysis at 350 nm, 85–95% yield.[1]

1. D. Gravel, S. Murray, and G. Ladouceur, *J. Chem. Soc., Chem. Commun.*, 1828 (1985).

5. Diacetyl Acetals and Ketals: $R_2C(OAc)_2$

Formation

1. Ac_2O, 1 drop concd. H_2SO_4, 20°, 1 h, 95% yield.[1]
2. Ac_2O, $FeCl_3$, rt, <30 min, 60–93% yield.[2] These conditions will selectively protect an aldehyde in the presence of a ketone.[3] This combination also converts *t*-butyldimethylsilyl (TBDMS) ethers to acetates.
3. Ac_2O, PCl_3, 20°, 1–24 h, 30–90% yield.[4] Aromatic aldehydes bearing electron-withdrawing groups tend to give low yields under these conditions.
4. Ac_2O, Nafion H, 50–99% yield.[5]

Cleavage

1. NaOH or K_2CO_3, THF, H_2O or MeOH.[2] This protective group is stable to MeOH (18 h); 10% HCl (MeOH, 30 min); 10% Na_2CO_3 (H_2O, Et_2O, 70 min); $NaHCO_3$ (THF, H_2O, 4 h).

1. M. Tomita, T. Kikuchi, K. Bessho, T. Hori, and Y. Inubushi, *Chem. Pharm. Bull.*, **11,** 1484 (1963).
2. K. S. Kochhar, B. S. Bal, R. P. Deshpande, S. N. Rajadhyaksha, and H. W. Pinnick, *J. Org. Chem.*, **48,** 1765 (1983).
3. J. Kula, *Synth. Commun.*, **16,** 833 (1986).
4. J. K. Michie and J. A. Miller, *Synthesis*, 824 (1981).
5. G. A. Olah and A. K. Mehrotra, *Synthesis*, 962 (1982).

Cyclic Acetals and Ketals

Kinetic studies of acetal/ketal formation from cyclohexanone, and hydrolysis (3×10^{-3} N HCl/dioxane–H_2O, 20°), indicate the following orders of reactivity:[1]

Formation

$$HOCH_2C(CH_3)_2CH_2OH > HO(CH_2)_2OH > HO(CH_2)_3OH$$

Cleavage

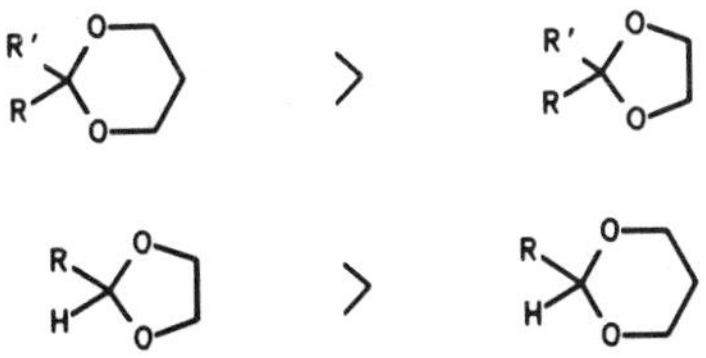

R′, R-1,3-dioxane > R′, R-1,3-dioxolane

R, H-1,3-dioxolane > R, H-1,3-dioxane

The relative rate of acid-catalyzed hydrolysis of some dioxolanes [dioxolane : aq. HCl (1 : 1)] are: 2,2-dimethyldioxolane : 2-methyldioxolane : dioxolane, 50,000: 5000 : 1.[2]

A review[3] discusses the condensation of aldehydes and ketones with glycerol to give 1,3-dioxanes and 1,3-dioxolanes.

1. M. S. Newman and R. J. Harper, *J. Am. Chem. Soc.*, **80,** 6350 (1958); S. W. Smith and M. S. Newman, *J. Am. Chem. Soc.*, **90,** 1249, 1253 (1968).
2. P. Salomaa and A. Kankaanperä, *Acta Chem. Scand.*, **15,** 871 (1961).
3. A. J. Showler and P. A. Darley, *Chem. Rev.*, **67,** 427–440 (1967).

6. 1,3-Dioxanes (Chart 5):

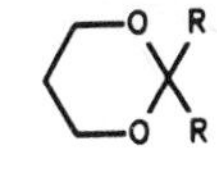

Formation

1. $HO(CH_2)_3OH$, TsOH, benzene, reflux.[1–3]

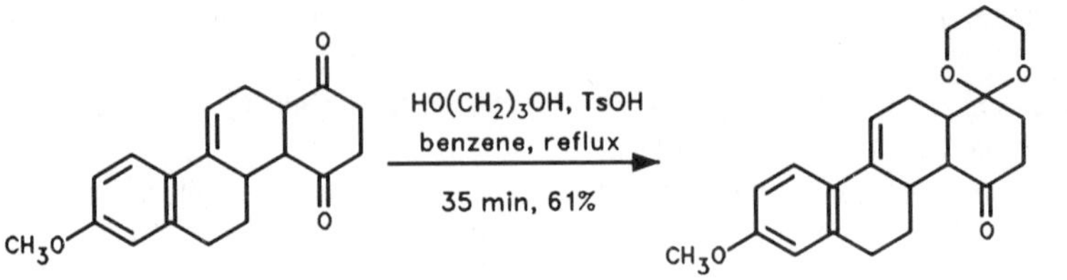

$HO(CH_2)_3OH$, TsOH, benzene, reflux, 35 min, 61% — Ref. 1

TsOH, 91% — Ref. 2

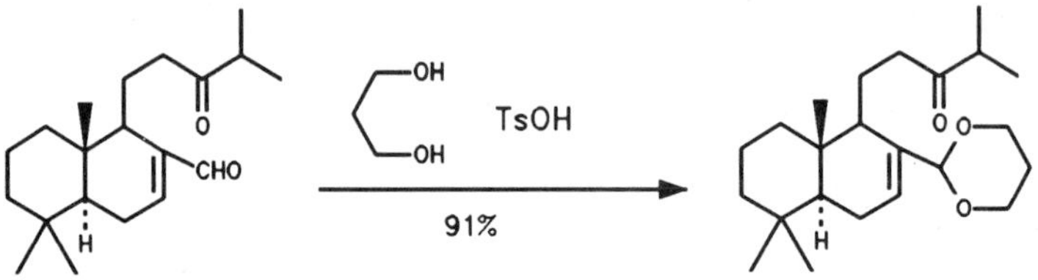

In the first example selective protection was more successful with 1,3-propanediol than with ethylene glycol.[1]

2. 1,3-Propanediol, THF, Amberlyst-15, 5 min, 50–70% yield.[4] This method is also effective for the preparation of 1,3-dioxolanes.

Cleavage

The section on the cleavage of 1,3-dioxolanes (Section 10, below) should be consulted.

1. J. E. Cole, W. S. Johnson, P. A. Robins, and J. Walker, *J. Chem. Soc.*, 244 (1962).
2. H. Okawara, H. Nakai, and M. Ohno, *Tetrahedron Lett.*, **23,** 1087 (1982).
3. For examples on the use of the related 4,4-dimethyl-1,3-dioxane, see: E. Piers, J. Banville, C. K. Lau, and I. Nagakura, *Can. J. Chem.*, **60,** 2965 (1982); M. A. Avery, C. Jennings-White, and W. K. M. Chong, *Tetrahedron Lett.*, **28,** 4629 (1987).
4. A. E. Dann, J. B. Davis, and M. J. Nagler, *J. Chem. Soc., Perkin Trans. I*, 158 (1979).

7. **5-Methylene-1,3-dioxane** (Chart 5):

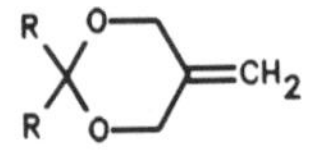

Formation[1]

1. $CH_2{=}C(CH_2OH)_2$, TsOH, benzene, reflux, 90% yield.

Cleavage[1]

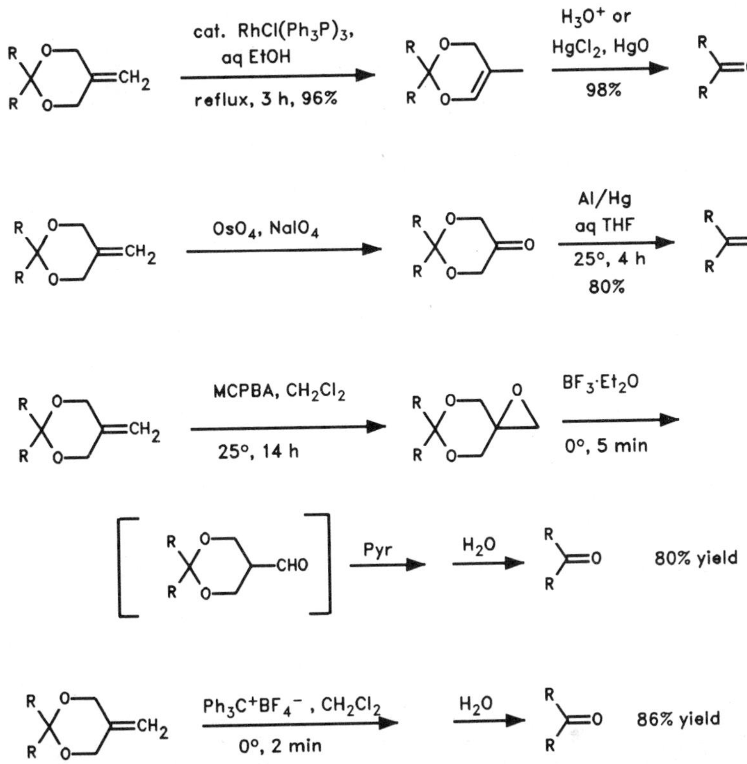

cat. $RhCl(Ph_3P)_3$, aq EtOH, reflux, 3 h, 96%; H_3O^+ or $HgCl_2$, HgO, 98%

OsO_4, $NaIO_4$; Al/Hg, aq THF, 25°, 4 h, 80%

MCPBA, CH_2Cl_2, 25°, 14 h; $BF_3 \cdot Et_2O$, 0°, 5 min; [CHO]; Pyr; H_2O; 80% yield

$Ph_3C^+BF_4^-$, CH_2Cl_2, 0°, 2 min; H_2O; 86% yield

1. E. J. Corey and J. W. Suggs, *Tetrahedron Lett.*, 3775 (1975).

8. 5,5-Dibromo-1,3-dioxane (Chart 5):

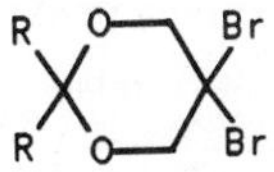

Formation

1. $Br_2C(CH_2OH)_2$, TsOH, benzene, heat for several hours, 84–94% yield.[1]

Cleavage

1. Zn–Ag, THF, AcOH, 25°, 1 h, ~90% yield.[1]

1. E. J. Corey, E. J. Trybulski, and J. W. Suggs, *Tetrahedron Lett.*, 4577 (1976).

9. 5-(2-Pyridyl)-1,3-dioxane:

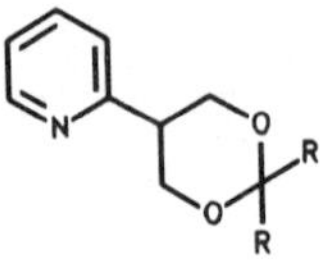

Formation/Cleavage[1]

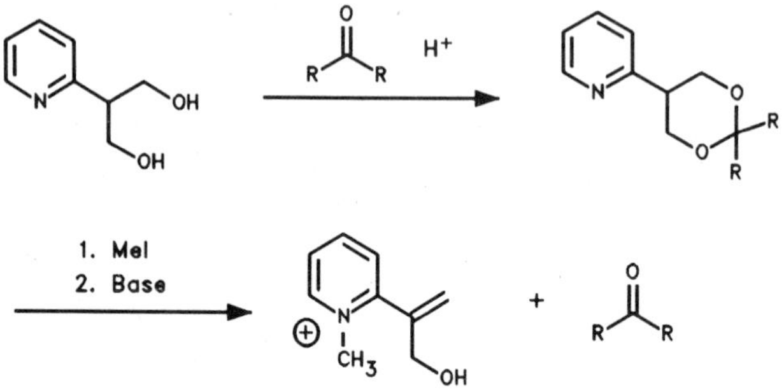

This group is stable to 0.1 M HCl.

1. A. R. Katritzky, W.-Q. Fan, and Q.-L. Li, *Tetrahedron Lett.*, **28**, 1195 (1987).

10. 1,3-Dioxolanes (Chart 5):

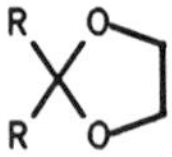

The 1,3-dioxolane group is probably the most widely used carbonyl protective group. For the protection of carbonyls containing other acid-sensitive functionality, one should use acids of low acidity or pyridinium salts. In general, a molecule containing two similar ketones can be selectively protected at the less hindered carbonyl, assuming that neither or both of the carbonyls are conjugated to an alkene.[1]

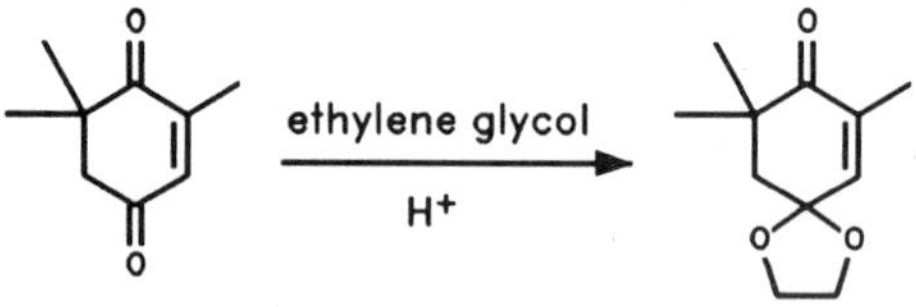

ethylene glycol, H+ — Ref. 1b

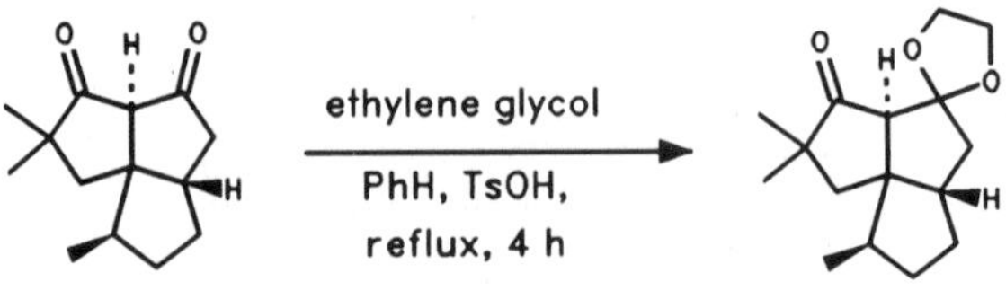

ethylene glycol, PhH, TsOH, reflux, 4 h — Ref. 1a

If one carbonyl is conjugated with an olefin, the unconjugated carbonyl is selectively protected. This generalization appears to be independent of ring size.[2] Simple aldehydes are generally selectively protected over simple ketones.[3] In the formation of 1,3-dioxolanes of enones, control of the olefin regiochemistry is determined by the acidity of the acid catalyst. Acids of high acidity ($pK_a \sim 1$) may cause the double bond to migrate to the β,γ-position whereas acids of low acidity ($pK_a \sim 3$) do not cause olefin isomerization (see Table 1).[4] In addition, the use of the bistrimethylsilyl derivative of ethylene glycol and Me_3SiOTf (CH_2Cl_2, −78°, 20 h, Pyr quench, 92%) for the protection of enones proceeds without double bond migration.[5,6]

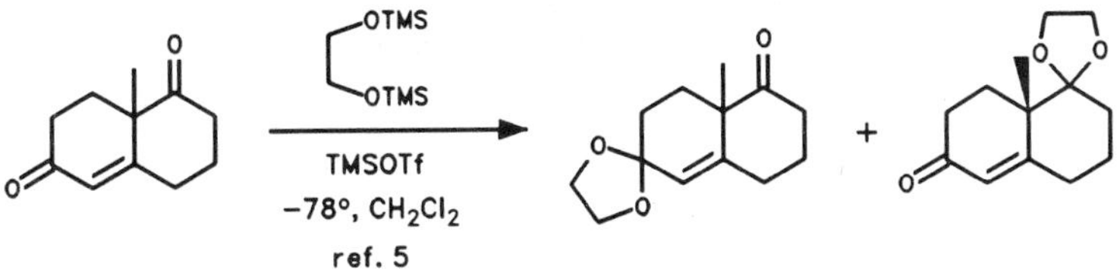

OTMS / OTMS, TMSOTf, −78°, CH_2Cl_2, ref. 5

Ratio = 27:1

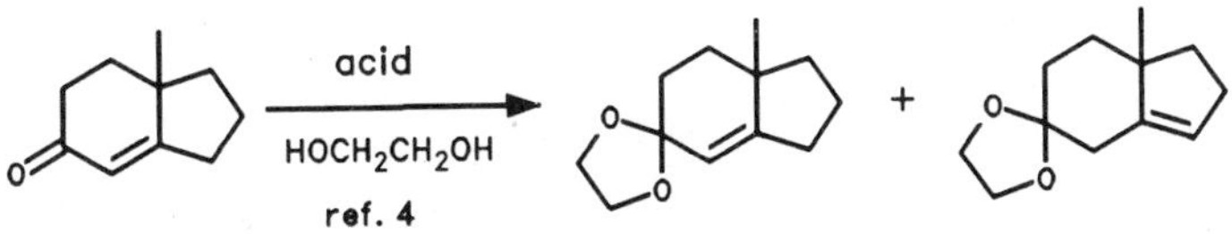

acid, $HOCH_2CH_2OH$, ref. 4

Table 1[4] Olefin Isomerization

Acid	pK_a	Percent α,β	Percent β,γ	Percent conversion
Fumaric acid	3.03	100	0	90
Phthalic acid	2.89	70	30	90
Oxalic acid[7]	1.23	80	20	93
TsOH acid	<1.0	0	100	100

A polymer-supported 1,2-diol has also been developed.[8]

Formation

1. $HO(CH_2)_2OH$, TsOH, C_6H_6, reflux, 75–85% yield.[9]
2. $HO(CH_2)_2OH$, TsOH, $(EtO)_3CH$, 25°, 65% yield.[10]
3. $HO(CH_2)_2OH$, $BF_3 \cdot Et_2O$, HOAc, 35–40°, 15 min, 90% yield.[11]
4. $HO(CH_2)_2OH$, HCl, 25°, 12 h, 55–90% yield.[12]
5. $HO(CH_2)_2OH$, Me_3SiCl, MeOH or CH_2Cl_2.[13]

6. $HO(CH_2)_2OH$, Al_2O_3, $PhCH_3$ or CCl_4, heat, 24 h, 80–100% yield.[3] These conditions are selective for the formation of acetals from aldehydes in the presence of ketones.
7. $Me_3SiOCH_2CH_2OSiMe_3$, Me_3SiOTf, 15 kbar (1.5 GPa), 40°, 48 h.[14] These conditions were used to prepare the ketal of fenchone, which cannot be done under normal acid-catalyzed conditions.
8. $HO(CH_2)_2OH$, 0.1 eq. $CuCl_2 \cdot H_2O$, 80°, 30 min, 82–100% yield.[15] The use of 5 eq. of $CuCl_2$ results in the formation of the α-chloro ketal.

OH OH
0.1 eq $CuCl_2$, 80°
89%
5 eq $CuCl_2$
2 h, 80°
OH OH
Cl
80%

9. $HO(CH_2)_2OH$, oxalic acid, CH_3CN, 25°, 95% yield.[16] Note that ketals prepared with oxalic acid from enones retain the olefin regiochemistry.[7]
10. $HO(CH_2)_2OH$, adipic acid, C_6H_6, reflux, 17–24 h, 10–85% yield.[17]
11. $HO(CH_2)_2OH$, SeO_2, $CHCl_3$, 28°, 4 h, 60% yield.[18]
12. $HO(CH_2)_2OH$, $C_5H_5N^+H\ Cl^-$, C_6H_6, reflux, 6 h, 85% yield.[19]
13. $HO(CH_2)_2OH$, $C_5H_5N^+H\ TsO^-$, C_6H_6, reflux, 1–3 h, 90–95% yield.[20]
14. $HO(CH_2)_nOH$ ($n = 2, 3$)/$MeOCH^+NMe_2\ MeOSO_3^-$, 0–25°, 2 h, 40–95% yield.[21]
15. $HO(CH_2)_nOH$ ($n = 2, 3$)/column packed with an acid ion-exchange resin, 5 min, 50–90% yield.[22]
16. $HOCH_2CH_2OH$, $(EtO)_3CH$, *p*-TsOH, 83% yield.[23]
17. 2-Methoxy-1,3-dioxolane/TsOH, C_6H_6, 40–50°, 4 h, 85% yield.[24]
18. 2-Ethoxy-1,3-dioxolane, pyridinium tosylate (PPTS), benzene, heat, 8 h, 89% yield.[25] In this case protection of an enone proceeds without olefin isomerization.
19. 2-Ethyl-2-methyl-1,3-dioxolane/TsOH, reflux, 75% yield.[26,27] These conditions selectively protect a ketone in the presence of an enone.
20. 2-Dimethylamino-1,3-dioxolane/cat. HOAc, CH_2Cl_2, 83% yield.[28] 2-Dimethylamino-1,3-dioxolane protects a reactive ketone under mild conditions: it reacts selectively with a C_3-keto steroid in the presence of a Δ^4-3-keto steroid. C_{12}- and C_{20}-keto steroids do not react.

21. Diethylene orthocarbonate, $C(-OCH_2CH_2O-)_2$/TsOH or wet $BF_3 \cdot Et_2O$, $CHCl_3$, 20°, 70–95% yield.[29]
22. 1,3-Dioxolanes have been prepared from a carbonyl compound and an epoxide (e.g., ketone/$SnCl_4$, CCl_4, 20°, 4 h, 53% yield[30] or aldehyde/$Et_4N^+Br^-$, 125–220°, 2–4 h, 20–85% yield[31]). Perhalo ketones can be protected by reaction with ethylene chlorohydrin under basic conditions (K_2CO_3, pentane, 25°, 2 h, 85% yield[32] or NaOH, EtOH—H_2O, 95% yield[33]).
23. When the carbonyl group is very electron-deficient, thus stabilizing the hemiacetal, a dioxolane can be prepared under basic conditions.[32,34]
24. $HOCH_2CH_2OH$, $(i\text{-PrO})_3CH$, $RhCl_3$(*triphos*), [*triphos* = $H_3CC(CH_2PPh_2)_3$], rt → reflux, 80–100% yield.[35]

Cleavage

1,3-Dioxolanes can be cleaved by acid-catalyzed exchange dioxolanation, acid-catalyzed hydrolysis, or oxidation. Some representative examples are shown below.

1. Pyridinium tosylate (PPTS), acetone, H_2O, heat, 100% yield.[36]

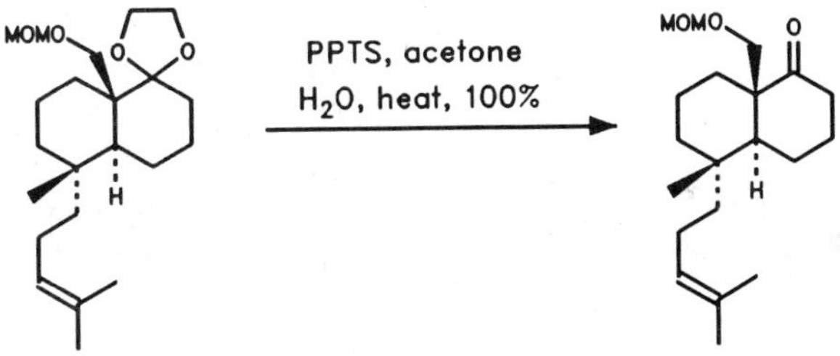

MOMO ... PPTS, acetone H_2O, heat, 100% ... MOMO

2. Acetone, TsOH, 20°, 12 h.[37] The reactant is a 3,6,17-tris(ethylenedioxy) steroid; the product has carbonyl groups at C-6 and C-17.
3. Acetone, H_2O, PPTS, reflux, 1–3 h, 90–95% yield.[20]
4. 5% HCl, THF, 25°, 20 h.[38]

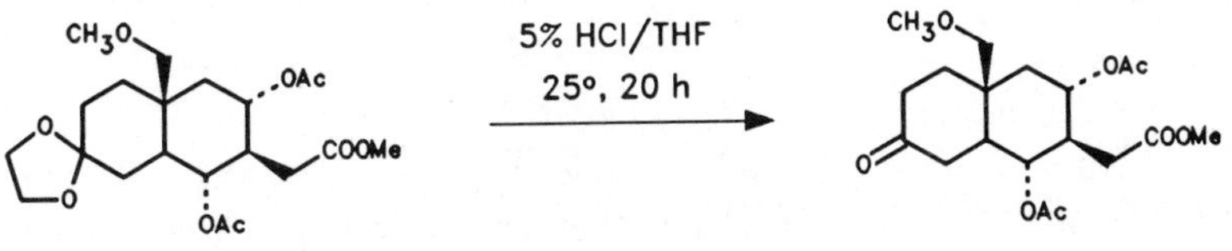

CH_3O ... OAc ... COOMe ... OAc — 5% HCl/THF, 25°, 20 h → CH_3O ... OAc ... COOMe ... OAc

5. 1 *M* HCl, THF, 0° → 25°, 13 h, 71% yield. Note that the acetonide survives these conditions.[39] Some variations have been reported in this system (including the use of 30% AcOH, 90°, high yield).[40]

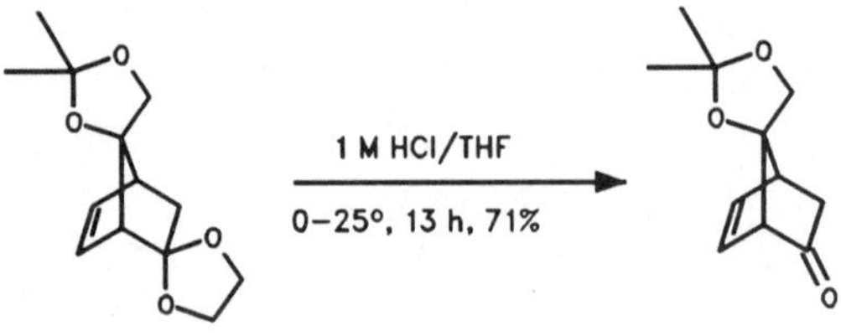

1 M HCl/THF

0–25°, 13 h, 71%

6. 80% AcOH, 65°, 5 min, 85% yield.[41]
7. Wet magnesium sulfate (C_6H_6, 20°, 1 h) effects selective, quantitative cleavage of an α,β-unsaturated 1,3-dioxolane in the presence of a 1,3-dioxolane.[17]
8. Perchloric acid (79% $HClO_4/CH_2Cl_2$, 0°, 1 h → 25°, 3 h, 87% yield)[42] and periodic acid (aq. dioxane, 3 h, quant. yield)[43] cleave 1,3-dioxolanes; the latter drives the reaction to completion by oxidation of the ethylene glycol that forms. Yields are substantially higher from cleavage with perchloric acid (3 *N* $HClO_4$/THF, 25°, 3 h, 80% yield) than with hydrochloric acid (HCl/HOAc, 65% yield).[44]
9. SiO_2, H_2O, CH_2Cl_2, oxalic acid, 90–95% yield.[45] These conditions selectively cleave α,β-unsaturated ketals.

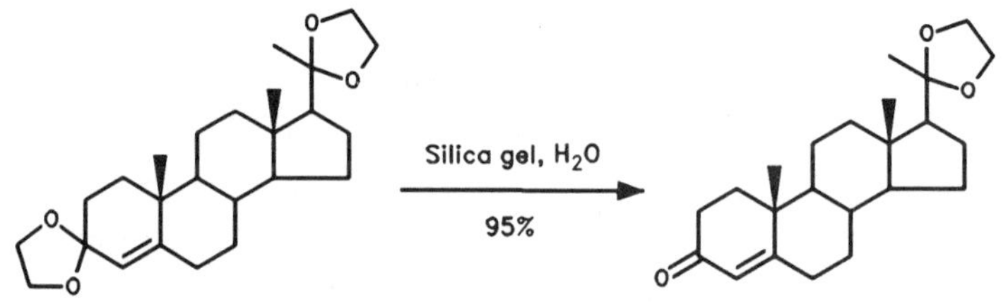

Silica gel, H_2O

95%

10. $Ph_3C^+BF_4^-$, CH_2Cl_2, 25°, 60–100% yield.[46,47]

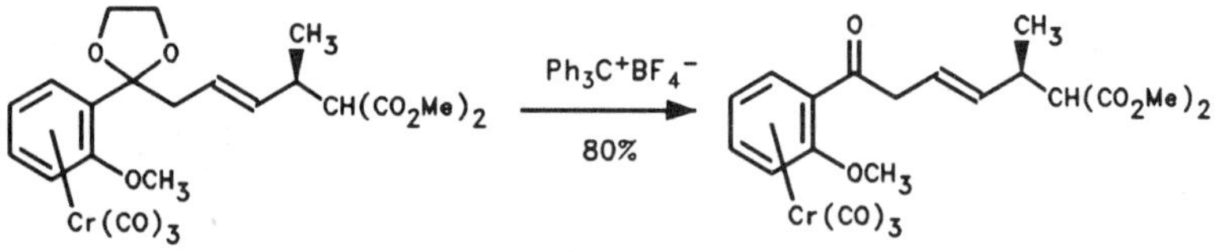

CH_3 $CH(CO_2Me)_2$ OCH_3 $Cr(CO)_3$ $Ph_3C^+BF_4^-$ 80% CH_3 $CH(CO_2Me)_2$ OCH_3 $Cr(CO)_3$

Ref. 47

1,3-Dithiolanes are not affected by these conditions, but a 1,3-oxathiolane is cleaved (100% yield).[48]

11. Me_2BBr, CH_2Cl_2, −78°, 90–97% yield.[49] This reagent also cleaves MTM, MEM, and MOM ethers (87–95% yield).
12. $PdCl_2(CH_3CN)_2$, acetone, H_2O, 82–100% yield.[50]
13. Me_3SiI.[51]

14. *t*-BuOOH, Pd(OOCCF$_3$)(OO-*t*-Bu), benzene, 50°, 12 h, 60–80% yield.[52] In this case an acetal is oxidized to the ester of ethylene glycol ($RCO_2CH_2CH_2OH$).
15. $LiBF_4$, wet CH_3CN.[53] Unsubstituted 1,3-dioxolanes are cleaved slowly under these conditions (40% in 5 h). The 4,5-dimethyl- and 4,4,5,5-tetramethyldioxolane and 1,3-dioxane are inert under these conditions. Dimethyl ketals are readily cleaved.
16. $TiCl_4$, Et_2O, LiI, rt, 61–91% yield.[54] A THP ether was stable to these conditions, but methyl ethers can be cleaved.
17. AlI_3, CH_3CN, benzene, 10 min, 70–92% yield.[55] Ethyl ketals are cleaved under these conditions, but thioketals are not affected.
18. Dimethyl sulfoxide, 180°, H_2O, 10 h, 89% yield.[56] A diethyl acetal can be cleaved in the presence of a 1,3-dioxolane under these conditions. TBDMS, THP, and MOM groups are stable.
19. NaTeH, EtOH, 25°, 30 min; air, 80–85% yield.[57]
20. H_2SiI_2, $CDCl_3$, −42°, 1–10 min, 100% yield.[58] Aromatic ketals are cleaved faster than the corresponding aliphatic derivatives, and cyclic ketals are cleaved more slowly than the acyclic analogues such as dimethyl ketals. Substituted ketals such as those derived from butane-2,3-diol, which react only slowly with Me_3SiI, can also be cleaved with H_2SiI_2. If the reaction is run at 22°, ketals and acetals are reduced to iodides in excellent yield.

1. For two examples, see: (a) M. T. Crimmins and J. A. DeLoach, *J. Am. Chem. Soc.*, **108,** 800 (1986); (b) M. G. Constantino, P. M. Donate, and N. Petragnani, *J. Org. Chem.*, **51,** 253 (1986).
2. For a variety of examples with varying ring sizes, see: Y. Ohtsuka and T. Oishi, *Tetrahedron Lett.*, **27,** 203 (1986); C. Iwata, Y. Takemoto, M. Doi, and T. Imanishi, *J. Org. Chem.*, **53,** 1623 (1988); S. D. Burke, C. W. Murtiashaw, J. O. Saunders, and M. S. Dike, *J. Am. Chem. Soc.*, **104,** 872 (1982); P. A. Wender, M. A. Eisenstat, and M. P. Filosa, *J. Am. Chem. Soc.*, **101,** 2196 (1979); A. A. Devreese, P. J. de Clercq, and M. Vandewalle, *Tetrahedron Lett.*, **21,** 4767 (1980); P. G. Baraldi, A. Barco, S. Benetti, G. P. Pollini, E. Polo, and D. Simoni, *J. Org. Chem.*, **50,** 23 (1985); M. P. Bosch, F. Camps, J. Coll, A. Guerrero, T. Tatsuoka, and J. Meinwald, *J. Org. Chem.*, **51,** 773 (1986).
3. Y. Kamitori, M. Hojo, R. Masuda, and T. Yoshida, *Tetrahedron Lett.*, **26,** 4767 (1985).
4. J. W. De Leeuw, E. R. De Waard, T. Beetz, and H. O. Huisman, *Recl. Trav. Chim. Pays-Bas*, **92,** 1047 (1973).
5. J. R. Hwu and J. M. Wetzel, *J. Org. Chem.*, **50,** 3946 (1985); J. R. Hwu, L.-C. Leu, J. A. Robl, D. A. Anderson, and J. M. Wetzel, *J. Org. Chem.*, **52,** 188 (1987).
6. T. Tsunoda, M. Suzuki, and R. Noyori, *Tetrahedron Lett.*, **21,** 1357 (1980).
7. G. H. Posner and G. L. Loomis, *Tetrahedron Lett.*, 4213 (1978).

8. P. Hodge and J. Waterhouse, *J. Chem. Soc., Perkin Trans. I*, 2319 (1983); Z. H. Xu, C. R. McArthur, and C. C. Leznoff, *Can. J. Chem.*, **61,** 1405 (1983).
9. R. A. Daignault and E. L. Eliel, *Org. Synth., Collect. Vol. V*, 303 (1973).
10. F. F. Caserio, Jr., and J. D. Roberts, *J. Am. Chem. Soc.*, **80,** 5837 (1958).
11. L. F. Fieser and R. Stevenson, *J. Am. Chem. Soc.*, **76,** 1728 (1954).
12. E. G. Howard and R. V. Lindsey, *J. Am. Chem. Soc.*, **82,** 158 (1960).
13. T. H. Chan, M. A. Brook, and T. Chaly, *Synthesis*, 203 (1983).
14. W. G. Dauben, J. M. Gerdes, and G. C. Look, *J. Org. Chem.*, **51,** 4964 (1986); H. Eibisch, *Z. Chem.*, **26,** 375 (1986).
15. J. Y. Satoh, C. T. Yokoyama, A. M. Haruta, K. Nishizawa, M. Hirose, and A. Hagitani, *Chem. Lett.*, 1521 (1974).
16. N. H. Anderson and H.-S. Uh, *Synth. Commun.*, **3,** 125 (1973).
17. J. J. Brown, R. H. Lenhard, and S. Bernstein *J. Am. Chem. Soc.*, **86,** 2183 (1964).
18. E. P. Oliveto, H. Q. Smith, C. Gerold, L. Weber, R. Rausser, and E. B. Hershberg, *J. Am. Chem. Soc.*, **77,** 2224 (1955).
19. F. T. Bond, J. E. Stemke, and D. W. Powell, *Synth. Commun.*, **5,** 427 (1975).
20. R. Sterzycki, *Synthesis*, 724 (1979).
21. W. Kantlehner and H.-D. Gutbrod, *Liebigs Ann. Chem.*, 1362 (1979).
22. A. E. Dann, J. B. Davis, and M. J. Nagler, *J. Chem. Soc., Perkin Trans. I*, 158 (1979).
23. M. Koreeda and L. Brown, *J. Org. Chem.*, **48,** 2122 (1983).
24. B. Glatz, G. Helmchen, H. Muxfeldt, H. Porcher, R. Prewo, J. Senn, J. J. Stezowski, R. J. Stojda, and D. R. White, *J. Am. Chem. Soc.*, **101,** 2171 (1979).
25. R. A. Holton, R. M. Kennedy, H.-B. Kim, and M. E. Krafft, *J. Am. Chem. Soc.*, **109,** 1597 (1987).
26. H. J. Dauben, B. Löken, and H. J. Ringold, *J. Am. Chem. Soc.*, **76,** 1359 (1954).
27. H. Hagiwara and H. Uda, *J. Org. Chem.*, **53,** 2308 (1988); Y. Tamai, H. Hagiwara, and H. Uda, *J. Chem. Soc., Perkin Trans. I*, 1311 (1986).
28. H. Vorbrueggen, *Steroids*, **1,** 45 (1963).
29. D. H. R. Barton, C. C. Dawes, and P. D. Magnus, *J. Chem. Soc., Chem. Commun.*, 432 (1975).
30. J. L. E. Erickson and F. E. Collins, *J. Org. Chem.*, **30,** 1050 (1965).
31. F. Nerdel, J. Buddrus, G. Scherowsky, D. Klamann, and M. Fligge, *Liebigs Ann. Chem.*, **710,** 85 (1967).
32. H. E. Simmons and D. W. Wiley, *J. Am. Chem. Soc.*, **82,** 2288 (1960).
33. R. J. Stedman, L. D. Davis, and L. S. Miller, *Tetrahedron Lett.*, 4915 (1967).
34. G. R. Newkome, J. D. Sauer, and C. L. McClure, *Tetrahedron Lett.*, 1599 (1973).
35. J. Ott, G. M. Ramos Tombo, B. Schmid, L. M. Venanzi, G. Wang and T. R. Ward, *Tetrahedron Lett.*, **30,** 6151 (1989).
36. H. Hagiwara and H. Uda, *J. Chem. Soc., Chem. Commun.*, 1351 (1987).
37. G. Bauduin, D. Bondon, Y. Pietrasanta, and B. Pucci, *Tetrahedron*, **34,** 3269 (1978).
38. P. A. Grieco, M. Nishizawa, T. Oguri, S. D. Burke, and N. Marinovic, *J. Am. Chem. Soc.*, **99,** 5773 (1977).
39. P. A. Grieco, Y. Yokoyama, G. P. Withers, F. J. Okuniewicz, and C.-L. J. Wang, *J. Org. Chem.*, **43,** 4178 (1978).

40. P. A. Grieco, Y. Ohfune, and G. Majetich, *J. Am. Chem. Soc.*, **99,** 7393 (1977).
41. J. H. Babler, N. C. Malek, and M. J. Coghlan, *J. Org. Chem.*, **43,** 1821 (1978).
42. P. A. Grieco, T. Oguri, S. Gilman, and G. R. DeTitta, *J. Am. Chem. Soc.*, **100,** 1616 (1978).
43. H. M. Walborsky, R. H. Davis, and D. R. Howton, *J. Am. Chem. Soc.*, **73,** 2590 (1951).
44. J. A. Zderic and D. C. Limon, *J. Am. Chem. Soc.*, **81,** 4570 (1959).
45. F. Huet, A. Lechevallier, M. Pellet, and J. M. Conia, *Synthesis*, 63 (1978).
46. D. H. R. Barton, P. D. Magnus, G. Smith, and D. Zurr, *J. Chem. Soc., Chem. Commun.*, 861 (1971).
47. M. Uemura, T. Minami, and Y. Hayashi, *Tetrahedron Lett.*, **29,** 6271 (1988).
48. D. H. R. Barton, P. D. Magnus, G. Smith, G. Streckert, and D. Zurr, *J. Chem. Soc., Perkin Trans. I*, 542 (1972).
49. Y. Guindon, H. E. Morton, and C. Yoakim, *Tetrahedron Lett.*, **24,** 3969 (1983).
50. B. H. Lipshutz, D. Pollart, J. Monforte, and H. Kotsuki, *Tetrahedron Lett.*, **26,** 705 (1985).
51. M. E. Jung, W. A. Andrus, and P. L. Ornstein, *Tetrahedron Lett.*, 4175 (1977).
52. T. Hosokawa, Y. Imada, and S.-i. Murahashi, *J. Chem. Soc., Chem. Commun.*, 1245 (1983).
53. B. H. Lipshutz and D. F. Harvey, *Synth. Commun.*, **12,** 267 (1982).
54. G. Balme and J. Goré, *J. Org. Chem.*, **48,** 3336 (1983).
55. P. Sarmah and N. C. Barua, *Tetrahedron Lett.*, **30,** 4703 (1989).
56. T. Kametani, H. Kondoh, T. Honda, H. Ishizone, Y. Suzuki and W. Mori, *Chem. Lett.*, 901 (1989).
57. P. Lue, W.-Q. Fan and X.-J. Zhou, *Synthesis*, 692 (1989).
58. E. Keinan, D. Perez, M. Sahai, and R. Shvily, *J. Org. Chem.*, **55,** 2927 (1990).

11. 4-Bromomethyl-1,3-dioxolane (Chart 5):

CH$_2$Br
O O
R R

Formation

1. HOCH$_2$CH(OH)CH$_2$Br, TsOH, benzene, reflux, 5 h, 93–98% yield.[1]

Cleavage[1]

1. Activated Zn, MeOH, reflux, 12 h, 89–96% yield. This ketal is stable to several reagents that react with carbonyl groups (e.g., *m*-ClC$_6$H$_4$CO$_3$H, NH$_3$, NaBH$_4$, and MeLi). It is cleaved under neutral conditions.

1. E. J. Corey and R. A. Ruden, *J. Org. Chem.*, **38,** 834 (1973).

12. 4-(3-Butenyl)-1,3-dioxolane:

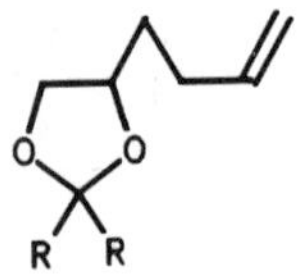

Formation/Cleavage[1]

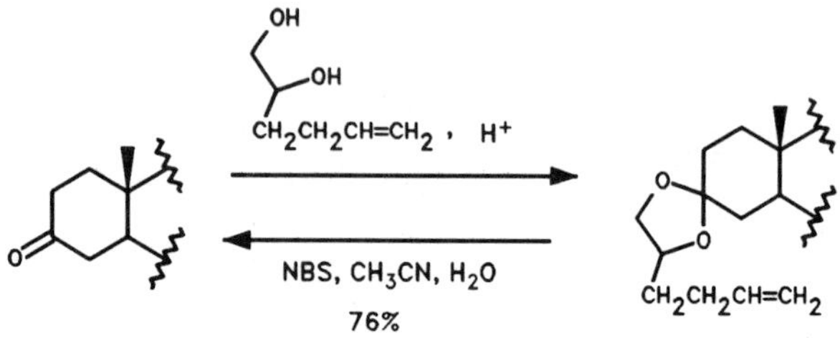

OH
OH
$CH_2CH_2CH{=}CH_2$, H^+
NBS, CH_3CN, H_2O
76%
$CH_2CH_2CH{=}CH_2$

1. Z. Wu, D. R. Mootoo, and B. Fraser-Reid, *Tetrahedron Lett.*, **29,** 6549 (1988).

13. 4-Phenyl-1,3-dioxolane:

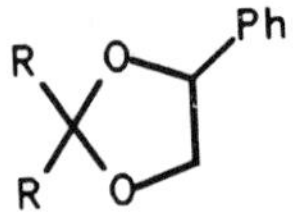

Cleavage[1]

Electrolysis: $LiClO_4$, H_2O, Pyr, CH_3CN, *N*-hydroxyphthalimide, 0.85 V SCE, 22–90% yield.

1. M. Masui, T. Kawaguchi, and S. Ozaki, *J. Chem. Soc., Chem. Commun.*, 1484 (1985).

14. 4-(2-Nitrophenyl)-1,3-dioxolane (Chart 5):

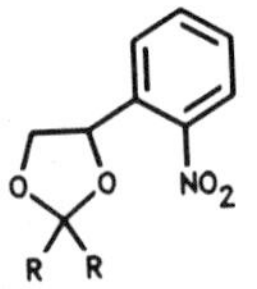

This dioxolane is readily formed from the glycol (TsOH, benzene, reflux, 70–95% yield); it is cleaved by irradiation (350 nm, benzene, 25°, 6 h, 75–90% yield). This group is stable to 5% HCl/THF; 10% AcOH/THF; 2% oxalic acid/THF; 10% aq. H_2SO_4/THF; 3% aq. TsOH/THF.[1]

1. J. Hébert and D. Gravel, *Can. J. Chem.*, **52,** 187 (1974); D. Gravel, J. Hébert, and D. Thoraval, *Can. J. Chem.*, **61,** 400 (1983).

15. 4,5-Dimethoxymethyl-1,3-dioxolane:

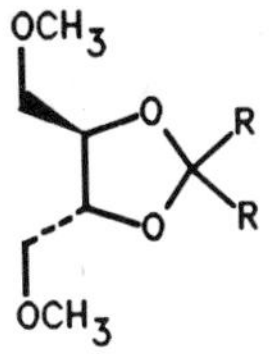

Formation/Cleavage[1]

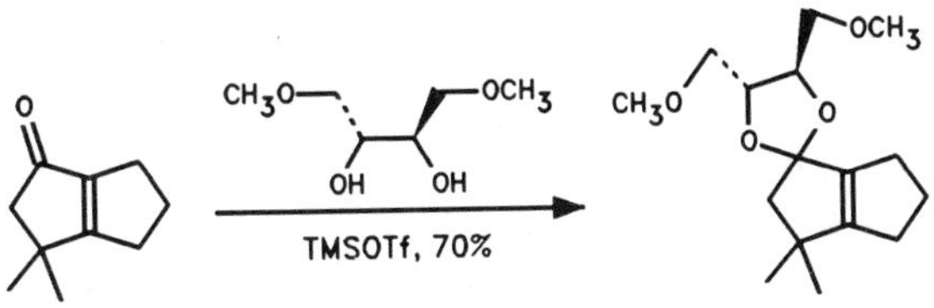

TMSOTf, 70%

This protective group was used to direct the selective cyclopropanation of a variety of enones. Hydrolysis (HCl, MeOH, H_2O, rt, 94% yield) affords optically active cyclopropyl ketones.

1. E. A. Mash, S. K. Math, and C. J. Flann, *Tetrahedron Lett.*, **29**, 2147 (1988).

16. *O,O′*-Phenylenedioxy Ketal:

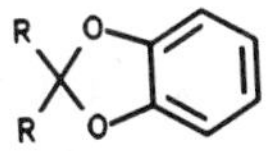

The phenylenedioxy ketal is prepared from catechol (TsOH, 90°, 30 h, 85% yield) and is cleaved with 5 *N* HCl (dioxane, reflux, 6 h). It is more stable to acid than the ethylene ketal.[1,2]

1. M. Rosenberger, D. Andrews, F. DiMaria, A. J. Duggan, and G. Saucy, *Helv. Chim. Acta*, **55**, 249 (1972).
2. M. Rosenberger, A. J. Duggan, and G. Saucy, *Helv. Chim. Acta*, **55**, 1333 (1972).

17. 1,5-Dihydro-3*H*-2,4-benzodioxepin:

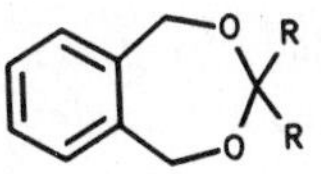

Formation/Cleavage[1,2]

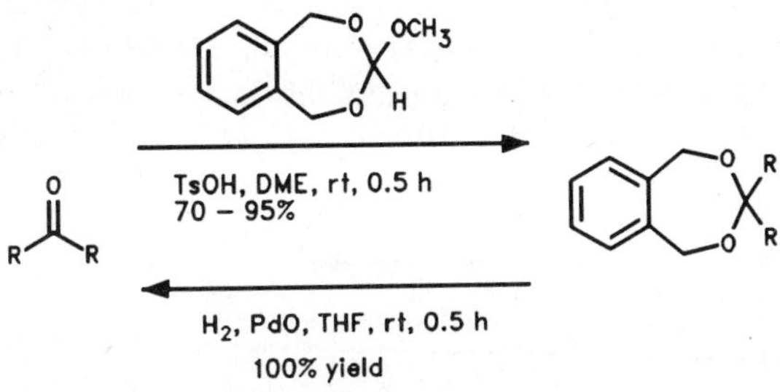

TsOH, DME, rt, 0.5 h
70 – 95%

Ref. 1

H_2, PdO, THF, rt, 0.5 h
100% yield

Camphor cannot be protected with this reagent, indicating that steric factors will prevent its use in very hindered systems.

1. N. Machinaga and C. Kibayashi, *Tetrahedron Lett.*, **30,** 4165 (1989).
2. K. Mori, T. Yoshimura, and T. Sugai, *Liebigs Ann. Chem.*, 899 (1988).

Dithio Acetals and Ketals

A carbonyl group can be protected as a sulfur derivative—for example, a dithio acetal or ketal, 1,3-dithiane, or 1,3-dithiolane—by reaction of the carbonyl compound in the presence of an acid catalyst with a thiol or dithiol. The derivatives are in general cleaved by reaction with Hg(II) salts or oxidation; acidic hydrolysis is unsatisfactory. The acyclic derivatives are formed and hydrolyzed much more readily than their cyclic counterparts. Representative examples of formation and cleavage are shown below.

ACYCLIC DITHIO ACETALS AND KETALS

18. ***S,S'*-Dimethyl Acetals and Ketals:** RR'C$(SCH_3)_2$ (Chart 5)

19. ***S,S'*-Diethyl Acetals and Ketals:** RR'C$(SC_2H_5)_2$

20. ***S,S'*-Dipropyl Acetals and Ketals:** RR'C$(SC_3H_7)_2$

21. ***S,S'*-Dibutyl Acetals and Ketals:** RR'C$(SC_4H_9)_2$

22. ***S,S'*-Dipentyl Acetals and Ketals:** RR'C$(SC_5H_{11})_2$

23. ***S,S'*-Diphenyl Acetals and Ketals:** RR'C$(SC_6H_5)_2$

24. ***S,S'*-Dibenzyl Acetals and Ketals:** RR'C$(SCH_2C_6H_5)_2$

General Methods of Formation

1. RSH, cond. HCl, 20°, 30 min.[1] These conditions were used to protect an aldose as the methyl or ethyl thioketal.
2. $RSSiMe_3$ [R = Me, Et, $(-CH_2-)_3$], ZnI_2, Et_2O, 0–25°, 70–95% yield.[2] This method is satisfactory for a variety of aldehydes and ketones and is also suitable for the preparation of 1,3-dithianes. Methacrolein gives the product of Michael addition rather than the thioacetal. The less hindered of two ketones is readily protected using this methodology.[3]

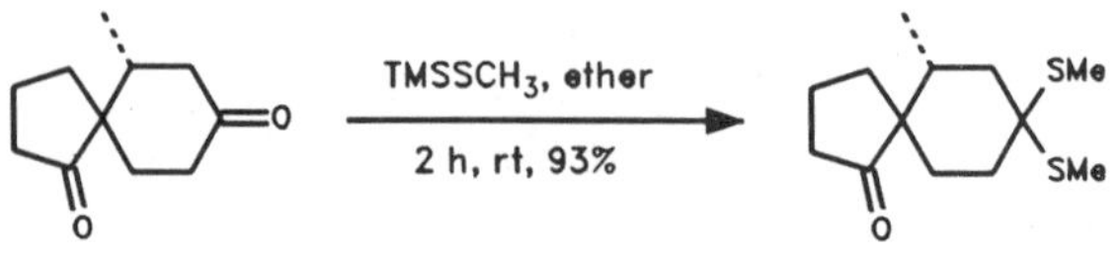

$TMSSCH_3$, ether

2 h, rt, 93%

SMe

SMe

3. RSH (R = Et, Pr, Ph), Me_3SiCl, $CHCl_3$, 20°, 1 h, >80% yield.[4]
4. $B(SR)_3$ (R = Et, Bu, C_5H_{11}), reflux, 2 h or 25°, 18 h, 75–85% yield.[5]
5. $Al(SPh)_3$, 25°, 1 h, 65% yield.[6] This method also converts esters to thioesters.
6. PhSH, $BF_3 \cdot Et_2O$, $CHCl_3$, 0°, 10 min, 86% yield.[7] $ZnCl_2$[8] and $MgBr_2$[9] have also been used as catalysts. With $MgBr_2$ acetals can be converted to thioketals in the presence of ketones.[9]
7. RSH, SO_2, benzene, 54–81% yield.[10]
8. EtSH, $TiCl_4$, $CHCl_3$, 6–12 h, rt, 90–98% yield.[11]
9. Ⓟ-$PPh_2 \cdot I_2$, RSH, Et_3N, CH_3CN; K_2CO_3, H_2O, 80–98% yield.[12] This method is also effective for the formation of dioxolanes and dithiolanes.
10. RSSR (R = Me, Ph, Bu), Bu_3P, rt, 15–83% yield.[13] This reagent also reacts with epoxides to form 1,2-dithioethers.

General Methods of Cleavage

1. $AgNO_3/Ag_2O$, CH_3CN–H_2O, 0°, 2 h, 85% yield.[14]

OH BuS BuS O C_5H_{11} OTHP —[$AgNO_3$, Ag_2O, CH_3CN, H_2O, 0°, 2 h, 85%]→ OH O O C_5H_{11} OTHP

This method has also been used to cleave dithianes and dithiolanes.[15] The *S,S′*-dibutyl group is stable to acids (e.g., HOAc/H_2O–THF, 45°, 3 h; TsOH/CH_2Cl_2, 0°, 0.5 h).[14]

2. $AgClO_4$, H_2O, C_6H_6, 25°, 4 h, 80–100% yield.[16]
3. $HgCl_2$, $CdCO_3$, aq. acetone[17] or $HgCl_2$, $CaCO_3$, CH_3CN, H_2O.[18]
4. $Me_2CH(CH_2)_2ONO$, CH_2Cl_2, 25°, 15 min; H_2O, 63–93% yield.[19] Isoamyl nitrite cleaves aromatic dithioacetals in preference to aliphatic dithioacetals, and dithioacetals in preference to dithioketals. It also cleaves 1,3-oxathiolanes (1 h, 65–90% yield).
5. $Tl(NO_3)_3$, CH_3OH, H_2O, 25°, 5 min, 73–98% yield.[7] These conditions are also effective for the cleavage of dithiolanes and dithianes.
6. SO_2Cl_2, SiO_2–H_2O, CH_2Cl_2, 25°, 2–3 h, 90–100% yield.[20,21]
7. I_2, $NaHCO_3$, dioxane, H_2O, 25°, 4.5 h, 80–95% yield.[22]
8. I_2, MeOH, reflux, 2 h, 79%; $HClO_4$, H_2O, 25°, 16 h, 87% yield.[23] These conditions also cleave acetonides and benzylidene acetals.[24]
9. H_2O_2, aq. acetone or $NaIO_4/H_2O$, 25°; g $HCl/CHCl_3$, 0°, 50–70% yield.[25]
10. O_2, $h\nu$, hexane, Ph_2CO, 2–5 h, 60–80% yield.[26] 1,3-Oxathiolanes and dithiolanes are also cleaved by these conditions.

11. CuCl, CuO, H_2O, acetone, 2 h, 20°, 61–73% yield.[27]
12. $HgCl_2$, HgO, 80% CH_3CN, H_2O, 30 min, rt, 96% yield.[28]
13. MCPBA, CF_3COOH, CH_2Cl_2, 0°.[29]
14. Ph_3CClO_4, Ph_3COMe, CH_2Cl_2, −45°, 2.5 h; aq. $NaHCO_3$, 84–96% yield.[30] A diethyl thioketal could be cleaved in the presence of a diphenyl thioketal.

1. H. Zinner, *Chem. Ber.*, **83,** 275 (1950).
2. D. A. Evans, L. K. Truesdale, K. G. Grimm, and S. L. Nesbitt, *J. Am. Chem. Soc.*, **99,** 5009 (1977).
3. D. A. Evans, K. G. Grimm, and L. K. Truesdale, *J. Am. Chem. Soc.*, **97,** 3229 (1975).
4. B. S. Ong, and T. H. Chan, *Synth. Commun.*, **7,** 283 (1977).
5. F. Bessette, J. Brault, and J. M. Lalancette, *Can. J. Chem.*, **43,** 307 (1965).
6. T. Cohen and R. E. Gapinski, *Tetrahedron Lett.*, 4319 (1978).
7. E. Fujita, Y. Nagao, and K. Kaneko, *Chem. Pharm. Bull.*, **26,** 3743 (1978).
8. W. E. Truce and F. E. Roberts, *J. Org. Chem.*, **28,** 961 (1963).
9. J. H. Park and S. Kim, *Chem. Lett.*, 629 (1989).
10. B. Burczyk and Z. Kortylewicz, *Synthesis*, 831 (1982).
11. V. Kumar and S. Dev, *Tetrahedron Lett.*, **24,** 1289 (1983).
12. R. Caputo, C. Ferreri, and G. Palumbo, *Synthesis*, 386 (1987).
13. M. Tazaki and M. Takagi, *Chem. Lett.*, 767 (1979).
14. E. J. Corey, M. Shibasaki, J. Knolle, and T. Sugahara, *Tetrahedron Lett.*, 785 (1977).
15. C. H. Heathcock, M. J. Taschner, T. Rosen, J. A. Thomas, C. R. Hadley, and G. Popják, *Tetrahedron Lett.*, **23,** 4747 (1982); R. Zamboni and J. Rokach, *Tetrahedron Lett.*, **23,** 4751 (1982).
16. T. Mukaiyama, S. Kobayashi, K. Kamio, and H. Takei, *Chem. Lett.*, 237 (1972).
17. J. English, Jr., and P. H. Griswold, Jr., *J. Am. Chem. Soc.*, **67,** 2039 (1945).
18. A. I. Meyers, D. L. Comins, D. M. Roland, R. Henning, and K. Shimizu, *J. Am. Chem. Soc.*, **101,** 7104 (1979).
19. K. Fuji, K. Ichikawa, and E. Fujita, *Tetrahedron Lett.*, 3561 (1978).
20. M. Hojo and R. Masuda, *Synthesis*, 678 (1976).
21. Y. Kamitori, M. Hojo, R. Masuda, T. Kimura, and T. Yoshida, *J. Org. Chem.*, **51,** 1427 (1986).
22. G. A. Russell and L. A. Ochrymowycz, *J. Org. Chem.*, **34,** 3618 (1969).
23. B. M. Trost, T. N. Salzmann, and K. Hiroi, *J. Am. Chem. Soc.*, **98,** 4887 (1976).
24. W. A. Szarek. A. Zamojski, K. N. Tiwari, and E. R. Ison, *Tetrahedron Lett.*, **27,** 3827 (1986).
25. H. Nieuwenhuyse and R. Louw, *Tetrahedron Lett.*, 4141 (1971).
26. T. T. Takahashi, C. Y. Nakamura, and J. Y. Satoh, *J. Chem. Soc., Chem. Commun.*, 680 (1977).

27. B. Cazes and S. Julia, *Tetrahedron Lett.*, 4065 (1978).
28. V. E. Amoo, S. De Bernardo, and M. Weigele, *Tetrahedron Lett.*, **29**, 2401 (1988).
29. J. Cossy, *Synthesis*, 1113 (1987).
30. M. Ohshima, M. Murakami, and T. Mukaiyama, *Chem. Lett.*, 1593 (1986).

25. *S,S'*-Diacetyl Acetals and Ketals: $R_2C(SCOCH_3)_2$

Formation[1]

CH_3COSH, 6 N H_2SO_4, 25°, 84%

Cleavage[1]

1. NaOMe, MeOH, 25°, 40 h, 96%
2. $CHCl_3$, cat. H_2O or SiO_2, facile

The formyl group was lost during attempted protection with ethylene glycol, TsOH.

1. T. Kametani, Y. Kigawa, K. Takahashi, H. Nemoto, and K. Fukumoto, *Chem. Pharm. Bull.*, **26**, 1918 (1978).

CYCLIC DITHIO ACETALS AND KETALS

26. 1,3-Dithiane Derivative ($n = 3$) (Chart 5)

27. 1,3-Dithiolane Derivative ($n = 2$) (Chart 5):

General Methods of Formation

1. $HS(CH_2)_nSH$, $BF_3 \cdot Et_2O$, CH_2Cl_2, 25°, 12 h, high yield, $n = 2$,[1] $n = 3$.[2] In α,β-unsaturated ketones the olefin does not isomerize to the β,γ-position as occurs when an ethylene ketal is prepared.[3] Aldehydes are selectively protected in the presence of ketones except when steric factors force the ketone to be protected as in the example below.[4] A TBDMS group is not stable to these conditions.[5]

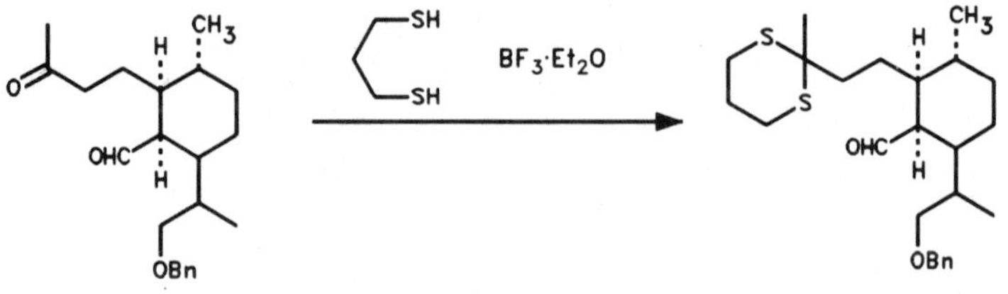

$BF_3 \cdot Et_2O$

2. 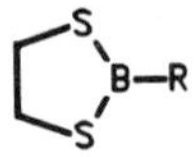 $CHCl_3$, 25°, 2 h, 90–100% yield.[6]

R = Cl or Ph

When R = Ph, the reaction is selective for unhindered ketones. Diaryl ketones, generally unreactive compounds, react rapidly when R = Cl.

3. $Me_3SiSCH_2CH_2SSiMe_3$, ZnI_2, Et_2O, 0–25°, 12–24 h, high yields.[7] Less hindered ketones can be selectively protected in the presence of more hindered ketones. α,β-Unsaturated ketones are selectively protected (94:1, 94:4) in the presence of saturated ketones by this reagent.[8]
4. $HS(CH_2)_nSH$, $SOCl_2$–SiO_2, 88–100% yield.[9] Aldehydes are selectively protected in the presence of ketones.
5. $HS(CH_2)_2SH$, $TiCl_4$, −10° → 25°, 96% yield.[10]
6. $HSCH_2CH_2SH$, $Zn(OTf)_2$ or $Mg(OTf)_2$, $ClCH_2CH_2Cl$, heat, 16 h, 85–99% yield.[11,12] α,β-Unsaturated ketones such as carvone are not cleanly converted to ketals because of Michael addition of the thiol.[11]

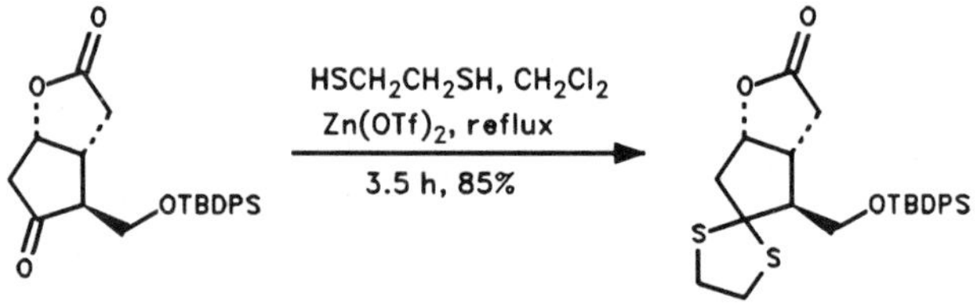

$HSCH_2CH_2SH$, CH_2Cl_2, $Zn(OTf)_2$, reflux, 3.5 h, 85%

In this case other methods failed because of β-elimination.

7. 1,3-Dioxolanes[13,15] and 1,3-dioxanes[14] are readily converted to 1,3-dithianes and 1,3-dithiolanes in good to excellent yields.

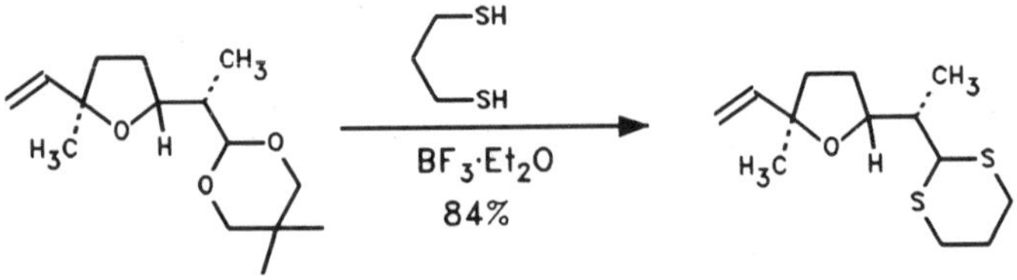

$BF_3 \cdot Et_2O$, 84% — Ref. 14

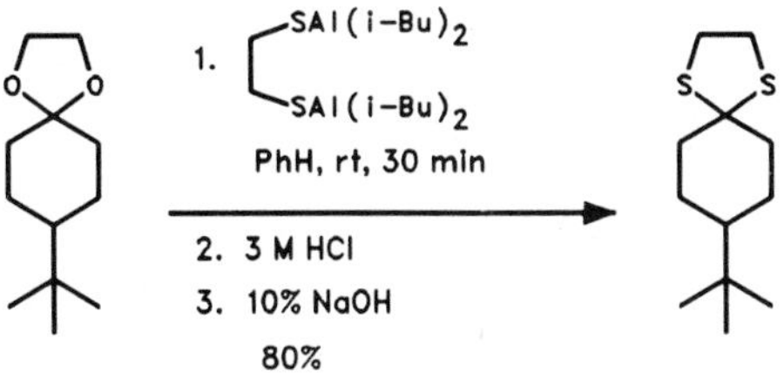

1. $HS(CH_2)_2SAl(i\text{-}Bu)_2$, PhH, rt, 30 min; 2. 3 M HCl; 3. 10% NaOH; 80%

Ref. 15

8. 2,2-Dimethyl-2-sila-1,3-dithiane, $BF_3 \cdot Et_2O$, CH_2Cl_2, 0°, 82–99% yield.[16] This method was reported to be superior to the conventional synthesis because cleaner products are formed. Aldehydes are selectively protected in the presence of ketones, which do not react competitively with this reagent.
9. 2,2-Dibutyl-2-stanna-1,3-dithiane, $Bu_2Sn(OTf)_2$, $ClCH_2CH_2Cl$, 35°, 1 h, 77–94% yield.[17] TBDMS, TBDPS, THP, and OAc groups are not affected by these conditions.

General Methods of Cleavage

1. $Hg(ClO_4)_2$, MeOH, $CHCl_3$, 25°, 5 min, 93% yield.[18,19]

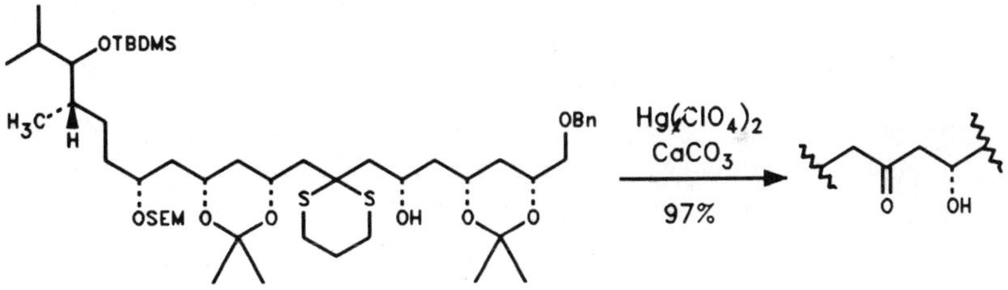

OTBDMS, H_3C, H, OSEM, OBn, OH; $Hg(ClO_4)_2$, $CaCO_3$, 97%

Ref. 19

2. A 1,3-dithiane is stable to the conditions ($HgCl_2$, $CaCO_3$, CH_3CN–H_2O, 25°, 1–2 h) used to cleave a methylthiomethyl (MTM) ether (i.e., a monothio acetal).[20]
3. $CuCl_2$, CuO, acetone, reflux, 90 min, 85% yield.[21]

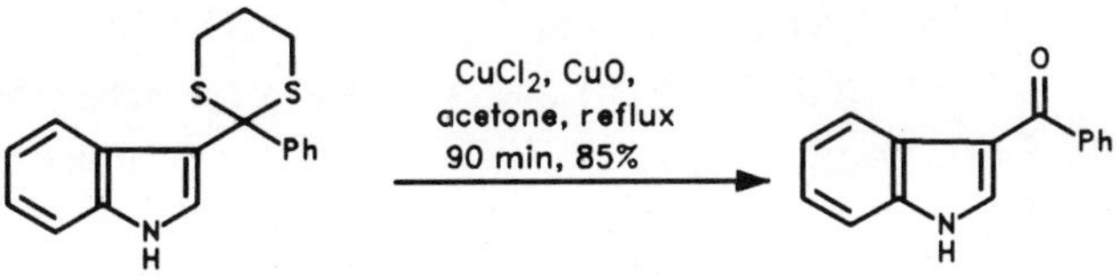

$CuCl_2$, CuO, acetone, reflux, 90 min, 85%

4. $AgNO_3$, EtOH, H_2O, 50°, 20 min, 55% yield.[22]

$AgNO_3$, EtOH, H_2O
50°, 20 min, 55%

Attempted cleavage using Hg(II) salts gave material that could not be distilled. 1,3-Dithiolanes can also be cleaved with Ag_2O (MeOH, H_2O, reflux, 16 h–4 days, 75–85% yield).[23]

5. For (n = 2): NBS, aq. acetone, 0°, 20 min, 80% yield.[24]
6. For (n = 3): NCS, $AgNO_3$, CH_3CN, H_2O, 25°, 5–10 min, 70–100% yield.[25,26]
7. For (n = 2, 3): $Tl(NO_3)_3$, CH_3OH, 25°, 5 min, 73–99% yield.[19] These conditions have been used to effect selective cleavage of α,β-unsaturated thioketals.[27]

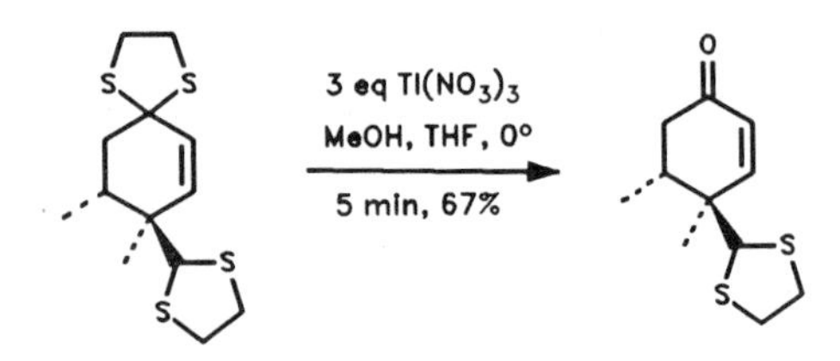

8. For (n = 2, 3): $Tl(OCOCF_3)_3$, THF, 25°, 1 min, 83–95% yield.[28] α,β-Unsaturated 1,3-dithiolanes are selectively cleaved in the presence of saturated 1,3-dithiolanes [$Tl(NO_3)_3$, 5 min, 97% yield].[29]
9. For (n = 2, 3): SO_2Cl_2, SiO_2, CH_2Cl_2, H_2O, 0–25°, 90–100% yield.[30]
10. For (n = 2): I_2, DMSO, 90°, 1 h, 75–85% yield.[31]
11. For (n = 2,[32] 3[33]): p-$MeC_6H_4SO_2N(Cl)Na$, aq. MeOH, 75–100% yield. 1,3-Oxathiolanes are also cleaved by Chloramine-T.[33]
12. For (n = 2, 3): $(PhSeO)_2O$, THF or CH_2Cl_2, 25°, 30 min–50 h, 63–78% yield.[34]
13. For (n = 3): $Me_2CH(CH_2)_2ONO$, CH_2Cl_2, reflux, 2.5 h, 65% yield.[35] 1,3-Oxathiolanes are also cleaved by isoamyl nitrite.
14. For (n = 2, 3): N-chlorobenzotriazole, CH_2Cl_2, −80°; NaOH, 50% yield.[36] 1,3-Dithianes and 1,3-dithiolanes, used in this example to protect C_3-keto steroids, were not cleaved by $HgCl_2$–$CdCO_3$.
15. For (n = 2, 3): $Ce(NH_4)_2(NO_3)_6$, aq. CH_3CN, 3 min, 70–87% yield.[37]
16. For (n = 2): O_2, $h\nu$, 4.5 h, 60–80% yield.[38] 1,3-Oxathiolanes are also cleaved by $O_2/h\nu$.
17. Electrolysis: 1.5 V, CH_3CN, H_2O, $LiClO_4$ or $Bu_4N^+ClO_4^-$, 50–75% yield.[39,40] 1,3-Dithiolanes were not cleaved efficiently by electrolytic oxidation.

18. For (n = 2, 3): $MeOSO_2F$, C_6H_6, 25°, 1 h, 62–88% yield[41] or liq. SO_2, 70–85% yield.[42]
19. For (n = 2): MeI, aq. MeOH, reflux, 2–20 h, 60–80% yield.[42]
20. For (n = 3): MeI, aq. CH_3CN, 25°.[43]
21. For (n = 2): $Et_3O^+BF_4^-$, followed by 3% aq. $CuSO_4$, 81% yield.[44]
22. For (n = 2): $Me_2S^+Br\ Br^-$, CH_2Cl_2, 25°, 1 h → reflux, 8 h, followed by H_2O, 55–91% yield.[45]
23. OHCCOOH, HOAc, 25°, 15 min–20 h, 60–90% yield.[46]
24. $NO^+HSO_4^-$, CH_2Cl_2, 25°, 45 min; H_2O, 56–82% yield.[47]
25. Electrolysis: 1 V, (p-$CH_3C_6H_4)_3N$, CH_3CN, H_2O, $NaHCO_3$, 70–95% yield.[48]
26. Diiodohydantoin, −20°, 5 : 5 : 1 acetone : THF : H_2O.[8]
27. $(CF_3CO_2)_2IPh$, H_2O, CH_3CN, 85–99% yield.[49] In the presence of ethylene glycol the dithiane can be converted to a dioxolane (91% yield). The reaction conditions are not compatible with primary amides. Thioesters are not affected.
28. MCPBA; Ac_2O, Et_3N, H_2O, THF, 28–37% yield.[50]

1. MCPBA
2. Ac_2O, TEA
H_2O, THF
28–37%

29. Pyr·HBr·Br_2, CH_2Cl_2, pyridine, $Bu_4N^+Br^-$, 0°–rt, 2 h, 80–90% yield.[51] The deprotection proceeds without olefin or aromatic ring bromination.
30. $PhOP(O)Cl_2$, DMF, NaI, 1 h, rt, 71–94% yield.[52]
31. $MeP(Ph)_3{}^+Br^-$, CH_2Cl_2, H_2O, NaH_2PO_4, Na_2HPO_4, 0–100% yield.[53]
32. For (n = 2): Me_3SiI or Me_3SiBr, DMSO, 65–99% yield.[54]
33. For (n = 3): DMSO, dioxane, 1.8 M HCl, 90–96% yield.[55]
34. For (n = 3): $Me_3S^+SbCl_6^-$, −77°; Na_2CO_3, H_2O, 95–97% yield.[55]
35. For (n = 3): $(CF_3CO_2)_2IPh$, H_2O, CH_3CN or MeOH, 1–10 min, 85–100% yield.[56] A phenylthio ester is stable to these conditions, but amides are not.
36. For (n = 3): MCPBA, TFA, CH_2Cl_2, 0°, 75–96% yield.[57]

1. R. P. Hatch, J. Shringarpure, and S. M. Weinreb, *J. Org. Chem.*, **43,** 4172 (1978).
2. J. A. Marshall and J. L. Belletire, *Tetrahedron Lett.*, 871 (1971).
3. F. Sondheimer and D. Rosenthal, *J. Am. Chem. Soc.*, **80,** 3995 (1958).
4. W.-S. Zhou, *Pure Appl. Chem.*, **58,** 817 (1986).
5. T. Nakata, S. Nagao, N. Mori, and T. Oishi, *Tetrahedron Lett.*, **26,** 6461 (1985).

6. D. R. Morton and S. J. Hobbs, *J. Org. Chem.*, **44,** 656 (1979).
7. D. A. Evans, L. K. Truesdale, K. G. Grimm, and S. L. Nesbitt, *J. Am. Chem. Soc.*, **99,** 5009 (1977).
8. E. J. Corey, M. A. Tius, and J. Das, *J. Am. Chem. Soc.*, **102,** 7612 (1980).
9. Y. Kamitori, M. Hojo, R. Masuda, T. Kimura, and T. Yoshida, *J. Org. Chem.*, **51,** 1427 (1986).
10. V. Kumar and S. Dev, *Tetrahedron Lett.*, **24,** 1289 (1983).
11. E. J. Corey and K. Shimoji, *Tetrahedron Lett.*, **24,** 169 (1983).
12. M. E. Kuehne, W. G. Bornmann, W. G. Earley, and I. Marko, *J. Org. Chem.*, **51,** 2913 (1986).
13. R. A. Moss and C. B. Mallon, *J. Org. Chem.*, **40,** 1368 (1975).
14. Y. Honda, A. Ori, and G. Tsuchihashi, *Chem. Lett.*, 1259 (1987).
15. T. Satoh, S. Uwaya, and K. Yamakawa, *Chem. Lett.*, 667 (1983).
16. J. A. Soderquist and E. I. Miranda, *Tetrahedron Lett.*, **27,** 6305 (1986).
17. T. Sato, E. Yoshida, T. Kobayashi, J. Otera, and H. Nozaki, *Tetrahedron Lett.*, **29,** 3971 (1988).
18. E. Fujita, Y. Nagao, and K. Kaneko, *Chem. Pharm. Bull.*, **26,** 3743 (1978).
19. B. H. Lipshutz, R. Moretti, and R. Crow, *Tetrahedron Lett.*, **30,** 15 (1989).
20. E. J. Corey and M. G. Bock, *Tetrahedron Lett.*, 2643 (1975).
21. P. Stütz and P. A. Stadler, *Org. Synth.*, **56,** 8 (1977); *Collect. Vol. VI*, 109 (1988).
22. C. A. Reece, J. O. Rodin, R. G. Brownlee, W. G. Duncan, and R. M. Silverstein, *Tetrahedron*, **24,** 4249 (1968).
23. D. Gravel, C. Vaziri, and S. Rahal, *J. Chem. Soc., Chem. Commun.*, 1323 (1972).
24. E. N. Cain and L. L. Welling, *Tetrahedron Lett.*, 1353 (1975).
25. E. J. Corey and B. W. Erickson, *J. Org. Chem.*, **36,** 3553 (1971).
26. A. V. Rama Rao, G. Venkatswamy, S. M. Javeed, V. H. Deshpande, and B. R. Rao, *J. Org. Chem.*, **48,** 1552 (1983).
27. P. S. Jones, S. V. Ley, N. S. Simpkins, and A. J. Whittle, *Tetrahedron*, **42,** 6519 (1986).
28. T.-L. Ho and C. M. Wong, *Can. J. Chem.*, **50,** 3740 (1972).
29. R. A. J. Smith and D. J. Hannah, *Synth. Commun.*, **9,** 301 (1979).
30. M. Hojo and R. Masuda, *Synthesis*, 678 (1976).
31. J. B. Chattopadhyaya and A. V. Rama Rao, *Tetrahedron Lett.*, 3735 (1973).
32. W. F. J. Huurdeman, H. Wynberg, and D. W. Emerson, *Tetrahedron Lett.*, 3449 (1971).
33. D. W. Emerson and H. Wynberg, *Tetrahedron Lett.*, 3445 (1971).
34. D. H. R. Barton, N. J. Cussans, and S. V. Ley, *J. Chem. Soc., Chem. Commun.*, 751 (1977).
35. K. Fuji, K. Ichikawa, and E. Fujita, *Tetrahedron Lett.*, 3561 (1978).
36. P. R. Heaton, J. M. Midgley, and W. B. Whalley, *J. Chem. Soc., Chem. Commun.*, 750 (1971).
37. T.-L. Ho, H. C. Ho, and C. M. Wong, *J. Chem. Soc., Chem. Commun.*, 791 (1972).
38. T. T. Takahashi, C. Y. Nakamura, and J. Y. Satoh, *J. Chem. Soc., Chem. Commun.*, 680 (1977).

39. Q. N. Porter and J. H. P. Utley, *J. Chem. Soc., Chem. Commun.*, 255 (1978).
40. H. J. Cristau, B. Chabaud, and C. Niangoran, *J. Org. Chem.*, **48,** 1527 (1983).
41. T.-L. Ho and C. M. Wong, *Synthesis*, 561 (1972).
42. M. Fetizon and M. Jurion, *J. Chem. Soc., Chem. Commun.*, 382 (1972).
43. S. Takano, S. Hatakeyama, and K. Ogasawara, *J. Chem. Soc., Chem. Commun.*, 68 (1977).
44. T. Oishi, K. Kamemoto, and Y. Ban, *Tetrahedron Lett.*, 1085 (1972).
45. G. A. Olah, Y. D. Vankar, M. Arvanaghi, and G. K. S. Prakash, *Synthesis*, 720 (1979).
46. H. Muxfeldt, W.-D. Unterweger, and G. Helmchen, *Synthesis*, 694 (1976).
47. G. A. Olah, S. C. Narang, G. F. Salem, and B. G. B. Gupta, *Synthesis*, 273 (1979).
48. M. Platen and E. Steckhan, *Tetrahedron Lett.*, **21,** 511 (1980); *idem., Chem. Ber.*, **117,** 1679 (1984).
49. G. Stork and K. Zhao, *Tetrahedron Lett.*, **30,** 287 (1989).
50. A. B. Smith, III, B. D. Dorsey, M. Visnick, T. Maeda, and M. S. Malamas, *J. Am. Chem. Soc.*, **108,** 3110 (1986).
51. G. S. Bates and J. O'Doherty, *J. Org. Chem.*, **46,** 1745 (1981).
52. H.-J. Liu and V. Wiszniewski, *Tetrahedron Lett.*, **29,** 5471 (1988).
53. H. -J. Cristau, A. Bazbouz, P. Morand, and E. Torreilles, *Tetrahedron Lett.*, **27,** 2965 (1986).
54. G. A. Olah, S. C. Narang, and A. K. Mehrotra, *Synthesis*, 965 (1982).
55. M. Prato, U. Quintily, G. Scorrano, and A. Sturaro, *Synthesis*, 679 (1982).
56. G. Stork and K. Zhao, *Tetrahedron Lett.*, **30,** 287 (1989).
57. J. Cossy, *Synthesis*, 1113 (1987).

28. 1,5-Dihydro-3*H*-2,4-benzodithiepin Derivative:

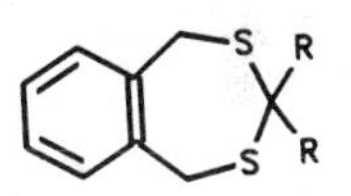

Dithiepin derivatives, prepared in high yield from 1,2-bis(mercaptomethyl)benzenes, are cleaved by $HgCl_2$ (80% yield). Neither reagents nor products have unpleasant odors.[1]

1. I. Shahak and E. D. Bergmann, *J. Chem. Soc. C*, 1005 (1966).

Monothio Acetals and Ketals

ACYCLIC MONOTHIO ACETALS AND KETALS

Acyclic monothio acetals and ketals can be prepared directly from a carbonyl compound or by transketalization, a reaction that does not involve a free carbonyl group, from a 1,3-dithiane or 1,3-dithiolane. They are cleaved by acidic hydrolysis or Hg(II) salts.

29. *O*-Trimethylsilyl-*S*-alkyl Acetals and Ketals: $R_2C(SR')OSiMe_3$

Formation

1. $RSSiMe_3$, ZnI_2, 25°, 30 min, 80–90% yield.[1]
2. Me_3SiCl, R'SH, Pyr, 25°, 3 h, 75–90% yield.[2]

Cleavage

1. Dilute HCl.[2]

In ether or tetrahydrofuran organolithium reagents cleave the silicon–oxygen bond; in hexamethylphosphoramide, they react at the carbon atom.[2]

1. D. A. Evans, L. K. Truesdale, K. G. Grimm, and S. L. Nesbitt, *J. Am. Chem. Soc.*, **99**, 5009 (1977).
2. T. H. Chan and B. S. Ong, *Tetrahedron Lett.*, 319 (1976).

30. *O*-Methyl-*S*-alkyl or -*S*-phenyl Acetals and Ketals: $R_2C(OMe)SR'$

Formation

1. From a dimethyl acetal: Et_2AlSPh, 0°, 78% yield.[1]
2. From a dimethyl acetal: $BCl_3 \cdot Et_2O$, −45°, CH_3SH, 73% yield.[2]
3. From a dialkyl acetal: Bu_3SnSPh, $BF_3 \cdot Et_2O$, toluene, −78° → 0°, 64–100% yield.[3] These conditions also convert MOM and MEM groups to the corresponding phenylthiomethyl groups in 64–77% yield.
4. From a dialkyl acetal: $MgBr_2$, Et_2O, rt, PhSH, 91% yield.[4] MOM groups are converted to phenylthiomethyl groups, 75% yield.

Cleavage

1. Electrolysis: Pt electrode, KOAc, AcOH, 10 V, 18–20°; K_2CO_3, MeOH, 81–91% yield.[5] These cleavage conditions could, in principle, be used to cleave the MTM group.
2. Mercury salts should also be effective for the cleavage of this protective group. The section on MTM ethers should be consulted.

1. Y. Masaki, Y. Serizawa, and K. Kaji, *Chem. Lett.*, 1933 (1985).
2. F. Nakatsubo, A. J. Cocuzza, D. E. Keely, and Y. Kishi, *J. Am. Chem. Soc.*, **99,** 4835 (1977).
3. T. Sato, T. Kobayashi, T. Gojo, E. Yishida, J. Otera, and H. Nozaki, *Chem. Lett.*, 1661 (1987).

4. S. Kim, J. H. Park, and S. Lee, *Tetrahedron Lett.*, **30,** 6697 (1989).
5. T. Mandai, H. Irei, M. Kuwada, and J. Otera, *Tetrahedron Lett.*, **25,** 2371 (1984).

31. *O*-Methyl-*S*-2-(methylthio)ethyl Acetals and Ketals: $R_2C(OMe)SCH_2CH_2SMe$

Formation[1]

$$R_2C[S(CH_2)_nS] \xrightarrow[\text{2. MeOH, } CH_2Cl_2,\ 23°,\ 2\ h,\ 72\%]{\text{1. } MeOSO_2F,\ CH_2Cl_2,\ 0°,\ 10\ min \rightarrow 25°,\ 2\ h} R_2C(OCH_3)S(CH_2)_nSCH_3$$

n = 2,3

Cleavage[1]

1. $HgCl_2$, $CaCO_3$, THF, H_2O, 0°, rapid. These derivatives are less susceptible to oxidation and hydrogenolysis than are the 1,3-dithiane and 1,3-dithiolane precursors.

1. E. J. Corey and T. Hase, *Tetrahedron Lett.*, 3267 (1975).

CYCLIC MONOTHIO ACETALS AND KETALS

32. 1,3-Oxathiolanes (Chart 5):

(structure: R_2C in a five-membered ring with O and S)

Formation

1. $HSCH_2CH_2OH$, $ZnCl_2$ AcONa, dioxane, 25°, 20 h, 60–90% yield.[1,2]

Cleavage

The section on the cleavage of 1,3-dithianes and 1,3-dithiolanes (pp. 203–205), should be consulted since many of the methods described there are also applicable to the cleavage of oxathiolanes.

1. $HgCl_2$, AcOH, AcOK, 100°, 1 h, 83% yield.[3]
2. $HgCl_2$, NaOH, EtOH, H_2O, 25°, 30 min, 91% yield.[3]
3. Raney Ni, AcOH, AcOK, 100°, 90 min, 92% yield.[3]
4. HCl, AcOH, reflux, 22 h, 60% yield.[4]
5. $AgNO_3$, NCS, 80% CH_3CN, H_2O.[5]
6. Benzyne, $ClCH_2CH_2Cl$, 49–100% yield.[6]

1. J. Romo, G. Rosenkranz, and C. Djerassi, *J. Am. Chem. Soc.*, **73,** 4961 (1951).
2. V. K. Yadav and A. G. Fallis, *Tetrahedron Lett.*, **29,** 897 (1988).
3. C. Djerassi, M. Shamma, and T. Y. Kan, *J. Am. Chem. Soc.*, **80,** 4723 (1958).
4. R. H. Mazur and E. A. Brown, *J. Am. Chem. Soc.*, **77,** 6670 (1955).
5. S. V. Frye and E. L. Eliel, *Tetrahedron Lett.*, **26,** 3907 (1985).
6. J. Nakayama, H. Sugiura, A. Shiotsuki, and M. Hoshino, *Tetrahedron Lett.*, **26,** 2195 (1985).

33. Diseleno Acetals and Ketals: $R_2C(SeR')_2$

Formation

1. RSeH, $ZnCl_2$, N_2, CCl_4, 20°, 3 h, 70–95% yield.[1]
2. From a ketal: $(PhSe)_3B$, CF_3COOH, $CHCl_3$, 20°, 20 min–24 h.[2]

Cleavage[1]

1. $HgCl_2$, $CaCO_3$, CH_3CN, H_2O, 20°, 2–4 h, 65–80% yield.
2. $CuCl_2$, CuO, acetone, H_2O, 20°, 5 min–2 h, 73–99% yield.
3. H_2O_2, THF, 0°, 15 min → 20°, 3 h, 60–65% yield.
4. $(PhSeO)_2O$, THF, 20° or 60°, 5 min → 6 h, 60–90% yield.
5. Clay-supported ferric nitrate (Clayfen) or clay-supported cupric nitrate (Claycop), pentane, rt, 60–97% yield.[3]

Diseleno acetals and ketals are cleaved more rapidly than their dithio counterparts; a methyl derivative is cleaved more rapidly than a phenyl derivative. Methyl iodide or ozone converts diseleno acetals and ketals to vinyl selenides.[1]

1. A. Burton, L. Hevesi, W. Dumont, A. Cravador, and A. Krief, *Synthesis*, 877 (1979).
2. D. L. J. Clive and S. M. Menchen, *J. Org. Chem.*, **44,** 4279 (1979).
3. P. Laszlo, P. Pennetreau, and A. Krief, *Tetrahedron Lett.*, **27,** 3153 (1986).

MISCELLANEOUS DERIVATIVES

O-Substituted Cyanohydrins

34. *O*-Acetyl Cyanohydrin: $R_2C(CN)OAc$

Formation

1. $Me_2C(CN)OH$, Et_3N, 25°, 2 h, 82% yield; Ac_2O, Pyr, 25°, 40 h, 82% yield.[1]

2. From a cyanohydrin: Ac_2O, $FeCl_3$, 25–92% yield.[2] Other anhydrides are also effective in this conversion.

Cleavage

1. $Li(O\text{-}t\text{-}Bu)_3AlH$, THF; KOH, CH_3OH, H_2O, 25°, 5 min, 84% yield.[1]

35. *O*-Trimethylsilyl Cyanohydrin: $R_2C(CN)OSiMe_3$ (Chart 5)

Formation

1. Me_3SiCN, cat. KCN or $Bu_4N^+F^-$, 18-crown-6, 75–95% yield.[3]
2. Me_3SiCN, Ph_3P, CH_3CN, 0°, 1 h, 100% yield.[4]

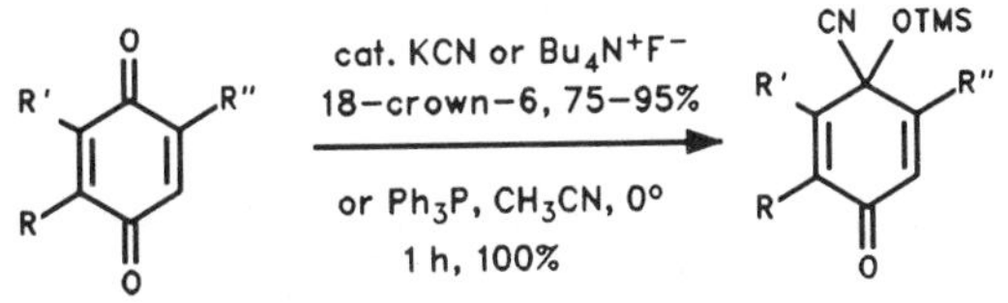

cat. KCN or $Bu_4N^+F^-$ 18-crown-6, 75–95% → or Ph_3P, CH_3CN, 0° 1 h, 100%

3. $Me_2C(CN)OSiMe_3$, KCN, 130°.[5]
4. Me_3SiCl, KCN, Amberlite XAD-4, CH_3CN, 60°, 8 h, 81–97% yield.[6]
5. Me_3SiCl, KCN, NaI, Pyr, CH_3CN, 50–77% distilled yields, 100% by NMR.[7]
6. R_3SiCl, KCN, ZnI_2, CH_3CN, 86–98% yield.[8] This method was used to prepare the t-$BuPh_2Si$, t-$BuMe_2Si$, and i-Pr_3Si cyanohydrins.

Cleavage

1. AgF, THF, H_2O, 25°, 2.5 h, 77% yield.[3]
2. Dilute acid or base.[9]

36. *O*-1-Ethoxyethyl Cyanohydrin: $R_2C(CN)OCH(OC_2H_5)CH_3$

The ethoxyethyl cyanohydrin was prepared (NaCN, HCl, THF, 0°, 75% yield, followed by $EtOCH{=}CH_2$, HCl, 50% yield) to convert an aldehyde ultimately to a protected ketone. It was cleaved by hydrolysis (0.01 N HCl, MeOH, 25°, followed by NaOH, 0°, 85% yield).[10]

37. *O*-Tetrahydropyranyl Cyanohydrin: $R_2C(CN)O$–THP

The tetrahydropyranyl cyanohydrin was prepared from a steroid cyanohydrin (dihydropyran, TsOH, reflux, 1.5 h) and cleaved by hydrolysis (cat, concd. HCl, acetone, reflux, 15 min, followed by aq. pyridine, reflux, 1 h).[11]

1. P. D. Klimstra and F. B. Colton, *Steroids*, **10,** 411 (1967).
2. T. Hiyama, H. Oishi, and H. Saimoto, *Tetrahedron Lett.*, **26,** 2459 (1985).
3. D. A. Evans, J. M. Hoffman, and L. K. Truesdale, *J. Am. Chem. Soc.*, **95,** 5822 (1973).
4. D. A. Evans and R. Y. Wong, *J. Org. Chem.*, **42,** 350 (1977).
5. D. A. Evans and L. K. Truesdale, *Tetrahedron Lett.*, 4929 (1973).
6. K. Sukata, *Bull. Chem. Soc. Jpn.*, **60,** 3820 (1987).
7. F. Duboudin, Ph. Cazeau, F. Moulines, and O. Laporte, *Synthesis*, 212 (1982).
8. V. H. Rawal, J. A. Rao, and M. P. Cava, *Tetrahedron Lett.*, **26,** 4275 (1985).
9. D. A. Evans, L. K. Truesdale, and G. L. Carroll, *J. Chem. Soc., Chem. Commun.*, 55 (1973).
10. G. Stork and L. Maldonado, *J. Am. Chem. Soc.*, **93,** 5286 (1971).
11. P. deRuggieri and C. Ferrari, *J. Am. Chem. Soc.*, **81,** 5725 (1959).

Substituted Hydrazones

38. *N,N*-Dimethylhydrazone: RR′C=NN$(CH_3)_2$ (Chart 5)

Formation

1. H_2NNMe_2, EtOH–HOAc, reflux, 24 h, 90–94% yield.[1]

Cleavage

1. $NaIO_4$, MeOH, pH 7, 2–3 h, 90% yield.[2]
2. $Cu(OAc)_2$, H_2O, THF, pH 5.4, 25°, 15 min, 97% yield.[3]
3. $CuCl_2$, THF, HPO_4^-, → pH 7, 85–100% yield.[3]
4. CH_3I, 95% EtOH, reflux, 80–90% yield.[4]
5. O_3, CH_2Cl_2, −78°, 60–100% yield.[5]
6. O_2, *hν*, Rose Bengal, MeOH, −78° → −20°, followed by Ph_3P or Me_2S, 48–88% yield.[6]
7. CoF_3 ($CHCl_3$, reflux, 67–93% yield);[7] $MoOCl_3$ or MoF_6 (H_2O, THF, 25°, 4 h, 80–90% yield);[8] WF_6 ($CHCl_3$, 0 → 25°, 1 h, 84–95% yield;[9] UF_6 (50–95% yield).[10]
8. N_2O_4, −40 → 0°, CH_3CN, THF, $CHCl_3$, CCl_4, ~10 min, 75–95% yield.[11] This method is also effective for the regeneration of ketones from oximes (45–95% yield).
9. MCPBA, DMF, −63°, 100% yield.[12] An axial α-methyl group on a cyclohexanone does not epimerize under these conditions.
10. $NaBO_3 \cdot 4H_2O$, *t*-BuOH, pH 7, 60°, 24 h, 70–95% yield.[13]
11. AcOH, THF, H_2O, AcONa, 25°, 24 h, 95% yield.[14]

CH_3O ... H_3C ... N–NMe_2 ... CH_3O → (AcOH, THF; H_2O, NaOAc; 25°, 24 h; 95%) → CH_3O ... H_3C ... O ... CH_3O

N,N-Dimethylhydrazones are stable to CrO_3/H_2SO_4 (0°, 3 min), to $NaBH_4$ (EtOH, 25°), to $LiAlH_4$ (THF, 25°), and to B_2H_6 followed by H_2O_2/OH^-. They are cleaved by CrO_3/Pyr and by *p*-$NO_2C_6H_4CO_3H/CHCl_3$, 25°.[4]

12. Silica gel, THF, H_2O, rt, 3–10 h, 60–74% yield.[15]

1. G. R. Newkome and D. L. Fishel, *Org. Synth.*, **50,** 102 (1970); *Collect. Vol. VI*, 12 (1988).
2. E. J. Corey and D. Enders, *Tetrahedron Lett.*, 3 (1976).
3. E. J. Corey and S. Knapp, *Tetrahedron Lett.*, 3667 (1976).
4. M. Avaro, J. Levisalles, and H. Rudler, *J. Chem. Soc., Chem. Commun.*, 445 (1969).
5. R. E. Erickson, P. J. Andrulis, J. C. Collins, M. L. Lungle, and G. D. Mercer, *J. Org. Chem.*, **34,** 2961 (1969).
6. E. Friedrich, W. Lutz, H. Eichenauer, and D. Enders, *Synthesis*, 893 (1977).
7. G. A. Olah, J. Welch, and M. Henninger, *Synthesis*, 308 (1977).
8. G. A. Olah, J. Welch, G. K. S. Prakash, and T.-L. Ho, *Synthesis*, 808 (1976).
9. G. A. Olah and J. Welch, *Synthesis*, 809 (1976).
10. G. A. Olah, J. Welch, and T.-L. Ho, *J. Am. Chem. Soc.*, **98,** 6717 (1976).
11. S. B. Shim, K. Kim, and Y. H. Kim, *Tetrahedron Lett.*, **28,** 645 (1987).
12. M. Duraisamy and H. M. Walborsky, *J. Org. Chem.*, **49,** 3410 (1984).
13. D. Enders and V. Bhushan, *Z. Naturforsh. B: Chem. Sci.*, **42,** 1595 (1987).
14. E. J. Corey and H. L. Pearce, *J. Am. Chem. Soc.*, **101,** 5841 (1979).
15. R. B. Mitra and G. B. Reddy, *Synthesis*, 694 (1989).

39. 2,4-Dinitrophenylhydrazone (2,4-DNP group): $R_2C{=}NNHC_6H_3$-2,4-$(NO_2)_2$ (Chart 5)

Formation

1. 2,4-$(NO_2)_2C_6H_3NHNH_2 \cdot H_2SO_4$, EtOH, H_2O, 25°, 10 min, 80% yield.[1]

In a synthesis of sativene a carbonyl group was protected as a 2,4-DNP while a double bond was hydrated with $BH_3/H_2O_2/OH^-$. Attempted protection of the carbonyl group as a ketal caused migration of the double bond; protection as an oxime or oxime acetate was unsatisfactory since they would be reduced with BH_3.

Cleavage

2,4-Dinitrophenylhydrazones are cleaved by various oxidizing and reducing agents, and by exchange reactions.

1. O_3, EtOAc, −78°, 70% yield.[1]
2. $TiCl_3$, DME, H_2O, N_2, reflux, 80–95% yield.[2]
3. Acetone, sealed tube, 75°, 20 h, 80–85% yield.[3]

1. J. E. McMurry, *J. Am. Chem. Soc.*, **90,** 6821 (1968).
2. J. E. McMurry and M. Silvestri, *J. Org. Chem.*, **40,** 1502 (1975).
3. S. R. Maynez, L. Pelavin, and G. Erker, *J. Org. Chem.*, **40,** 3302 (1975).

40. Oxime Derivatives: $R_2C{=}NOH$

Formation

1. $H_2NOH \cdot HCl$, Pyr, 60°. This is the standard method for the preparation of oximes. Ethanol or methanol can be used as cosolvents.
2. $H_2NOH \cdot HCl$, DABCO, MeOH, rt, 87% for a camphor derivative.[1] This method was reported to be better than when pyridine was used as the solvent and base.

Cleavage

Oximes are cleaved by oxidation, reduction, or hydrolysis in the presence of another carbonyl compound. Some synthetically useful methods are shown below.

1. $CH_3CO(CH_2)_2COOH$, 1 *N* HCl, 25°, 3 h, 94% yield.[2] Pyruvic acid (HOAc, reflux, 1–3 h, 77% yield)[3] and acetone (80–100 h, 72% yield)[4] effect cleavage in a similar manner.
2. $(PhSeO)_2O$/THF, 50°, 1–3 h, 80–95% yield.[5] An *O*-methyl oxime is stable to phenylselenic anhydride.
3. $Na_2S_2O_4$, H_2O, 25°, 12 h or 40°, a few hours, ~95% yield.[6]
4. $NaHSO_3$, EtOH–H_2O, reflux, 2–16 h; dil. HCl, 30 min, 85% yield.[7]
5. Ac_2O, 20°; $Cr(OAc)_2$/THF–H_2O, 25–65°, 75–95% yield.[8] Chromous acetate also cleaves unsubstituted oximes, but the reaction is slow and requires high temperatures.
6. $NaNO_2$, 1 *N* HCl, CH_3OH, H_2O, 0°, 3 h, 76% yield.[9] In the last step of a synthesis of erythronolide A, acid-catalyzed hydrolysis of an acetonide failed because the carbonyl-containing precursor was unstable to acidic hydrolysis (3% MeOH, HCl, 0°, 30 min, conditions developed for the synthesis of erythronolide B). Consequently the carbonyl group was protected

as an oxime, the acetonide was cleaved, and the carbonyl group was regenerated.

7. NOCl, Pyr, −20°; H_2O, reflux, 70–90% yield.[10] Olefins were not affected under these conditions. The related nitrosyl tetrafluoroborate has also been used.[11]
8. $TiCl_3$, H_2O, rt, 1 h, 85% yield.[12] This is an excellent reagent that works when cleavage of a methoxy oxime with chromous ion fails.
9. VCl_2, H_2O, THF, 8 h, rt, 75–92% yield.[13]
10. Et_3NH^+ $ClCrO_3^-$, $ClCH_2CH_2Cl$, 2 h, rt, 60–90% yield.[14] This reagent was reported to work better than PCC (pyridinium chlorochromate).
11. Bu_3P, PhSSPh, THF, 85% yield.[15]
12. *t*-BuONO, *t*-BuOK; H_2O, NaOH; acidify, 40°.[16]

1. R. V. Stevens, F. C. A. Gaeta, and D. S. Lawrence, *J. Am. Chem. Soc.*, **105,** 7713 (1983).
2. C. H Depuy and B. W. Ponder, *J. Am. Chem. Soc.*, **81,** 4629 (1959).
3. E. B. Hershberg, *J. Org. Chem.*, **13,** 542 (1948).
4. S. R. Maynez, L. Pelavin, and G. Erker, *J. Org. Chem.*, **40,** 3302 (1975).
5. D. H. R. Barton, D. J. Lester, and S. V. Ley, *J. Chem. Soc., Chem. Commun.*, 445 (1977).
6. P. M. Pojer, *Aust. J. Chem.*, **32,** 201 (1979).
7. S. H. Pines, J. M. Chemerda, and M. A. Kozlowski, *J. Org. Chem.*, **31,** 3446 (1966).
8. E. J. Corey and J. E. Richman, *J. Am. Chem. Soc.*, **92,** 5276 (1970).
9. E. J. Corey, P. B. Hopkins, S. Kim, S. Yoo, K. P. Nambiar, and J. R. Falck, *J. Am. Chem. Soc.*, **101,** 7131 (1979).
10. C. R. Narayanan, P. S. Ramaswamy, and M. S. Wadia, *Chem. Ind. (London)*, 454 (1977).
11. G. A. Olah and T. L. Ho, *Synthesis*, 609 (1976).
12. G. H. Timms and E. Wildsmith, *Tetrahedron Lett.*, 195 (1971).
13. G. A. Olah, M. Arvanaghi, and G. K. S. Prakash, *Synthesis*, 220 (1980).
14. C. Gundu Rao, A. S. Radhakrishna, B. Bali Singh, and S. P. Bhatnagar, *Synthesis*, 808 (1983).
15. D. H. R. Barton, W. B. Motherwell, E. S. Simon, and S. Z. Zard, *J. Chem. Soc., Chem. Commun.*, 337 (1984).
16. E. J. Corey, M. Narisada, T. Hiraoka, and R. A. Ellison, *J. Am. Chem. Soc.*, **92,** 396 (1970).

41. *O*-Methyl Oxime: $R_2C{=}NOCH_3$

Formation[1]

1. $MeONH_2 \cdot HCl$, Pyr, MeOH, 23°, 30 min, 81% yield.

Cleavage[1]

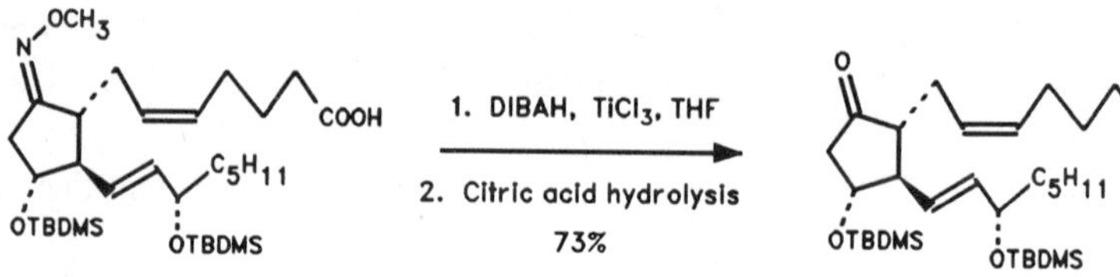

1. DIBAH, $TiCl_3$, THF
2. Citric acid hydrolysis
73%

This method was developed because conventional procedures failed to cleave the oxime.

1. E. J. Corey, K. Niimura, Y. Konishi, S. Hashimoto, and Y. Hamada, *Tetrahedron Lett.*, **27**, 2199 (1986).

42. *O*-Benzyl Oxime: $R_2C{=}NOCH_2Ph$

The reactions shown below were used in a synthesis of perhydrohistrionicotoxin; the carbonyl groups were protected as an oxime and an *O*-benzyl oxime.[1]

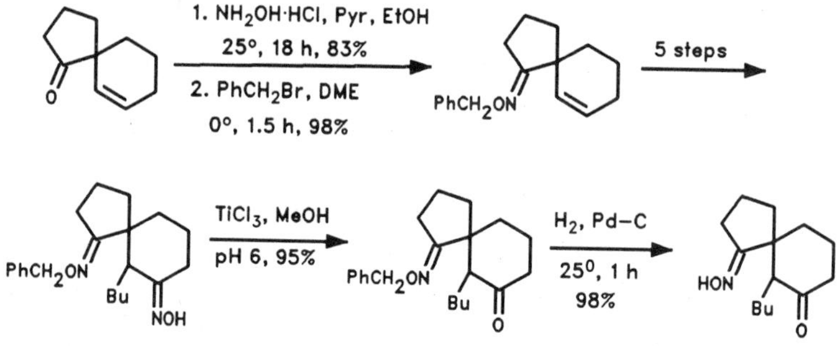

1. $NH_2OH \cdot HCl$, Pyr, EtOH, 25°, 18 h, 83%
2. $PhCH_2Br$, DME, 0°, 1.5 h, 98%

5 steps

$TiCl_3$, MeOH, pH 6, 95%

H_2, Pd–C, 25°, 1 h, 98%

1. E. J. Corey, M. Petrzilka, and Y. Ueda, *Helv. Chim. Acta*, **60**, 2294 (1977).

43. *O*-Phenylthiomethyl Oxime: $R_2C{=}NOCH_2SC_6H_5$ (Chart 5)

In a prostaglandin synthesis a carbonyl group was protected as an oxime in which the hydroxyl group was protected against Collins oxidation by the phenylthiomethyl group. The phenylthiomethyl group is readily removed to give an oxime that is then cleaved to the carbonyl compound.[1]

Formation

1. $PhSCH_2ONH_2$, Pyr, 25°, 24 h, 100% yield.

Cleavage

1. $HgCl_2$, HgO, AcOH, AcOK, 25–50°, 0.5–48 h, 75% yield; K_2CO_3, MeOH, 25°, 5 min, 100% yield. These conditions remove the $PhSCH_2$— group from the oxime, which is then cleaved with AcOH/$NaNO_2$ (10°, 1 h). This group was also stable to acid, base, and $LiAlH_4$.

1. I. Vlattas, L. Della Vecchia, and J. J. Fitt, *J. Org. Chem.*, **38,** 3749 (1973).

Imines

In general, imines are too reactive to be used to protect carbonyl groups. In a synthesis of juncusol,[1] however, a bromo- and an iodocyclohexylimine of two identical aromatic aldehydes were coupled by an Ullman coupling reaction modified by Ziegler.[2] The imines were cleaved by acidic hydrolysis (aq. oxalic acid, THF, 20°, 1 h, 95% yield). Imines of aromatic aldehydes have also been prepared to protect the aldehyde during ring metalation with *s*-BuLi.[3]

1. A. S. Kende and D. P. Curran, *J. Am. Chem. Soc.*, **101,** 1857 (1979).
2. F. E. Ziegler, K. W. Fowler, and S. Kanfer, *J. Am. Chem. Soc.*, **98,** 8282 (1976).
3. B. A. Keay and R. Rodrigo, *J. Am. Chem. Soc.*, **104,** 4725 (1982).

44. Substituted Methylene Derivatives: RR′C=C(CN)R″ (Chart 5)

RR′ = substituted pyrrole; R″ = —CN,[1] —CO_2Et[2]

The substituted methylene derivative, prepared from a 2-formylpyrrole and a malonic acid derivative, was used in a synthesis of chlorophyll.[1] It is cleaved under drastic conditions (concd alkali).[1,2]

1. R. B. Woodward and 17 co-workers, *J. Am. Chem. Soc.*, **82,** 3800 (1960).
2. J. B. Paine, R. B. Woodward, and D. Dolphin, *J. Org. Chem.*, **41,** 2826 (1976).

Cyclic Derivatives

45. Oxazolidines: 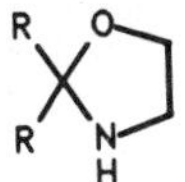

An oxazolidine was used to protect the carbonyl group in an α,β-unsaturated aldehyde during reduction of the carbon–carbon double bond by H_2/Raney Ni. It was prepared in ~60% yield from 2-aminoethanol and cleaved by acidic hydrolysis (10% HCl, 25°, 12 h, 60% yield).[1]

1. E. P. Goldberg and H. R. Nace, *J. Am. Chem. Soc.*, **77**, 359 (1955).

46. 1-Methyl-2-(1′-hydroxyalkyl)imidazoles:

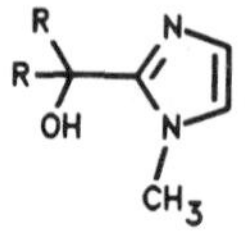

Formation/Cleavage[1]

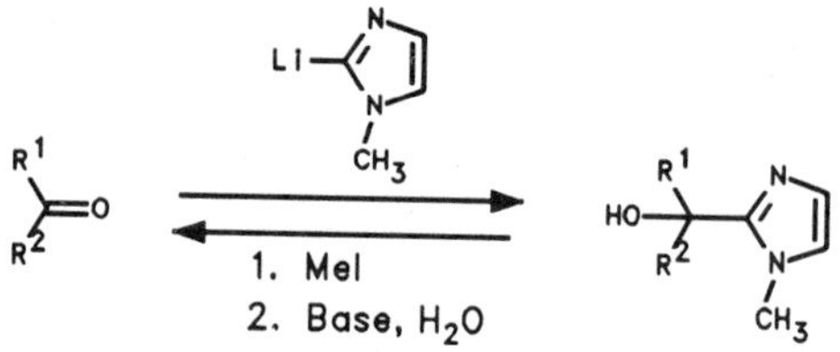

1. MeI
2. Base, H_2O

This protective group is stable to 1 *N* KOH/MeOH, 70°, 7 h; 20% H_2SO_4, 70°, 7 h; H_2, Pd–C, EtOH, 1 atm, 18 h; $NaBH_4$, $LiAlH_4$, CF_3COOH, Al_2O_3/MeOH.

1. S. Ohta, S. Hayakawa, K. Nishimura, and M. Okamoto, *Tetrahedron Lett.*, **25**, 3251 (1984).

47. *N,N′*-Dimethylimidazolidine:

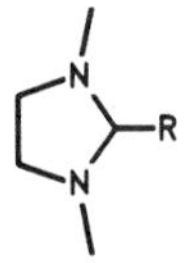

The imidazolidine was prepared from an aldehyde with *N,N′*-dimethyl-1,2-ethylenediamine (benzene, heat, 78% yield) and cleaved with MeI (Et_2O; H_2O, 92% yield). Derivatization is chemoselective for aldehydes. The imidazolidine is stable to BuLi and LDA.[1] The diphenylimidazolidine has been prepared analogously and can be cleaved with aqueous HCl.[2,3]

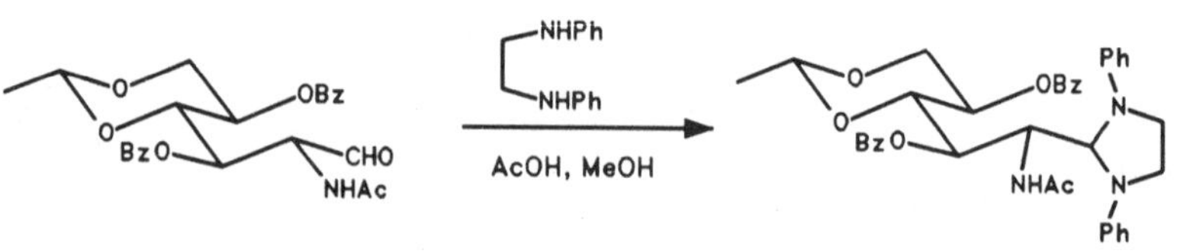

AcOH, MeOH

Ref. 3

1. A. J. Carpenter and D. J. Chadwick, *Tetrahedron*, **41,** 3803 (1985).
2. H.-W. Wanzlick and W. Löchel, *Chem. Ber.*, **86,** 1463 (1953).
3. A. Giannis, P. Münster, K. Sandhoff, and W. Steglich, *Tetrahedron*, **44,** 7177 (1988).

48. 2,3-Dihydro-1,3-benzothiazole:

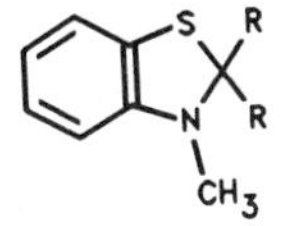

The benzothiazole group is introduced by heating 2-methylaminobenzenethiol with a carbonyl compound in ethanol (70–93% yield).[1] An enone is selectively protected over a ketone and aldehydes react faster than ketones. Cleavage is effected with $AgNO_3$ (CH_3CN, H_2O, pH 7, 83–93% yield) or by heating in Ac_2O followed by aqueous hydrolysis (HCl, $CHCl_3$, 50°, 1 h, 40% yield) of the resulting enamide.[2] Nonaromatic thiazolidines have also been used as protective groups. They can be cleaved by basic hydrolysis (NaOH, 25°, 95% yield).[3]

1. H. Chikashita, N. Ishimoto, S. Komazawa, and K. Itoh, *Heterocycles*, **23,** 2509 (1985).
2. G. Trapani, A. Reho, A. Latrofa, and G. Liso, *Synthesis*, 84, (1988).
3. K. Ueno, F. Ishikawa, and T. Naito, *Tetrahedron Lett.*, 1283 (1969).

49. Diethylamine Adduct: $R_2C[OTi(NEt_2)_3]NEt_2$

Titanium tetrakis(diethylamide) selectively adds to aldehydes in the presence of ketones and to the least hindered ketone in compounds containing more than one ketone. The protection is *in situ*, which thus avoids the usual protection–deprotection sequence. Selective aldol and Grignard additions are readily performed employing this protection methodology.[1]

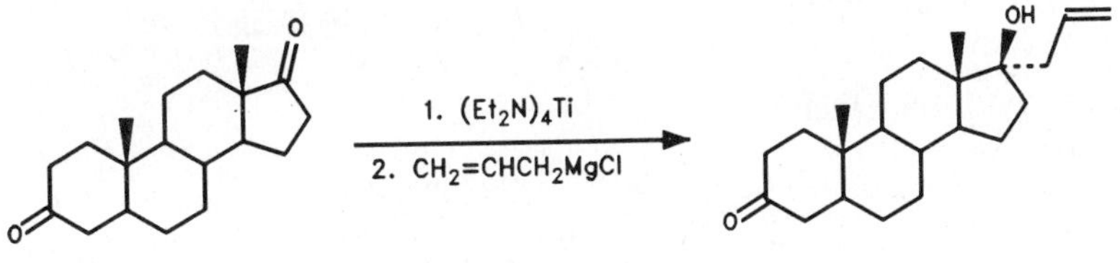

1. $(Et_2N)_4Ti$
2. $CH_2{=}CHCH_2MgCl$

1. M. T. Reetz, B. Wenderoth, and R. Peter, *J. Chem. Soc., Chem. Commun.*, 406 (1983).

50. Methylaluminum Bis(2,6-di-*t*-butyl-4-methylphenoxide) (MAD) Complex

This approach to carbonyl protection uses the relative differences in the basicity and the differences in steric effects to protect selectively either the more basic carbonyl group or the less hindered carbonyl group from reactions with nucleophiles such as DIBAH.[1]

1. K. Maruoka, Y. Araki, and H. Yamamoto, *J. Am. Chem. Soc.*, **110,** 2650 (1988).

MONOPROTECTION OF DICARBONYL COMPOUNDS

Selective Protection of α- and β-Diketones

α- and β-Diketones can be protected as enol ethers, thioenol ethers, enol acetates, and enamines.

51. Enamines:

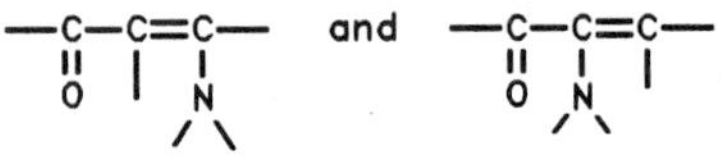

—C(=O)—C=C(N<)— and —C(=O)—C(N<)=C—

52. Enol Acetates:

—C(=O)—C=C(OAc)— and —C(=O)—C(OAc)=C—

Enol Ethers:

—C(=O)—C=C(OR)— and —C(=O)—C(OR)=C—

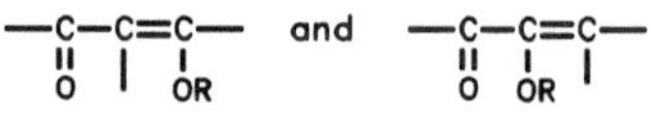

53. Methyl Enol Ether:

54. Ethyl Enol Ether:

55. *i*-Butyl Enol Ether:

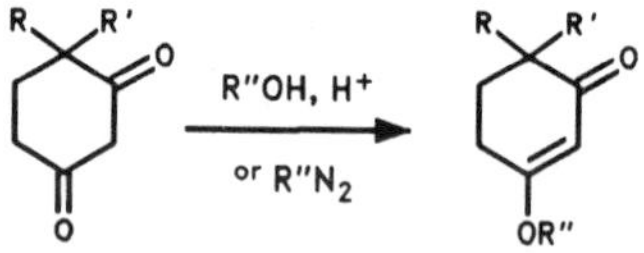

$$\xrightarrow[\text{or } R''N_2]{R''OH,\ H^+}$$

R″OH: R″ = Me (HCl, 25°, 8 h, 83% yield).[1]
R″ = Et (TsOH, benzene, reflux, 6–8 h, 70–75% yield).[2]
R″ = $(CH_3)_2CHCH_2$ (*i*-BuOH, benzene, reflux, TsOH, 16 h, 100% yield).[3] In this case 2-methyl-1,3-cyclopentanedione was monoprotected.

56. Piperidinyl Enol Ether:

57. Morpholinyl Enol Ether:

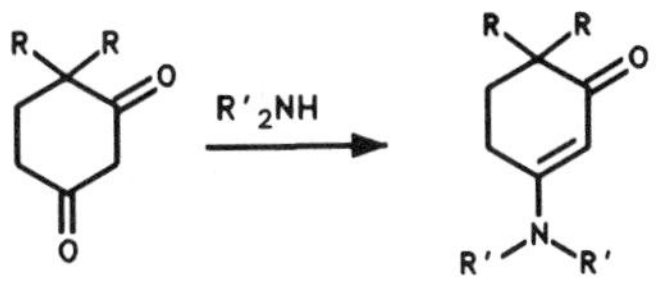

R'₂NH

R'_2NH = piperidine, TsOH, benzene, reflux, 92% yield.[4]
R'_2NH = morpholine, TsOH, $PhCH_3$, reflux, 4–5 h, 72–80% yield.[5]

58. 4-Methyl-1,3-dioxolanyl Enol Ether:

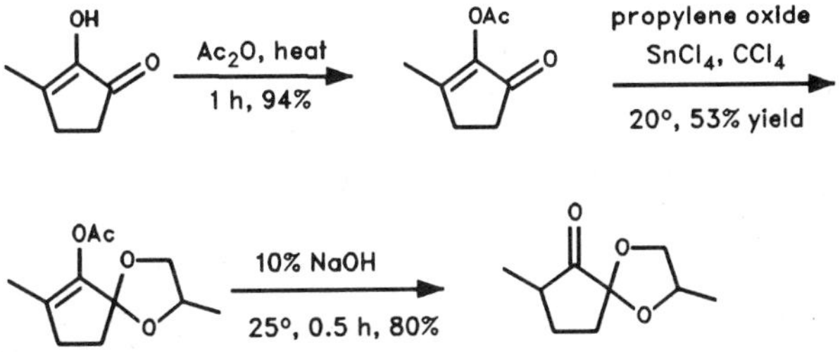

Ac_2O, heat, 1 h, 94%

propylene oxide, $SnCl_4$, CCl_4, 20°, 53% yield

10% NaOH, 25°, 0.5 h, 80%

Ref. 6

59. Pyrrolidinyl Enol Ether:

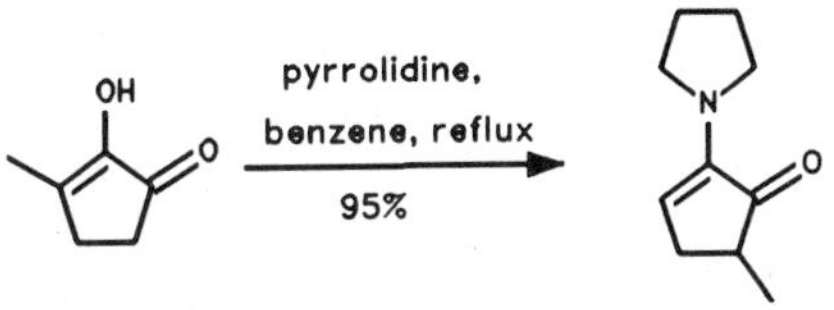

pyrrolidine, benzene, reflux, 95%

Ref. 7

60. Benzyl Enol Ether:

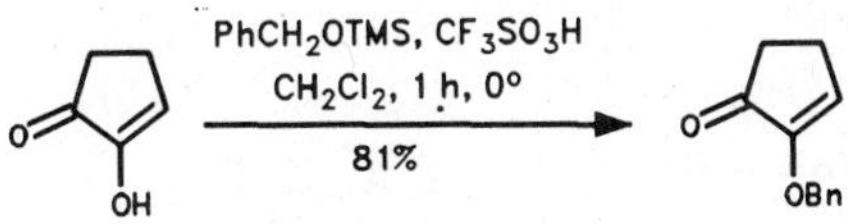

$PhCH_2OTMS$, CF_3SO_3H, CH_2Cl_2, 1 h, 0°, 81%

Ref. 8

61. *S*-Butyl Enol Ether:

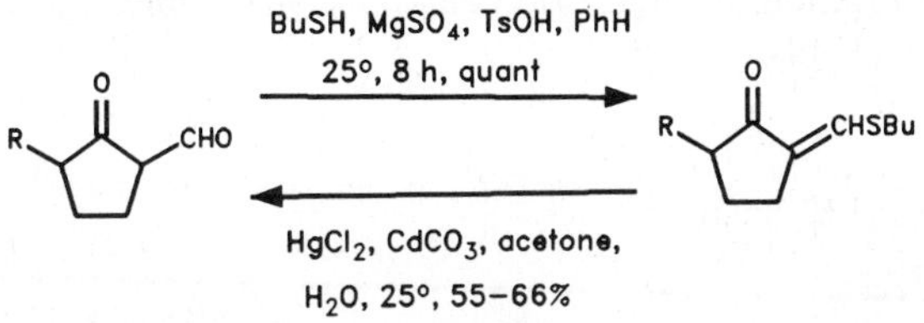

BuSH, $MgSO_4$, TsOH, PhH, 25°, 8 h, quant

$HgCl_2$, $CdCO_3$, acetone, H_2O, 25°, 55–66%

Ref. 9

1. H. O. House and G. H. Rasmusson, *J. Org. Chem.*, **28,** 27 (1963).
2. W. F. Gannon and H. O. House, *Org. Synth., Collect. Vol. V,* 539 (1973).
3. M. Rosenberger and P. J. McDougal, *J. Org. Chem.*, **47,** 2134 (1982).
4. P. Kloss, *Chem. Ber.*, **97,** 1723 (1964).
5. S. Hünig, E. Lücke, and W. Brenninger, *Org. Synth., Collect. Vol. V,* 808 (1973).
6. J. L. E. Erickson and F. E. Collins, Jr., *J. Org. Chem.*, **30,** 1050 (1965).
7. E. Gordon, F. Martens, and H. Gault, *C. R. Hebd. Seances Acad. Sci., Ser. C,* **261,** 4129 (1965).
8. A. A. Ponaras and Md. Y. Meah, *Tetrahedron Lett.*, **27,** 4953 (1986).
9. P. R. Bernstein, *Tetrahedron Lett.*, 1015 (1979).

62. Trimethylsilyl Enol Ethers: R(R)C=C(R)OTMS

Trimethylsilyl enol ethers can be used to protect ketones, but in general are not used for this purpose because they are reactive under both acidic and basic conditions. More highly hindered silyl enol ethers are much less susceptible to acid and base. A less hindered silyl enol can be hydrolyzed in the presence of a more hindered one.[1]

Bu_3SnF, PhH, $PdCl_2(TPP)_2$, reflux, 91%

The preparation of silyl enol ethers has been reviewed.[2,3]

1. H. Urabe, Y. Takano, and I. Kuwajima, *J. Am. Chem. Soc.*, **105,** 5703 (1983).
2. E. Colvin, *Silicon in Organic Synthesis,* Butterworths, Boston, 1981, pp. 198–287.
3. W. P. Weber, *Silicon Reagents for Organic Synthesis,* Springer-Verlag, New York, 1983, pp. 255–272.

Cyclic Ketals, Monothio and Dithio Ketals

Cyclohexane-1,2-dione reacts with ethylene glycol (TsOH, benzene, 6 h) to form the diprotected compound. Monoprotected 1,3-oxathiolanes and 1,3-dithiolanes are isolated on reaction under similar conditions with 2-mercaptoethanol and ethanedithiol, respectively.[1]

1. R. H. Jaeger and H. Smith, *J. Chem. Soc.*, 160, 646 (1955).

63. Bismethylenedioxy Derivatives (Chart 5):

Formation/Cleavage:[1]

CH_2O, concd HCl
$CHCl_3$, 48 h,
50–70%

60% HCOOH, 90°,
30 min or
50% AcOH, 100°,
7 h, 50–70%

This derivative is stable to TsOH/benzene at reflux and to CrO_3/H^+.[2] It is stable to NBS/*hν*.[3] In the formation of this derivative formaldehyde from formalin can react with a C_{11}-hydroxyl group to form a methoxymethyl ether. Paraformaldehyde can be used to avoid formation of the ethers.[4]

1. R. E. Beyler, F. Hoffman, R. M. Moriarty, and L. H. Sarett, *J. Org. Chem.*, **26,** 2421 (1961).
2. J. F. W. Keana, in *Steroid Reactions,* C. Djerassi, Ed., Holden-Day, San Francisco, 1963, pp. 56–61.
3. D. Duval, R. Condom, and R. Emiliozzi, *C. R. Hebd. Seances Acad. Sci., Ser C,* **285,** 281 (1977).
4. J. A. Edwards, M. C. Calzada, and A. Bowers, *J. Med. Chem.*, **7,** 528 (1964).

64. Tetramethylbismethylenedioxy Derivatives:

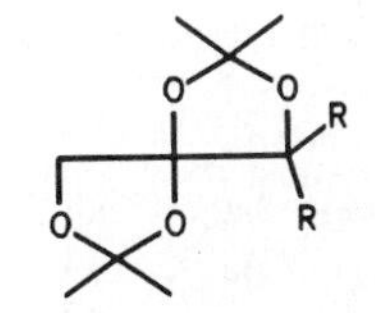

A bismethylenedioxy group in a 4-chloro or 11-keto steroid is stable to cleavage by formic acid or glacial acetic acid (100°, 6 h). The tetramethyl derivative is readily hydrolyzed (50% AcOH, 90°, 3–4 h, 80–90% yield).[1]

1. A. Roy, W. D. Slaunwhite, and S. Roy, *J. Org. Chem.*, **34,** 1455 (1969).

5

PROTECTION FOR THE CARBOXYL GROUP

Carboxylic acids are protected for a number of reasons: (1) to mask the acidic proton so that it does not interfere with base-catalyzed reactions, (2) to mask the carbonyl group to prevent nucleophilic addition reactions, and (3) to improve the handling of the molecule in question (e.g., to make the compound less water-soluble, to improve its NMR characteristics, or to render it more volatile so that it can be analyzed by gas chromatography). Besides stability to a planned set of reaction conditions, the protective group must also be removed without affecting other functionality in the molecule. For this reason a large number of protective groups for acids have been developed that are removed under a variety of conditions even though most can be readily cleaved by simple hydrolysis. Hydrolysis is an important means of deprotection and the rate of hydrolysis is, of course, dependent on steric and electronic factors that then help to achieve differential deprotection in polyfunctional substrates. These factors are also important in the selective protection of compounds containing two or more carboxylic acids. Hydrolysis using HOO^- is about 400 times faster than simple hydrolysis with hydroxide (phenyl acetate = substrate).[1]

Polymer-supported esters[2] are widely used in solid-phase peptide synthesis, and extensive information for this specialized protection is reported annually.[3] Some activated esters that have been used as macrolide precursors and some that have been used in peptide synthesis are also described in this chapter; the many activated esters that are used in peptide synthesis are discussed elsewhere.[3] A useful list, with references, of many protected amino acids (e.g., $-NH_2$, COOH, and side-chain-protected compounds) has been compiled.[4] Some general methods for the preparation of esters are provided at the beginning of this chapter;[5] conditions that are unique to a protective group are described with that group.[6] Some esters that have been used as protective groups are included in Reactivity Chart 6.

1. W. P. Jencks and M. Gilchrist, *J. Am. Chem. Soc.*, **90,** 2622 (1968).
2. See: P. Hodge, "Polymer-Supported Protecting Groups," *Chem. Ind. (London)*, 624 (1979); R. B. Merrifield, G. Barany, W. L. Cosand, M. Engelhard, and S. Mojsov, "Some Recent Developments in Solid Phase Peptide Synthesis," in *Peptides: Proceedings of the Fifth American Peptide Symposium*, M. Goodman and J. Meienhofer, Eds., Wiley, New York, 1977, pp. 488–502; J. M. J. Fréchet, "Synthesis and Applications of Organic Polymers as Supports and Protecting Groups," *Tetrahedron*, **37,** 663 (1981).
3. *Specialist Periodical Reports*: "Amino-Acids, Peptides, and Proteins," Royal Society of Chemistry, London, Vols. 1–16 (1969–1983); "Amino Acids and Peptides," Vols. 17–21 (1984–1990).
4. E. Gross and J. Meinenhofer, Eds., "The Peptides: Analysis, Synthesis, Biology: Vol. 3: Protection of Functional Groups in Peptide Synthesis," Academic Press, New York, 1981.
5. For classical methods, see: C. A. Buehler and D. E. Pearson, *Survey of Organic Syntheses*, Wiley-Interscience, New York, 1970, Vol. 1, pp. 801–830; 1977, Vol. 2, pp. 711–726.
6. See also: E. Haslam, "Recent Developments in Methods for the Esterification and Protection of the Carboxyl Group," *Tetrahedron*, **36,** 2409–2433 (1980); E. Haslam, "Activation and Protection of the Carboxyl Group," *Chem. Ind. (London)*, 610–617 (1979); E. Haslam, "Protection of Carboxyl Groups," in *Protective Groups in Organic Chemistry*, J. F. W. McOmie, Ed., Plenum Press, New York and London, 1973, pp. 183–215.

ESTERS

General Preparation of Esters

The preparation of esters can be classified into two main categories: (1) carboxylate activation with a good leaving group and (2) nucleophilic displacement of a carboxylate on an alkyl halide or sulfonate. The latter approach is generally not suitable for the preparation of esters if the halide or tosylate is sterically hindered, but there has been some success with simple secondary halides[1] and tosylates (ROTs, DMF, K_2CO_3, 69–93% yield).[2]

1. RCO_2H + R′OH, MeTHF, Me_3SiCl (or Me_2SiCl_2, $MeSiCl_3$, or $SiCl_4$), rt, 15 min–100 h, 90–97% yield.[3] In this case both R and R′ can be hindered. Since the reaction conditions generate HCl the substrates should be stable to strong acid.

2. RCO_2H, R'OH, DCC/DMAP, Et_2O, 25°, 1–24 h, 70–95% yield. This method is suitable for a large variety of hindered and unhindered acids and alcohols.[4]
3. RCO_2H, R'OH, BOP–Cl, Et_3N, CH_2Cl_2, 23°, 2 h, 71–99% yield.[5] This is an excellent general method for the preparation of esters.

BOP–Cl =

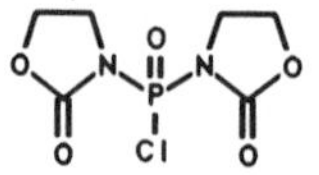

4. $RCO_2H \xrightarrow[\text{2. R'OH, DMAP, >95\%}]{\text{1. 2,4,6-Cl}_3\text{C}_6\text{H}_2\text{COCl, Et}_3\text{N, THF}^6} RCO_2R'$

   This method is best suited to the preparation of relatively unhindered esters; otherwise some esterification of the benzoic acid may occur at the expense of the acid to be esterified.
5. $RCO_2H + R'X \xrightarrow[\text{25–80°, 1–10 h}]{\text{DBU, benzene}^7} RCO_2R'$, 70–95%

   RCO_2H = alkyl, aryl, hindered acids
   R' = Et, *n*- and *s*-Bu, CH_3SCH_2 . . .
   X = Cl, Br, I

   The reaction also proceeds well in acetonitrile, allowing lower temperatures (25°) and shorter times.[8]
6. $RCH(NHPG)CO_2H \xrightarrow[\text{pH 7}]{Cs_2CO_3} \xrightarrow[\text{6 h}]{\text{R'X, DMF}^9} RCH(NHPG)CO_2R'$

   R' = Me, 80%; $PhCH_2$, 70–90%; *o*-$NO_2C_6H_4CH_2$, 90%; *p*-$MeOC_6H_4CH_2$, 70%; Ph_3C, 40–60%; *t*-Bu, 14%; PhCOCH(Me), 80%; *N*-phthalimidomethyl, 80% yield

   A study of relative rates of this reaction indicates that $Cs^+ > K^+ > Na^+ > Li^+$; $I^- \gg Br^- \gg Cl^-$; HMPA > DMSO > DMF.[10]
7. $RCH(NHPG)CO_2H + R'X \xrightarrow[\text{25°, 24 h, 90–95\%}]{\text{NaHCO}_3\text{, DMF}^{11}} RCH(NHPG)CO_2R'$

   R' = Et, *n*-Bu, *s*-Bu
   X = Br, I
8. $RCH(NHPG)CO_2H + R'X \xrightarrow[\text{25°, 3–24 h, 70–95\%}]{(C_8H_{17})_3N^+MeCl^-\text{, aq. NaHCO}_3\text{, CH}_2\text{Cl}_2{}^{12}} RCH(NHPG)CO_2R'$
9. $RCO_2H + R'_3O^+BF_4^- \xrightarrow[\text{20°, 1–24 h}]{\text{EtN-}i\text{-Pr}_2\text{, CH}_2\text{Cl}_2{}^{13}} RCO_2R'$, 70–95% yield

   RCO_2H = hindered acids
   R' = Me, Et
10. $RCO_2H + Me_2NCH(OR')_2 \xrightarrow{\text{25–80°, 1–36 h}^{14}} RCO_2R'$, 80–95%

   RCO_2H = Ph, 2,4,6-$Me_3C_6H_2$—, *N*-protected amino acids
   R' = Me, Et, $PhCH_2$, *s*-Bu

11. $RCO_2H + R'OH \xrightarrow[0\text{–}20°,\ 24\ h]{t\text{-BuNC}^{15}} RCO_2R'$, 36–98%

RCO_2H = amino, dicarboxylic acids; ≠ $PhCO_2H$

R′ = Me, Et, *t*-Bu

12. $RCO_2H + R'OH \xrightarrow[25°,\ 12\ h,\ 75\text{–}85\%]{Ph_3P(OSO_2CF_3)_2,\ CH_2Cl_2{}^{16}} RCO_2R'$

R = aryl

R′ = Et

13. $RCO_2H \xrightarrow[Et_3N,\ DMAP]{ClCO_2R',\ CH_2Cl_2,\ 0°{}^{17}} RCO_2R'$, 89–98%

The reaction is not suitable for hindered carboxylic acids, since considerable symmetrical anhydride formation (52% with pivalic acid) results. Symmetrical anhydride formation can sometimes be suppressed by the use of stoichiometric quantities of DMAP.

14. $RCO_2H + R'X \xrightarrow[R_4''N^+X^-,\ rt,\ 80\text{–}99\%\ yield]{electrolysis:\ pyrrolidone,\ DMF^{18}} RCO_2R'$

This method is based on the generation of the tetraalkylammonium salt of pyrrolidone, which acts as a base. The method is compatible with a large variety of carboxylic acids and alkylating agents. The method is effective for the preparation of macrolides.

15. RCH(NHPG)$CO_2H \xrightarrow[DMAP,\ CH_2Cl_2,\ 0°,\ R'OH]{Isopropenyl\ chloroformate^{19}}$ RCH(NHPG)CO_2R', 60–96%

16. $RCO_2H + R'OH \xrightarrow[0.1\text{–}48\ h,\ 67\text{–}98\%]{TMSCl,\ 22\text{–}78°{}^{20}} RCO_2R'$

17. $RCO_2H \xrightarrow[2.\ R'OH,\ reflux,\ 5\text{–}120\ min,\ 80\text{–}90\%]{1.\ TsCl,\ K_2CO_3,\ TEBAC^{21},\ 40°\text{–reflux},\ 5\text{–}60\ min} RCO_2R'$

TEBAC = $Et_3N^+CH_2Ph\ Cl^-$

General Cleavage of Esters

1. $RCO_2R' + Nu^- \xrightarrow{aprotic\ solvent^{22}} RCO_2H$

Nu^- = LiS-*n*-Pr: HMPA, 25°, 1 h, ~ quant. yield[23]

= NaSePh: HMPA-THF, reflux, 7 h, 90–100% yield[24]

= LiCl: DMF or Pyr, reflux, 1–18 h, 60–90% yield[25]

= KO-*t*-Bu: DMSO, 50–100°, 1–24 h, 65–95% yield[26]

= NaCN (for decarboxylation of malonic esters): DMSO, 160°, 4 h, 70–80% yield[27]

= NaTeH from Te, DMF, *t*-BuOH, $NaBH_4$, 80–90°, 15 min, 85–98% yield[28]

= KO_2: 18-crown-6, benzene, 25°, 8–72 h, 80–95% yield[29]

2. $RCO_2R' \xrightarrow[reflux,\ 5\text{–}35\ h,\ 70\text{–}90\%]{TMSCl,\ NaI,\ CH_3CN^{30\text{–}32}} RCO_2H$

RCO_2H = alkyl, aryl, hindered acids

R′ = Me, Et, *i*-Pr, *t*-Bu, $PhCH_2$

This method generates Me_3SiI *in situ*. The reagent also cleaves a number of other protective groups.

3. $RCO_2R' \xrightarrow[80\text{–}100\%]{\text{KO-}t\text{-Bu/H}_2\text{O (4:1), 25°, 2–48 h}^{33}} RCO_2H$

   RCO_2H = Ph, aryl, hindered acids
   R′ = Me, *t*-Bu, alkyl

   "Anhydrous hydroxide" also cleaves tertiary amides.

4. $RCH(NHPG)CO_2R' \xrightarrow[-10°,\ 1\ h \rightarrow 25°,\ 2\ h]{BBr_3,\ CH_2Cl_2{}^{34}} RCH(NH_2)CO_2H$, 60–85%

   R′ = Me, Et, *t*-Bu, $PhCH_2$
   PG = $-CO_2CH_2Ph$, $-CO_2$-*t*-Bu; OMe, OEt, O-*t*-Bu, OCH_2Ph side-chain ethers

5. $RCO_2R' \xrightarrow[70\text{–}95\%]{AlX_3,\ R''SH,\ 25°,\ 5\text{–}50\ h^{35,36}} RCO_2H$

   R = Ph, steroid side chain, . . .
   R′ = Me, Et, $PhCH_2$
   R″ = Et, $HO(CH_2)_2-$
   X = Cl, Br

6. $RCO_2R' \xrightarrow[\text{benzene, 1–30 h}]{\text{xs }(Bu_3Sn)_2O,\ 80°^{37}} RCO_2H$, 40–95% yield

   R′ = $CH_2O_2CC(CH_3)_3$, Me, Et, Ph

7. Transesterification can be used to convert one type of ester to another type that can then be removed under a new set of conditions.

   $$RCO_2R' + R''OH \xrightarrow[\text{toluene, >88\% yield}]{Bu_2Sn(OH)OSn(NCS)Bu_2\ \text{cat.}^{38}} RCO_2R'' + R'OH$$

   This method is not effective for tertiary alcohols. It has a strong rate dependence on solvent polarity, with the less polar solvents giving the faster rates.

1. T. Shono, O. Ishige, H. Uyama, and S. Kashimura, *J. Org. Chem.*, **51,** 546 (1986).
2. W. L. Garbrecht, G. Marzoni, K. R. Whitten, and M. L. Cohen, *J. Med. Chem.*, **31,** 444 (1988).
3. R. Nakao, K. Oka, and T. Fukomoto, *Bull. Chem. Soc. Jpn.*, **54,** 1267 (1981).
4. A. Hassner and V. Alexanian, *Tetrahedron Lett.*, 4475 (1978).
5. J. Diago-Meseguer, A. L. Palomo-Coll, J. R. Fernández-Lizarbe, and A. Zugaza-Bilbao, *Synthesis*, 547 (1980).
6. J. Inanaga, K. Hirata, H. Saeki, T. Katsuki, and M. Yamaguchi, *Bull. Chem. Soc. Jpn.*, **52,** 1989 (1979).
7. N. Ono, T. Yamada, T. Saito, K. Tanaka, and A. Kaji, *Bull. Chem. Soc. Jpn.*, **51,** 2401 (1978).
8. C. G. Rao, *Org. Prep. Proc. Int.*, **12,** 225 (1980).
9. S.-S. Wang, B. F. Gisin, D. P. Winter, R. Makofske, I. D. Kulesha, C. Tzougraki, and J. Meienhofer, *J. Org. Chem.*, **42,** 1286 (1977).
10. P. E. Pfeffer and L. S. Silbert, *J. Org. Chem.*, **41,** 1373 (1976).

11. V. Bocchi, G. Casnati, A. Dossena, and R. Marchelli, *Synthesis*, 961 (1979).
12. V. Bocchi, G. Casnati, A. Dossena, and R. Marchelli, *Synthesis*, 957 (1979).
13. D. J. Raber, P. Gariano, A. O. Brod, A. Gariano, W. C. Guida, A. R. Guida, and M. D. Herbst, *J. Org. Chem.*, **44,** 1149 (1979).
14. H. Brechbühler, H. Büchi, E. Hatz, J. Schreiber, and A. Eschenmoser, *Helv. Chim. Acta*, **48,** 1746 (1965).
15. D. Rehn and I. Ugi, *J. Chem. Res., Synop.*, 119 (1977).
16. J. B. Hendrickson and S. M. Schwartzman, *Tetrahedron Lett.*, 277 (1975).
17. S. Kim, Y. C. Kim, and J. I. Lee, *Tetrahedron Lett.*, **24,** 3365 (1983).
18. T. Shono, O. Ishige, H. Uyama, and S. Kashimura, *J. Org. Chem.*, **51,** 546 (1986).
19. P. Jouin, B. Castro, C. Zeggaf, A. Pantaloni, J. P. Senet, S. Lecolier, and G. Sennyey, *Tetrahedron Lett.*, **28,** 1661 (1987).
20. M. A. Brook and T. H. Chan, *Synthesis*, 201 (1983).
21. Z. M. Jászay, I. Petneházy, and L. Töke, *Synthesis*, 745 (1989).
22. J. McMurry, "Ester Cleavages via S_N2-Type Dealkylation," *Org. React.*, **24,** 187–224 (1976).
23. P. A. Bartlett and W. S. Johnson, *Tetrahedron Lett.*, 4459 (1970).
24. D. Liotta, W. Markiewicz, and H. Santiesteban, *Tetrahedron Lett.*, 4365 (1977).
25. F. Elsinger, J. Schreiber, and A. Eschenmoser, *Helv. Chim. Acta*, **43,** 113 (1960).
26. F. C. Chang and N. F. Wood, *Tetrahedron Lett.*, 2969 (1964).
27. A. P. Krapcho, G. A. Glynn, and B. J. Grenon, *Tetrahedron Lett.*, 215 (1967).
28. J. Chen and X. J. Zhou, *Synthesis*, 586 (1987).
29. J. San Filippo, L. J. Romano, C.-I. Chern, and J. S. Valentine, *J. Org. Chem.*, **41,** 586 (1976).
30. M. E. Jung and M. A. Lyster, *J. Am. Chem. Soc.*, **99,** 968 (1977).
31. T. Morita, Y. Okamoto, and H. Sakurai, *J. Chem. Soc., Chem. Commun.*, 874 (1978).
32. G. A. Olah, S. C. Narang, B. G. B. Gupta, and R. Malhotra, *J. Org. Chem.*, **44,** 1247 (1979).
33. P. G. Gassman and W. N. Schenk, *J. Org. Chem.*, **42,** 918 (1977).
34. A. M. Felix, *J. Org. Chem.*, **39,** 1427 (1974).
35. M. Node, K. Nishide, M. Sai, and E. Fujita, *Tetrahedron Lett.*, 5211 (1978).
36. M. Node, K. Nishide, M. Sai, K. Fuji, and E. Fujita, *J. Org. Chem.*, **46,** 1991 (1981).
37. E. G. Mata and O. A. Mascaretti, *Tetrahedron Lett.*, **29,** 6893 (1988).
38. J. Otera, T. Yano, A. Kawabata, and H. Nozaki, *Tetrahedron Lett.*, **27,** 2383 (1986); J. Otera, S. Ioka, and H. Nozaki, *J. Org. Chem.*, **54,** 4013 (1989).

1. Methyl Ester: RCO_2CH_3 (Chart 6)

Formation

The section on general preparation of esters should also be consulted.

1. $H_2NCON(NO)Me$, KOH, DME, H_2O, 0°, 75% yield. This method generates diazomethane *in situ*.[1] *N*-Methyl-*N*-nitrosourea is a proven carcinogen.

2. Me_3SiCHN_2, MeOH, benzene, 20°.[2,3] This reagent does not react with phenols. This is a safe alternative to the use of diazomethane. A detailed, large-scale preparation of this useful reagent has been described.[4]
3. $Me_2C(OMe)_2$, cat. HCl, 25°, 18 h, 80–95% yield.[5] These reaction conditions were used to prepare methyl esters of amino acids.
4. MeOH, H_2SO_4, 0°, 1 h; 5°, 18 h, 98% yield.[6]

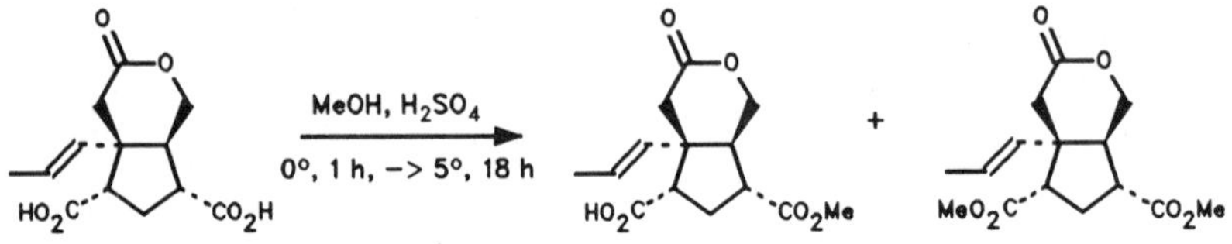

MeOH, H_2SO_4, 0°, 1 h, –> 5°, 18 h

Ratio = 4:1

5. MeOH, HBF_4, Na_2SO_4, 25–60°, 15 h, 45–94% yield.[7] The selectivity observed here is also observed for Et, *i*-Pr, Bn, and cyclohexyl esters (n = 1,2).

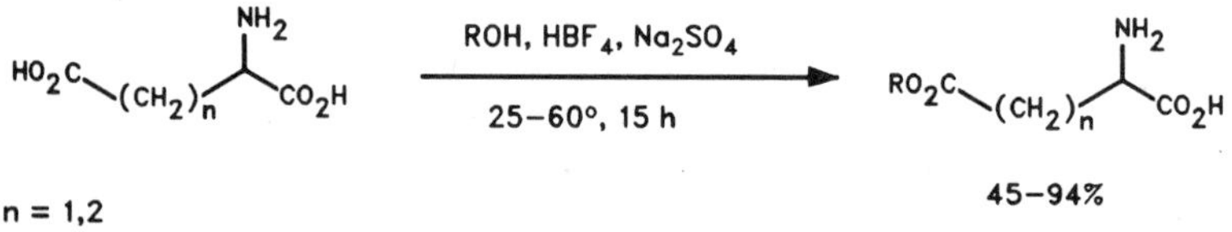

ROH, HBF_4, Na_2SO_4, 25–60°, 15 h

45–94%

n = 1,2

R = CH_3, Et, *i*-Pr, Bn, cyclohexyl

Cleavage

1. LiOH, CH_3OH, H_2O, (3 : 1), 5°, 15 h.[8]
2. NaCN, HMPA, 75°, 24 h, 75–92% yield.[9] Ethyl esters are not cleaved under these conditions.
3. LiI, Pyr, reflux, 91% yield.[10]

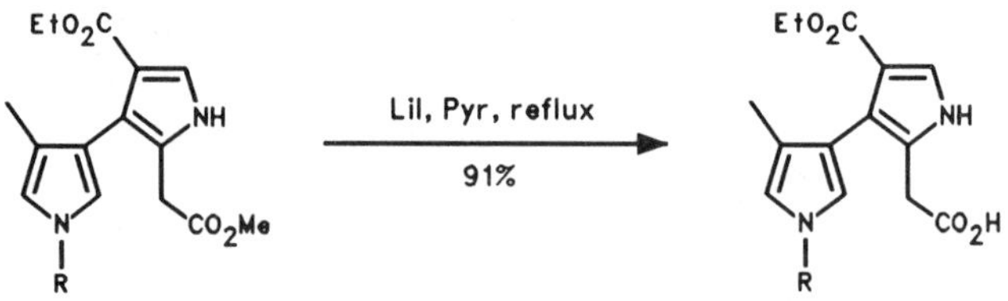

LiI, Pyr, reflux

91%

4. $(CH_3)_3SiOK$, ether, 4 h, 61–95% yield as the acid salt.[11]
5. $Ba(OH)_2 \cdot 8H_2O$, MeOH, rt, 7 h, 72% yield.[12]

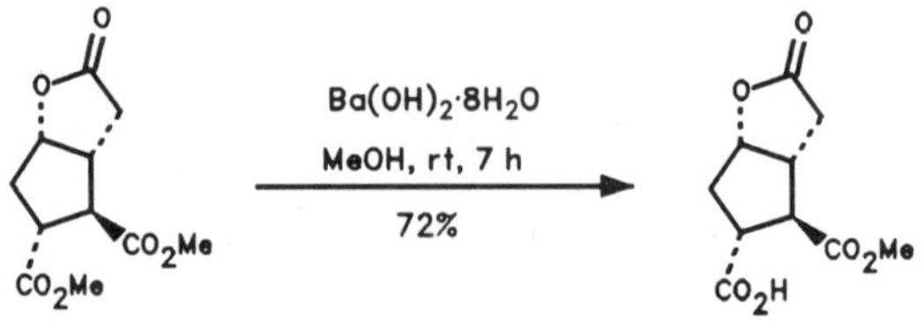

$Ba(OH)_2 \cdot 8H_2O$, MeOH, rt, 7 h

72%

6. 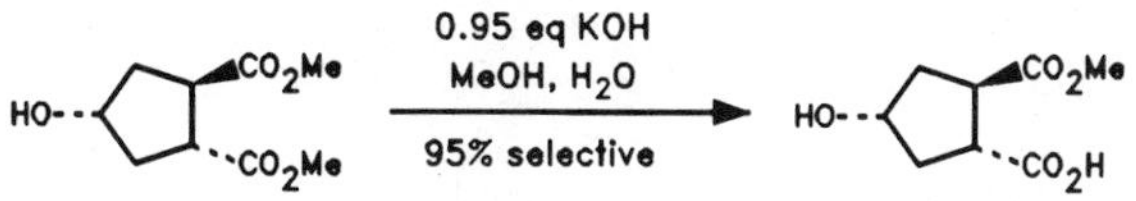 HO-cyclopentane(CO_2Me)(CO_2Me) → (0.95 eq KOH, MeOH, H_2O; 95% selective) HO-cyclopentane(CO_2Me)(CO_2H) Ref. 13

The authors propose that the selectivity is due to participation of the hydroxyl group.

7. $H_2NC_6H_4SH$, Cs_2CO_3, DMF, 85°, 1–3 h.[14]
8. Pig liver esterase is particularly effective in cleaving one ester of a symmetrical pair.[15–17]
9. 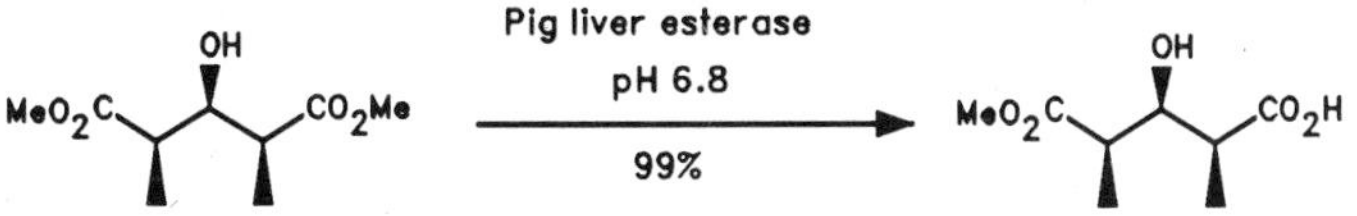 MeO_2C–CH(Me)–CH(OH)–CH(Me)–CO_2Me → (Pig liver esterase, pH 6.8; 99%) MeO_2C–CH(Me)–CH(OH)–CH(Me)–CO_2H Ref. 18

10. 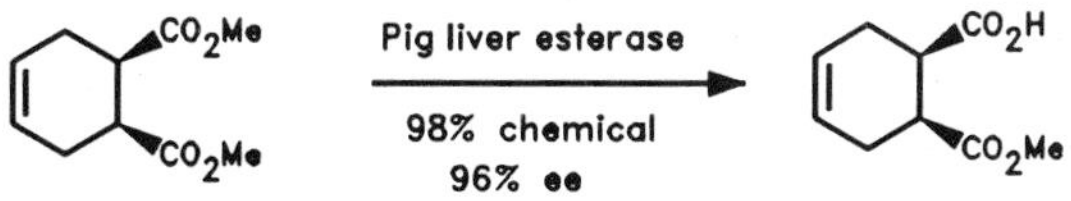 cyclohexene-diester (CO_2Me, CO_2Me) → (Pig liver esterase; 98% chemical, 96% ee) (CO_2H, CO_2Me) Ref. 19

11. 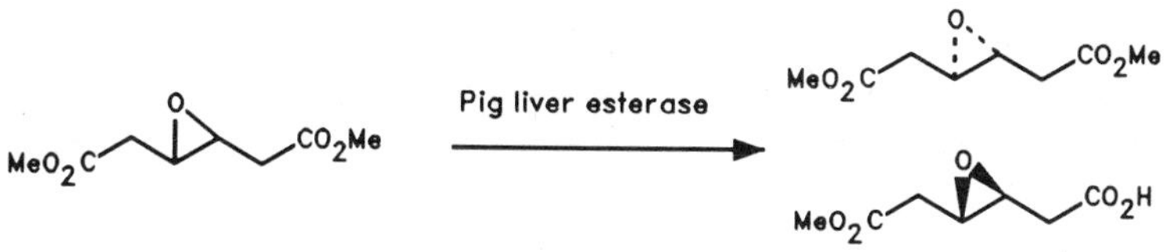 MeO_2C-epoxide-CO_2Me → (Pig liver esterase) MeO_2C-epoxide-CO_2Me + MeO_2C-epoxide-CO_2H

E = 21.5 (enantiomeric ratio)

Ref. 20

12. BCl_3, 0°, 5–6 h, 90% yield.[21] In this example a phenolic methyl group, normally cleaved with boron trichloride, was not affected.
13. $NaBH_4$, I_2, 3 h, rt.[22]
14. 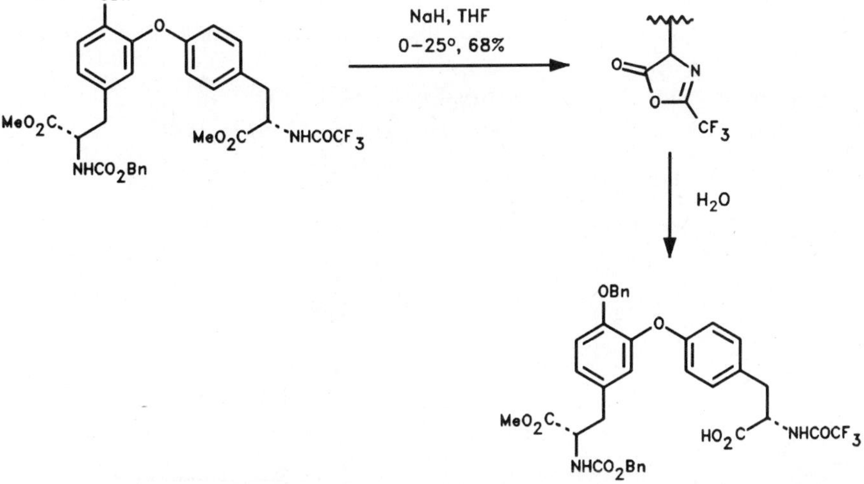 OBn, O, MeO_2C, $NHCO_2Bn$, MeO_2C, $NHCOCF_3$ → (NaH, THF, 0–25°, 68%) oxazolone O, N, O, CF_3 → (H_2O) OBn, O, MeO_2C, $NHCO_2Bn$, HO_2C, $NHCOCF_3$

Ref. 23

15. $(Bu_3Sn)_2O$, benzene, 80°, 2–24 h, 73–100% yield.[24]

1. For example, see: S. M. Hecht and J. W. Kozarich, *Tetrahedron Lett.*, 1397 (1973).
2. N. Hashimoto, T. Aoyama, and T. Shioiri, *Chem. Pharm. Bull.*, **29,** 1475 (1981).
3. Y. Hirai, T. Aida, and S. Inoue, *J. Am. Chem. Soc.*, **111,** 3062 (1989).
4. T. Shioiri, T. Aoyama, and S. Mori, *Org. Synth.*, **68,** 1 (1989).
5. J. R. Rachele, *J. Org. Chem.*, **28,** 2898 (1963).
6. S. Danishefsky, M. Hirama, K. Gombatz, T. Harayama, E. Berman, and P. Schuda, *J. Am. Chem. Soc.*, **100,** 6536 (1978); *idem*, **101,** 7020 (1979).
7. R. Albert, J. Danklmaier, H. Hönig, and H. Kandolf, *Synthesis*, 635 (1987).
8. E. J. Corey, I. Székely, and C. S. Shiner, *Tetrahedron Lett.*, 3529 (1977).
9. P. Müller and B. Siegfried, *Helv. Chim. Acta*, **57,** 987 (1974).
10. P. Magnus and T. Gallagher, *J. Chem. Soc., Chem. Commun.*, 389 (1984).
11. E. D. Laganis and B. L. Chenard, *Tetrahedron Lett.*, **25,** 5831 (1984).
12. K. Inoue and K. Sakai, *Tetrahedron Lett.*, 4063 (1977).
13. M. Honda, K. Hirata, H. Sueoka, T. Katsuki, and M. Yamaguchi, *Tetrahedron Lett.*, **22,** 2679 (1981).
14. E. Keinan and D. Eren, *J. Org. Chem.*, **51,** 3165 (1986).
15. M. Ohno, Y. Ito, M. Arita, T. Shibata, K. Adachi, and H. Sawai, *Tetrahedron*, **40,** 145 (1984).
16. E. Alvarez, T. Cuvigny, C. Hervé du Penhoat, and M. Julia, *Tetrahedron*, **44,** 119 (1988).
17. K. Adachi, S. Kobayashi, and M. Ohno, *Chimia*, **40,** 311 (1986).
18. D. S. Holmes, U. C. Dyer, S. Russell, J. A. Sherringham, and J. A. Robinson, *Tetrahedron Lett.*, **29,** 6357 (1988).
19. S. Kobayashi, K. Kamiyama, T. Iimori, and M. Ohno, *Tetrahedron Lett.*, **25,** 2557 (1984).
20. P. Mohr, L. Rösslein, and C. Tamm, *Tetrahedron Lett.*, **30,** 2513 (1989).
21. P. S. Manchand, *J. Chem. Soc., Chem. Commun.*, 667 (1971).
22. D. H. R. Barton, L. Bould, D. L. J. Clive, P. D. Magnus, and T. Hase, *J. Chem. Soc. C*, 2204 (1971).
23. D. L. Boger and D. Yohannes, *J. Org. Chem.*, **54,** 2498 (1989).
24. E. G. Mata and O. A. Mascaretti, *Tetrahedron Lett.*, **29,** 6893 (1988).

Substituted Methyl Esters

2. 9-Fluorenylmethyl (Fm) Ester:

O_2CR

9-Fluorenylmethyl esters of *N*-protected amino acids were prepared using the DCC/DMAP method (50–89% yield)[1] or by imidazole-catalyzed transesterification of protected amino acid active esters with FmOH.[2] Cleavage is accomplished with either diethylamine or piperidine in CH_2Cl_2 at room temperature for 2 h. No racemization was observed during formation or cleavage of the Fm esters.[1] The Fm ester is cleaved slowly by hydrogenolysis, but complete selectivity for the benzyloxycarbonyl group could not be obtained. Fm esters also improved the solubility of protected peptides in organic solvents.[2]

1. H. Kessler and R. Siegmeier, *Tetrahedron Lett.*, **24,** 281 (1983).
2. M. A. Bednarek and M. Bodanszky, *Int. J. Pept. Protein Res.*, **21,** 196 (1983).

3. Methoxymethyl Ester (MOM Ester): $RCOOCH_2OCH_3$ (Chart 6)

Formation

The section on the formation of MOM ethers should be consulted since many of the methods described there should also be applicable to the formation of MOM esters.

1. CH_3OCH_2Cl, Et_3N, DMF, 25°, 1 h.[1]
2. $CH_3OCH_2OCH_3$, Zn/$BrCH_2CO_2Et$, 0°; CH_3COCl, 0–20°, 2 h, 75–85%.[2] A number of methoxymethyl esters were prepared by this method, which avoids the use of the carcinogen chloromethyl methyl ether.

Cleavage

1. R'_3SiBr, trace MeOH. Methoxymethyl ethers are stable to these cleavage conditions.[3] Methoxymethyl esters are unstable to silica gel chromatography, but are stable to mild acid (0.01 *N* HCl, EtOAc, MeOH, 25°, 16 h).[4]

1. A. B. A. Jansen and T. J. Russell, *J. Chem. Soc.*, 2127 (1965).
2. F. Dardoize, M. Gaudemar, and N. Goasdoue, *Synthesis*, 567 (1977).
3. S. Masamune, *Aldrichimica Acta*, **11,** 23–30 (1978); see p. 30.
4. L. M. Weinstock, S. Karady, F. E. Roberts, A. M. Hoinowski, G. S. Brenner, T. B. K. Lee, W. C. Lumma, and M. Sletzinger, *Tetrahedron Lett.*, 3979 (1975).

4. Methylthiomethyl Ester (MTM Ester): $RCOOCH_2SCH_3$ (Chart 6)

Formation

1. From RCO_2K: CH_3SCH_2Cl, NaI, 18-crown-6, C_6H_6, reflux, 6 h, 85–97% yield.[1]
2. $Me_2S^+ClX^-$, Et_3N, 0.5 h, −70° → 25°, 80–85% yield.[2]
3. *t*-BuBr, DMSO, $NaHCO_3$, 62–98% yield.[3,4] This method was used to prepare the MTM esters of *N*-protected amino acids.

Cleavage

1. $HgCl_2$, CH_3CN, H_2O, reflux, 6 h; H_2S, 20°, 30 min, 82–98% yield.[1]
2. MeI, acetone, reflux, 24 h; 1 *N* NaOH, 87–97% yield.[5]
3. CF_3COOH, 25°, 15 min, 80–90% yield.[6]
4. H_2O_2, $(NH_4)_6Mo_7O_{24}$; NaOH, pH 11, 97% yield.[5]
5. HCl, Et_2O, 6 h, 83–88% yield.[4] Acidic deprotection of the BOC group could not be achieved with complete selectivity in the presence of an MTM ester. The trityl and NPS (2-nitrophenylsulfenyl) groups were the preferred nitrogen protective groups.

1. L. G. Wade, J. M. Gerdes, and R. P. Wirth, *Tetrahedron Lett.*, 731 (1978).
2. T.-L. Ho, *Synth. Commun.*, **9,** 267 (1979).
3. A. Dossena, R. Marchelli, and G. Casnati, *J. Chem. Soc., Perkin Trans. I*, 2737 (1981).
4. A. Dossena, G. Palla, R. Marchelli, and T. Lodi, *Int. J. Pept. Protein Res.*, **23,** 198 (1984).
5. J. M. Gerdes and L. G. Wade, *Tetrahedron Lett.*, 689 (1979).
6. T.-L. Ho and C. M. Wong, *J. Chem. Soc., Chem. Commun.*, 224 (1973).

5. Tetrahydropyranyl Ester (THP Ester): RCOO–2-tetrahydropyranyl (Chart 6)

Formation

1. Dihydropyran, TsOH, CH_2Cl_2, 20°, 1.5 h, quant.[1]

Cleavage

1. AcOH, THF, H_2O (4:2:1), 45°, 3.5 h.[1]

1. K. F. Bernady, M. B. Floyd, J. F. Poletto, and M. J. Weiss, *J. Org. Chem.*, **44,** 1438 (1979).

6. Tetrahydrofuranyl Ester: RCO_2-2-tetrahydrofuranyl

Formation/Cleavage[1]

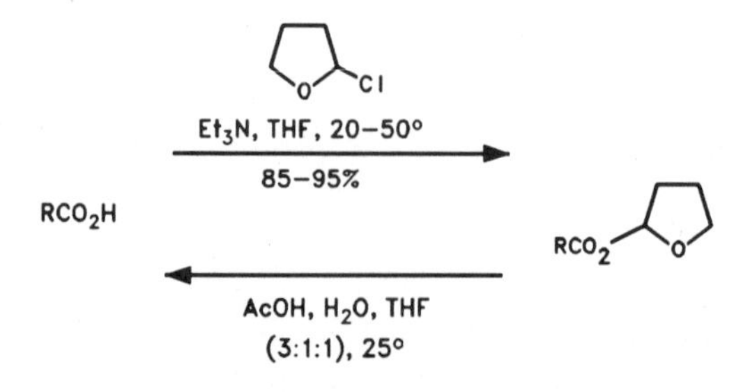

1. C. G. Kruse, N. L. J. M. Broekhof, and A. van der Gen, *Tetrahedron Lett.*, 1725 (1976).

7. Methoxyethoxymethyl Ester (MEM Ester): $RCO_2CH_2OCH_2CH_2OCH_3$

Formation/Cleavage[1]

$$RCOOH \xrightarrow[\text{0°, 2 h, high yield}]{MeOCH_2CH_2OCH_2Cl,\ i\text{-}Pr_2NEt,\ CH_2Cl_2} RCO_2MEM$$

$$\xleftarrow[\text{40°, 12 h}]{3\ N\ HCl,\ THF}$$

In an attempt to synthesize the macrolide antibiotic chlorothricolide, an unhindered —COOH group was selectively protected, in the presence of a hindered —COOH group, as a MEM ester that was then reduced to an alcohol group.[2]

1. A. I. Meyers and P. J. Reider, *J. Am. Chem. Soc.*, **101,** 2501 (1979).
2. R. E. Ireland and W. J. Thompson, *Tetrahedron Lett.*, 4705 (1979).

8. 2-(Trimethylsilyl)ethoxymethyl Ester (SEM Ester): $RCO_2CH_2OCH_2CH_2Si(CH_3)_3$

The SEM ester was used to protect a carboxyl group where DCC-mediated esterification caused destruction of the substrate. It was formed from the acid and SEM chloride (THF, 0°, 80% yield) and was removed solvolytically. The ease of removal in this case was attributed to anchimeric assistance by the phosphate group.[1] Normally SEM groups are cleaved by treatment with fluoride ion. Note that in this case the SEM group is removed considerably faster than the phenyl groups from the phosphate.

CO_2SEM / $OP(O)(OPh)_2$ —MeOH, half-life = 12 h, 65%→ CO_2H / $OP(O)(OH)_2$

1. E. W. Logusch, *Tetrahedron Lett.*, **25,** 4195 (1984).

9. Benzyloxymethyl Ester (BOM Ester): $RCOOCH_2OCH_2C_6H_5$ (Chart 6)

Formation[1]

$$RCOONa + PhCH_2OCH_2Cl \xrightarrow{\text{HMPA, 25°, 70\%}} RCO_2BOM$$

Cleavage[1]

1. H_2/Pd–C, EtOH, 25°, 70–100% yield.
2. Aqueous HCl, THF, 25°, 2 h, 75–95% yield.

1. P. A. Zoretic, P. Soja, and W. E. Conrad, *J. Org. Chem.*, **40,** 2962 (1975).

10. Phenacyl Ester: $RCOOCH_2COC_6H_5$ (Chart 6)

Formation

1. $\underset{\displaystyle NHCOOCH_2Ph}{RCHCOOH} \xrightarrow[20°,\ 12\ h,\ 83\%]{PhCOCH_2Br,\ Et_3N,\ EtOAc^1} \underset{\displaystyle NHCOOCH_2Ph}{RCHCOOCH_2COPh}$

2. $RCOOH \xrightarrow[25°,\ 10\ min,\ 90\text{–}99\%]{PhCOCH_2Br,\ KF/DMF^2} RCO_2CH_2COPh$

   R = alkyl, aryl, or hindered acids (at 100°)

Cleavage

1. Zn/HOAc, 25°, 1 h, 90% yield.[3]
2. H_2/Pd–C, aq. MeOH, 20°, 1 h, 72% yield.[1]
3. PhSNa, DMF, 20°, 30 min, 72% yield.[1]
4. PhSeH, DMF, rt, 48 h, 79% yield.[4]

Under basic coupling conditions an aspartyl peptide that has a β-phenacyl ester is converted to a succinimide.[5] The use of PhSeH prevents the α,β-rearrangement of the aspartyl residue during deprotection.

A phenacyl ester is much more readily cleaved by nucleophiles than are other esters such as the benzyl ester. Phenacyl esters are stable to acidic hydrolysis (e.g., concd. HCl;[1] HBr/HOAc;[1] 50% CF_3COOH/CH_2Cl_2;[6] HF, 0°, 1 h[6]).

1. G. C. Stelakatos, A. Paganou, and L. Zervas, *J. Chem. Soc. C*, 1191 (1966).
2. J. H. Clark and J. M. Miller, *Tetrahedron Lett.*, 599 (1977).
3. J. B. Hendrickson and C. Kandall, *Tetrahedron Lett.*, 343 (1970).
4. J. L. Morell, P. Gaudreau, and E. Gross, *Int. J. Pept. Protein Res.*, **19,** 487 (1982).
5. M. Bodanszky and J. Martinez, *J. Org. Chem.*, **43,** 3071 (1978).
6. C. C. Yang and R. B. Merrifield, *J. Org. Chem.*, **41,** 1032 (1976).

11. *p*-Bromophenacyl Ester: $RCOOCH_2COC_6H_4$-*p*-Br

In a penicillin synthesis the carboxyl group was protected as a *p*-bromophenacyl ester that was cleaved by nucleophilic displacement (PhSK, DMF, 20°, 30 min,

64% yield). Hydrogenolysis of a benzyl ester was difficult (perhaps because of catalyst poisoning by sulfur); basic hydrolysis of methyl or ethyl esters led to attack at the β-lactam ring.[1]

1. P. Bamberg, B. Eckström, and B. Sjöberg, *Acta Chem. Scand.*, **21,** 2210 (1967).

12. α-Methylphenacyl Ester: $RCO_2CH(CH_3)COC_6H_5$

13. *p*-Methoxyphenacyl Ester: $RCO_2CH_2COC_6H_4$-*p*-OCH_3

Phenacyl esters can be prepared from the phenacyl bromide, a carboxylic acid, and potassium fluoride as base.[1] These phenacyl esters can be cleaved by irradiation (313 nm, dioxane or EtOH, 20°, 6 h, 80–95% yield, R = amino acids;[2] >300 nm, 30°, 8 h, R = a gibberellic acid, 36–62% yield[3]). Another phenacyl derivative, $RCO_2CH(COC_6H_5)C_6H_3$-3,5-$(OCH_3)_2$, cleaved by irradiation, has also been reported.[4]

1. F. S. Tjoeng and G. A. Heavner, *Synthesis*, 897 (1981).
2. J. C. Sheehan and K. Umezawa, *J. Org. Chem.*, **38,** 3771 (1973).
3. E. P. Serebryakov, L. M. Suslova, and V. K. Kucherov, *Tetrahedron*, **34,** 345 (1978).
4. J. C. Sheehan, R. M. Wilson, and A. W. Oxford, *J. Am. Chem. Soc.*, **93,** 7222 (1971).

14. Carboxamidomethyl Ester (Cam Ester): $RCO_2CH_2CONH_2$

The carboxamidomethyl ester was prepared for use in peptide synthesis. It is formed from the cesium salt of an *N*-protected amino acid and α-chloroacetamide (60–85% yield). It is cleaved with 0.5 *M* NaOH or $NaHCO_3$ in DMF/H_2O. It is stable to the conditions required to remove BOC, Cbz, Fmoc, and *t*-butyl esters. It cannot be selectively cleaved in the presence of a benzyl ester of aspartic acid.[1]

1. J. Martinez, J. Laur, and B. Castro, *Tetrahedron*, **41,** 739 (1985); *idem*, *Tetrahedron Lett.*, **24,** 5219 (1983).

15. *N*-Phthalimidomethyl Ester (Chart 6):

NCH_2O_2CR

Formation

1. XCH_2-*N*-phthalimido + RCO_2H.
   X = OH: Et_2NH, EtOAc, 37°, 12 h, 70–80% yield.[1]

X = Cl: (c-$C_6H_{11})_2NH$, DMF or DMSO, 60°, a few minutes, 70–80% yield.[1]
X = Cl, Br: KF, DMF, 80°, 2 h, 65–75% yield.[2]

Cleavage

1. H_2NNH_2/MeOH, 20°, 3 h, 90% yield.[1]
2. Et_2NH/MeOH, H_2O, 25°, 24 h or reflux, 2 h, 82% yield.[1]
3. NaOH/MeOH, H_2O, 20°, 45 min, 77% yield.[1]
4. Zn/HOAc, 25°, 12 h, 80% yield.[3]
5. g HCl/EtOAc, 20°, 16 h, 83% yield.[1]
6. HBr/HOAc, 20°, 10–15 min, 80% yield.[1]

1. G. H. L. Nefkens, G. I. Tesser, and R. J. F. Nivard, *Recl. Trav. Chim. Pays-Bas*, **82,** 941 (1963).
2. K. Horiki, *Synth. Commun.*, **8,** 515 (1978).
3. D. L. Turner and E. Baczynski, *Chem. Ind.* (*London*), 1204 (1970).

2-Substituted Ethyl Esters

16. 2,2,2,-Trichloroethyl Ester: $RCO_2CH_2CCl_3$ (Chart 6)

Formation

1. CCl_3CH_2OH, DCC, Pyr.[1]
2. CCl_3CH_2OH, TsOH, toluene, reflux.[1,2]
3. CCl_3CH_2OCOCl, THF, Pyr, >60% yield.[3]

Cleavage

1. Zn, AcOH, 0°, 2.5 h.[1] Trichloroethyl esters are cleaved with zinc/THF buffer at pH 4.2–7.2 (20°, 10 min, 75–95% yield).[4]
2. Electrolysis: −1.65 V, $LiClO_4$, MeOH, 87–91% yield.[5] A tribromoethyl ester is cleaved by electrolytic reduction at −0.70 V (85% yield); a dichloroethyl ester is cleaved at −1.85 V (78% yield).[5]
3. Cat. Se, $NaBH_4$, DMF, 40–50°, 1 h, 77–93% yield.[6]

1. R. B. Woodward, K. Heusler, J. Gosteli, P. Naegeli, W. Oppolzer, R. Ramage, S. Ranganathan, and H. Vorbrüggen, *J. Am. Chem. Soc.*, **88,** 852 (1966).
2. J. F. Carson, *Synthesis*, 24 (1979).

3. R. R. Chauvette, P. A. Pennington, C. W. Ryan, R. D. G. Cooper, F. L. José, I. G. Wright, E. M. Van Heyningen, and G. W. Huffman, *J. Org. Chem.*, **36,** 1259 (1971).
4. G. Just and K. Grozinger, *Synthesis*, 457 (1976).
5. M. F. Semmelhack and G. E. Heinsohn, *J. Am. Chem. Soc.*, **94,** 5139 (1972).
6. Z.-Z. Huang and X.-J. Zhou, *Synthesis*, 693 (1989).

17. 2-Haloethyl Ester: $RCOOCH_2CH_2Cl$ (Chart 6)

2-Haloethyl esters have been cleaved by a variety of nucleophiles.

1. Li^+ or Na^+ Co(I)phthalocyanine/MeOH, 0–20°, 40 min–60 h, 60–98% yield.[1]
2. Electrolysis: Co(I)phthalocyanine, $LiClO_4$, EtOH, H_2O, −1.95 V, 95% yield.[2]
3. $NaS(CH_2)_2SNa/CH_3CN$, reflux, 2 h, 80–85% yield.[3]
4. NaSeH/EtOH, 25°, 1 h → reflux, 6 min, 92–99% yield.[4]
5. $(NaS)_2CS/CH_3CN$, reflux, 1.5 h, 75–86% yield.[5]
6. Me_3SnLi/THF, 3 h → $Bu_4N^+F^-$, reflux, 15 min, 78–86% yield.[6]
7. NaHTe, EtOH, 2–60 min, 80–92% yield.[7]
8. Na_2S, 40–68% yield.[8]

1. H. Eckert and I. Ugi, *Angew Chem., Int. Ed. Engl.*, **15,** 681 (1976).
2. R. Scheffold and E. Amble, *Angew. Chem., Int. Ed. Engl.*, **19,** 629 (1980).
3. T.-L. Ho, *Synthesis*, 510 (1975).
4. T.-L. Ho, *Synth. Commun.*, **8,** 301 (1978).
5. T.-L. Ho, *Synthesis*, 715 (1974).
6. T.-L. Ho, *Synth. Commun.*, **8,** 359 (1978).
7. J. Chen and X. Zhou, *Synth. Commun.*, **17,** 161 (1987).
8. M. Joaquina, S. A. Amaral Trigo, and M. I. A. Oliveira Sartos, in *Peptides 1986*, D. Theodoropoulos, Ed., Walter de Gruyter & Co., Berlin, 1987, p. 61.

18. ω-Chloroalkyl Ester: $RCOO(CH_2)_nCl$

ω-Chloroalkyl esters (n = 4,5) have been cleaved by sodium sulfide (reflux, 4 h, 58–85% yield). The reaction proceeds by sulfide displacement of the chloride ion followed by intramolecular displacement of the carboxylate group by the (now) sulfhydryl group.[1]

1. T.-L. Ho and C. M. Wong, *Synth. Commun.*, **4,** 307 (1974).

19. 2-(Trimethylsilyl)ethyl Ester: $RCO_2CH_2CH_2Si(CH_3)_3$ ($RCO_2CH_2CH_2TMS$)

Formation

1. $Me_3SiCH_2CH_2OH$, DCC, Pyr, CH_3CN, 0°, 5–15 h, 66–97% yield.[1]
2. From an acid chloride: $Me_3SiCH_2CH_2OH$, Pyr, 25°, 3 h.[2]
3. $Me_3SiCH_2CH_2OH$, Me_3SiCl, THF, reflux, 12–36 h.[3] This method of esterification is also effective for the preparation of other esters.
4. From an anhydride: $Me_2AlOCH_2CH_2SiMe_3$, benzene, heat, >85% yield.[4]

OMOM
$Me_2AlOCH_2CH_2TMS$
benzene, heat
>85%
OMOM
HO_2C $CO_2CH_2CH_2TMS$

5. $Me_3SiCH_2CH_2OH$, 2-chloro-1-methylpyridinium iodide, Et_3N, 90% yield.[5]

Cleavage

1. $Et_4N^+F^-$ or $Bu_4N^+F^-$, DMF or DMSO, 20–30°, 5–60 min, quant. yield.[1,6]
2. DMF, $Bu_4N^+Cl^-$, $KF \cdot 2H_2O$, 42–62% yield (substrate = polypeptide).[7]

1. P. Sieber, *Helv. Chim. Acta*, **60,** 2711 (1977).
2. H. Gerlach, *Helv. Chim. Acta*, **60,** 3039 (1977).
3. M. A. Brook and T. H. Chan, *Synthesis*, 201 (1983).
4. E. Vedejs and S. D. Larsen, *J. Am. Chem. Soc.*, **106,** 3030 (1984).
5. J. D. White and L. R. Jayasinghe, *Tetrahedron Lett.*, **29,** 2139 (1988).
6. P. Sieber, R. H. Andreatta, K. Eisler, B. Kamber, B. Riniker, and H. Rink, *Peptides: Proceedings of the Fifth American Peptide Symposium*, M. Goodman and J. Meienhofer, Eds., Halsted Press, New York, 1977, pp. 543–545.
7. R. A. Forsch and A. Rosowsky, *J. Org. Chem.*, **49,** 1305 (1984).

20. 2-Methylthioethyl Ester: $RCO_2CH_2CH_2SCH_3$

The 2-methylthioethyl ester is prepared from a carboxylic acid and methylthioethyl alcohol or methylthioethyl chloride ($MeSCH_2CH_2OH$, TsOH, benzene, reflux, 55 h, 55% yield; $MeSCH_2CH_2Cl$, Et_3N, 65°, 12 h, 50–70% yield).[1] It is cleaved by oxidation [H_2O_2, $(NH_4)_6Mo_7O_{24}$, acetone, 25°, 2 h, 80–95% yield → pH 10–11, 25°, 12–24 h, 85–95% yield][2] and by alkylation followed by hydrolysis (MeI, 70–95% yield → pH 10, 5–10 min, 70–95% yield).[1]

1. M. J. S. A. Amaral, G. C. Barrett, H. N. Rydon, and J. E. Willet, *J. Chem. Soc. C*, 807 (1966).
2. P. M. Hardy, H. N. Rydon, and R. C. Thompson, *Tetrahedron Lett.*, 2525 (1968).

21. 1,3-Dithianyl-2-methyl Ester (Dim Ester): 1,3-dithian-2-yl–CH_2O_2CR

The Dim ester was developed for the protection of the carboxyl function during peptide synthesis. It is prepared by transesterification of amino acid methyl esters with 2-(hydroxymethyl)-1,3-dithiane and Al(*i*-PrO)$_3$ (reflux, 4 h, 75°, 12 torr, 75% yield). It is removed by oxidation [H_2O_2, $(NH_4)_2MoO_4$; pH 8, H_2O, 60 min, 83% yield]. Since it must be removed by oxidation it is not compatible with sulfur-containing amino acids such as cysteine and methionine. Its suitability for other, easily oxidized amino acids (e.g., tyrosine and tryptophan) must also be questioned. It is stable to CF_3CO_2H and HCl/ether.[1,2]

1. H. Kunz and H. Waldmann, *Angew. Chem., Int. Ed. Engl.*, **22,** 62 (1983).
2. H. Waldmann and H. Kunz, *J. Org. Chem.*, **53,** 4172 (1988).

22. 2-(*p*-Nitrophenylsulfenyl)ethyl Ester: $RCO_2CH_2CH_2SC_6H_4$-*p*-NO_2

This ester is similar to the 2-methylthioethyl ester in that it is prepared from a thioethyl alcohol and cleaved by oxidation [H_2O_2, $(NH_4)_6Mo_7O_{24}$].[1]

1. M. J. S. A. Amaral, *J. Chem. Soc. C*, 2495 (1969).

23. 2-(*p*-Toluenesulfonyl)ethyl Ester (Tse Ester): $RCO_2CH_2CH_2SO_2C_6H_4$-*p*-CH_3 (Chart 6)

Formation

1. $TsCH_2CH_2OH$, DCC, Pyr, 0°, 1 h → 20°, 16 h, 70–90% yield.[1]

Cleavage

1. Na_2CO_3, dioxane, H_2O, 20°, 2 h, 95% yield.[1]
2. 1 *N* NaOH, dioxane, H_2O, 20°, 3 min, 60–95% yield.[1]
3. KCN, dioxane, H_2O, 20°, 2.5 h, 60–85% yield.[1]
4. DBN, benzene, 25°, quant.[2]
5. DBU, benzene, 11 h, 100% yield.[3]

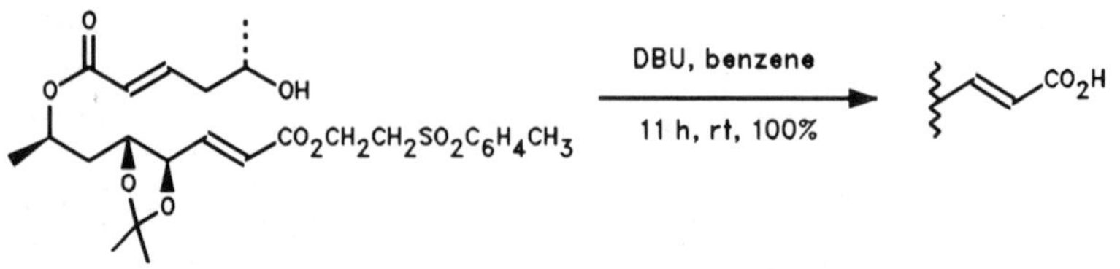

DBU, benzene

11 h, rt, 100%

6. $Bu_4N^+F^-$, THF, 0°, 1 h, 52–95% yield.[4] A primary alcohol protected as the *t*-butyldimethylsilyl ether is cleaved under these conditions, but a similarly protected secondary alcohol is stable.

1. A. W. Miller and C. J. M. Stirling, *J. Chem. Soc. C*, 2612 (1968).
2. E. W. Colvin, T. A. Purcell, and R. A. Raphael, *J. Chem. Soc., Chem. Commun.*, 1031 (1972).
3. H. Tsutsui and O. Mitsunobo, *Tetrahedron Lett.*, **25,** 2163 (1984).
4. H. Tsutsui, M. Muto, K. Motoyoshi, and O. Mitsunobo, *Chem. Lett.*, 1595 (1987).

24. 2-(2′-Pyridyl)ethyl Ester (Pet Ester): $RCO_2CH_2CH_2$–2-C_5H_4N

Formation

1. DCC, HOBt, $HOCH_2CH_2$–2-C_5H_4N, 0° → rt, CH_2Cl_2 or DMF, overnight, 50–92% yield.[1,2]
2. DCC, DMAP, $HOCH_2CH_2$–2-C_5H_4N, CH_2Cl_2, 61–92% yield.[3]
3. The related 2-(4′-pyridyl)ethyl ester has also been prepared from the acid chloride and the alcohol.[4]

Cleavage

1. MeI, CH_3CN; morpholine or diethylamine, methanol, 76–95% yield.[1,3] These conditions also cleave the 4′-pyridyl derivative.[4] The Pet ester is stable to the acidic conditions required to remove the BOC and *t*-butyl ester groups, to the basic conditions required to remove the Fmoc and Fm groups, and to hydrogenolysis. It is not recommended for use in peptides that contain methionine or histidine since these are susceptible to alkylation with methyl iodide.

1. H. Kessler, G. Becker, H. Kogler, and M. Wolff, *Tetrahedron Lett.*, **25,** 3971 (1984).
2. H. Kessler, G. Becker, H. Kogler, J. Friesse, and R. Kerssebaum, *Int. J. Pept. Protein Res.*, **28,** 342 (1986).
3. H. Kunz and M. Kneip, *Angew. Chem., Int. Ed. Engl.*, **23,** 716 (1984).
4. A. R. Katritsky, G. R. Khan, and O. A. Schwarz, *Tetrahedron Lett.*, **25,** 1223 (1984).

25. 2-(Diphenylphosphino)ethyl ester (Dppe ester): $(C_6H_5)_2PCH_2CH_2O_2CR$

The Dppe group was developed for carboxyl protection in peptide synthesis. It is formed from an *N*-protected amino acid and the alcohol (DCC, DMAP, 3–12 h, 0°, rt). It is most efficiently cleaved by quaternization with MeI followed by treatment with fluoride ion or K_2CO_3. The ester is stable to HBr/AcOH, $BF_3 \cdot Et_2O$, and CF_3CO_2H.[1]

1. D. Chantreux, J.-P. Gamet, R. Jacquier, and J. Verducci, *Tetrahedron*, **40**, 3087 (1984).

26. 1-Methyl-1-phenylethyl Ester (Cumyl Ester): $RCO_2C(CH_3)_2C_6H_5$

Cleavage

PhC(Me)$_2$O$_2$C ... CO$_2$CMe$_3$ → (dry HCl, CH_2Cl_2; 0°, 3 min, 65%) → HO$_2$C ... CO$_2$CMe$_3$

Note that a cumyl ester can be selectively cleaved in the presence of a *t*-butyl ester and a β-lactam.[1]

1. D. M. Brunwin and G. Lowe, *J. Chem. Soc., Perkin Trans. I*, 1321 (1973).

27. *t*-Butyl Ester: $RCO_2C(CH_3)_3$ (Chart 6)

Formation

The *t*-butyl ester is a relatively hindered ester, and many of the methods reported below should be—and in many cases are—equally effective for the preparation of other hindered esters. The related 1- and 2-adamantyl esters have been used for the protection of aspartic acid.[1]

1. Isobutylene, concd. H_2SO_4, Et_2O, 25°, 2–24 h, 50–60% yield.[2] This method works for the preparation of *t*-Bu esters of alkyl acids, amino acids,[3,4] and penicillins.[5]
2. Isobutylene, CH_2Cl_2, H_3PO_4 (P_2O_5), $BF_3 \cdot Et_2O$, −78°, 2 h, → 0°, 24 h.[6]
3. $(COCl)_2$, benzene, DMF, 7–10°, 45 min; *t*-BuOH, Et_3N, CH_2Cl_2, 0°, 3 h, 75% yield.[7]
4. From an aromatic acid chloride: LiO-*t*-Bu, 25°, 15 h, 79–82% yield.[8]

5. 2,4,6-$Cl_3C_6H_2COCl$, Et_3N, THF; *t*-BuOH, DMAP, benzene, 25°, 20 min, 90% yield.[9]
6. *t*-BuOH, Pyr, $(Me_2N)(Cl)C{=}N^+Me_2Cl^-$, 77% yield.[10] This method is also effective for the preparation of other esters.
7. $(Im)_2CO$ (*N,N'*-carbonyldiimidazole), *t*-BuOH, DBU, 54–91% yield.[11]
8. Bu_3PI_2, Et_2O, HMPA; *t*-BuOH, 73% yield.[12]
9. *t*-BuOH, EDCI (EDCI = 1-ethyl-3-[3-(dimethylamino)propyl]carbodiimide hydrochloride), DMAP, CH_2Cl_2, 88% yield.[13] Cbz-Proline was protected without racemization.
10. $(t\text{-BuO})_2CHNMe_2$, toluene, 80°, 30 min, 82% yield.[14]
11. $Cl_3C(t\text{-BuO})\,C{=}NH$, $BF_3\cdot Et_2O$, CH_2Cl_2, cyclohexane, 70–92% yield.[15] This reagent also forms *t*-butyl ethers from alcohols.
12. From an acid chloride: *t*-BuOH, AgCN, benzene, 20–80°, 60–100% yield.[16]
13. 2-Cl-3,5-$(NO_2)C_5H_2N$, Pyr, rt → 115°, *t*-BuOH.[17] Other esters are also prepared effectively using this methodology.
14. *t*-BuOCOF, Et_3N, DMAP, CH_2Cl_2, *t*-BuOH, rt, 82–96% yield.[18]

Cleavage

t-Butyl esters are stable to mild basic hydrolysis, hydrazine, and ammonia; they are cleaved by moderately acidic hydrolysis.

1. HCO_2H, 20°, 3 h.[19]
2. CF_3COOH, CH_2Cl_2, 25°, 1 h.[20]
3. AcOH, HBr, 10°, 10 min, 70% yield.[3] Phthaloyl or trifluoroacetyl groups on amino acids are stable to these conditions; benzyloxycarbonyl (Cbz) or *t*-butoxycarbonyl (BOC) groups are cleaved.
4. TsOH, benzene, reflux, 30 min, 76% yield.[3]

A *t*-butyl ester is stable to the conditions needed to convert an α,β-unsaturated ketone to a dioxolane ($HOCH_2CH_2OH$, TsOH, benzene, reflux).[21]

5. KOH, 18-crown-6, toluene, 100°, 5 h, 94% yield.[22] These conditions were used to cleave the *t*-butyl ester from an aromatic ester; they are probably too harsh to be used on more highly functionalized substrates.
6. 190–200°, 15 min, 100% yield.[23] Thermolytic conditions will also cleave the BOC group from amines.
7. Bromocatecholborane.[24] Ethyl esters are not affected by this reagent, but it does cleave other groups; see the section on methoxymethyl (MOM) ethers.

1. Y. Okada and S. Iguchi, *J. Chem. Soc., Perkin Trans. I*, 2129 (1988).

2. A. L. McCloskey, G. S. Fonken, R. W. Kluiber, and W. S. Johnson, *Org. Synth., Collect. Vol. IV*, 261 (1963).
3. G. W. Anderson and F. M. Callahan, *J. Am. Chem. Soc.*, **82,** 3359 (1960).
4. R. M. Valerio, P. F. Alewood, and R. B. Johns, *Synthesis*, 786 (1988).
5. R. J. Stedman, *J. Med. Chem.*, **9,** 444 (1966).
6. C.-Q. Han, D. DiTullio, Y.-F. Wang, and C. J. Sih, *J. Org. Chem.*, **51,** 1253 (1986).
7. C. F. Murphy and R. E. Koehler, *J. Org. Chem.*, **35,** 2429 (1970).
8. G. P. Crowther, E. M. Kaiser, R. A. Woodruff, and C. R. Hauser, *Org. Synth.*, **51,** 96 (1971); *Collect. Vol. VI*, 259 (1988).
9. J. Inanaga, K. Hirata, H. Saeki, T. Katsuki, and M. Yamaguchi, *Bull. Chem. Soc. Jpn.*, **52,** 1989 (1979).
10. T. Fujisawa, T. Mori, K. Fukumoto, and T. Sato, *Chem. Lett.*, 1891 (1982).
11. S. Ohta, A. Shimabayashi, M. Aona, and M. Okamoto, *Synthesis*, 833 (1982).
12. R. K. Haynes and M. Holden, *Aust. J. Chem.*, **35,** 517 (1982).
13. M. K. Dhaon, R. K. Olsen, and K. Ramasamy, *J. Org. Chem.*, **47,** 1962 (1982).
14. U. Widmer, *Synthesis*, 135 (1983).
15. A. Armstrong, I. Brackenridge, R. F. W. Jackson, and J. M. Kirk, *Tetrahedron Lett.*, **29,** 2483 (1988).
16. S. Takimoto, J. Inanaga, T. Katsuki, and M. Yamaguchi, *Bull. Chem. Soc. Jpn.*, **49,** 2335 (1976).
17. S. Takimoto, N. Abe, Y. Kodera, and H. Ohta, *Bull. Chem. Soc. Jpn.*, **56,** 639 (1983).
18. A. Loffet, N. Galeotti, P. Jouin, and B. Castro, *Tetrahedron Lett.*, **30,** 6859 (1989).
19. S. Chandrasekaran, A. F. Kluge, and J. A. Edwards, *J. Org. Chem.*, **42,** 3972 (1977).
20. D. B. Bryan, R. F. Hall, K. G. Holden, W. F. Huffman, and J. G. Gleason, *J. Am. Chem. Soc.*, **99,** 2353 (1977).
21. A. Martel, T. W. Doyle, and B.-Y. Luh, *Can. J. Chem.*, **57,** 614 (1979).
22. C. J. Pedersen, *J. Am. Chem. Soc.*, **89,** 7017 (1967).
23. L. H. Klemm, E. P. Antoniades, and D. C. Lind, *J. Org. Chem.*, **27,** 519 (1962).
24. R. K. Boeckman, Jr., and J. C. Potenza, *Tetrahedron Lett.*, **26,** 1411 (1985).

28. Cyclopentyl Ester: RCO_2-*c*-C_5H_9

29. Cyclohexyl Ester: RCO_2-*c*-C_6H_{11}

Cycloalkyl esters have been used to protect the β-CO_2H group in aspartyl peptides to minimize aspartimide formation during acidic or basic reactions. Aspartimide formation is limited to 2–3% in TFA (20 h, 25°), 5–7% with HF at 0°, and 1.5–4% TfOH (thioanisole in TFA). Cycloalkyl esters are also stable to Et_3N, whereas use of the benzyl ester leads to 25% aspartimide formation during Et_3N treatment. Cycloalkyl esters are stable to CF_3COOH, but are readily cleaved with HF or TfOH.[1-3]

1. J. Blake, *Int. J. Pept. Protein Res.*, **13,** 418 (1979).

2. J. P. Tam. T.-W. Wong, M. W. Riemen, F.-S. Tjoeng, and R. B. Merrifield, *Tetrahedron Lett.*, 4033 (1979).
3. N. Fujii, M. Nomizu, S. Futaki, A. Otaka, S. Funakoshi, K. Akaji, K. Watanabe, and H. Yajima, *Chem. Pharm. Bull.*, **34,** 864 (1986).

30. Allyl Ester: $RCO_2CH_2CH{=}CH_2$

Formation

1. Allyl bromide, Aliquat 336, $NaHCO_3$, CH_2Cl_2, 83% yield.[1] The carboxylic acid group of Z-serine (Z = Cbz = benzyloxycarbonyl) is selectively esterified without affecting the alcohol.
2. $R'R''C{=}CHCH_2OH$, NaH, THF, 1–3 days, 80–95% yield.[2] A methyl ester is exchanged for an allyl ester under these conditions.
3. Allyl bromide, Cs_2CO_3, 84% yield.[3]
4. Allyl alcohol, TsOH, benzene, $-H_2O$.[4] These conditions were used to prepare esters of amino acids.

Cleavage

1. $Pd(OAc)_2$, sodium 2-methylhexanoate, Ph_3P, acetone.[5]
2. $(Ph_3P)_3RhCl$ or $Pd(Ph_3P)_4$, 70°, EtOH, H_2O, 91% yield.[6]
3. $Pd(Ph_3P)_4$, pyrrolidine, 0°, 5–15 min, CH_3CN, 70–90% yield.[7] Morpholine has also been used as an allyl scavenger in this process.[1,3]
4. $PdCl_2(Ph_3P)_2$, dimedone, THF, 95% yield.[8] This method is also effective for removing the allyloxycarbonyl group from alcohols and amines.
5. $Pd(Ph_3P)_4$, 2-ethylhexanoic acid.[9]
6. Me_2CuLi, Et_2O, 0°, 1 h; H_3O^+, 75–85% yield.[10]

31. 3-Buten-1-yl Ester: $CH_2{=}CHCH_2CH_2O_2CR$

The ester, formed from the acid ($COCl_2$, toluene; then $CH_2{=}CHCH_2CH_2OH$, acetone, −78° warm to rt, 70–94% yield), can be cleaved by ozonolysis followed by Et_3N or DBU treatment (79–99% yield). The ester is suitable for the protection of enolizable and base-sensitive carboxylic acids.[11]

32. 4-(Trimethylsilyl)-2-buten-1-yl Ester: $RCO_2CH_2CH{=}CHCH_2Si(CH_3)_3$

This ester is formed by standard procedures and is readily cleaved with $Pd(Ph_3P)_4$ in CH_2Cl_2 to form trimethylsilyl esters that readily hydrolyze on treatment with water or alcohol or on chromatography on silica gel (73–98% yield). Amines can be protected using the related carbamate.[12]

33. Cinnamyl Ester: $RCO_2CH_2CH{=}CHC_6H_5$ (Chart 6)

The cinnamyl ester can be prepared from an activated carboxylic acid derivative and cinnamyl alcohol; it is cleaved under nearly neutral conditions [$Hg(OAc)_2$, MeOH, 23°, 2–4 h; KSCN, H_2O, 23°, 12–16 h, 90% yield].[13]

34. α-Methylcinnamyl Ester (MEC): $PhCH{=}CHCH(CH_3)O_2CR$

Formation

1. $PhCH{=}CHCH(CH_3)OH$, DCC, DMAP, THF, 98% yield.[14]
2. From an acid chloride: $PhCH{=}CHCH(CH_3)OH$, Pyr, DMAP, 75–88% yield.[14]

Cleavage

1. $Me_2Sn(SMe)_2$, $BF_3\cdot Et_2O$, $PhCH_3$, 0°, 3–24 h; AcOH, 75–100% yield.[11] An ethyl ester can be hydrolyzed in the presence of an MEC ester with 1 *N* aqueous NaOH–DMSO (1 : 1), and MEC esters can be cleaved in the presence of ethyl, benzyl, cinnamyl, and *t*-butyl esters as well as the acetate, TBDMS and MEM ethers.

1. S. F.-Bochnitschek, H. Waldmann, and H. Kunz, *J. Org. Chem.*, **54,** 751 (1989).
2. N. Engel, B. Kübel, and W. Steglich, *Angew. Chem., Int. Ed. Engl.*, **16,** 394 (1977).
3. H. Kunz, H. Waldmann, and C. Unverzagt, *Int. J. Pept. Protein Res.*, **26,** 493 (1985).
4. H. Waldmann and H. Kunz, *Liebigs Ann. Chem.*, 1712 (1983).
5. L. N. Jungheim, *Tetrahedron Lett.*, **30,** 1889 (1989).
6. H. Kunz and H. Waldmann, *Helv. Chim. Acta*, **68,** 618 (1985).
7. R. Deziel, *Tetrahedron Lett.*, **28,** 4371 (1987).
8. H. X. Zhang, F. Guibé, and G. Balavoine, *Tetrahedron Lett.*, **29,** 623 (1988).
9. P. D. Jeffrey and S. W. McCombie, *J. Org. Chem.*, **47,** 587 (1982).
10. T.-L. Ho, *Synth. Commun.*, **8,** 15 (1978).
11. A. G. M. Barrett, S. A. Lebold, and X.-an Zhang, *Tetrahedron Lett.*, **30,** 7317 (1989).
12. H. Mastalerz, *J. Org. Chem.*, **49,** 4092 (1984).
13. E. J. Corey and M. A. Tius, *Tetrahedron Lett.*, 2081 (1977).
14. T. Sato, J. Otera, and H. Nozaki, *Tetrahedron Lett.*, **30,** 2959 (1989).

35. Phenyl Ester: $RCO_2C_6H_5$

Phenyl esters can be prepared from *N*-protected amino acids (PhOH, DCC, CH_2Cl_2, $-20° \rightarrow 20°$, 12 h, 86% yield;[1] PhOH, BOP, Et_3N, CH_2Cl_2, 25°, 2 h,

73–97% yield[2]). Phenyl esters are readily cleaved under basic conditions (H_2O_2, H_2O, DMF, pH 10.5, 20°, 15 min).[3]

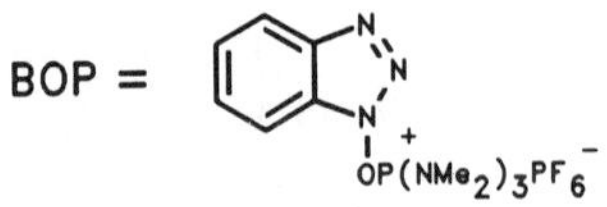

BOP = [benzotriazolyl-$\overset{+}{O}P(NMe_2)_3PF_6^-$]

1. I. J. Galpin, P. M. Hardy, G. W. Kenner, J. R. McDermott, R. Ramage, J. H. Seely, and R. G. Tyson, *Tetrahedron*, **35**, 2577 (1979).
2. B. Castro, G. Evin, C. Selve, and R. Seyer, *Synthesis*, 413 (1977).
3. G. W. Kenner and J. H. Seely, *J. Am. Chem. Soc.*, **94**, 3259 (1972).

36. *p*-(Methylmercapto)phenyl Ester: $RCO_2C_6H_4$-*p*-SCH_3

The *p*-(methylmercapto)phenyl ester has been prepared from an *N*-protected amino acid and 4-$CH_3SC_6H_4OH$ (DCC, CH_2Cl_2, 0°, 1 h → 20°, 12 h, 60–70% yield). The *p*-(methylmercapto)phenyl ester serves as an unactivated ester that is activated on oxidation to the sulfone (H_2O_2, AcOH, 20°, 12 h, 60–80% yield) which then serves as an activated ester in peptide synthesis.[1]

1. B. J. Johnson and T. A. Ruettinger, *J. Org. Chem.*, **35**, 255 (1970).

37. Benzyl Ester: $RCO_2CH_2C_6H_5$, RCO_2Bn (Chart 6)

Formation

Benzyl esters are readily prepared by many of the classical methods (see introduction to this chapter), as well as by many newer methods, since benzyl alcohol is unhindered and relatively stable to acid.

1. BnOCOCl, Et_3N, 0°, DMAP, CH_2Cl_2, 30 min, 97% yield.[1]

In the case of very hindered acids the yields are poor and formation of the symmetrical anhydride is observed. Useful selectivity can be achieved for a less hindered acid in the presence of a more hindered one.[2]

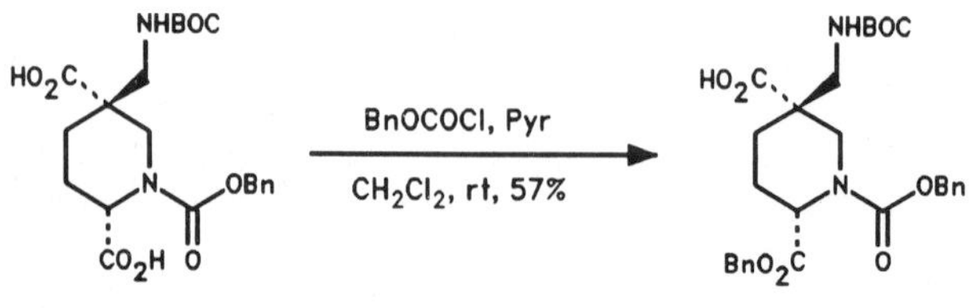

NHBOC, HO_2C, N, OBn, CO_2H, O → (BnOCOCl, Pyr; CH_2Cl_2, rt, 57%) NHBOC, HO_2C, N, OBn, BnO_2C, O

2. A methyl ester can be exchanged for a benzyl ester thermally (185°, 1.25 h, −MeOH).[3]

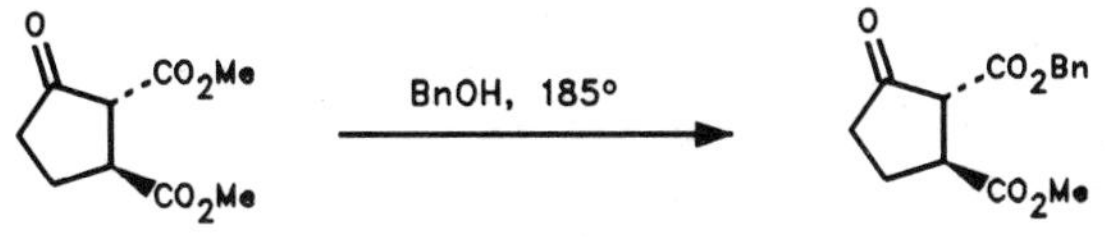

BnOH, 185°

3. For amino acids: DCC, DMAP, BnOH, 92% yield.[4]

Cleavage

The most useful property of benzyl esters is that they are readily cleaved by hydrogenolysis.

1. H_2/Pd–C, 25°, 45 min–24 h, high yields.[5]

Catalytic transfer hydrogenation (entries 2 and 3 below) can be used to cleave benzyl esters in some compounds that contain sulfur, a poison for hydrogenolysis catalysts.

2. Pd–C, cyclohexene[6] or 1,4-cyclohexadiene,[7] 25°, 1.5–6 h, good yields.
3. Pd–C, 4.4% HCOOH, MeOH, 25°, 5–10 min in a column, 100% yield.[8]
4. K_2CO_3, H_2O, THF, 0° → 25°, 1 h, 75% yield.[9]

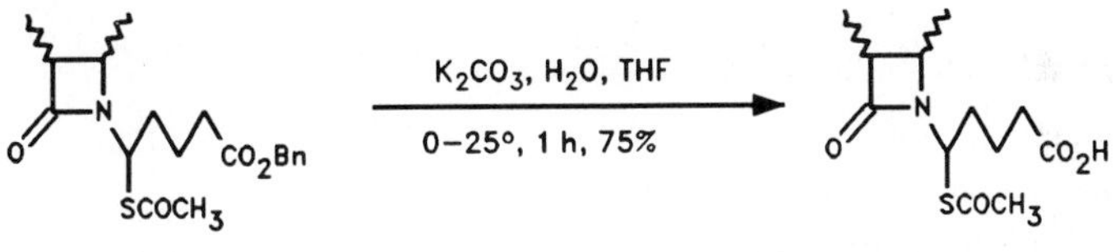

K_2CO_3, H_2O, THF

0–25°, 1 h, 75%

5. $AlCl_3$, anisole, CH_2Cl_2, CH_3NO_2, 0° → 25°, 5 h, 80–95% yield.[10] These conditions were used to cleave the benzyl ester in a variety of penicillin derivatives.
6. Na, ammonia, 50% yield.[11] These conditions were used to cleave the benzyl ester of an amino acid; the Cbz and benzylsulfenamide derivatives were also cleaved.
7. Aqueous $CuSO_4$, EtOH, pH 8, 32°, 60 min; pH 3; EDTA (ethylenediaminetetraacetic acid), 75% yield.[12]

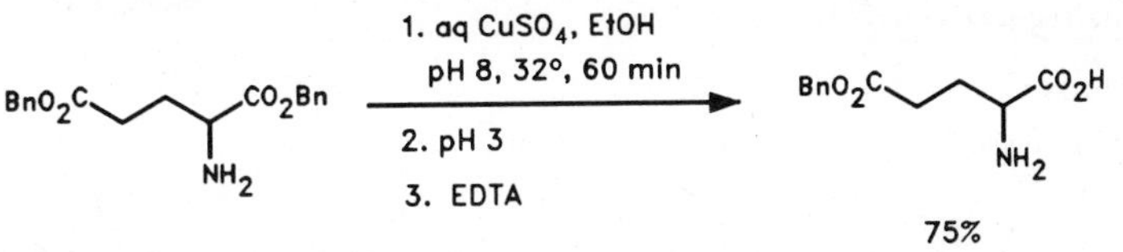

1. aq $CuSO_4$, EtOH pH 8, 32°, 60 min
2. pH 3
3. EDTA

75%

8. Benzyl esters can be cleaved by electrolytic reduction at −2.7 V.[13]

9. *t*-$BuMe_2SiH$, $Pd(OAc)_2$, CH_2Cl_2, Et_3N, 100% yield.[14] Cbz groups and Alloc groups are also cleaved, but benzyl ethers are stable.

$$\text{CH}_2\text{=CHCH}_2\text{CH(NHBOC)CO}_2\text{Bn} \xrightarrow[\text{Pd(OAc)}_2,\ \text{CH}_2\text{Cl}_2]{t\text{-BuMe}_2\text{SiH},\ \text{Et}_3\text{N}} \text{CH}_2\text{=CHCH}_2\text{CH(NHBOC)CO}_2\text{TBDMS}$$

10. NaHTe, DMF, *t*-BuOH, 80–90°, 5 min, 98% yield.[15] Methyl and propyl esters are also cleaved (13–97% yield).
11. W2 Raney nickel, EtOH, Et_3N, rt, 0.5 h, 75–85% yield.[16] A disubstituted olefin was not reduced.

1. S. Kim, Y. C. Kim, and J. I. Lee, *Tetrahedron Lett.*, **24,** 3365 (1983); S. Kim, J. I. Lee, and Y. C. Kim, *J. Org. Chem.*, **50,** 560 (1985).
2. J. E. Baldwin, M. Otsuka, and P. M. Wallace, *Tetrahedron*, **42,** 3097 (1986).
3. W. L. White, P. B. Anzeveno, and F. Johnson, *J. Org. Chem.*, **47,** 2379 (1982).
4. B. Neises, T. Andries, and W. Steglich, *J. Chem. Soc., Chem. Commun.*, 1132 (1982).
5. W. H. Hartung and R. Simonoff, *Org. React.*, **VII,** 263–326 (1953).
6. G. M. Anantharamaiah and K. M. Sivanandaiah, *J. Chem. Soc., Perkin Trans. I*, 490 (1977).
7. A. M. Felix, E. P. Heimer, T. J. Lambros, C. Tzougraki, and J. Meienhofer, *J. Org. Chem.*, **43,** 4194 (1978).
8. B. ElAmin, G. M. Anantharamaiah, G. P. Royer, and G. E. Means, *J. Org. Chem.*, **44,** 3442 (1979).
9. W. F. Huffman, R. F. Hall, J. A. Grant, and K. G. Holden, *J. Med. Chem.*, **21,** 413 (1978).
10. T. Tsuji, T. Kataoka, M. Yoshioka, Y. Sendo, Y. Nishitani, S. Hirai, T. Maeda, and W. Nagata, *Tetrahedron Lett.*, 2793 (1979).
11. C. W. Roberts, *J. Am. Chem. Soc.*, **76,** 6203 (1954).
12. R. L. Prestidge, D. R. K. Harding, J. E. Battersby, and W. S. Hancock, *J. Org. Chem.*, **40,** 3287 (1975).
13. W. G. Mairanovsky, *Angew. Chem., Int. Ed. Engl.*, **15,** 281 (1976).
14. M. Sakaitani, N. Kurokawa, and Y. Ohfune, *Tetrahedron Lett.*, **27,** 3753 (1986).
15. J. Chen and X. J. Zhou, *Synthesis*, 586 (1987).
16. S.-i. Hashimoto, Y. Miyazaki, T. Shinoda, and S. Ikegami, *Tetrahedron Lett.*, **30,** 7195 (1989).

Substituted Benzyl Esters

38. Triphenylmethyl Ester: $RCO_2C(C_6H_5)_3$ (Chart 6)

Triphenylmethyl esters are unstable in aqueous solution, but are stable to oxymercuration.[1]

Formation

$$RCO_2^-M^+ + Ph_3CBr \xrightarrow[\text{3-5 h, 85-95\%}]{\text{benzene, reflux}^2} RCO_2CPh_3$$

$M^+ = Ag^+, K^+, Na^+$

$$RCO_2SiMe_3 \xrightarrow[\text{0°, 0.5 h, 86\%}]{\text{Ph}_3\text{COTMS, TMSOTf, CH}_2\text{Cl}_2{}^3} RCO_2CPh_3$$

Cleavage

1. $HCl \cdot H_2NCH_2CO_2CPh_3 \xrightarrow{\text{MeOH or H}_2\text{O/dioxane}} HCl \cdot H_2NCH_2CO_2H$
   18°, 5 h, 72%; 18°, 24 h, 98%; 100°, 1 min, 98%[4]
2. Trityl esters have been cleaved by electrolytic reduction at −2.6 V.[5]

1. W. A. Slusarchyk, H. E. Applegate, C. M. Cimarusti, J. E. Dolfini, P. Funke, and M. Puar, *J. Am. Chem. Soc.*, **100,** 1886 (1978).
2. K. D. Berlin, L. H. Gower, J. W. White, D. E. Gibbs, and G. P. Sturm, *J. Org. Chem.*, **27,** 3595 (1962).
3. S. Murata and R. Noyori, *Tetrahedron Lett.*, **22,** 2107 (1981).
4. G. C. Stelakatos, A. Paganou, and L. Zervas, *J. Chem. Soc. C*, 1191 (1966).
5. V. G. Mairanovsky, *Angew. Chem., Int. Ed., Engl.*, **15,** 281 (1976).

39. Diphenylmethyl Ester (Dpm Ester): $RCO_2CH(C_6H_5)_2$

Diphenylmethyl esters are similar in acid lability to *t*-butyl esters and can be cleaved by acidic hydrolysis from *S*-containing peptides that poison hydrogenolysis catalysts.

Formation

1. Ph_2CN_2, acetone, 0°, 30 min → 20°, 4 h, 70%.[1,2]
2. $Ph_2C{=}NNH_2$, I_2, AcOH, >90% yield.[3]
3. $(Ph_2CHO)_3PO$, CF_3COOH, CH_2Cl_2, reflux, 1–5 h, 70–87% yield.[4] Free alcohols are converted to the corresponding Dpm ethers. This reaction has also been used for the selective protection of amino acids as their tosylate salts (CCl_4, 15 min–3 h, 63–91% yield).[5]
4. $Ph_2C{=}NNH_2$, $PhI(OAc)_2$, CH_2Cl_2, cat. I_2, −10° → 0°, 1 h, 73–93% yield.[6]

Cleavage

1. H_2/Pd black, MeOH, THF, 3 h, 90% yield.[1,7]
2. CF_3COOH, PhOH, 20°, 30 min, 82% yield.[1]

3. AcOH, reflux, 6 h.[8]
4. $BF_3 \cdot Et_2O$, AcOH, 40°, 0.5 h → 10°, several hours, 65% yield.[9] The sulfur–sulfur bond in cystine is stable to these conditions.
5. H_2NNH_2, MeOH, reflux, 60 min, 100% yield.[10] In this case the ester is converted to a hydrazide.
6. Diphenylmethyl esters are cleaved by electrolytic reduction at −2.6 V.[11]
7. HF, CH_3NO_2, AcOH (12:2:1), 91% yield.[12]

HF, CH_3NO_2, AcOH
12:2:1 91%

8. HCl, CH_3NO_2, <5 min, 25°.[13]
9. 98% HCOOH, 40–50°, 70–97% yield.[2]
10. 1 *N* NaOH, MeOH, rt.[5]
11. $AlCl_3$, CH_3NO_2, anisole, 3–6 h, 73–95% yield.[14,15] These conditions also cleaved the *p*-$MeOC_6H_4CH_2$ ester and ether in penam and cephalosporin-type intermediates.

1. G. C. Stelakatos, A. Paganou, and L. Zervas, *J. Chem. Soc. C*, 1191 (1966).
2. T. Kametani, H. Sekine, and T. Hondo, *Chem. Pharm. Bull.*, **30,** 4545 (1982).
3. R. Bywood, G. Gallagher, G. K. Sharma, and D. Walker, *J. Chem. Soc., Perkin Trans. I*, 2019 (1975).
4. L. Lapatsanis, *Tetrahedron Lett.*, 4697 (1978).
5. C. Froussios and M. Kolovos, *Synthesis*, 1106 (1987).
6. L. Lapatsanis, G. Milias, and S. Paraskewas, *Synthesis*, 513 (1985).
7. S. De Bernardo, J. P. Tengi, G. J. Sasso, and M. Weigele, *J. Org. Chem.*, **50,** 3457 (1985).
8. E. Haslam, R. D. Haworth, and G. K. Makinson, *J. Chem. Soc.*, 5153 (1961).
9. R. G. Hiskey and E. L. Smithwick, *J. Am. Chem. Soc.*, **89,** 437 (1967).
10. R. G. Hiskey and J. B. Adams, *J. Am. Chem. Soc.*, **87,** 3969 (1965).
11. V. G. Mairanovsky, *Angew. Chem., Int. Ed. Engl.*, **15,** 281 (1976).
12. L. R. Hillis and R. C. Ronald, *J. Org. Chem.*, **50,** 470 (1985).
13. R. C. Kelly, I. Schletter, S. J. Stein, and W. Wierenga, *J. Am. Chem. Soc.*, **101,** 1054 (1979).
14. T. Tsuji, T. Kataoka, M. Yoshioka, Y. Sendo, Y. Nishitani, S. Hirai, T. Maeda, and W. Nagata, *Tetrahedron Lett.*, 2793 (1979).
15. M. Ohtani, F. Watanabe, and M. Narisada, *J. Org. Chem.*, **49,** 5271 (1984).

40. Bis(o-nitrophenyl)methyl Ester: $RCOOCH(C_6H_4\text{-}o\text{-}NO_2)_2$ (Chart 6)

Bis(*o*-nitrophenyl)methyl esters are formed and cleaved by the same methods used for diphenylmethyl esters. They can also be cleaved by irradiation ($h\nu$ = 320 nm, dioxane, THF, . . ., 1–24 h, quant. yield).[1]

1. A. Patchornik, B. Amit, and R. B. Woodward, *J. Am. Chem. Soc.*, **92,** 6333 (1970).

41. 9-Anthrylmethyl Ester: $RCOOCH_2$-9-anthryl (Chart 6)

Formation

1. RCOOH + 9-anthrylmethyl chloride $\xrightarrow[\text{reflux, 4–6 h, 70–90\%}]{\text{Et}_3\text{N, MeCN}^1}$ RCO_2CH_2-9-anthryl
2. RCOOH + N_2CH-9-anthryl $\xrightarrow[\text{25°, 10 min, 80\%}]{\text{hexane}^2}$ RCO_2CH_2-9-anthryl

Cleavage

9-Anthrylmethyl esters are cleaved by acidic (2 *N* HBr/HOAc, 25°, 10–30 min, 100% yield) and basic (0.1 *N* NaOH/dioxane, 25°, 15 min, 97% yield) hydrolysis,[1] and by nucleophiles (MeSNa, THF–HMPA, −20°, 1 h, 90–100% yield).[3]

1. F. H. C. Stewart, *Aust. J. Chem.*, **18,** 1699 (1965).
2. M. G. Krakovyak, T. D. Amanieva, and S. S. Skorokhodov, *Synth. Commun.*, **7,** 397 (1977).
3. N. Kornblum and A. Scott, *J. Am. Chem. Soc.*, **96,** 590 (1974).

42. 2-(9,10-Dioxo)anthrylmethyl Ester (Chart 6):

O R' O_2CR O R = H, Ph

This derivative is prepared from an *N*-protected amino acid and the anthrylmethyl alcohol in the presence of DCC/hydroxybenzotriazole.[1] It can also be prepared from 2-(bromomethyl)-9,10-anthraquinone (Cs_2CO_3).[2] It is stable to moderately acidic conditions (e.g., CF_3COOH, 20°, 1 h; HBr/HOAc, $t_{1/2}$ = 65 h; HCl/CH_2Cl_2, 20°, 1 h).[1] Cleavage is effected by reduction of the quinone to the hydroquinone **i**; in the latter, electron release from the —OH group of the hydroquinone results in facile cleavage of the methylene–carboxylate bond. The related 2-phenyl-2-(9,10-dioxo)anthrylmethyl ester has also been prepared, but is cleaved by electrolysis (−0.9 V, DMF, 0.1 *M* $LiClO_4$, 80% yield).[3]

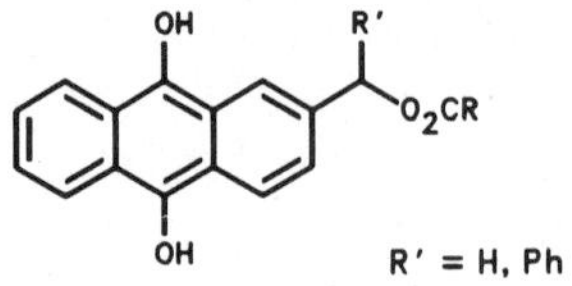

OH R' O_2CR OH R' = H, Ph

i

Cleavage[1]

This derivative is cleaved by hydrogenolysis and by the following conditions:

1. $Na_2S_2O_4$, dioxane–H_2O, pH 7–8, 8 h, 100% yield.
2. Irradiation, *i*-PrOH, 4 h, 99% yield.
3. 9-Hydroxyanthrone, Et_3N/DMF, 5 h, 99% yield.
4. 9,10-Dihydroxyanthracene/polystyrene resin, 1.5 h, 100% yield.

1. D. S. Kemp and J. Reczek, *Tetrahedron Lett.*, 1031 (1977).
2. P. Hoogerhout, C. P. Guis, C. Erkelens, W. Bloemhoff, K. E. T. Kerling, and J. H. Boom, *Recl. Trav. Chim. Pays-Bas*, **104**, 54 (1985).
3. R. L. Blankespoor, A. N. K. Lau, and L. L. Miller, *J. Org. Chem.*, **49**, 4441 (1984).

43. 5-Dibenzosuberyl Ester:

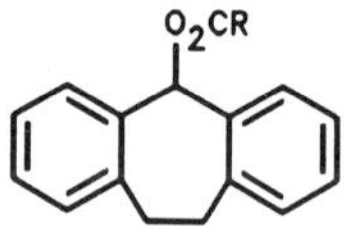

O_2CR

The dibenzosuberyl ester is prepared from dibenzosuberyl chloride (which is also used to protect —OH, —NH, and —SH groups) and a carboxylic acid (Et_3N, reflux, 4 h, 45% yield). It can be cleaved by hydrogenolysis and, like *t*-butyl esters, by acidic hydrolysis (aq. HCl/THF, 20°, 30 min, 98% yield).[1]

1. J. Pless, *Helv. Chim. Acta*, **59**, 499 (1976).

44. 1-Pyrenylmethyl Ester (R′ = H, Me, Ph):

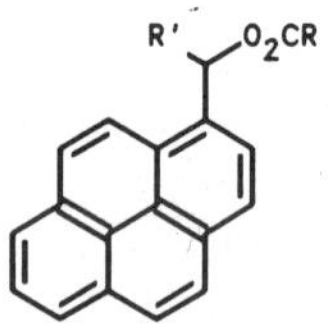

R′ O_2CR

These esters are prepared from the diazomethylpyrenes and carboxylic acids in DMF (R′ = H, 60% yield, R′ = Me, 80% yield, R′ = Ph, 20% yield for 4-methylbenzoic acid). They are cleaved by photolysis at 340 nm (80–100% yield, R′ = H).[1] The esters are very fluorescent.

1. M. Iwamura, T. Ishikawa, Y. Koyama, K. Sakuma, and H. Iwamura, *Tetrahedron Lett.*, **28,** 679 (1987).

45. 2-(Trifluoromethyl)-6-chromylmethyl Ester (Tcrom ester):

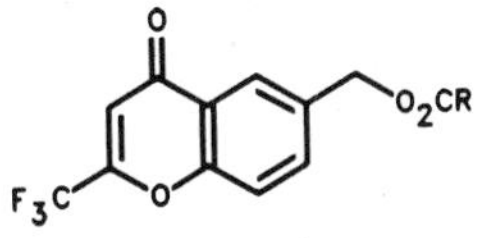

The Tcrom ester is prepared from the cesium salt of an *N*-protected amino acid by reaction with 2-(trifluoromethyl)-6-chromylmethyl bromide (DMF, 25°, 4 h, 53–89% yield). Cleavage of the Tcrom group is effected by brief treatment with *n*-propylamine (2 min, 25°, 96% yield). It is stable to HCl/dioxane, used to cleave a BOC group.[1]

1. D. S. Kemp and G. Hanson, *J. Org. Chem.*, **46,** 4971 (1981).

46. 2,4,6-Trimethylbenzyl Ester: $RCOOCH_2C_6H_2$-2,4,6-$(CH_3)_3$

The 2,4,6-trimethylbenzyl ester has been prepared from an amino acid and the benzyl chloride (Et_3N, DMF, 25°, 12 h, 60–80% yield); it is cleaved by acidic hydrolysis (CF_3COOH, 25°, 60 min, 60–90% yield; 2 *N* HBr/HOAc, 25°, 60 min, 80–95% yield) and by hydrogenolysis. It is stable to methanolic hydrogen chloride used to remove *N*-*o*-nitrophenylsulfenyl groups or triphenylmethyl esters.[1]

1. F. H. C. Stewart, *Aust. J. Chem.*, **21,** 2831 (1968).

47. *p*-Bromobenzyl Ester: $RCOOCH_2C_6H_4$-*p*-Br

The *p*-bromobenzyl ester has been used to protect the β-COOH group in aspartic acid. It is cleaved by strong acidic hydrolysis (HF, 0°, 10 min, 100% yield), but is stable to 50% CF_3COOH/CH_2Cl_2 used to cleave *t*-butyl carbamates. It is 5–7 times more stable than a benzyl ester.[1]

1. D. Yamashiro, *J. Org. Chem.*, **42,** 523 (1977).

48. *o*-Nitrobenzyl Ester: $RCOOCH_2C_6H_4$-*o*-NO_2

49. *p*-Nitrobenzyl Ester: $RCOOCH_2C_6H_4$-*p*-NO_2

The *o*-nitrobenzyl ester, used in this example to protect penicillin precursors, can be cleaved by irradiation (H_2O/dioxane, pH 7). Reductive cleavage of benzyl or *p*-nitrobenzyl esters occurred in lower yields.[1,2]

p-Nitrobenzyl esters have been prepared from the Hg(I) salt of penicillin precursors and the phenyldiazomethane.[3] They are much more stable to acidic hydrolysis (e.g., HBr) than are *p*-chlorobenzyl esters and are recommended for terminal —COOH protection in solid-phase peptide synthesis.[4]

p-Nitrobenzyl esters of penicillin and cephalosporin precursors have been cleaved by alkaline hydrolysis with Na_2S (0°, aq. acetone, 25–30 min, 75–85% yield).[5] They are also cleaved by electrolytic reduction at −1.2 V.[6]

1. L. D. Cama and B. G. Christensen, *J. Am. Chem. Soc.*, **100,** 8006 (1978).
2. For a review covering the photolytic removal of protective groups, see: V. N. R. Pillai, *Synthesis*, 1 (1980).
3. W. Baker, C. M. Pant, and R. J. Stoodley, *J. Chem. Soc., Perkin Trans. I*, 668 (1978).
4. R. L. Prestidge, D. R. K. Harding, and W. S. Hancock, *J. Org. Chem.*, **41,** 2579 (1976).
5. S. R. Lammert, A. I. Ellis, R. R. Chauvette, and S. Kukolja, *J. Org. Chem.*, **43,** 1243 (1978).
6. V. G. Mairanovsky, *Angew. Chem., Int. Ed. Engl.*, **15,** 281 (1976).

50. *p*-Methoxybenzyl Ester: $RCOOCH_2C_6H_4$-*p*-OCH_3

p-Methoxybenzyl esters have been prepared from the Ag(I) salt of amino acids and the benzyl halide (Et_3N, $CHCl_3$, 25°, 24 h, 60% yield)[1] and from cephalosporin precursors and the benzyl alcohol [$Me_2NCH(OCH_2$-*t*-$Bu)_2$, CH_2Cl_2, 90% yield].[2] They are readily prepared by activation of the acid with isopropenyl oxychloroformate ($MeOC_6H_4CH_2OH$, DMAP, 0°, CH_2Cl_2, 91%).[3] They are cleaved by acidic hydrolysis (CF_3COOH/PhOMe, 25°, 3 min, 98% yield;[4] HCOOH, 22°, 1 h, 81% yield[1]), by treatment with $AlCl_3$ (anisole, CH_2Cl_2 or CH_3NO_2, −50°; $NaHCO_3$, −50°, 73–95% yield),[5,6] or by $CF_3CO_2H/B(OTf)_3$.[7]

1. G. C. Stelakatos and N. Argyropoulos, *J. Chem. Soc. C*, 964 (1970).
2. J. A. Webber, E. M. Van Heyningen, and R. T. Vasileff, *J. Am. Chem. Soc.*, **91,** 5674 (1969).
3. P. Jouin, B. Castro, C. Zeggaf, A. Pantaloni, J. P. Senet, S. Lecolier, and G. Sennyey, *Tetrahedron Lett.*, **28,** 1661 (1987).
4. F. H. C. Stewart, *Aust. J. Chem.*, **21,** 2543 (1968).
5. M. Ohtani, F. Watanabe, and M. Narisada, *J. Org. Chem.*, **49,** 5271 (1984).
6. T. Tsuji, T. Kataoka, M. Yoshioka, Y. Sendo, Y. Nishitani, S. Hirai, T. Maeda, and W. Nagata, *Tetrahedron Lett.*, 2793 (1979).
7. S. D. Young and P. P. Tamburini, *J. Am. Chem. Soc.*, **111,** 1933 (1989).

51. 2,6-Dimethoxybenzyl Esters: $2,6\text{-}(CH_3O)_2C_6H_3CH_2OOCR$

2,6-Dimethoxybenzyl esters prepared from the acid chloride and the benzyl alcohol are readily cleaved oxidatively by DDQ (CH_2Cl_2, H_2O, rt, 18 h, 90–95% yield). A 4-methoxybenzyl ester was found not to be cleaved by DDQ. The authors have also explored the oxidative cleavage (ceric ammonium nitrate, CH_3CN, H_2O, 0°, 4 h, 65–97% yield) of a variety of 4-hydroxy- and 4-amino-substituted phenolic esters.[1]

1. C. U. Kim and P. F. Misco, *Tetrahedron Lett.*, **26,** 2027 (1985).

52. 4-(Methylsulfinyl)benzyl (Msib) Ester: $4\text{-}CH_3S(O)C_6H_4CH_2O_2CR$

The 4-(methylsulfinyl)benzyl ester was recommended as a selectively cleavable carboxyl protective group for peptide synthesis. It is readily prepared from 4-(methylsulfinyl)benzyl alcohol (EDCI, HOBt, $CHCl_3$, 78–100% yield) or from 4-methylthiobenzyl alcohol followed by oxidation of the derived ester with MCPBA or H_2O_2/AcOH. The Msib ester is exceptionally stable to CF_3COOH (cleavage rate = 0.000038% ester cleaved per minute) and only undergoes 10% cleavage in HF (anisole, 0°, 1 h). Anhydrous HCl/dioxane rapidly reduces the sulfoxide to the sulfide (Mtb ester), which is completely cleaved in 30 min with CF_3CO_2H. A number of reagents readily reduce the Msib ester to the Mtb ester with $(CH_3)_3SiCl$/Ph_3P as the reagent of choice.[1]

1. J. M. Samanen and E. Brandeis, *J. Org. Chem.*, **53,** 561 (1988).

53. 4-Sulfobenzyl Ester: $Na^+ \ ^-O_3SC_6H_4CH_2O_2CR$

4-Sulfobenzyl esters were prepared (cesium salt or dicyclohexylammonium salt, $NaO_3SC_6H_4CH_2Br$, DMF, 37–95% yield) from *N*-protected amino acids. They are cleaved by hydrogenolysis (H_2/Pd), or hydrolysis (NaOH, dioxane/water). Treatment with ammonia or hydrazine results in formation of the amide or hydrazide. The ester is stable to 2 *M* HBr/AcOH and to CF_3SO_3H in CF_3CO_2H. The relative rates of hydrolysis and hydrazinolysis for different esters are as follows:

Hydrolysis: $NO_2C_6H_4CH_2O{-} \gg C_6H_4CH_2O{-} > {^-O_3SC_6H_4CH_2O{-}} > MeO{-}$

Hydrazinolysis: $NO_2C_6H_4CH_2O{-} > {^-O_3SC_6H_4CH_2O{-}} > C_6H_5CH_2O{-} > MeO{-}$

A benzyl ester can be cleaved in the presence of the 4-sulfobenzyl ester by CF_3SO_3H.[1,2]

1. R. Bindewald, A. Hubbuch, W. Danho, E. E. Büllesbach, J. Föhles and H. Zahn, *Int. J. Pept. Protein Res.*, **23,** 368 (1984).
2. A. Hubbuch, R. Bindewald, J. Föhles, V. K. Naithani, and H. Zahn, *Angew. Chem., Int. Ed. Engl.*, **19,** 394 (1980).

54. Piperonyl Ester (Chart 6):

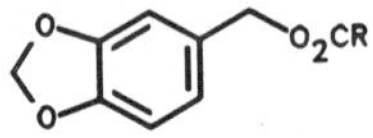

The piperonyl ester can be prepared from an amino acid ester and the benzyl alcohol (imidazole/dioxane, 25°, 12 h, 85% yield) or from an amino acid and the benzyl chloride (Et_3N, DMF, 25°, 57–95% yield). It is cleaved, more readily than a *p*-methoxybenzyl ester, by acidic hydrolysis (CF_3COOH, 25°, 5 min, 91% yield).[1]

1. F. H. C. Stewart, *Aust. J. Chem.*, **24,** 2193 (1971).

55. 4-Picolyl Ester: RCO_2CH_2-4-pyridyl

The picolyl ester has been prepared from amino acids and picolyl alcohol (DCC/CH_2Cl_2, 20°, 16 h, 60% yield) or picolyl chloride (DMF, 90–100°, 2 h, 50% yield). It is cleaved by reduction (H_2/Pd–C, aq. EtOH, 10 h, 98% yield; Na/NH_3, 1.5 h, 93% yield) and by basic hydrolysis (1 *N* NaOH, dioxane, 20°, 1 h, 93% yield). The basic site in a picolyl ester allows its ready separation by extraction into an acidic medium.[1]

1. R. Camble, R. Garner, and G. T. Young, *J. Chem. Soc. C*, 1911 (1969).

56. *p*-Ⓟ-Benzyl Ester: $RCOOCH_2C_6H_4$-*p*-Ⓟ

The first,[1] and still widely used, polymer-supported ester is formed from an amino acid and a chloromethylated copolymer of styrene–divinylbenzene. Originally it was cleaved by basic hydrolysis (2 *N* NaOH, EtOH, 25°, 1 h). Subsequently it has been cleaved by hydrogenolysis (H_2/Pd–C, DMF, 40°, 60 psi, 24 h, 71% yield),[2] and by HF, which concurrently removes many amine protective groups.[3]

Monoesterification of a symmetrical dicarboxylic acid chloride can be effected by reaction with a hydroxymethyl copolymer of styrene–divinylbenzene to give an ester; a mono salt of a diacid was converted into a dibenzyl polymer.[4]

1. R. B. Merrifield, *J. Am. Chem. Soc.*, **85,** 2149 (1963).
2. J. M. Schlatter and R. H. Mazur, *Tetrahedron Lett.*, 2851 (1977).
3. J. Lenard and A. B. Robinson, *J. Am. Chem. Soc.*, **89,** 181 (1967).
4. D. D. Leznoff and J. M. Goldwasser, *Tetrahedron Lett.*, 1875 (1977).

Silyl Esters

Silyl esters are stable to nonaqueous reaction conditions. A trimethylsilyl ester is cleaved by refluxing in alcohol; the more substituted and therefore more stable silyl esters are cleaved by mildly acidic or basic hydrolysis.

57. Trimethylsilyl Ester: $RCOOSi(CH_3)_3$ (Chart 6)

Some of the more common reagents for the conversion of carboxylic acids to trimethylsilyl esters are listed below. For additional methods that can be used to silylate acids, the section on alcohol protection should be consulted since many of the methods presented there are also applicable to carboxylic acids. Trimethylsilyl esters are cleaved in aqueous solutions.

Formation

1. Me_3SiCl/Pyr, CH_2Cl_2, 30°, 2 h.[1]
2. $MeC(OSiMe_3){=}NSiMe_3$, HBr, dioxane, α-picoline, 6 h, 80% yield.[2]
3. $MeCH{=}C(OMe)OSiMe_3/CH_2Cl_2$, 15–25°, 5–40 min, quant.[3]
4. $Me_3SiNHSO_2OSiMe_3/CH_2Cl_2$, 30°, 0.5 h, 92–98% yield.[4]

1. B. Fechti, H. Peter, H. Bickel, and E. Vischer, *Helv. Chim. Acta*, **51,** 1108 (1968).
2. J. J. de Koning, H. J. Kooreman, H. S. Tan, and J. Verweij, *J. Org. Chem.*, **40,** 1346 (1975).
3. Y. Kita, J. Haruta, J. Segawa, and Y. Tamura, *Tetrahedron Lett.*, 4311 (1979).
4. B. E. Cooper and S. Westall, *J. Organomet. Chem.*, **118,** 135 (1976).

58. Triethylsilyl Ester: $RCOOSi(C_2H_5)_3$

Formation/Cleavage[1]

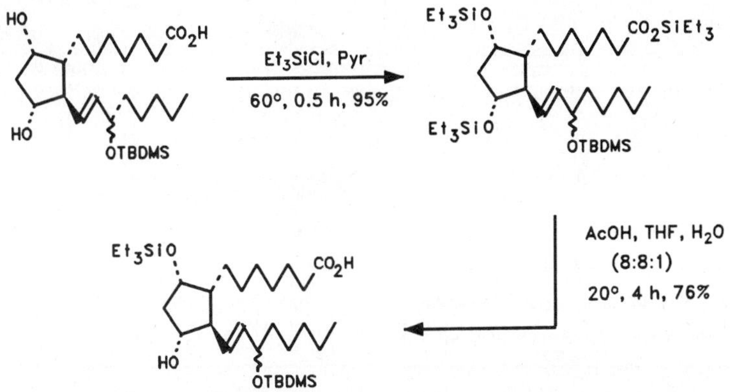

HO, HO, CO_2H, OTBDMS
Et_3SiCl, Pyr
60°, 0.5 h, 95%
Et_3SiO, Et_3SiO, CO_2SiEt_3, OTBDMS
AcOH, THF, H_2O
(8:8:1)
20°, 4 h, 76%
Et_3SiO, HO, CO_2H, OTBDMS

1. T. W. Hart, D. A. Metcalfe, and F. Scheinmann, *J. Chem. Soc., Chem. Commun.*, 156 (1979).

59. *t*-Butyldimethylsilyl Ester (TBDMS Group): $RCOOSi(CH_3)_2C(CH_3)_3$ (Chart 6)

Formation

1. *t*-$BuMe_2SiCl$, imidazole, DMF, 25°, 48 h, 88%.[1]
2. Morpholine, TBDMSCl, THF, 2 min, 20°, >80% yield.[2] In this case the ester was formed in the presence of a phenol.

Cleavage

1. AcOH, H_2O, THF, (3 : 1 : 1), 25°, 20 h.[1]

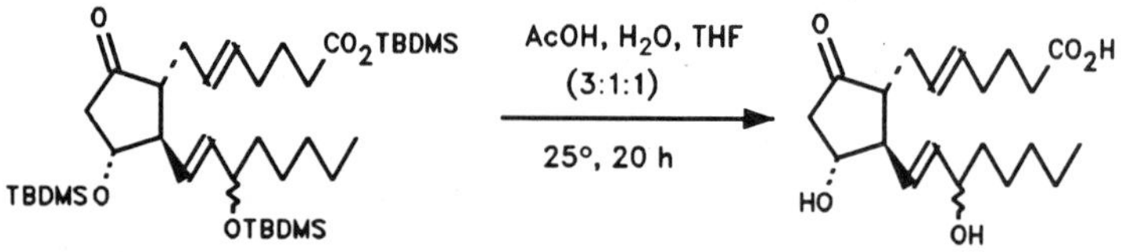

2. $Bu_4N^+F^-$, DMF, 25°.[1]
3. K_2CO_3, MeOH, H_2O, 25°, 1 h, 88% yield.[3]

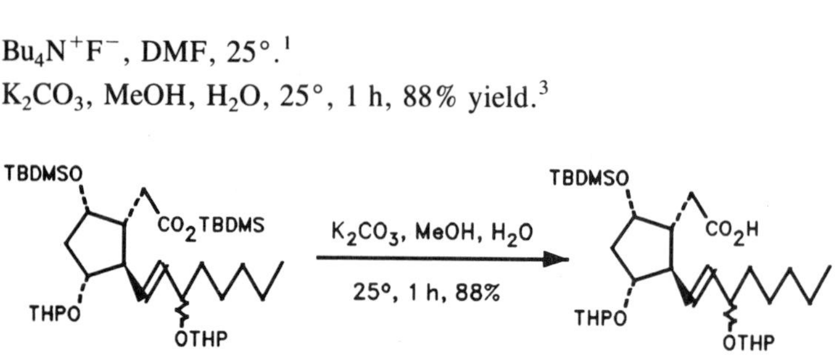

4. The TBDMS ester can be converted directly to an acid chloride [DMF, $(COCl)_2$, rt, CH_2Cl_2] and then converted to another ester, with different properties, by standard means. This procedure avoids the generation of HCl during the acid chloride formation and is thus suitable for acid-sensitive substrates.[4]

1. E. J. Corey and A. Venkateswarlu, *J. Am. Chem. Soc.*, **94,** 6190 (1972).
2. J. W. Perich and R. B. Johns, *Synthesis*, 701 (1989).
3. D. R. Morton and J. L. Thompson, *J. Org. Chem.*, **43,** 2102 (1978).
4. A. Wissner and G. V. Grudzinskas, *J. Org. Chem.*, **43,** 3972 (1978).

60. *i*-Propyldimethylsilyl Ester: $RCOOSi(CH_3)_2CH(CH_3)_2$

The *i*-propyldimethylsilyl ester is prepared from a carboxylic acid and the silyl chloride (Et_3N, 0°). It is cleaved at pH 4.5 by conditions that do not cleave a tetrahydropyranyl ether (HOAc–NaOAc, acetone–H_2O, 0°, 45 min → 25°, 30 min, 91% yield).[1]

1. E. J. Corey and C. U. Kim, *J. Org. Chem.*, **38,** 1233 (1973).

61. Phenyldimethylsilyl Ester: $RCOOSi(CH_3)_2C_6H_5$

The phenyldimethylsilyl ester has been prepared from an amino acid and phenyldimethylsilane (Ni/THF, reflux, 3–5 h, 62–92% yield).[1]

1. M. Abe, K. Adachi, T. Takiguchi, Y. Iwakura, and K. Uno, *Tetrahedron Lett.*, 3207 (1975).

62. Di-*t*-butylmethylsilyl Ester (DTBMS Ester): $(t\text{-Bu})_2CH_3SiO_2CR$

The DTBMS ester was prepared (THF, DTBMSOTf, Et_3N, rt) to protect an ester so that a lactone could be reduced to an aldehyde. The ester is cleaved with aq. HF/THF or $Bu_4N^+F^-$ in wet THF. A THP derivative can be deprotected (pyridinium *p*-toluenesulfonate, warm ethanol) in the presence of a DTBMS ester.[1]

1. R. S. Bhide, B. S. Levison, R. B. Sharma, S. Ghosh, and R. G. Salomon, *Tetrahedron Lett.*, **27**, 671 (1986).

Activated Esters

63. Thiol Esters

Thiol esters, which are more reactive to nucleophiles than are the corresponding oxygen esters, have been prepared to activate carboxyl groups for both lactonization and peptide bond formation. For lactonization *S*-*t*-butyl[1] and *S*-2-pyridyl[2] esters are widely used. Some methods used to prepare thiol esters are shown below. The *S*-*t*-butyl ester is included in Reactivity Chart 6.

1. $RCOOH + R'SH \xrightarrow[0°,\ 5\ \text{min} \rightarrow 20°,\ 3\ \text{h}]{\text{DCC, DMAP, } CH_2Cl_2{}^{3}} RCOSR'$, 85–92%

   R′ = Et, *t*-Bu

DMAP = 4-dimethylaminopyridine (10^4 times more effective than pyridine)

2. 2-fluoro-1-methylpyridinium tosylate (CH_3, TsO^-) + RCOOH $\xrightarrow[-15°,\ 1\ h]{Et_3N,\ CH_2Cl_2}$ $\xrightarrow[2\ h,\ 75\text{–}95\%]{R'SH,\ Et_3N,\ CH_2Cl_2}$ RCOSR′[4]

   R′ = *n*-Bu, *s*-Bu, *t*-Bu, Ph, 2-pyridyl

3. RCOOH + R′SH $\xrightarrow[25°,\ 1\ h,\ 70\text{–}100\%]{Me_2NPOCl_2,\ Et_3N,\ DME}$ RCOSR′

   R′ = Et, *i*-Pr, *t*-Bu, *c*-C_6H_{11}, Ph

   These neutral conditions can be used to prepare thiol esters of acid- or base-sensitive compounds including penicillins.[5]

4. RCH(NHPG)COOH + Ph_2POCl $\xrightarrow[0°,\ 30\ min]{Et_3N,\ CH_2Cl_2}$ $\xrightarrow{R'SH,\ Et_3N}$ or $\xrightarrow[25°,\ 1\ h]{R'STl}$ RCH(NHPG)COSR′, 70–100%[6]

   R′ = *t*-Bu, Ph, $PhCH_2$

5. RCOOH + R′SH $\xrightarrow[Et_3N,\ DMF,\ 25°,\ 3\ h,\ 70\text{–}85\%]{(EtO)_2POCN\ or\ (PhO)_2PON_3}$ RCOSR′[7]

   R = alkyl, aryl, benzyl, amino acids; penicillins
   R′ = Et, *i*-Pr, *n*-Bu, Ph, $PhCH_2$

6. RCOCl + *n*-Bu_3SnSR' $\xrightarrow{CHCl_3}$ RCOSR′[8]

   R′ = *t*-Bu: 60°, 0.5 h, 90–95% yield
   = Ph: 25°, 12 h, 92–95% yield
   = $PhCH_2$: 25°, 0.5–1 h, 87–96% yield

7. RCOOR′ + Me_2AlS-*t*-Bu $\xrightarrow[25°,\ 75\text{–}100\%]{CH_2Cl_2}$ RCOS-*t*-Bu[9]

   R′ = Me, Et
   This reaction avoids the use of toxic thallium compounds.

8. RCOOH + PhSCN $\xrightarrow[25°,\ 30\ min,\ 80\text{–}95\%]{Bu_3P,\ CH_2Cl_2}$ RCOSPh[10]

9. RCOOH + ClCOS-2-pyridyl $\xrightarrow[0.5\ h,\ 95\text{–}100\%]{Et_3N,\ 0°}$ RCOS-2-pyridyl + $Et_3N\cdot HCl$[11]

10. RCO_2H + hydroxybenzotriazole $\xrightarrow{DCC}$ $\xrightarrow{R'SH,\ Et_3N}$ or $\xrightarrow{R'STl}$ RCOSR′, 70–100%[6]

    R′ = *t*-Bu, Ph, $PhCH_2$

Cleavage

1. $AgNO_3$, H_2O, dioxane, (1:4), 2 h.[12]
2. ROH, $Hg(O_2CCF_3)_2$ 90% yield.[1]
3. Electrolysis, $Bu_4N^+Br^-$, H_2O, CH_3CN, $NaHCO_3$.[13] This method is unsatisfactory for primary and secondary alcohols, aldehydes, olefins, or amines.
4. MeI, ROH (R = *t*-Bu, PhSH, etc.), 68–97% yield.[14]

1. S. Masamune, S. Kamata, and W. Schilling, *J. Am. Chem. Soc.*, **97,** 3515 (1975).
2. T. Mukaiyama, R. Matsueda, and M. Suzuki, *Tetrahedron Lett.*, 1901 (1970); E. J. Corey, P. Ulrich, and J. M. Fitzpatrick, *J. Am. Chem. Soc.*, **98,** 222 (1976).
3. B. Neises and W. Steglich, *Angew. Chem., Int. Ed. Engl.*, **17,** 522 (1978).
4. Y. Watanabe, S.-i. Shoda, and T. Mukaiyama, *Chem. Lett.*, 741 (1976).
5. H.-J. Liu, S. P. Lee, and W. H. Chan, *Synth. Commun.*, **9,** 91 (1979).
6. K. Horiki, *Synth. Commun.*, **7,** 251 (1977).
7. S. Yamada, Y. Yokoyama, and T. Shiori, *J. Org. Chem.*, **39,** 3302 (1974).
8. D. N. Harpp, T. Aida, and T. H. Chan, *Tetrahedron Lett.*, 2853 (1979).
9. R. P. Hatch and S. M. Weinreb, *J. Org. Chem.*, **42,** 3960 (1977).
10. P. A. Grieco, Y. Yokoyama, and E. Williams, *J. Org. Chem.*, **43,** 1283 (1978).
11. E. J. Corey and D. A. Clark, *Tetrahedron Lett.*, 2875 (1979).
12. A. B. Shenvi and H. Gerlach, *Helv. Chim. Acta*, **63,** 2426 (1980).
13. M. Kimura, S. Matsubara, and Y. Sawaki, *J. Chem. Soc., Chem. Commun.*, 1619 (1984).
14. D. Ravi and H. B. Mereyala, *Tetrahedron Lett.*, **30,** 6089 (1989).

Miscellaneous Derivatives

64. Oxazoles:

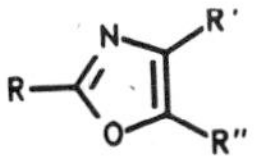

Oxazoles, prepared from carboxylic acids (benzoin, DCC; NH_4OAc, AcOH, 80–85% yield), have been used as carboxylic acid protective groups in a variety of synthetic applications.[1] They are readily cleaved by singlet oxygen followed by hydrolysis (ROH, TsOH, benzene[1] or K_2CO_3, MeOH[2]).

65. 2-Alkyl-1,3-oxazolines (Chart 6):

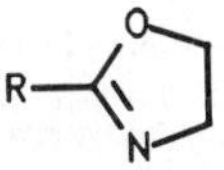

2-Alkyl-1,3-oxazolines are prepared to protect both the carbonyl and hydroxyl groups of an acid. They are stable to Grignard reagents[3] and to lithium aluminum hydride (25°, 2 h).[4]

Formation

1. $HOCH_2C(CH_3)_2NH_2$, $PhCH_3$, reflux, 70–80% yield.[5]
2. From an acid chloride: $HOCH_2C(CH_3)_2NH_2$, $SOCl_2$, CH_2Cl_2, 25°, 30 min, >80% yield.[6]
3. Dimethylaziridine, DCC, 3% H_2SO_4, Et_2O or CH_2Cl_2, rt, 6–16 h, 50–80% yield.[4]

4. $H_2NCH_2CH_2OH$, Ph_3P, Et_3N, CCl_4, CH_3CN, Pyr, rt, 70% yield.[7]
5. From an acid chloride: $BrCH_2CH_2NH_3^+Br^-$, Et_3N, benzene, reflux, 24 h, 46–67% yield.[8]

Cleavage

1. 3 *N* HCl, EtOH, 90% yield.[3]
2. MeI, 25°, 12 h; 1 *N* NaOH, 25°, 15 h, 94% yield.[9]

1. H. H. Wasserman, K. E. McCarthy, and K. S. Prowse, *Chem. Rev.*, **86,** 845 (1986).
2. M. A. Tius and D. P. Astrab, *Tetrahedron Lett.*, **30,** 2333 (1989).
3. A. I. Meyers and D. L. Temple, *J. Am. Chem. Soc.*, **92,** 6644 (1970).
4. D. Haidukewych and A. I. Meyers, *Tetrahedron Lett.*, 3031 (1972).
5. H. L. Wehrmeister, *J. Org. Chem.*, **26,** 3821 (1961).
6. S. R. Schow, J. D. Bloom, A. S. Thompson, K. N. Winzenberg, and A. B. Smith, III, *J. Am. Chem. Soc.*, **108,** 2662 (1986).
7. H. Vorbrüggen and K. Krolikiewicz, *Tetrahedron Lett.*, **22,** 4471 (1981).
8. C. Kashima and H. Arao, *Synthesis*, 873 (1989).
9. A. I. Meyers, D. L. Temple, R. L. Nolen, and E. D. Mihelich, *J. Org. Chem.*, **39,** 2778 (1974).

66. 4-Alkyl-5-oxo-1,3-oxazolidine:

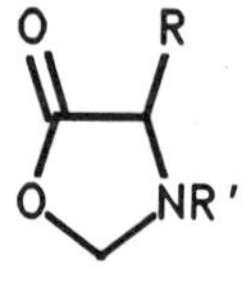

O R
O NR'

1,3-Oxazolidines are prepared to allow selective protection of the α- or ω-CO_2H groups in aspartic and glutamic acids.

Formation/Cleavage[1]

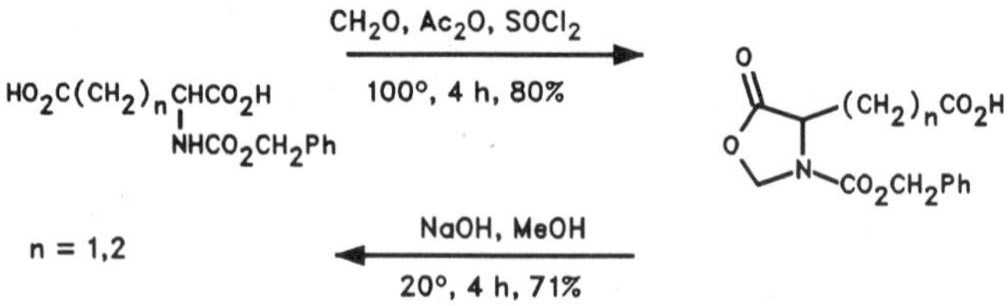

$HO_2C(CH_2)_nCHCO_2H$ ($NHCO_2CH_2Ph$) —[CH_2O, Ac_2O, $SOCl_2$; 100°, 4 h, 80%]→ oxazolidinone with $(CH_2)_nCO_2H$ and N–CO_2CH_2Ph

←[NaOH, MeOH; 20°, 4 h, 71%]—

n = 1,2

The related 2-*t*-butyl derivative has also been prepared and used to advantage as a temporary protective group for the stereogenic center of amino acids during alkylations.[2]

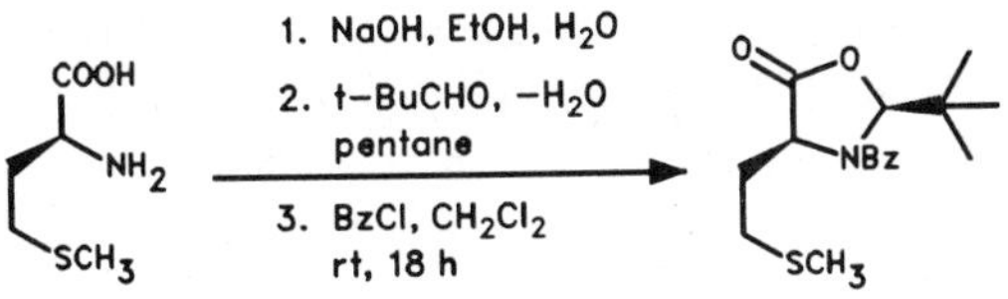

1. NaOH, EtOH, H_2O
2. t-BuCHO, $-H_2O$ pentane
3. BzCl, CH_2Cl_2 rt, 18 h

1. M. Itoh, *Chem. Pharm. Bull.*, **17,** 1679 (1969).
2. D. Seebach and A. Fadel, *Helv. Chim. Acta*, **68,** 1243 (1985).

67. 5-Alkyl-4-oxo-1,3-dioxolane:

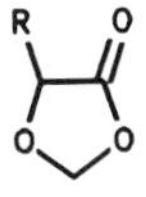

These derivatives are prepared to protect α-hydroxy carboxylic acids; they are cleaved by acidic hydrolysis of the acetal structure (HCl, DMF, 50°, 7 h, 71% yield), or basic hydrolysis of the lactone.[1]

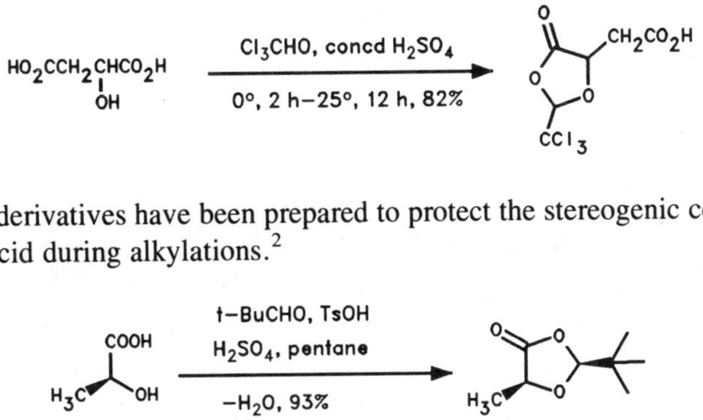

$HO_2CCH_2CHCO_2H$ (OH) → Cl_3CHO, concd H_2SO_4; 0°, 2 h–25°, 12 h, 82%

The 2-alkyl derivatives have been prepared to protect the stereogenic center of the α-hydroxy acid during alkylations.[2]

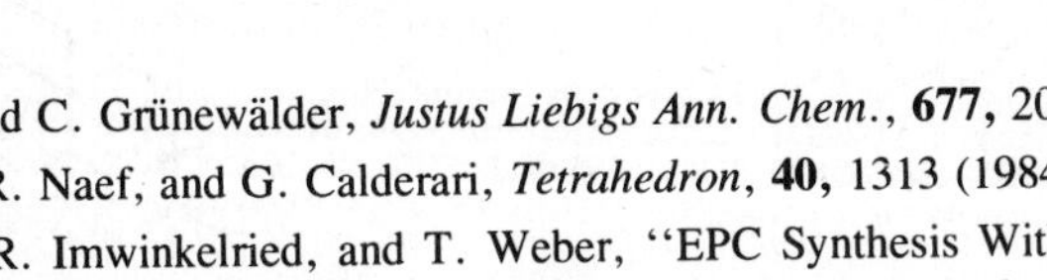

t-BuCHO, TsOH
H_2SO_4, pentane
$-H_2O$, 93%
61% crystalized

This methodology is also effective for protection of β-hydroxy acids.[3]

1. H. Eggerer and C. Grünewälder, *Justus Liebigs Ann. Chem.*, **677,** 200 (1964).
2. D. Seebach, R. Naef, and G. Calderari, *Tetrahedron*, **40,** 1313 (1984).
3. D. Seebach, R. Imwinkelried, and T. Weber, "EPC Synthesis With C,C Bond Formation via Acetals and Enimines," in *Modern Synthetic Methods 1986*, Vol. 4, R. Scheffold, Ed., Springer-Verlag, New York, 1986, p. 125.

68. Ortho Esters: RC(OR′)$_3$

Ortho esters are among the few derivatives that can be prepared from acids and esters that protect the carbonyl against nucleophilic attack by hydroxide or other strong nucleophiles such as Grignard reagents. In general, ortho esters are difficult to prepare directly from acids and are therefore more often prepared from the nitrile.[1,2] Simple ortho esters derived from normal alcohols are the least stable in terms of acid stability and stability toward Grignard reagents, but as the ortho ester becomes more constrained its stability increases.

Formation

1. RCOCl + (3-methyl-3-oxetanemethanol) —Pyr, 0°, CH_2Cl_2, 12 h, 75–85%→ (oxetane ester, O_2CR) —$BF_3 \cdot Et_2O$, −15°, CH_2Cl_2, 75–90%→ (bicyclic ortho ester, R) Ref. 3

This is one of the few methods available for the direct and efficient conversion of an acid, via the acid chloride, to an ortho ester.

2. $Br(CH_2)_5CO_2H$ —TsOH, xylene, reflux, $-H_2O$; cis,cis-1,3,5-cyclohexanetriol (OH, HO, OH)→ ortho ester $(CH_2)_5Br$, 14% Ref. 4

3. $Br(CH_2)_5CN$ —1. HCl, MeOH; 2. cis,cis-1,3,5-cyclohexanetriol (OH, HO, OH)→ ortho ester $(CH_2)_5Br$, 68% Ref. 4

4. (lactone) —$HOCH_2CH_2OH$, H^+, benzene, reflux, >88%→ (ketal) Refs. 5, 6

Note that this method does not work on simple esters.

5. (lactone) —1. $Me_2AlS(CH_2)_2SAlMe_2$; 2. TsOH 94%→ (dithiolane ortho ester) ; reverse: Hg^{++}, H_2O, $BF_3 \cdot Et_2O$, THF, 25°, 40 min Ref. 7

Cleavage

Oxygen ortho esters are readily cleaved by mild aqueous acid (TsOH·Pyr, H_2O;[8] $NaHSO_4$, 5:1 DME, H_2O, 0°, 20 min[9]) to form esters that are then hydrolyzed with aqueous base to give the acid. Note that a trimethyl ortho ester is readily hydrolyzed in the presence of an acid-sensitive ethoxyethyl acetal.[7] The order of acid stability is

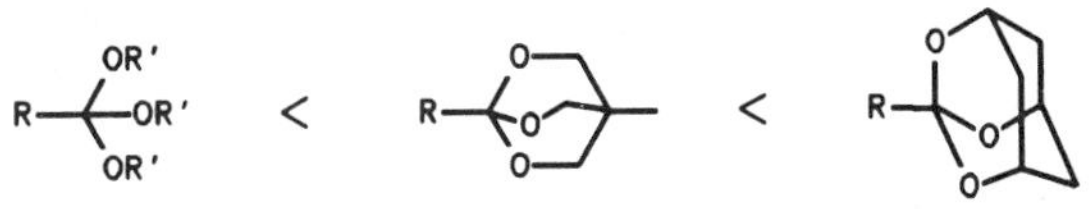

1. S. M. McElvain and J. W. Nelson, *J. Am. Chem. Soc.*, **64,** 1825 (1942).
2. The synthesis and interconversion of simple ortho esters has been reviewed: R. H. DeWolfe, *Synthesis*, 153 (1974).
3. E. J. Corey and N. Raju, *Tetrahedron Lett.*, **24,** 5571 (1983).
4. G. Voss and H. Gerlach, *Helv. Chim. Acta*, **66,** 2294 (1983).
5. T. Wakamatsu, H. Hara, and Y. Ban, *J. Org. Chem.*, **50,** 108 (1985).
6. J. D. White, S.-c. Kuo, and, T. R. Vedananda, *Tetrahedron Lett.*, **28,** 3061 (1987).
7. E. J. Corey and D. J. Beames, *J. Am. Chem. Soc.*, **95,** 5829 (1973).
8. G. Just, C. Luthe, and M. T. P. Viet, *Can. J. Chem.*, **61,** 712 (1983).
9. E. J. Corey, K. Niimura, Y. Konishi, S. Hashimoto, and Y. Hamada, *Tetrahedron Lett.*, **27,** 2199 (1986).

69. Phenyl Group: C_6H_5—

The phenyl group became a practical "protective" group for carboxylic acids when Sharpless published a mild, effective one-step method for its conversion to a carboxylic acid.[1] It has recently been used in a synthesis of the amino acid statine, where it served as a masked or carboxylic acid equivalent.[2]

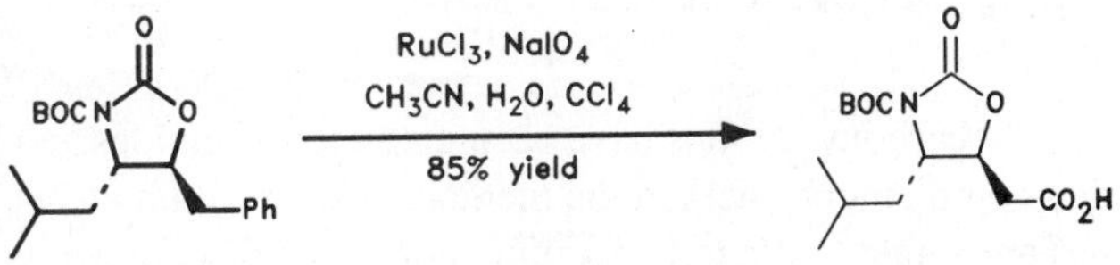

1. P. H. J. Carlsen, T. Katsuki, V. S. Martin, and K. B. Sharpless, *J. Org. Chem.*, **46,** 3936 (1981).
2. S. Kano, Y. Yuasa, T. Yokomatsu, and S. Shibuya, *J. Org. Chem.*, **53,** 3865 (1988).

70. Pentaaminecobalt(III) Complex: $[RCO_2Co(NH_3)_5](BF_4)_3$

The pentaaminecobalt(III) complex has been prepared from amino acids to protect the carboxyl group during peptide synthesis [$(H_2O)Co(NH_3)_5(ClO_4)_3$, 70°, H_2O,

6 h; cool to 0°; filter; HBF_4, 60–80% yield]. It is cleaved by reduction [$NaBH_4$, NaSH, or $(NH_4)_2S$, Fe(II)EDTA]. These complexes do not tend to racemize and are stable to CF_3CO_2H, which is used to remove BOC groups.[1-3]

1. S. Bagger, I. Kristjansson, I. Soetofte, and A. Thorlacius, *Acta Chem. Scand. Ser A*, **A39,** 125 (1985).
2. S. S. Isied, A. Vassilian, and J. M. Lyon, *J. Am. Chem. Soc.*, **104,** 3910 (1982).
3. S. S. Isied, J. Lyon, and A. Vassilian, *Int. J. Pept. Protein Res.*, **19,** 354 (1982).

Stannyl Esters

71. Triethylstannyl Ester: $RCOOSn(C_2H_5)_3$

72. Tri-*n*-butylstannyl Ester: $RCOOSn(n\text{-}C_4H_9)_3$

Stannyl esters have been prepared to protect a —COOH group in the presence of an $-NH_2$ group [$(n\text{-}Bu_3Sn)_2O$ or $n\text{-}Bu_3SnOH$, C_6H_6, reflux, 88%].[1] Stannyl esters of *N*-acylamino acids are stable to reaction with anhydrous amines, and to water and alcohols;[2] aqueous amines convert them to ammonium salts.[2] Stannyl esters of amino acids are cleaved in quantitative yield by water or alcohols (PhSK, DMF, 25°, 15 min, 63% yield or HOAc, EtOH, 25°, 30 min, 77% yield).[2]

1. P. Bamberg, B. Ekström, and B. Sjöberg, *Acta Chem. Scand.*, **22,** 367 (1968).
2. M. Frankel, D. Gertner, D. Wagner, and A. Zilkha, *J. Org. Chem.*, **30,** 1596 (1965).

AMIDES AND HYDRAZIDES

To a limited extent carboxyl groups have been protected as amides and hydrazides, derivatives that complement esters in the methods used for their cleavage. Amides and hydrazides are stable to the mild alkaline hydrolysis that cleaves esters. Esters are stable to nitrous acid, effective in cleaving amides, and to the oxidizing agents [including $Pb(OAc)_4$, MnO_2, SeO_2, CrO_3, and $NaIO_4$;[1] $Ce(NH_4)_2(NO_3)_6$;[2] Ag_2O;[3] and $Hg(OAc)_2$[4]] that have been used to cleave hydrazides.

Classically, amides and hydrazides have been prepared from an ester or an acid chloride and an amine or hydrazine, respectively; they can also be prepared directly from the acid as shown in eqs. 1–3.

1. $$\underset{\text{NHPG}}{\text{RCHCOOH}} + \text{R'NH}_2 \xrightarrow[20°,\ 4\ \text{h},\ 70\text{–}90\%]{\text{DCC, THF or CH}_2\text{Cl}_2{}^{5}} \underset{\text{NHPG}}{\text{RCHCONHR'}}$$

2. $\underset{\text{NHPG}}{\text{RCHCOOH}} \xrightarrow{H_2NNH_2,\ N\text{-hydroxybenzotriazole}^{6}} \underset{\text{NHPG}}{\text{RCHCONHNH}_2}$

3. $RCOOH + R'R''NH \xrightarrow{Ph_3P^{7} \text{ or } Bu_3P/o\text{-}NO_2-C_6H_4SCN^{8}} RCONR'R''$

Equations 4–8 illustrate some mild methods that can be used to cleave amides. Equations 4 and 5 indicate the conditions that were used by Woodward[9] and Eschenmoser,[10] respectively, in their synthesis of vitamin B_{12}. Butyl nitrite,[11] nitrosyl chloride,[12] and nitrosonium tetrafluoroborate ($NO^+BF_4^-$)[13] have also been used to cleave amides. Since only tertiary amides are cleaved by potassium *t*-butoxide (eq. 6), this method can be used to effect selective cleavage of tertiary amides in the presence of primary or secondary amides.[14] (Esters, however, are cleaved by similar conditions.)[15] Photolytic cleavage of nitro amides (eq. 7) is discussed in a review.[16]

4. $RCONH_2 \xrightarrow{N_2O_4/CCl_4{}^{9}} RCOOH$

5. $RCONH_2 \xrightarrow{[ClCH_2CH{=}N(\rightarrow O)\text{-}c\text{-}C_6H_{11} + AgBF_4]^{10}} \xrightarrow{H_3O^+} RCOOH$

6. $RCONR'R'' \xrightarrow[24°,\ 2\text{–}48\ h,\ 88\text{–}96\%]{KO\text{-}t\text{-}Bu/H_2O\ (6:2),\ Et_2O^{14}} RCOOH$

$R', R'' \neq H$

7. $a, b, \text{ or } c \xrightarrow[5\text{–}10\ h,\ 70\text{–}100\%]{350\ nm^{16}} RCOOH$

a = *o*-nitroanilides[17]
b = *N*-acyl-7-nitroindoles[18]
c = *N*-acyl-8-nitrotetrahydroquinolines[19]

8. $RC(O)NH_2 \xrightarrow[\text{Amberlyst A-21, } 45\text{–}66°,\ CH_3CN]{CH_3O\text{–}CH(Cl)CH_2CH_2CH(Cl)\text{–}OCH_3} RC(O)\text{–}N\text{-pyrrole} \xrightarrow{\text{NaOMe, quant.}} RCO_2Me$

Treatment of acyl pyrroles with primary and secondary amines affords amides.[20]

9. The following cleavage proceeds via intramolecular assistance from the alkoxide formed on base treatment.[21]

$RC(O)NHR' \xrightarrow[\text{2. LiOH, rt, 71–87\%}]{\text{1. } AcOCH_2C(CH_3)_2COCl,\ CH_2Cl_2,\ DMAP,\ TEA,\ 88\text{–}99\%} RCO_2H + R'NH_2$

Hydrazides have been used in penicillin and peptide syntheses. In the latter syntheses they are converted by nitrous acid to azides to facilitate coupling.

Some amides and hydrazides that have been prepared to protect carboxyl groups are included in Reactivity Chart 6.

1. M. J. V. O. Baptista, A. G. M. Barrett, D. H. R. Barton, M. Girijavallabhan, R. C. Jennings, J. Kelly, V. J. Papadimitriou, J. V. Turner, and N. A. Usher, *J. Chem. Soc., Perkin Trans. I*, 1477 (1977).
2. T.-L. Ho, H. C. Ho, and C. M. Wong, *Synthesis*, 562 (1972).
3. Y. Wolman, P. M. Gallop, A. Patchornik, and A. Berger, *J. Am. Chem. Soc.*, **84,** 1889 (1962).
4. J. B. Aylward and R. O. C. Norman, *J. Chem. Soc. C*, 2399 (1968).
5. J. C. Sheehan and G. P. Hess, *J. Am. Chem. Soc.*, **77,** 1067 (1955).
6. For example, see S. S. Wang, I. D. Kulesha, D. P. Winter, R. Makofske, R. Kutny, and J. Meienhofer, *Int. J. Pept. Protein Res.*, **11,** 297 (1978).
7. L. E. Barstow and V. J. Hruby, *J. Org. Chem.*, **36,** 1305 (1971).
8. P. A. Grieco, D. S. Clark, and G. P. Withers, *J. Org. Chem.*, **44,** 2945 (1979).
9. R. B. Woodward, *Pure Appl. Chem.*, **33,** 145 (1973).
10. U. M. Kempe, T. K. Das Gupta, K. Blatt, P. Gygax, D. Felix, and A. Eschenmoser, *Helv. Chim. Acta*, **55,** 2187 (1972).
11. N. Sperber, D. Papa, and E. Schwenk, *J. Am. Chem. Soc.*, **70,** 3091 (1948).
12. M. E. Kuehne, *J. Am. Chem. Soc.*, **83,** 1492 (1961).
13. G. A. Olah and J. A. Olah, *J. Org. Chem.*, **30,** 2386 (1965).
14. P. G. Gassman, P. K. G. Hodgson, and R. J. Balchunis, *J. Am. Chem. Soc.*, **98,** 1275 (1976).
15. P. G. Gassman and W. N. Schenk, *J. Org. Chem.*, **42,** 918 (1977).
16. B. Amit, U. Zehavi, and A. Patchornik, *Isr. J. Chem.*, **12,** 103 (1974).
17. B. Amit and A. Patchornik, *Tetrahedron Lett.*, 2205 (1973).
18. B. Amit, D. A. Ben-Efraim, and A. Patchornik, *J. Am. Chem. Soc.*, **98,** 843 (1976).
19. B. Amit, D. A. Ben-Efraim, and A. Patchornik, *J. Chem. Soc., Perkin Trans. I*, 57 (1976).
20. S. D. Lee, M. A. Brook, and T. H. Chan, *Tetrahedron Lett.*, **24,** 1569 (1983).
21. T. Tsunoda, O. Sasaki, and S. Itô, *Tetrahedron Lett.*, **31,** 731 (1990).

Amides

73. *N,N*-Dimethylamide: $RCON(CH_3)_2$ (Chart 6)

Formation/Cleavage[1]

$$RCO_2H \xrightarrow{SOCl_2,\ 70°,\ 3\ h} \xrightarrow{Me_2NH} RCONMe_2$$

$$RCO_2H \xleftarrow{KOH,\ HOCH_2CH_2OH,\ 170°,\ 6\ h} RCONMe_2$$

In these papers the carboxylic acid to be protected was a stable, unsubstituted compound. Harsh conditions were acceptable for both formation and cleavage of the amide.

1. D. E. Ames and P. J. Islip, *J. Chem. Soc.*, 351 (1961); *idem*, *ibid.*, 4363 (1963).

74. Pyrrolidinamide: RCONR′R″, [R′R″ = $(-CH_2-)_4$]

Formation/Cleavage[1]

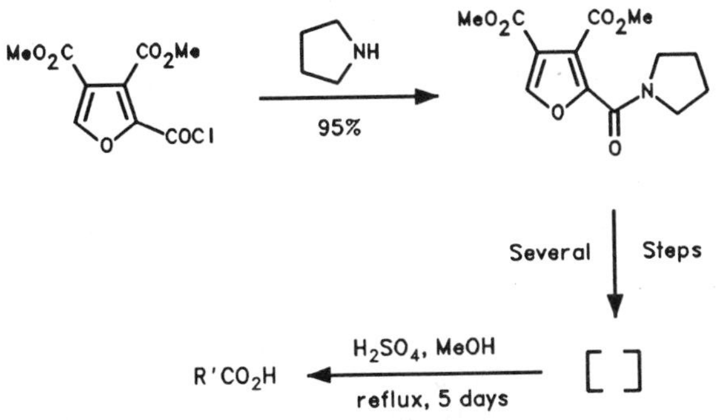

Several Steps

H_2SO_4, MeOH, reflux, 5 days → $R'CO_2H$

R′COOH = precursor of DL-camptothecin

1. A. S. Kende, T. J. Bentley, R. W. Draper, J. K. Jenkins, M. Joyeux, and I. Kubo, *Tetrahedron Lett.*, 1307 (1973).

75. Piperidinamide: RCONR′R″, [R′R″ = $(-CH_2-)_5$]

Formation/Cleavage[1]

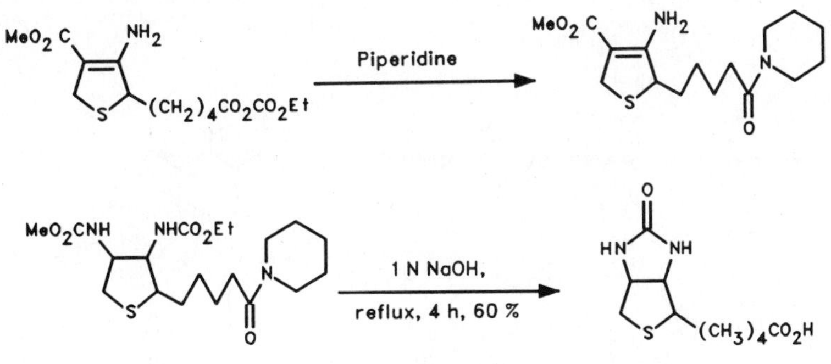

Piperidine

1 N NaOH, reflux, 4 h, 60 %

biotin

1. P. N. Confalone, G. Pizzolato, and M. R. Uskoković, *J. Org. Chem.*, **42,** 1630 (1977).

76. 5,6-Dihydrophenanthridinamide:

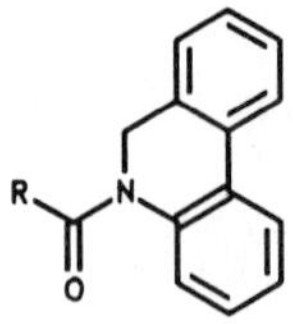

Formation/Cleavage[1]

The amide is stable to HCl or KOH (THF, MeOH, H_2O, 70°, 10 h) and MeMgI, THF, HMPA, −78°. It can also be formed directly from the acid chloride.

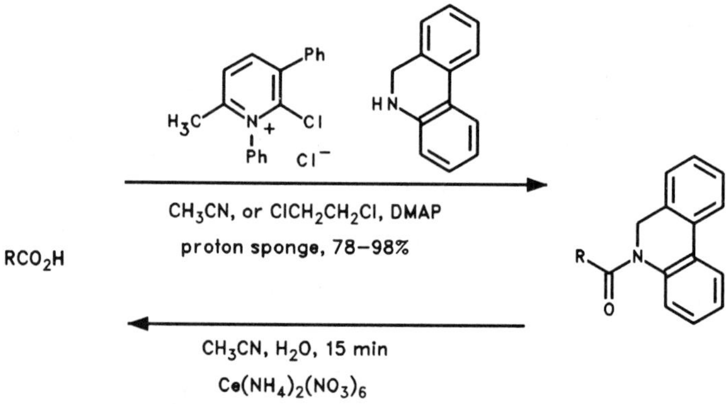

RCO_2H

CH_3CN, or $ClCH_2CH_2Cl$, DMAP
proton sponge, 78–98%

CH_3CN, H_2O, 15 min
$Ce(NH_4)_2(NO_3)_6$

1. T. Uchimaru, K. Narasaka, and T. Mukaiyama, *Chem. Lett.*, 1551 (1981).

77. *o*-Nitroanilide: $RCONR'C_6H_4$-*o*-NO_2, R′ ≠ H

78. *N*-7-Nitroindolylamide (Chart 6):

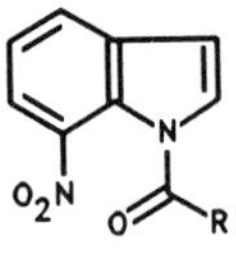

79. *N*-8-Nitro-1,2,3,4-tetrahydroquinolylamide:

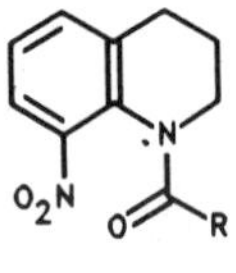

o-Nitroanilides (R′ = Me, *n*-Bu, *c*-C_6H_{11}, Ph, $PhCH_2$; ≠ H),[1] nitroindolylamides,[2] and tetrahydroquinolylamides[3] are cleaved in high yields under mild conditions by irradiation at 350 nm (5–10 h).

1. B. Amit and A. Patchornik, *Tetrahedron Lett.*, 2205 (1973).
2. B. Amit, D. A. Ben-Efraim, and A. Patchornik, *J. Am. Chem. Soc.*, **98,** 843 (1976).
3. B. Amit, D. A. Ben-Efraim, and A. Patchornik, *J. Chem. Soc., Perkin Trans. I*, 57 (1976).

80. *p*-Ⓟ-Benzenesulfonamide: $RCONHSO_2C_6H_4$-*p*-Ⓟ

A polymer-supported sulfonamide, prepared from an amino acid activated ester and a polystyrene–sulfonamide, is stable to acidic hydrolysis (CF_3COOH; HBr/HOAc). It is cleaved by the "safety-catch" method shown below.[1]

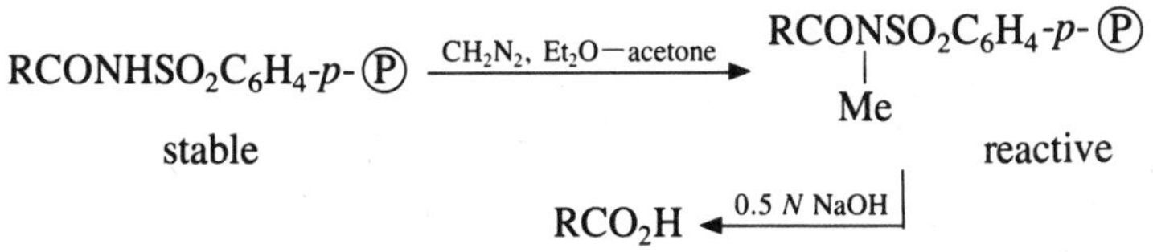

$$\underset{\text{stable}}{RCONHSO_2C_6H_4\text{-}p\text{-}Ⓟ} \xrightarrow{CH_2N_2,\ Et_2O-\text{acetone}} \underset{\text{reactive}}{RCON(Me)SO_2C_6H_4\text{-}p\text{-}Ⓟ} \xrightarrow{0.5\ N\ NaOH} RCO_2H$$

1. G. W. Kenner, J. R. McDermott, and R. C. Sheppard, *J. Chem. Soc., Chem. Commun.*, 636 (1971).

Hydrazides

81. Hydrazides: $RCONHNH_2$ (Chart 6)

See also pp. 270–272.

Cleavage

1. 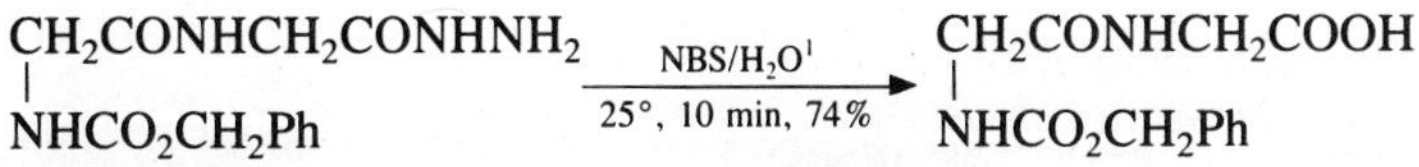 $CH_2(NHCO_2CH_2Ph)CONHCH_2CONHNH_2 \xrightarrow[25°,\ 10\ \text{min},\ 74\%]{NBS/H_2O^{1}} CH_2(NHCO_2CH_2Ph)CONHCH_2COOH$
2. $RCH(NHCOPh)CONHNH_2 \xrightarrow[48°,\ 24\ \text{h},\ 100\%]{60\%\ HClO_4^{2}} RCH(NHCOPh)COOH$

   R = Ph; optically pure
3. $POCl_3$, H_2O, 94% yield.[2]
4. HBr/HOAc or HCl/HOAc, 94% yield.[2]
5. $CuCl_2$, H_2O, THF.[3]

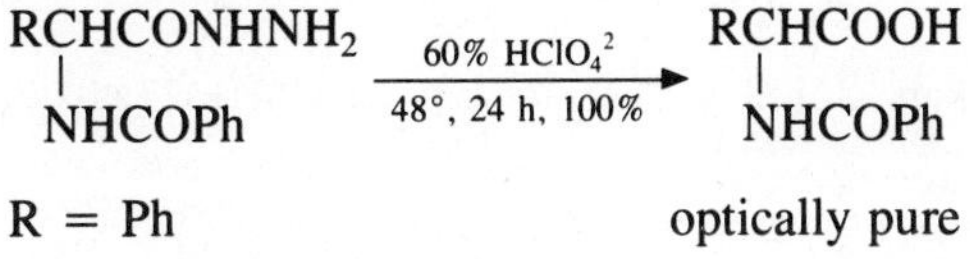

If an alcohol such as ethanol is substituted for H_2O in this reaction, the ester is produced instead of the acid.

1. H. T. Cheung and E. R. Blout, *J. Org. Chem.*, **30,** 315 (1965).
2. J. Schnyder and M. Rottenberg, *Helv. Chim. Acta*, **58,** 521 (1975).
3. O. Attanasi and F. Serra-Zannetti, *Synthesis*, 314 (1980).

82. *N*-Phenylhydrazide: $RCONHNHC_6H_5$ (Chart 6)

Phenylhydrazides have been prepared from amino acid esters and phenylhydrazine in 70% yield;[1] they are cleaved by oxidation [$Cu(OAc)_2$, 95°, 10 min, 67% yield;[2] $FeCl_3$/1 *N* HCl, 96°, 14 min, 85% yield[3]].

1. R. B. Kelly, *J. Org. Chem.*, **28,** 453 (1963).
2. E. W.-Leitz and K. Kühn, *Chem. Ber.*, **84,** 381 (1951).
3. H. B. Milne, J. E. Halver, D. S. Ho, and M. S. Mason, *J. Am. Chem. Soc.*, **79,** 637 (1957).

83. *N,N'*-Diisopropylhydrazide: $RCON(i\text{-}C_3H_7)NH\text{-}i\text{-}C_3H_7$ (Chart 6)

The *N,N'*-diisopropylhydrazide, prepared to protect penicillin derivatives, is cleaved oxidatively by the following methods.[1]

1. $Pb(OAc)_4$/Pyr, 25°, 10 min, 90% yield.
2. $NaIO_4/H_2O$–THF, H_2SO_4, 20°, 5 min, 89% yield.
3. Aq NBS/THF–Pyr, 20°, 10 min, 90% yield.
4. CrO_3/HOAc, 25°, 10 min, 65% yield.

A number of di- and trisubstituted hydrazides of penicillin and cephalosporin derivatives were prepared to study the effect of *N*-substitution on ease of oxidative cleavage.[2]

1. D. H. R. Barton, M. Girijavallabhan, and P. G. Sammes, *J. Chem. Soc., Perkin Trans. I*, 929 (1972).
2. D. H. R. Barton and 8 co-workers, *J. Chem. Soc., Perkin Trans. I*, 1477 (1977).

6

PROTECTION FOR THE THIOL GROUP

Protection for the thiol group is important in many areas of organic research, particularly in peptide and protein syntheses that often involve the amino acid cysteine, $HSCH_2CH(NH_2)CO_2H$, CySH. The synthesis[1] of coenzyme A, which converts a carboxylic acid into a thioester, an acyl transfer agent in the biosynthesis or oxidation of fatty acids, also requires the use of thiol protective groups. A free —SH group can be protected as a thioether or a thioester, or oxidized to a symmetrical disulfide, from which it is regenerated by reduction. Thioethers are in general formed by reaction of the thiol, in a basic solution, with a halide; they are cleaved by reduction with sodium/ammonia, by acid-catalyzed hydrolysis, or by reaction with a heavy-metal ion such as silver(I) or mercury(II), followed by hydrogen sulfide treatment. Some groups, including *S*-diphenylmethyl and *S*-triphenylmethyl thioethers, and *S*-2-tetrahydropyranyl and *S*-isobutoxymethyl monothioacetals, can be oxidized by thiocyanogen, $(SCN)_2$, iodine, or a sulfenyl chloride to a disulfide that is subsequently reduced to the thiol. Thioesters are formed and cleaved in the same way as oxygen esters; they are more reactive to nucleophilic substitution, as indicated by their use as "activated esters." Several miscellaneous protective groups, including thiazolidines, unsymmetrical disulfides, and *S*-sulfenyl derivatives, have been used to a more limited extent. This chapter discusses some synthetically useful thiol protective groups.[2,3] Some of the more useful groups are included in Reactivity Chart 7.

1. J. G. Moffatt and H. G. Khorana, *J. Am. Chem. Soc.*, **83,** 663 (1961).
2. See also: Y. Wolman, "Protection of the Thiol Group," in *The Chemistry of the Thiol Group*, S. Patai, Ed., Wiley-Interscience, New York, 1974, Vol. 15/2, pp. 669–684; R. G. Hiskey, V. R. Rao, and W. G. Rhodes, "Protection of Thiols," in *Protective Groups in Organic Chemistry*, J. F. W. McOmie, Ed., Plenum Press, New York and London, 1973, pp. 235–308; J. F. W. McOmie, "Protective Groups," *Adv. Org. Chem.*, **3,** 251–255 (1963).
3. R. G. Hiskey, "Sulfhydryl Group Protection in Peptide Synthesis," in *The Peptides*, E. Gross and J. Meienhofer, Eds. Academic Press, 1981, Vol. 3, pp. 137–167.

THIOETHERS

S-Benzyl and substituted *S*-benzyl derivatives, readily cleaved with sodium/ammonia, are the most frequently used thioethers. *n*-Alkyl thioethers are difficult to cleave and have not been used as protective groups. Alkoxymethyl or alkylthiomethyl mono or dithioacetals ($RSCH_2OR'$ or $RSCH_2SR'$) can be cleaved by acidic hydrolysis, or by reaction with silver or mercury salts, respectively. Mercury(II) salts also cleave dithioacetals, RS–CH_2SR', *S*-triphenylmethyl thioethers, RS–CPh_3, *S*-diphenylmethyl thioethers, RS–$CHPh_2$, *S*-acetamidomethyl derivatives, RS–$CH_2NHCOCH_3$, and *S*-(*N*-ethylcarbamates), RS–CONHEt. *S*-*t*-Butyl thioethers, RS-*t*-Bu, are cleaved if refluxed with mercury(II); *S*-benzyl thioethers, RS–CH_2Ph, are cleaved if refluxed with mercury(II)/1 *N* HCl. Some β-substituted *S*-ethyl thioethers are cleaved by reactions associated with the β-substituent.

1. *S*-Benzyl Thioether: $RSCH_2Ph$ (Chart 7)

For the most part cysteine and its derivatives have been protected by the following reactions.

Formation

1. $PhCH_2Cl$, 2 *N* NaOH or NH_3, EtOH, 30 min, 25°, 90% yield.[1]
2. $PhCH_2Cl$, Cs_2CO_3, DMF, 20°.[2]

Cleavage

1. Na, NH_3, 10 min.[3] Sodium in boiling butyl alcohol[4] or boiling ethyl alcohol[5] can be used if the benzyl thioether is insoluble in ammonia.
2. HF, anisole, 25°, 1 h.[6]
3. 5% cresol, 5% thiocresol, 90% HF.[7]

In the HF deprotection of thioethers and many other protective groups, anisole serves as a scavenger for the liberated cation formed during the deprotection process. If cations liberated during this deprotection are not scavenged, they can react with other amino acid residues, especially tyrosine. The authors[6] list 15 protective groups that are cleaved by this method, including some branched-chain carbonates and esters, benzyl esters and ethers, the nitro-protective group in arginine, and *S*-benzyl and *S*-*t*-butyl thioethers. They report that 12 protective groups are stable under these conditions, including some straight-chain carbonates and esters, *N*-benzyl derivatives, and *S*-methyl, *S*-ethyl, and *S*-isopropyl thioethers.[6] Dimethyl sulfide, thiocresol, cresol, and thioanisole have also been used as scavengers when strong acids are used for deprotection. A mixture of 5% cresol, 5% *p*-thiocresol, and 90% HF is recommended for benzyl thioether deprotection.[7] These conditions cause cleavage by an S_N1 mechanism. The use of low concentrations of HF in dimethyl sulfide (1:3), which has been recommended for deprotection of other peptide protective groups, does not cleave the *S*-4-methylbenzyl group. Reactions that use low HF concentrations are considered to proceed via an S_N2 mechanism. The use of low HF concentrations with thioanisole results in some methylation of free thiols. The use of HF in anisole can also result in alkylation of methionine.

4. Electrolysis, NH_3, 90 min.[8]
5. Electrolysis, −2.8 V, DMF, $R_4N^+X^-$, 82% yield.[9]

1. M. Frankel, D. Gertner, H. Jacobson, and A. Zilkha, *J. Chem. Soc.*, 1390 (1960).
2. F. Vögtle and B. Klieser, *Synthesis*, 294 (1982).
3. J. E. T. Corrie, J. R. Hlubucek, and G. Lowe, *J. Chem. Soc., Perkin Trans. I*, 1421 (1977).
4. W. I. Patterson and V. du Vigneaud, *J. Biol. Chem.*, **111,** 393 (1935).
5. K. Hofmann, A. Bridgwater, and A. E. Axelrod, *J. Am. Chem. Soc.*, **71,** 1253 (1949).
6. S. Sakakibara, Y. Shimonishi, Y. Kishida, M. Okada, and H. Sugihara, *Bull. Chem. Soc. Jpn.*, **40,** 2164 (1967).
7. W. F. Heath, J. P. Tam, and R. B. Merrifield, *Int. J. Pept. Protein Res.*, **28,** 498 (1986).
8. D. A. J. Ives, *Can. J. Chem.*, **47,** 3697 (1969).
9. V. G. Mairanovsky, *Angew. Chem., Int. Ed. Engl.*, **15,** 281 (1976).

2. *S-p*-Methoxybenzyl Thioether: $RSCH_2C_6H_4$-*p*-OCH_3 (Chart 7)

Formation

1. 4-$MeOC_6H_4CH_2Cl$, NH_3, 78% yield.[1]

Cleavage

An *S*-4-methoxybenzyl thioether is stable to HBr/AcOH[1] and I_2/MeOH.[2] The latter reagent cleaves *S*-trityl and *S*-diphenylmethyl groups.

1. $Hg(OAc)_2$, CF_3COOH, 0°, 10–30 min or $Hg(OCOCF)_3)_2$, aq. AcOH, 20°, 2–3 h, followed by H_2S or $HSCH_2CH_2OH$, 100% yield.[3,4] An *S*-*t*-butyl thioether is cleaved in quantitive yield under these conditions.
2. $Hg(OCOCF_3)_2$, CF_3COOH, anisole.[5]

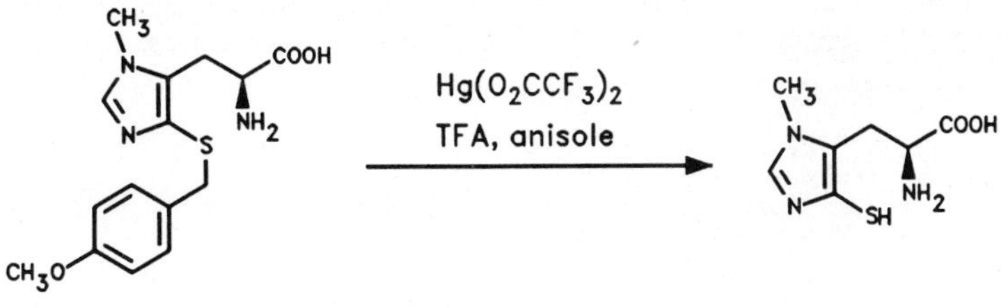

$Hg(O_2CCF_3)_2$, TFA, anisole

3. CF_3COOH, reflux.[1]
4. Anhydrous HF, anisole, 25°, 1 h, quant.[6]

5.

BOC–Cys–OMe | 4-MeOC$_6$H$_4$CH$_2$ —[(4-BrC$_6$H$_4$)$_3$N$^{\cdot+}$ SbCl$_6^-$, 75%,[2] or *in situ* electrolysis, CH$_3$CN, NaHCO$_3$ cat. (4-BrC$_6$H$_4$)$_3$N$^{\cdot+}$SbCl$_6^-$, 91%]→ BOC–Cys–OMe | BOC–Cys–OMe

During the synthesis of peptides that contain 4-methoxybenzyl-protected cysteine residues, sulfoxide formation may occur. These sulfoxides, when treated with HF/anisole, form thiophenyl ethers that cannot be deprotected; therefore, the peptides should be subjected to a reduction step prior to deprotection.[7]

NaBO$_3$ → ; ← PhSH, 100%; PhOMe, HF or MSA →

MeOZNH, CO$_2$H, OCH$_3$; H$_2$N, CO$_2$H

Note the missing methylene

MSA = methanesulfonic acid

1. S. Akabori, S. Sakakibara, Y. Shimonishi, and Y. Nobuhara, *Bull. Chem. Soc. Jpn.*, **37,** 433 (1964).
2. M. Platen and E. Steckhan, *Liebigs Ann. Chem.*, 1563 (1984).
3. O. Nishimura, C. Kitada, and M. Fujino, *Chem. Pharm. Bull.*, **26,** 1576 (1978).
4. E. M. Gordon, J. D. Godfrey, N. G. Delaney, M. M. Asaad, D. Von Langen, and D. W. Cushman, *J. Med. Chem.*, **31,** 2199 (1988).
5. T. P. Holler, A. Spaltenstein, E. Turner, R. E. Klevit, B. M. Shapiro, and P. B. Hopkins, *J. Org. Chem.*, **52,** 4420 (1987).
6. S. Sakakibara and Y. Shimonishi, *Bull. Chem. Soc. Jpn.*, **38,** 1412 (1965).
7. S. Funakoshi, N. Fujii, K. Akaji, H. Irie, and H. Yajima, *Chem. Pharm. Bull.*, **27,** 2151 (1979).

3. *S-o-* or *p*-Hydroxy- or Acetoxybenzyl Thioether: RSCH$_2$C$_6$H$_4$-*o*-(or *p*-)-OR′: R′ = H or Ac

Formation/Cleavage[1]

HS–(1-phenyltetrazol-5-yl) —[4-AcO–C$_6$H$_4$CH$_2$Cl, acetone, reflux, 2 h, 80%]→ 4-AcOC$_6$H$_4$CH$_2$S–(1-phenyltetrazol-5-yl)

← 5% aq NaOH, 25°, 2 h

The cleavage process occurs by *p*-quinonemethide formation after acetate hydrolysis.

1. L. D. Taylor, J. M. Grasshoff, and M. Pluhar, *J. Org. Chem.*, **43,** 1197 (1978).

4. *S-p*-Nitrobenzyl Thioether: $RSCH_2C_6H_4$-*p*-NO_2 (Chart 7)

Formation

1. 4-$NO_2C_6H_4CH_2Cl$, 1 *N* NaOH, 0°, 1 h → 25°, 0.5 h[1] or NaH, $PhCH_3$, 68% yield.[2]

Cleavage

1. H_2, Pd–C, HCl or AcOH, 7–8 h, 60–68%; $HgSO_4$, H_2SO_4, 20 h, 60%; H_2S, 15 min, 60%[2] or RSSR, 76% after air oxidn.[1] Hydrogenation initially produces the *p*-amino derivative, which is then cleaved with Hg(II).

1. M. D. Bachi and K. J. Ross-Peterson, *J. Org. Chem.*, **37,** 3550 (1972).
2. M. D. Bachi and K. J. Ross-Petersen, *J. Chem. Soc., Chem. Commun.*, 12 (1974).

5. *S*-4-Picolyl Thioether: $RSCH_2$-4-pyridyl (Chart 7)

Formation

1. 4-Picolyl chloride, 60% yield.[1]

Cleavage

1. Electrolytic reduction, 0.25 *M* H_2SO_4, 88% yield. *S*-4-Picolylcysteine is stable to CF_3COOH (7 days), to HBr/AcOH, and to 1 *M* NaOH. References for the electrolytic removal of seven other protective groups are included.[1]

1. A. Gosden, R. Macrae, and G. T. Young, *J. Chem. Res., Synop.*, 22 (1977).

6. *S*-2-Picolyl *N*-Oxide Thioether: $RSCH_2$-2-pyridyl *N*-Oxide (Chart 7)

Formation

1. 2-Picolyl chloride *N*-oxide, aq. NaOH, moderate yields.[1]

Cleavage

1. Ac_2O, reflux, 7 min or 25°, 1.5 h followed by hydrolysis; aq. NaOH, 25°, 3–12 h, 79% yield.[1]

1. Y. Mizuno and K. Ikeda, *Chem. Pharm. Bull.*, **22,** 2889 (1974).

7. *S*-9-Anthrylmethyl Thioether: $RSCH_2$-9-anthryl (Chart 7)

Formation

1. 9-Anthrylmethyl chloride, DMF, −20°, N_2.[1]

Cleavage

1. CH_3SNa, DMF or HMPA, 0–25°, 2–5 h, 68–92% yield.[1]

1. N. Kornblum and A. Scott, *J. Am. Chem. Soc.*, **96,** 590 (1974).

8. *S*-9-Fluorenylmethyl Ether (Fm-SR):

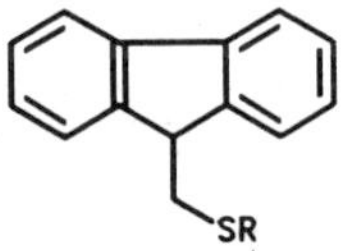

SR

Formation

1. Et(*i*-Pr)$_2$N, DMF, FmCl.[1]

Cleavage

1. 50% Piperidine, DMF or NH_4OH, 2 h.[2] The *S*-fluorenylmethyl group is stable to 95% HF/5% anisole for 1 h at 0°, to trifluoroacetic acid, to 12 *N* HCl, to 0.1 *M* I_2 in DMF, and to CF_3SO_3H in CF_3COOH.[2]

1. M. Bodanszky and M. A. Bednarek, *Int. J. Pept. Protein Res.*, **20,** 434 (1982).
2. M. Ruiz-Gayo, F. Albericio, E. Pedroso, and E. Giralt, *J. Chem. Soc., Chem. Commun.*, 1501 (1986).

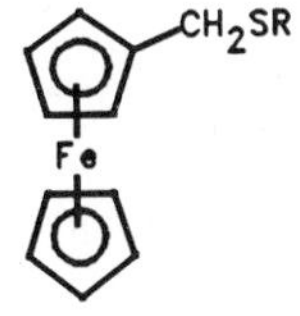

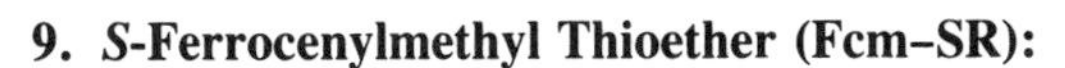

9. *S*-Ferrocenylmethyl Thioether (Fcm–SR):

Formation

1. Cp–Fe–$CpCH_2OH$, TFA, acetone, H_2O, 100% yield.[1]

Cleavage

The Fcm group can be removed with TFA or Hg(II). It is stable to mild acid and base.[1]

1. C. N. C. Drey and A. S. J. Stewart, in *Peptides 1986: Proceedings of the 19th European Peptide Symposium, Porto Carras, Chalkidiki, Greece, August 31–September 5, 1986*, D. Theodoropoulos, Ed., de Gruyter, Berlin and New York, 1987, p. 65.

S-Diphenylmethyl, Substituted *S*-Diphenylmethyl, and *S*-Triphenylmethyl Thioethers

S-Diphenylmethyl, substituted *S*-diphenylmethyl, and *S*-triphenylmethyl thioethers have often been formed or cleaved by the same conditions, although sometimes in rather different yields. As an effort has been made to avoid repetition in the sections that describe these three protective groups, the reader should glance at all the sections.

10. *S*-Diphenylmethyl Thioether: $RSCH(C_6H_5)_2$ (Chart 7)

Formation

1. Ph_2CHOH, CF_3COOH, 25°, 15 min or Ph_2CHOH, HBr, AcOH, 50°, 2 h, >90% yield.[1]

Boron trifluoride etherate (in HOAc, 60–80°, 15 min, high yields)[2] also catalyzes formation of *S*-diphenylmethyl and *S*-triphenylmethyl thioethers from aralkyl alcohols.

Yields of thioethers, formed under nonacidic conditions (Ph_2CHCl or Ph_3CCl, DMF, 80–90°, 2 h, N_2) are not as high ($RSCHPh_2$, 50% yield; $RSCPh_3$, 75% yield)[3] as the yields obtained under the acidic conditions described above.

Cleavage

1. CF_3COOH, 2.5% phenol, 30°, 2 h, 65% yield.[1] Zervas and co-workers tried many conditions for the acid-catalyzed formation and removal of the *S*-diphenylmethyl, *S*-4,4′-dimethoxydiphenylmethyl, and *S*-triphenylmethyl thioethers. The best conditions for the *S*-diphenylmethyl thioether are shown above. Phenol or anisole act as cation scavengers.
2. Na, NH_3, 97% yield.[3] Sodium/ammonia is an efficient but nonselective reagent (RS–Ph, RS–CH_2Ph, RS–CPh_3, and RS–SR are also cleaved).
3. 2-$NO_2C_6H_4SCl$, AcOH (results in disulfide formation), followed by $NaBH_4$ or $HS(CH_2)_2OH$ or dithioerythritol, quant.[4] *S*-Triphenylmethyl, *S*-4,4′-dimethoxydiphenylmethyl, and *S*-acetamidomethyl groups are also removed by this method.

1. I. Photaki, J. T.-Papadimitriou, C. Sakarellos, P. Mazarakis, and L. Zervas, *J. Chem. Soc. C.*, 2683 (1970).
2. R. G. Hiskey and J. B. Adams, Jr., *J. Org. Chem.*, **30,** 1340 (1965).
3. L. Zervas and I. Photaki, *J. Am. Chem. Soc.*, **84,** 3887 (1962).
4. A. Fontana, *J. Chem. Soc., Chem. Commun.*, 976 (1975).

11. *S*-Bis(4-methoxyphenyl)methyl Thioether: $RSCH(C_6H_4\text{–}4\text{-}OCH_3)_2$ (Chart 7)

Formation

1. $(4\text{-}MeOC_6H_4)_2CHCl$, DMF, 25°, 2 days, 96% yield.[1]

Cleavage

1. HBr, AcOH, 50–60°, 30 min, or CF_3COOH, phenol, reflux, 30 min, quant.[1]

1. R. W. Hanson and H. D. Law, *J. Chem. Soc.*, 7285 (1965).

12. *S*-5-Dibenzosuberyl Thioether:

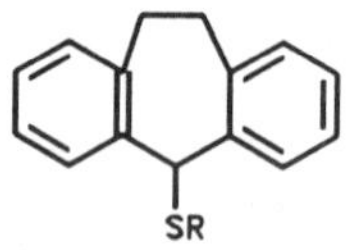

SR

5-Dibenzosuberyl alcohol reacts in 60% yield with cysteine to give a thioether that is cleaved by mercury(II) acetate or oxidized by iodine to cystine. The dibenzosuberyl group has also been used to protect —OH, —NH_2, and —CO_2H groups.[1]

1. J. Pless, *Helv. Chim. Acta*, **59,** 499 1976).

13. *S*-Triphenylmethyl Thioether: $RSC(C_6H_5)_3$ (Chart 7)

S-Triphenylmethyl thioethers have been formed by reaction of the thiol with triphenylmethyl alcohol/anhydrous CF_3COOH (85–90% yield) or with triphenylmethyl chloride (75% yield).

Cleavage

1. HCl, aq. AcOH, 90°, 1.5 h.[1]
2. $Hg(OAc)_2$, EtOH, reflux, 3 h, → 25°, 12 h; H_2S, 61% yield.[1]
3. $AgNO_3$, EtOH, Pyr, 90°, 1.5 h; H_2S, 47% yield.[1] An *S*-triphenylmethyl thioether can be selectively cleaved in the presence of an *S*-diphenylmethyl thioether by acidic hydrolysis or by heavy-metal ions. As a result of the structure of the substrate, the relative yields of cleavage by $AgNO_3$ and $Hg(OAc)_2$ can be reversed.[2]
4. Thiocyanogen [$(SCN)_2$, 5°, 4 h, 40% yield] selectively oxidizes an *S*-triphenylmethyl thioether to the disulfide (RSSR) in the presence of an *S*-diphenylmethyl thioether.[3]
5. *S*-Triphenylmethylcysteine is readily oxidized by iodine (MeOH, 25°) to cystine.[4] The *S*-triphenylmethylcysteine group can be selectively cleaved in the presence of a —Cys(Acm)— group (Acm = acetamidomethyl).[5] *S*-Benzyl and *S*-*t*-butyl thioethers are stable to the action of iodine.
6. Electrolysis, −2.6 V, DMF, $R_4N^+X^-$.[6]
7. Et_3SiH, 50% TFA, CH_2Cl_2, 1 h, rt.[7]

1. R. G. Hiskey, T. Mizoguchi, and H. Igeta, *J. Org. Chem.*, **31,** 1188 (1966).
2. R. G. Hiskey and J. B. Adams, *J. Org. Chem.*, **31,** 2178 (1966).
3. R. G. Hiskey, T. Mizoguchi, and E. L. Smithwick, *J. Org. Chem.*, **32,** 97 (1967).
4. B. Kamber, *Helv. Chim. Acta*, **54,** 398 (1971).
5. B. Kamber, A. Hartmann, K. Eisler, B. Riniker, H. Rink, P. Sieber, and W. Rittel, *Helv. Chim. Acta*, **63,** 899 (1980).
6. V. G. Mairanovsky, *Angew. Chem., Int. Ed. Eng.*, **15,** 281 (1976).
7. D. A. Pearson, M. Blanchette, M. L. Baker, and C. A. Guindon, *Tetrahedron Lett.*, **30,** 2739 (1989).

14. *S*-Diphenyl-4-pyridylmethyl Thioether: $RSC(C_6H_5)_2$-4-pyridyl

Formation

1. $Ph_2(4\text{-}C_5H_4N)COH$, $BF_3 \cdot Et_2O$, AcOH, 60°, 48 h.[1]

Cleavage

1. $Hg(OAc)_2$, AcOH, pH 4, 25°, 15 min.[1]
2. Zn, 80% AcOH, H_2O.[2]

The diphenylpyridylmethyl thioether is stable to acids (e.g., CF_3COOH, 21°, 48 h; 45% HBr/AcOH, 21°); it is oxidized by iodine to cystine (91%) or reduced by electrolysis at a mercury cathode.[1]

1. S. Coyle and G. T. Young, *J. Chem. Soc., Chem. Commun.*, 980 (1976).
2. S. Coyle, A. Hallett, M. S. Munns, and G. T. Young, *J. Chem. Soc., Perkin Trans. I*, 522 (1981).

15. *S*-Phenyl Thioether: RSC_6H_5

Although a sulfhydryl group generally is not converted to an *S*-phenyl thioether, thiophenol can be used to introduce sulfur into molecules, and thus the phenyl group serves as a suitable protective group that can be removed by electrolysis (-2.7 V, DMF, $R_4N^+X^-$).[1]

1. V. G. Mairanovsky, *Angew. Chem., Int. Ed. Engl.*, **15,** 281 (1976).

16. *S*-2,4-Dinitrophenyl Thioether: RSC_6H_3-2,4-$(NO_2)_2$ (Chart 7)

Formation

1. 2,4-$(NO_2)_2$-C_6H_3F, base.[1]

The sulfhydryl group in cysteine can be selectively protected in the presence of the amino group by reaction with 2,4-dinitrophenol at pH 5–6.[2]

Cleavage

1. $HSCH_2CH_2OH$, pH 8, 22°, 1 h, quant.[1]

1. S. Shaltiel, *Biochem. Biophys. Res. Commun.*, **29,** 178 (1967).
2. H. Zahn and K. Traumann, *Z. Naturforsch.*, **9b,** 518 (1954).

17. *S-t*-Butyl Thioether: $RSC(CH_3)_3$ (Chart 7)

Formation

1. Isobutylene, H_2SO_4, CH_2Cl_2, 25°, 12 h, 73% yield.[1] The *S-t*-butyl derivative of cysteine is stable to HBr/AcOH and to CF_3COOH.

Cleavage

1. $Hg(OAc)_2$, CF_3COOH, anisole, 0°, 15 min; H_2S, quant.[2]
2. $Hg(OCOCF_3)_2$, aq. AcOH, 25°, 1 h; H_2S, quant.[2]
3. HF, anisole, 20°, 30 min.[3]
4. $2\text{-}NO_2C_6H_4SCl$; $NaBH_4$.[4] Treatment of the thioether with the sulfenyl chloride initially produces a disulfide which is then reduced to afford the free thiol.

1. F. M. Callahan, G. W. Anderson, R. Paul, and J. E. Zimmerman, *J. Am. Chem. Soc.*, **85,** 201 (1963).
2. O. Nishimura, C. Kitada, and M. Fujino, *Chem. Pharm. Bull.*, **26,** 1576 (1978).
3. S. Sakakibara, Y. Shimonishi, Y. Kishida, M. Okada, and H. Sugihara, *Bull. Chem. Soc. Jpn.*, **40,** 2164 (1967).
4. J. J. Pastuszak and A. Chimiak, *J. Org. Chem.*, **46,** 1868 (1981).

18. *S*-1-Adamantyl Thioether: RS-1-adamantyl

Formation

1. 1-Adamantyl alcohol, CF_3COOH, 25°, 12 h, 90% yield.[1]

Cleavage

1. $Hg(OAc)_2$, CF_3COOH, 0°, 15 min, 100% yield.[1]
2. $Hg(OCOCF_3)_2$, aq. AcOH, 20°, 60 min, 100% yield.[1]
3. 1 *M* CF_3SO_3H, $PhSCH_3$ or $Tl(OCOCF_3)_3$.[2] The *S*-adamantyl group is less prone to sulfoxide formation than is the *S*-4-methoxybenzyl group. It is also more stable to CF_3COOH.

1. O. Nishimura, C. Kitada, and M. Fujino, *Chem. Pharm. Bull.*, **26,** 1576 (1978).
2. N. Fujii, A. Otaka, S. Funakoshi, H. Yajima, O. Nishimura, and M. Fujino, *Chem. Pharm. Bull.*, **34,** 869 (1986); N. Fujii, H. Yajima, A. Otaka, S. Funakoshi, M. Nomizu, K. Akaji, I. Yamamoto, K. Torizuka, K. Kitagawa, T. Akita, K. Ando, T. Kawamoto, Y. Shimonishi, and T. Takao, *J. Chem. Soc., Chem. Commun.*, 602 (1985).

Substituted *S*-Methyl Derivatives

MONOTHIO, DITHIO, AND AMINOTHIO ACETALS

19. *S*-Methoxymethyl Monothioacetal: $RSCH_2OCH_3$

Formation

1. Zn, $(CH_3O)_2CH_2$, $BrCH_2CO_2Et$, 80–82% yield. Formation of the methoxymethyl thioether with dimethoxymethane[1] avoids the use of the carcinogen chloromethyl methyl ether.[2] The reaction forms an intermediate zinc thiolate, which then forms the monothioacetal.

The related *S*-benzyloxymethyl monothioacetal (BOM) has been used for cysteine protection.[3]

1. F. Dardoize, M. Gaudemar, and N. Goasdoue, *Synthesis*, 567 (1977).
2. T. Fukuyama, S. Nakatsuka, and Y. Kishi, *Tetrahedron Lett.*, 3393 (1976).
3. A. Otaka, H. Morimoto, N. Fujii, T. Koide, S. Funakoshi, and H. Yajima, *Chem. Pharm. Bull.*, **37,** 526 (1989).

20. *S*-Isobutoxymethyl Monothioacetal: $RSCH_2OCH_2CH(CH_3)_2$ (Chart 7)

Formation

1. $ClCH_2OCH_2CH(CH_3)_2$, 82% yield.[1]

Cleavage

1. 2 *N* HBr, AcOH, rapid.[1] The *S*-isobutoxymethyl monothioacetal is stable to 2 *N* hydrochloric acid and to 50% acetic acid; some decomposition occurs in 2 *N* sodium hydroxide.[1]

The monothioacetal is also stable to 12 *N* hydrochloric acid in acetone (used to remove an *N*-triphenylmethyl group) and to hydrazine hydrate in refluxing ethanol (used to cleave an *N*-phthaloyl group). It is cleaved by boron trifluoride etherate in acetic acid, silver nitrate in ethanol, and trifluoroacetic acid. The monothioacetal is oxidized to a disulfide by thiocyanogen, $(SCN)_2$.[2]

1. P. J. E. Brownlee, M. E. Cox, B. O. Handford, J. C. Marsden, and G. T. Young, *J. Chem. Soc.*, 3832 (1964).
2. R. G. Hiskey and J. T. Sparrow, *J. Org. Chem.*, **35,** 215 (1970).

21. *S*-2-Tetrahydropyranyl Monothioacetal: RS-2-tetrahydropyranyl (Chart 7)

Formation

1. Dihydropyran, $BF_3 \cdot Et_2O$, Et_2O, 0°, 0.5 h → 25°, 1 h, satisfactory yields.[1]
2. Dihydropyran, PPTS (pyridinium *p*-toluenesulfonate), 4 h, 25°, 92% yield.[2]

Cleavage

1. Aqueous $AgNO_3$, 0°, 10 min, quant.[3]
2. HBr, CF_3COOH, 90 min, 100% yield.[4] An *S*-tetrahydropyranyl monothioacetal is stable to 4 *N* HCl/CH_3OH, 0° and to reduction with Na/NH_3. (An *O*-tetrahydropyranyl acetal is cleaved by 0.1 *N* HCl, 22°, $t_{1/2}$ = 4 min.)[5]

An *S*-2-tetrahydropyranyl monothioacetal is oxidized to a disulfide by iodine[3] or thiocyanogen, $(SCN)_2$.[6]

1. R. G. Hiskey and W. P. Tucker, *J. Am. Chem. Soc.*, **84,** 4789 (1962).
2. E. Block, V. Eswarakrishnan, M. Gernon, G. O.-Okai, C. Saha, K. Tang, and J. Zubieta, *J. Am. Chem. Soc.*, **111,** 658 (1989).
3. G. F. Holland and L. A. Cohen, *J. Am. Chem. Soc.*, **80,** 3765 (1958).
4. K. Hammerström, W. Lunkenheimer, and H. Zahn, *Makromol. Chem.*, **133,** 41 (1970).
5. B. E. Griffin, M. Jarman, and C. B. Reese, *Tetrahedron*, **24,** 639 (1968).
6. R. G. Hiskey and W. P. Tucker, *J. Am. Chem. Soc.*, **84,** 4794 (1962).

22. *S*-Benzylthiomethyl Dithioacetal: $RSCH_2SCH_2C_6H_5$

23. *S*-Phenylthiomethyl Dithioacetal: $RSCH_2SC_6H_5$

Formation

1. $ClCH_2SCH_2Ph$, NH_3, 91% yield.[1]

Cleavage

1. $Hg(OAc)_2$, H_2O, 80% AcOH, $HSCH_2CH_2SH$, 25°, 5–20 min; H_2S, 2 h, high yield.[1] The removal of an *S*-benzylthiomethyl protective group from a dithioacetal with mercury(II) acetate avoids certain side reactions that occur when an *S*-benzyl thioether is cleaved with sodium/ammonia. The dithioacetal is stable to hydrogen bromide/acetic acid used to cleave benzyl carbamates.

S-Phenylthiomethyl dithioacetals ($RSCH_2SC_6H_5$) were prepared and cleaved by similar methods.[1]

The dithioacetal is stable to catalytic reduction (H_2/Pd–C, CH_3OH–HOAc, 12 h, the conditions used to cleave a *p*-nitrobenzyl carbamate).[2]

1. P. J. E. Brownlee, M. E. Cox, B. O. Handford, J. C. Marsden, and G. T. Young, *J. Chem. Soc.*, 3832 (1964).
2. R. Camble, R. Purkayastha, and G. T. Young, *J. Chem. Soc. C*, 1219 (1968).

24. Thiazolidine Derivative:

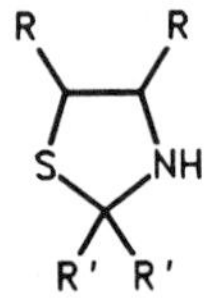

Thiazolidines have been prepared from β-aminothiols—for example, cysteine—to protect the —SH and —NH groups during syntheses of peptides, including glutathione.[1] Thiazolidines are oxidized to symmetrical disulfides with iodine;[2] they do not react with thiocyanogen in a neutral solution.[3]

Formation[4]

1. acetone, reflux, 6 h, 82%

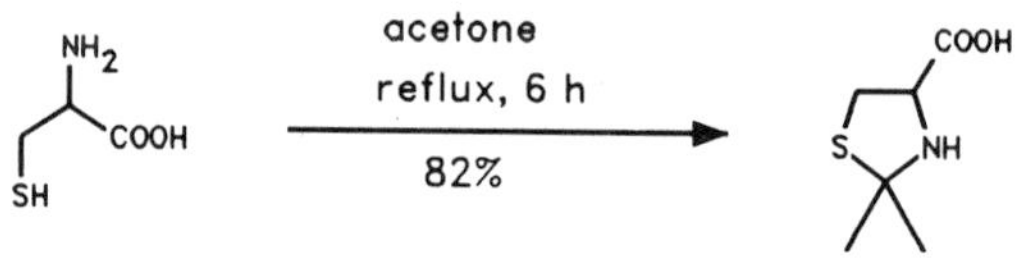

Cleavage

1. HCl, H_2O, CH_3OH, 25°, 3 days, high yield.[4]
2. $HgCl_2$, H_2O, 25°, 2 days or 60–70°, 15 min; H_2S, 20 min, 30–40% yield.[4]
3. *N*-BOC thiazolidines can be cleaved with ScmCl (methoxycarbonylsulfenyl chloride) (AcOH, DMF, H_2O) to afford the Scm derivative in >90% yield.[5]

1. F. E. King, J. W. Clark-Lewis, G. R. Smith, and R. Wade, *J. Chem. Soc.*, 2264 (1959).
2. S. Ratner and H. T. Clarke, *J. Am. Chem. Soc.*, **59**, 200 (1937).
3. R. G. Hiskey and W. P. Tucker, *J. Am. Chem. Soc.*, **84**, 4789 (1962).
4. J. C. Sheehan and D.-D. H. Yang, *J. Am. Chem. Soc.*, **80**, 1158 (1958).
5. D. S. Kemp and R. I. Carey, *J. Org. Chem.*, **54**, 3640 (1989).

25. *S*-Acetamidomethyl Aminothioacetal (Acm-SR): $RSCH_2NHCOCH_3$ (Chart 7)

Formation

1. $AcNHCH_2OH$, concd. HCl, pH 0.5, 25°, 1–2 days, 52% yield.[1]

Cleavage

1. $Hg(OAc)_2$, pH 4, 25°, 1 h; H_2S; air, 98% yield of cystine.[1] An *S*-acetamidomethyl group is hydrolyzed by the strongly acidic (6 *N* HCl, 110°, 6 h) or strongly basic conditions used to cleave amide bonds. It is stable to anhydrous trifluoroacetic acid and to hydrogen fluoride (0°, 1 h; 18°, 1 h, 10% cleaved). It is stable to zinc in acetic acid and to hydrazine in acetic acid or methanol.[1] If the Acm group is oxidized, there is no satisfactory method to liberate the cysteine. Cleavage of the sulfoxide with HF/anisole or CH_3SO_3H/anisole affords $Cys(C_6H_4OMe)$.[2]
2. $2\text{-}NO_2C_6H_4SCl$, AcOH; $HO(CH_2)_2SH$ or $NaBH_4$, quant.[3]
3. PhSH. This reagent affords the phenyl disulfide.[2]
4. $ClSCO_2Me$, MeOH, 80% yield.[4] These conditions convert the Acm group to a methyl *S*-sulfenylthiocarbonate group (Scm group), which can be cleaved with dithiothreitol.[5]

$$\underset{\text{BOCCys–OH}}{SCH_2NHAc} \xrightarrow{MeOC(O)SCl,\ CHCl_3} \underset{\text{BOCCys–OH}}{SSCO_2Me}$$

5. ClCOSCl, $CHCl_3$; PhNHMe.[5]

$$\underset{\text{BOC-CysOH}}{SCH_2NHAc} \xrightarrow[\text{2. } PhNHCH_3]{\text{1. } ClC(O)SCl,\ CHCl_3} \text{BOC-CysOH–S–S–C(=O)–N(CH_3)Ph}$$

The *S*-(*N*′-methyl-*N*′-phenylcarbamoyl)sulfenyl group (Snm group) produced under these conditions is stable to HF or CF_3SO_3H. Since there are few acid-stable —SH protective groups, the Snm group should prove to be useful where strong acids are encountered in synthesis.

1. D. F. Veber, J. D. Milkowski, S. L. Varga, R. G. Denkewalter, and R. Hirschmann, *J. Am. Chem. Soc.*, **94,** 5456 (1972); J. D. Milkowski, D. F. Veber, and R. Hirschmann, *Org. Synth.*, **59,** 190 (1980); *Collect. Vol. VI*, 5 (1988).

2. H. Yajima, K. Akaji, S. Funakoshi, N. Fujii, and H. Irie, *Chem. Pharm. Bull.*, **28,** 1942 (1980).
3. L. Moroder, F. Marchiori, G. Borin, and E. Schoffone, *Biopolymers*, **12,** 493 (1973); A. Fontana, *J. Chem. Soc., Chem. Commun.*, 976 (1975).
4. R. G. Hiskey, N. Muthukumaraswamy, and R. R. Vunnam, *J. Org. Chem.*, **40,** 950 (1975).
5. A. L. Schroll and G. Barany, *J. Org. Chem.*, **54,** 244 (1989).

26. *S*-Trimethylacetamidomethyl Aminothioacetal (Tacm–SR): $(CH_3)_3CCONHCH_2SR$

Formation[1]

1. $(CH_3)_3CCONHCH_2OH$, TFA, rt, 1 h, >85% yield.

Cleavage[1]

1. I_2, AcOH, EtOH, 25°, 1 h, 100% yield.
2. $Hg(OAc)_2$, TFA, 0°, 30 min. The Tacm group is stable to HF (0°, 1 h); to 1 *M* CF_3COOH, $PhSCH_3$ (0°, 1 h); to 0.5 *M* NaOH/MeOH (0°, 1 h); to NH_2NH_2, MeOH, and to Zn/AcOH. This group was reported to be more useful than the Acm group because it was less susceptible to byproduct formation and oxidation.[1]

1. Y. Kiso, M. Yoshida, T. Kimura, Y. Fujiwana, and M. Shimokura, *Tetrahedron Lett.*, **30,** 1979 (1989).

27. *S*-Benzamidomethyl Aminothioacetal: $RSCH_2NHCOC_6H_5$

S-Benzamidomethyl-*N*-methylcysteine has been prepared as a crystalline derivative ($HOCH_2NHCOC_6H_5$, anhydr. CF_3CO_2H, 25°, 45 min, 88% yield as the trifluoroacetate salt) and cleaved (100% yield) by treatment with mercury(II) acetate (pH 4, 25°, 1 h) followed by hydrogen sulfide. Attempted preparation of *S*-acetamidomethyl-*N*-methylcysteine resulted in noncrystalline material, shown by TLC to be a mixture.[1]

1. P. K. Chakravarty and R. K. Olsen, *J. Org. Chem.*, **43,** 1270 (1978).

28. *S*-Acetyl-, *S*-Carboxy-, and *S*-Cyanomethyl Thioethers: $ArSCH_2X$, $X = -COCH_3, -CO_2H, -CN$ (Chart 7)

In an attempt to protect thiophenols during electrophilic substitution reactions on the aromatic ring, the three substituted thioethers were prepared. After acetylation of the aromatic ring (moderate yields), the protective group was converted to the disulfide in moderate yields, 50–60%, by oxidation with hydrogen peroxide–boiling mineral acid, nitric acid, or acidic potassium permanganate.[1]

1. D. Walker, *J. Org. Chem.*, **31,** 835 (1966).

Substituted *S*-Ethyl Derivatives

A thiol, usually under basic catalysis, can undergo Michael addition to an activated double bond, resulting in protection of the sulfhydryl group as a substituted *S*-ethyl derivative.

29. *S*-2-Nitro-1-phenylethyl Thioether: $RSCH(C_6H_5)CH_2NO_2$ (Chart 7)

Formation

1. $PhCH{=}CHNO_2$, *N*-methylmorpholine, pH 7–8, 10 min, 70% yield.[1]

The protective group is removed by mildly alkaline conditions that do not cleave methyl or benzyl esters. The group is stable to CF_3COOH, HCl–AcOH, and HBr–AcOH. A polymer-bound version of this group has also been developed.[2]

1. G. Jung, H. Fouad, and G. Heusel, *Angew. Chem., Int. Ed. Engl.*, **14,** 817 (1975).
2. G. Heusel and G. Jung, *Liebigs Ann. Chem.*, 1173 (1979).

30. *S*-2-(4′-Pyridyl)ethyl Thioether: $C_4H_4NCH_2CH_2SR$

Formation[1]/Cleavage[2]

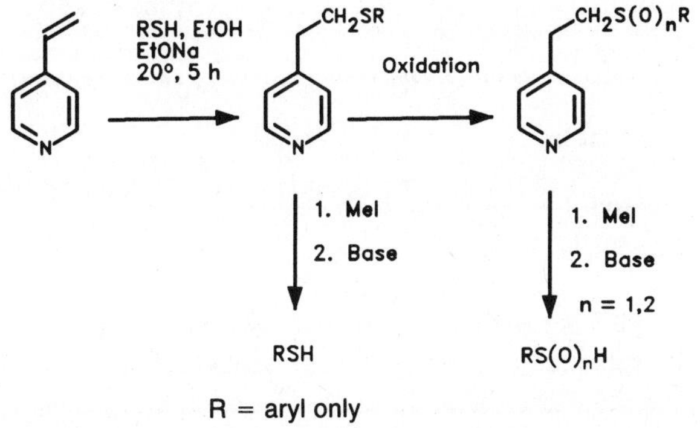

RSH, EtOH, EtONa, 20°, 5 h → CH_2SR (pyridine) → Oxidation → $CH_2S(O)_nR$ (pyridine)

1. MeI 2. Base → RSH

1. MeI 2. Base, n = 1,2 → $RS(O)_nH$

R = aryl only

The intermediate sulfides can be oxidized to the corresponding sulfoxides and sulfones and then liberated to give sulfenic and sulfinic acids.

1. A. R. Katritzky, I. Takahashi, and C. M. Marson, *J. Org. Chem.*, **51,** 4914 (1986).
2. A. R. Katritzky, G. R. Khan, and O. A. Schwarz, *Tetrahedron Lett.*, **25,** 1223 (1984).

31. *S*-2-Cyanoethyl Thioether: $NCCH_2CH_2SR$

The 2-cyanoethyl group was cleaved from an aromatic sulfide with $K_2CO_3/NaBH_4$ (DMF, 135°, 70% yield).[1]

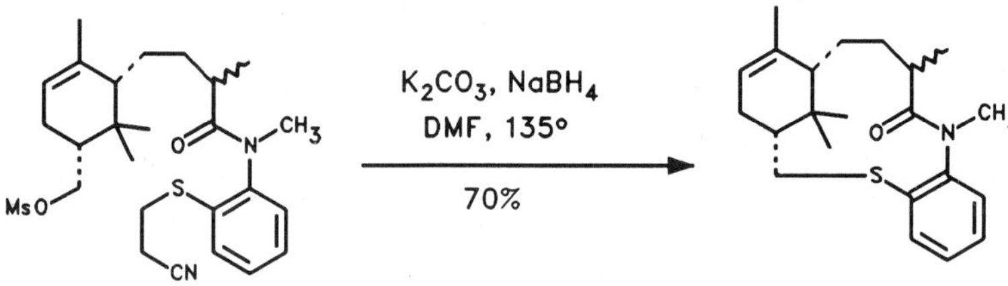

K_2CO_3, $NaBH_4$
DMF, 135°
70%

1. Y. Ohtsuka and T. Oishi, *Tetrahedron Lett.*, **27,** 203 (1986).

32. *S*-2,2-Bis(carboethoxy)ethyl Thioether: $RSCH_2CH(COOC_2H_5)_2$ (Chart 7)

Formation

1. $CH_2{=}C(CO_2Et)_2$, EtOH, 1 h, 74% yield.[1]

Cleavage

1. 1 *N* KOH, EtOH, 20°, 5–10 min, 80% yield. *S*-2,2-Bis(carboethoxy)ethyl thioether, stable to acidic reagents such as trifluoroacetic acid and hydrogen bromide/acetic acid, has been used in a synthesis of glutathione.[1]

1. T. Wieland and A. Sieber, *Justus Liebigs Ann. Chem.*, **722,** 222 (1969); *idem, ibid.*, **727,** 121 (1969).

33. *S*-1-*m*-Nitrophenyl-2-benzoylethyl Thioether: $ArSCH(C_6H_4\text{-}m\text{-}NO_2)CH_2COC_6H_5$

An *S*-1-*m*-nitrophenyl-2-benzoylethyl thioether was used to protect thiophenols during electrophilic substitution reactions of the benzene ring.[1]

Formation

1. $PhCOCH{=}CHC_6H_4$-*m*-NO_2, piperidine, benzene, 96% yield.

Cleavage

1. $Pb(OAc)_2$, EtOH, pH 8–10; dil. HCl, 77% yield.

1. A. H. Herz and D. S. Tarbell, *J. Am. Chem. Soc.*, **75,** 4657 (1953).

34. *S*-2-Phenylsulfonylethyl Thioether: $PhSO_2CH_2CH_2SR$

35. *S*-1-(4-Methylphenylsulfonyl)-2-methylprop-2-yl Thioether: $4\text{-}CH_3C_6H_4SO_2CH_2C(CH_3)_2SR$

Formation/Cleavage[1,2]

$$\text{RSH} \xrightarrow{\text{PhSO}_2\text{CH=CH}_2\text{, Et}_3\text{N, THF}} \text{or} \xrightarrow{\text{MeONa, MeOH}} \text{RSCH}_2\text{CH}_2\text{SO}_2\text{Ph}\ \ 84\text{–}100\%$$

$$\xleftarrow[t\text{-BuOK, THF, DME or } t\text{-BuOH, 80–100\%}]{}$$

1. Y. Kuroki and R. Lett, *Tetrahedron Lett.*, **25,** 197 (1984).
2. L. Horner and H. Lindel, *Phosphorus Sulfur*, **15,** 1 (1983).

Silyl Thioethers

Silyl-derived protective groups are also used to mask the thiol function. A complete compilation is not given here since silyl derivatives are described in the section on alcohol protection. The formation and cleavage of silyl thioethers proceed analogously to simple alcohols. The Si—S bond is weaker than the Si—O bond, and therefore sulfur derivatives are more susceptible to hydrolysis. For the most part silyl ethers are rarely used to protect the thiol function because of their instability. Silyl ethers have been used for *in situ* protection of the —SH group during amide formation.[1]

1. E. W. Abel, *J. Chem. Soc.*, 4933 (1961); L. Birkofer, W. Konkol, and A. Ritter, *Chem. Ber.*, **94,** 1263 (1961).

THIOESTERS

36. *S*-Acetyl Derivative: $RSCOCH_3$

37. *S*-Benzoyl Derivative: $RSCOC_6H_5$ (Chart 7)

Formation

1. Ac_2O, $KHCO_3$, 55% yield.[1]
2. BzCl, NaOH, $KHCO_3$, 0–5°, 30 min, 50% yield.[2]

$$\begin{array}{l} CH_2SH \\ | \\ CHOH \\ | \\ CH_2OH \end{array} \xrightarrow[\text{MeOH, 25°, 1.5 h, 54\%}]{\text{PhCSSMe, cat. NaOMe}} \begin{array}{l} CH_2SCSPh \\ | \\ CHOH \\ | \\ CH_2OH \end{array}$$

The base-catalyzed reaction of thiothreitol with methyl dithiobenzoate selectively protects a thiol group as an *S*-thiobenzoyl derivative in the presence of a hydroxyl group.[2]

Cleavage[1]

1. 0.2 *N* NaOH, N_2, 20°, 2–15 min, 100% yield.
2. Aqueous NH_3, N_2, 20°, 95–100% yield.
3. HBr, AcOH, 25°, 30 min, 5% to a substantial amount.
4. CF_3CO_2H, phenol, reflux, 30 min, 2–5% yield. In this case an *S*-Cbz group is removed.

Two disadvantages are associated with the use of *S*-acetyl or *S*-benzoyl derivatives in peptide syntheses: (a) base-catalyzed hydrolysis of *S*-acetyl- and *S*-benzoylcysteine occurs with β-elimination to give olefinic side products, $CH_2{=}C$-(NHPG)CO—;[3] (b) the yields of peptides formed by coupling an unprotected amino group in an *S*-acylcysteine are low because of prior S–N acyl migration.[4]

An *S*-acetyl group is stable to oxidation of a double bond by ozone (−20°, 5.5 h, 73% yield).[5]

1. L. Zervas, I. Photaki, and N. Ghelis, *J. Am. Chem. Soc.*, **85,** 1337 (1963).
2. E. J. Hedgley and N. H. Leon, *J. Chem. Soc. C*, 467 (1970).
3. R. G. Hiskey, R. A. Upham, G. M. Beverly, and W. C. Jones, Jr., *J. Org. Chem.*, **35,** 513 (1970).
4. R. G. Hiskey, T. Mizoguchi, and T. Inui, *J. Org. Chem.*, **31,** 1192 (1966).
5. I. Ernest, J. Gosteli, C. W. Greengrass, W. Holick, D. E. Jackman, H. R. Pfaendler, and R. B. Woodward, *J. Am. Chem. Soc.*, **100,** 8214 (1978).

38. *S-N*-[[(*p*-Biphenylyl)isopropoxy]carbonyl]-*N*-methyl-γ-aminobutyrate: $BpocN(CH_3)CH_2CH_2CH_2COSR$

39. *S-N-(t*-Butoxycarbonyl)-*N*-methyl-γ-aminobutyrate: $BOCN(CH_3)CH_2CH_2CH_2COSR$

Formation/Cleavage[1]

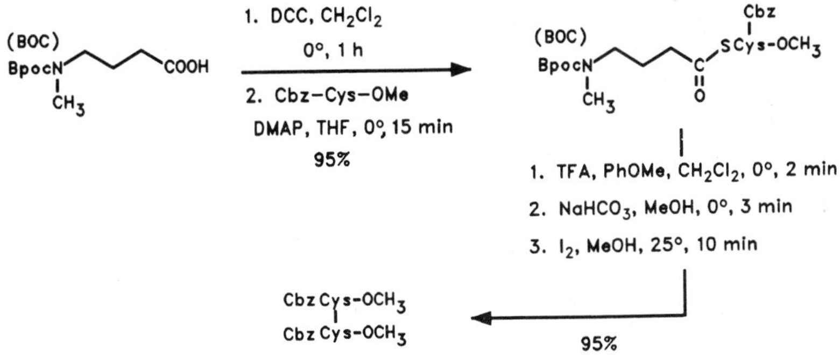

(BOC) BpocN(CH3)–(CH2)3–COOH
1. DCC, CH_2Cl_2, 0°, 1 h
2. Cbz–Cys–OMe, DMAP, THF, 0°, 15 min
95%
(BOC) BpocN(CH3)–(CH2)3–C(=O)–SCys(Cbz)–OCH3
1. TFA, PhOMe, CH_2Cl_2, 0°, 2 min
2. $NaHCO_3$, MeOH, 0°, 3 min
3. I_2, MeOH, 25°, 10 min
95%
Cbz Cys–OCH3 / Cbz Cys–OCH3 (disulfide)

Bpoc = *p*-C_6H_5–C_6H_4–$C(CH_3)_2$–OCO–

1. N. G. Galakatos and D. S. Kemp, *J. Org. Chem.*, **50,** 1302 (1985).

Thiocarbonate Derivatives

When cysteine reacts with an alkyl or aryl chloroformate, both the —SH and —NH groups are protected as a thiocarbonate and as a carbamate, respectively. Selective or simultaneous removal of the protective groups is possible.

40. *S*-2,2,2-Trichloroethoxycarbonyl Derivative: $RSCOOCH_2CCl_3$

Cleavage

1. Electrolysis, −1.5 V, $LiClO_4$, CH_3OH, 90% yield. The conditions can be adjusted to form either the sulfide or disulfide.[1]

1. M. F. Semmelhack and G. E. Heinsohn, *J. Am. Chem. Soc.*, **94,** 5139 (1972).

41. *S-t*-Butoxycarbonyl Derivative (BOC–SR): $RSCOOC(CH_3)_3$

t-Butyl chloroformate reacts with cysteine to protect both the amine and thiol groups; as with *N,S*-bis(benzyloxycarbonyl)cysteine, selective or simultaneous removal of the *N*- or *S*-protective groups can be effected.[1]

1. M. Muraki and T. Mizoguchi, *Chem. Pharm. Bull.*, **19,** 1708 (1971).

42. *S*-Benzyloxycarbonyl Derivative (RS-Cbz, R-SZ): $RSCOOCH_2C_6H_5$

Formation[1]

$$\text{S-Cbz-cysteine} \xleftarrow[\text{0°, 1 h; 10°, 1 h, 67\%}]{\text{aq } NaHCO_3,\ CbzCl} \text{cysteine (NH}_2\text{, SH, COOH)} \xrightarrow[\text{30 min, 71\%}]{\text{aq NaOH, CbzCl}} \text{N,S-di-Cbz-cysteine}$$

Cleavage

1. Concd. NH_4OH, 25°, 1 h, 90% yield.[1]
2. Na, NH_3, 62% yield.[1]
3. 0.1 *N* $NaOCH_3$, CH_3OH, N_2, 30 min–3 h, 100% yield.[2] An *S*-benzoyl group is removed (95–100% yield) in 5–10 min.
4. CF_3COOH, reflux, 30 min, ~quant.[2] An *N*-Cbz group is also removed under these conditions.
5. 2 *N* HBr, AcOH, 25°, 30 min.[2,3] The *S*-Cbz group is removed slowly under these conditions, but the *N*-Cbz group is completely cleaved, thus providing some selectivity in the protection scheme for cysteine.
6. Electrolysis, −2.6 V, $R_4N^+X^-$, DMF.[4] Both an *N*-Cbz group and an *S*-Cbz group are removed under these conditions.

1. A. Berger, J. Noguchi, and E. Katchalski, *J. Am. Chem. Soc.*, **78,** 4483 (1956).
2. L. Zervas, I. Photaki, and N. Ghelis, *J. Am. Chem. Soc.*, **85,** 1337 (1963).
3. M. Sokolovsky, M. Wilchek, and A. Patchornik, *J. Am. Chem. Soc.*, **86,** 1202 (1964).
4. V. G. Mairanovsky, *Angew Chem., Int. Ed. Engl.*, **15,** 281 (1976).

43. *S-p*-Methoxybenzyloxycarbonyl Derivative: $RSCOOCH_2C_6H_4$-*p*-OCH_3

S-p-Methoxybenzyloxycarbonylcysteine has been prepared in low yield (30%). It has been used in peptide syntheses, but is very labile to acids and bases.[1]

1. I. Photaki, *J. Chem. Soc. C*, 2687 (1970).

Thiocarbamate Derivatives

Thiocarbamates, formed by reaction of a thiol with an isocyanate, are stable in acidic and neutral solutions and are readily cleaved by basic hydrolysis. The β-elimination that can occur when an *S*-acyl group is removed with base from a cysteine derivative does not occur under the conditions needed to cleave a thiocarbamate.[1]

44. *S*-(*N*-Ethylcarbamate): $RSCONHC_2H_5$ (Chart 7)

Formation[1]

1. EtN=C=O, pH 1 → pH 6, 20°, 70 h, 67% yield.

Cleavage

1. 1 *N* NaOH, 20°, 20 min, 100% yield.[1]
2. NH_3 or NH_2NH_2, methanol, 20°, 2 h, 100% yield.[1]
3. Na/NH_3, −30°, 3 min, 100% yield.[1] This protective group is stable to acidic hydrolysis (4.5 *N* HBr/HOAc; 1 *N* HCl; CF_3CO_2H, reflux). There is no evidence of S → N acyl migration in *S*-(*N*-ethylcarbamates) (RS = cysteinyl).[1] Oxidation of *S*-(*N*-ethylcarbamoyl)cysteine with performic acid yields cysteic acid.[2]
4. $Hg(OAc)_2$, H_2O, CH_3OH, 30 min; H_2S, 4 h, 79% yield.[2]
5. $AgNO_3$, H_2O, CH_3OH; concd. HCl, 3 h, 62% yield.[2]

1. St. Guttmann, *Helv. Chim. Acta*, **49,** 83 (1966).
2. H. T. Storey, J. Beacham, S. F. Cernosek, F. M. Finn, C. Yanaihara, and K. Hofmann, *J. Am. Chem. Soc.*, **94,** 6170 (1972).

45. *S*-(*N*-Methoxymethylcarbamate): $RSCONHCH_2OCH_3$

Formation[1]

1. $CH_3OCH_2N{=}C{=}O$, pH 4–5, 2 min, 100% yield.
   At pH 4–5, the reaction is selective for protection of thiol groups in the presence of α- or ε-amino groups.

Cleavage[1]

1. At pH 9.6, a cysteine derivative is cleaved in 100% yield, glutathione in 80% yield.

1. H. Tschesche and H. Jering, *Angew. Chem., Int. Ed. Engl.*, **12,** 756 (1973).

MISCELLANEOUS DERIVATIVES

Unsymmetrical Disulfides

A thiol can be protected by oxidation (with O_2; H_2O_2; I_2. . .) to the corresponding symmetrical disulfide, which subsequently can be cleaved by reduction: [Sn/HCl; Na/xylene, Et_2O, or NH_3; $LiAlH_4$; $NaBH_4$; or thiols such as $HO(CH_2)_2SH$]. Unsymmetrical disulfides have also been prepared and are discussed.

46. *S*-Ethyl Disulfide: $RSSC_2H_5$ (Chart 7)

Formation

1. EtS(O)SEt, −70°, 1 h, 80–90% yield.[1]

Cleavage

1. PhSH, >50° or $HSCH_2CO_2H$, 45°, 15 h, quant.[2] The *S*-ethyl disulfide is stable to acid-catalyzed hydrolysis (CF_3CO_2H) of carbamates and to ammonolysis (25% NH_3/CH_3OH).[2]

1. D. A. Armitage, M. J. Clark, and C. C. Tso, *J. Chem. Soc., Perkin Trans. I*, 680 (1972).
2. N. Inukai, K. Nakano, and M. Murakami, *Bull. Chem. Soc. Jpn.*, **40,** 2913 (1967).

47. *S-t*-Butyl Disulfide: $RSSC(CH_3)_3$

Formation

1. $CH_3OC(O)SCl$, 0–5°, 1.5 h; *t*-BuSH, MeOH, 5 days, 97% crude, 46% pure.[1] The reaction proceeds through an *S*-sulfenyl thiocarbonate.
2. *t*-BuO_2CNHN(*S-t*-Bu)CO_2-*t*-Bu, H_2O.[2]

Cleavage

1. $NaBH_4$.[3]

1. L. Field and R. Ravichandran, *J. Org. Chem.*, **44,** 2624 (1979).
2. E. Wünsch, L. Moroder, and S. Romani, *Hoppe-Seyler's Z. Physiol. Chem.*, **363,** 1461 (1982).

3. E. Wünsch and R. Spangenberg, in *Peptides, 1969*, E. Schoffone, Ed., North Holland, Amsterdam, p. 1971.

48. Substituted *S*-Phenyl Disulfides: $RSSC_6H_4$–Y

Three substituted *S*-phenyl unsymmetrical disulfides have been prepared, **i**,[1] **ii**,[2] and **iii**[3]—compounds **i** and **ii** by reaction of a thiol with a sulfenyl halide, compound **iii** from a thiol and an aryl thiosulfonate ($ArSO_2SAr$). The disulfides are cleaved by reduction ($NaBH_4$) or by treatment with excess thiol ($HSCH_2CH_2OH$).

$RSSC_6H_3$-2-NO_2-4-R′ **i** (R′ = H, NO_2) $RSSC_6H_4$-2-N=N-C_6H_5 **ii** RSS-C_6H_4-2-COOH **iii**

1. A. Fontana, E. Scoffone, and C. A. Benassi, *Biochemistry*, **7**, 980 (1968); A. Fontana, *J. Chem. Soc., Chem. Commun.*, 976 (1975).
2. A. Fontana, F. M. Veronese, and E. Scoffone, *Biochemistry*, **7**, 3901 (1968).
3. L. Field and P. M. Giles, Jr., *J. Org. Chem.*, **36**, 309 (1971).

Sulfenyl Derivatives

49. *S*-Sulfonate Derivative: $RSSO_3^-$

Formation

1. Na_2SO_3, cat. cysteine, O_2, pH 7–8.5, 1 h, quant.[1]

Cleavage

1. $HSCH_2CH_2OH$, pH 7.5, 25°, 2 h, 100% yield.[1]
2. $NaBH_4$.[1] *S*-Sulfonates are stable at pH 1–9; they are unstable in hot acidic solutions and in 0.1 *N* sodium hydroxide.

1. W. W.-C. Chan, *Biochemistry*, **7**, 4247 (1968).

50. *S*-Sulfenylthiocarbonate: RSSCOOR′

A number of *S*-sulfenylthiocarbonates have been prepared to protect thiols. A benzyl derivative, R′=CH_2Ph, is stable to trifluoroacetic acid (25°, 1 h) and provides satisfactory protection during peptide syntheses;[1] a *t*-butyl derivative, R′ = *t*-Bu, is too labile in base to provide protection.[1] A methyl derivative, R′=CH_3, has been used to protect a cysteine fragment that is subsequently converted to a cystine.[2]

BOCNH–CH(CO₂H)–CH₂–STr → (BnOC(O)SCl, 0°, 2 h, −25°, 2 days, 60%) → BOCNH–CH(CO₂H)–CH₂–SSC(O)OBn → (40% HBr, AcOH) → BOCNH–CH(CO₂H)–CH₂–SH

1. K. Nokihara and H. Berndt, *J. Org. Chem.*, **43,** 4893 (1978).
2. R. G. Hiskey, N. Muthukumaraswamy, and R. R. Vunnam, *J. Org. Chem.*, **40,** 950 (1975).

51. *S*-3-Nitro-2-pyridinesulfenyl Sulfide (Npys-SR): 3-NO_2–C_5H_3NSSR

These sulfides are prepared from other sulfur protective groups by reaction with the sulfenyl chloride.[1] The Npys group can also be introduced directly by treatment of the thiol with NpysCl.[2]

Starting Material	Npys–X, Eq.	Conditions	Yield (%)
Boc–Cys(Bn)–OH	Cl, 1.2	rt, 24 h, CH_2Cl_2	No reaction
Boc–Cys(MeOBn)–OH[3]	Cl, 1.2	0°, 30 min, CH_2Cl_2	92
Boc–Cys(Me_2Bn)–OH	Cl, 1.2	0°, 30 min, CH_2Cl_2	90
Z–Cys(MeOBn)–Phe–Phe–Gln–Asn–O–*t*-Bu	Cl, 1.2	rt, 30 min, CH_2Cl_2/CF_3COOH (1:1)	85
Fmoc–Cys(*t*-Bu)–OH	Cl, 1.2	0°, 30 min, CH_2Cl_2	80
Boc–Cys(Tr)–OH	Cl, 1.2	−30°, 3 h, CH_2Cl_2	91
Boc–Cys(Acm)–OH	Cl, 1.2	0°, 30 min, AcOH	63
Z–Cys(Bn)–OH	Br, 2.0	rt, 10 h, CH_2Cl_2	21
Z–Cys(Bn)–OH	Cl, 2.0	rt, 5 h, CF_3CH_2OH	61
Z–Cys(Bn)–OH	Br, 2.4	rt, 3 h, CF_3CH_2OH/AcOH (10:1)	73
Z–Cys(Bn)–Pro–Leu–GlyNH_2	Br, 2.4	rt, 3 h, CF_3CH_2OH/AcOH (10:1)	70

The Npys group can be cleaved reductively with Bu_3P, H_2O or mercaptoethanol. It is stable to CF_3COOH (24 h), 4 *M* HCl/dioxane (24 h), and HF (1 h).[2] The related reagent, 2-pyridinesulfenyl chloride, has also been proposed as a useful reagent for the deprotection of the *S*-trityl, *S*-diphenylmethyl, *S*-acetamidomethyl, *S*-*t*-butyl, and *S*-*t*-butylsulfenyl groups, but this reagent is very susceptible to hydrolysis.[4]

1. R. Matsueda, S. Higashida, R. J. Ridge, and G. R. Matsueda, *Chem. Lett.*, 921 (1982).
2. R. Matsueda, T. Kimura, E. T. Kaiser, and G. R. Matsueda, *Chem. Lett.*, 737 (1981).
3. O. Ploux, M. Caruso, G. Chassaing, and A. Marquet, *J. Org. Chem.*, **53,** 3154 (1988).
4. J. V. Castell and A. Tun-Kyi, *Helv. Chim. Acta*, **62,** 2507 (1979).

Protection for Dithiols

DITHIO ACETALS AND KETALS

52. *S,S'*-Methylene (i), *S,S'*-Isopropylidene (ii), and *S,S'*-Benzylidene (iii) Derivatives

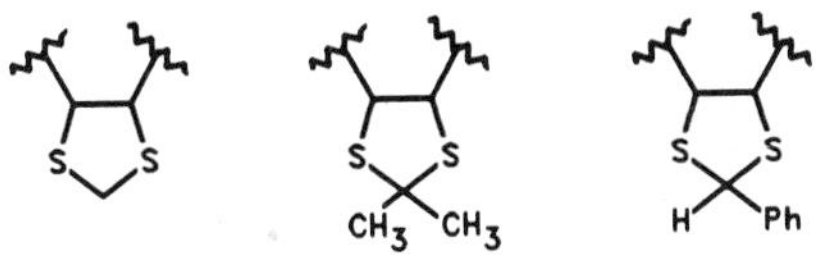

S S　　S S CH3 CH3　　S S H Ph

i　　ii　　iii

Dithiols, like diols, have been protected as *S,S'*-methylene,[1] *S,S'*-isopropylidene,[2] and *S,S'*-benzylidene[3] derivatives, formed by reaction of the dithiol with formaldehyde, acetone, or benzaldehyde, respectively. The methylene and benzylidene derivatives are cleaved by reduction with sodium/ammonia. The isopropylidene[2] and benzylidene[3] derivatives are cleaved by mercury(II) chloride; with sodium/ammonia the isopropylidene derivative is converted to a monothio ether, HSCHR–CHRSCHMe$_2$.[1]

1. E. D. Brown, S. M. Igbal, and L. N. Owen, *J. Chem. Soc. C*, 415 (1966).
2. E. P. Adams, F. P. Doyle, W. H. Hunter, and J. H. C. Nayler, *J. Chem. Soc.*, 2674 (1960).
3. L. W. C. Miles and L. N. Owen, *J. Chem. Soc.*, 2938 (1950).

53. *S,S'*-*p*-Methoxybenzylidene Derivative: (RS)$_2$CHC$_6$H$_4$–4-OCH$_3$

Formation

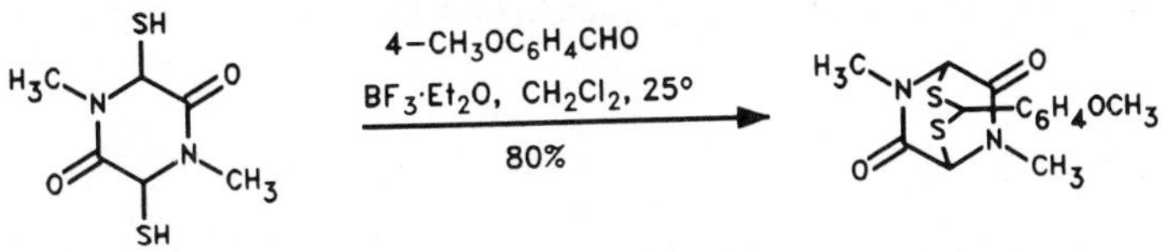

SH H3C N O O N CH3 SH

4–CH$_3$OC$_6$H$_4$CHO
BF$_3$·Et$_2$O, CH$_2$Cl$_2$, 25°
80%

H3C N O S S C6H4OCH3 O N CH3

Cleavage

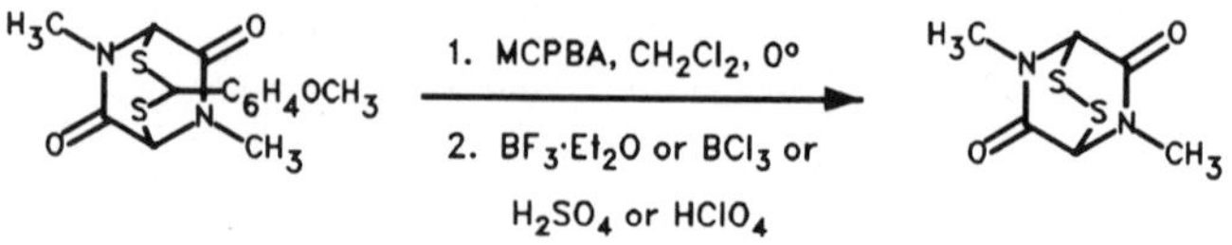

1. MCPBA, CH_2Cl_2, 0°
2. $BF_3 \cdot Et_2O$ or BCl_3 or H_2SO_4 or $HClO_4$

The epidithioketopiperazine shown above is present in natural products, including the gliotoxins and sporidesmins.[1]

1. Y. Kishi, T. Fukuyama, and S. Nakatusuka, *J. Am. Chem. Soc.*, **95,** 6490 (1973).

Protection for Sulfides

Since sulfides tend to react with electrophiles, a method for protection could be quite useful. Sulfoxides can be used to protect sulfides and are easily formed by a variety of oxidants. Sulfides can be regenerated with thiols,[1] $SiCl_4$ (0°, 15 min, TFA, anisole),[2] $LiBH_4/Me_3SiCl$,[3] $DMF \cdot SO_3/HSCH_2CH_2SH$ (DMF, Pyr, rt, 85% yield).[4]

Sulfides can also be protected as sulfonium salts.

54. *S*-Methylsulfonium Salt: $R_2S^+CH_3\ X^-$

Formation

1. $CH_3OSO_2CF_3$, CH_2Cl_2, 99% yield.[5]
2. MeOTs, EtOAc, rt, 4 days, 85% yield.[6]

Cleavage

1. DMF, Et_3N, $HSCH_2CH_2OH$, rt, 78% yield.[6]
2. $LiAlH_4$, THF.[5]

A methylsulfonium salt is stable to $NH_3/MeOH$ and to TFA, but not to hydrogenolysis (H_2/Pd–C).[6]

55. *S*-Benzyl- and *S*-4-Methoxybenzylsulfonium Salt: $R_2S^+CH_2Ph\ X^-$

Formation

1. $C_6H_5CH_2OTf$, CH_3CN.[7]
2. 4-$MeOC_6H_4CH_2Cl$, $AgBF_4$, CH_3CN, 97–99% yield.[8]

Cleavage

The benzylsulfonium salt is cleaved by hydrogenolysis (H_2/Pd–C, MeOH);[7] the 4-methoxybenzylsulfonium salt is cleaved by methylamine (100%).[8]

56. *S*-1-(4-Phthalimidobutyl)sulfonium Salt:

Formation/Cleavage[8]

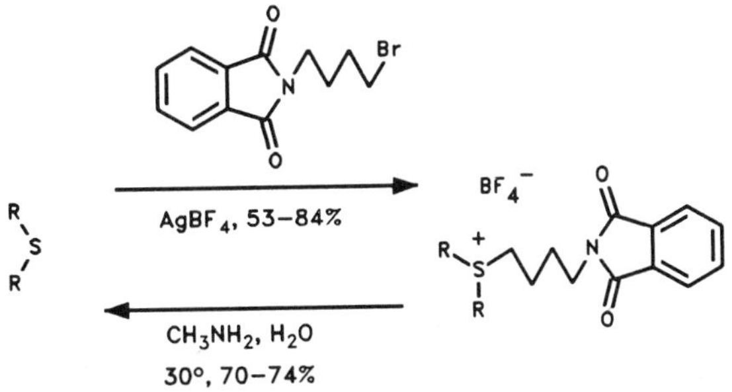

$AgBF_4$, 53–84%

CH_3NH_2, H_2O
30°, 70–74%

1. N. Fujii, A. Otaka, S. Funakoshi, H. Yajima, O. Nishimura, and M. Fujino, *Chem. Pharm. Bull.*, **34,** 869 (1986).
2. Y. Kiso, M. Yoshida, T. Fujisaki, T. Mimoto, T. Kimura, and M. Shimokura, *Pept. Chem. 1986*, **24th,** 205 (1987); *Chem. Abstr.*, **108,** 112924j (1988).
3. A. Giannis and K. Sandhoff, *Angew. Chem., Int. Ed. Engl.*, **28,** 218 (1989).
4. S. Futaki, T. Taike, T. Yagami, T. Akita, and K. Kitagawa, *Tetrahedron Lett.*, **30,** 4411 (1989).
5. V. Cere, A. Guenzi, S. Pollicino, E. Sandri, and A. Fava, *J. Org. Chem.*, **45,** 261 (1980).
6. M. Bodansky and M. A. Bednareck, *Int. J. Pept. Protein Res.*, **20,** 408 (1982).
7. R. C. Roemmele and H. Rapoport, *J. Org. Chem.*, **54,** 1866 (1989).
8. J. T. Doi and G. W. Luehr, *Tetrahedron Lett.*, **26,** 6143 (1985).

S–P Derivatives

57. *S*-(Dimethylphosphino)thioyl Group (Mpt–SR): $(CH_3)_2P(S)SR$

58. *S*-(Diphenylphosphino)thioyl Group (Ppt-SR): $Ph_2P(S)SR$

Formation

1. MptCl, (*i*-Pr)$_2$EtN, $CHCl_3$, 79% yield. The Mpt group on the nitrogen in cysteine can be selectively removed with HCl/Ph_3P leaving the S–Mpt group intact.[1]

Cleavage

1. $AgNO_3$, H_2O, Pyr, 0°, 1 h; H_2S, 100% yield.
2. KF, 18-crown-6 or $Bu_4N^+F^-$, CH_3CN, MeOH, 88% yield.[2]

The related *S*-diphenylphosphinothioyl group (Ppt group) has also been cleaved using these conditions.[3] The Mpt derivative of cysteine is not stable to DBU; it forms dehydroalanine. The Mpt group is stable to TFA and to 1 *M* HCl, but not to HBr/AcOH or 6 *M* HCl.[1]

1. M. Ueki and K. Shinozaki, *Bull. Chem. Soc. Jpn.*, **56,** 1187 (1983).
2. M. Ueki and K. Shinozaki, *Bull. Chem. Soc. Jpn.*, **57,** 2156 (1984).
3. L. Horner, R. Gehring, and H. Lindel, *Phosphorus Sulfur*, **11,** 349 (1981).

7

PROTECTION FOR THE AMINO GROUP

Many protective groups have been developed for the amino group, including carbamates ($>NCO_2R$), used for the protection of amino acids in peptide and protein syntheses, and amides ($>NCOR$), used more widely in syntheses of alkaloids and for the protection[1] of the nitrogen bases adenine, cytosine, and guanine in nucleotide syntheses.

Carbamates are formed from an amine with a wide variety of reagents, of which the chloroformate is the most common; amides are formed from the acid chloride. *n*-Alkyl carbamates are cleaved by acid-catalyzed hydrolysis; *N*-alkylamides are cleaved by acidic or basic hydrolysis at reflux, conditions that cleave peptide bonds.

In this chapter detailed information is provided for the more useful protective groups (some of which are included in Reactivity Charts 8–10); structures and references are given for protective groups that seem to have more limited use.[2]

CARBAMATES

Carbamates can be used as protective groups for amino acids to minimize racemization in peptide synthesis. Racemization occurs during the base-catalyzed coupling reaction of an *N*-protected, carboxyl-activated amino acid, and takes place in the intermediate oxazolone that forms readily from an *N*-acyl protected amino acid (R′ = alkyl, aryl):

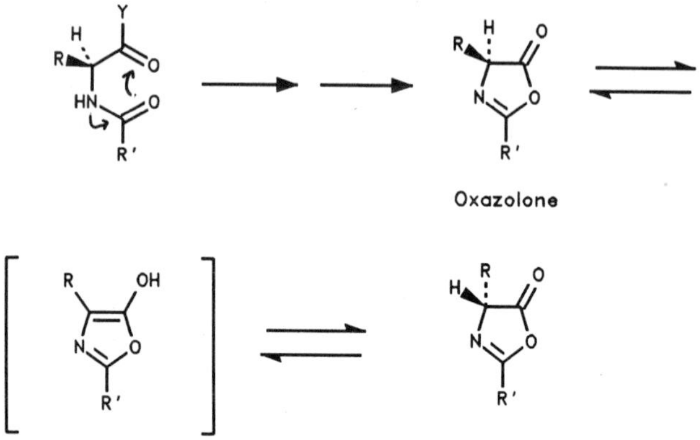

Oxazolone

To minimize racemization, the use of nonpolar solvents, a minimum of base, low reaction temperatures, and carbamate protective groups (R′ = *O*-alkyl or *O*-aryl) is effective. (A carbamate, R′ = O-*t*-Bu, has been reported to form an oxazolone that appears not to racemize during base-catalyzed coupling.)[3]

Many carbamates have been used as protective groups. They are arranged in this chapter in order of increasing complexity of structure. The most useful compounds do not necessarily have the simplest structures, but are *t*-butyl (BOC), readily cleaved by acidic hydrolysis; benzyl (Cbz or Z), cleaved by catalytic hydrogenolysis; 2,4-dichlorobenzyl, stable to the acid-catalyzed hydrolysis of benzyl and *t*-butyl carbamates; 2-(biphenylyl)isopropyl, cleaved more easily than *t*-butyl carbamate by dilute acetic acid; 9-fluorenylmethyl, cleaved by β-elimination with base; isonicotinyl, cleaved by reduction with zinc in acetic acid; 1-adamantyl, readily cleaved by trifluoroacetic acid; and allyl, readily cleaved by Pd-catalyzed isomerization.

1. C. B. Reese, *Tetrahedron*, **34,** 3143 (1978); V. Amarnath and A. D. Broom, *Chem. Rev.*, **77,** 183 (1977).
2. See also: E. Wünsch, "Blockierung und Schutz der α-Amino-Funktion," in *Methoden der Organischen Chemie (Houben-Weyl)*, Georg Thieme Verlag, Stuttgart, 1974, Vol. 15/1, pp. 46–305; J. W. Barton, "Protection of N—H Bonds and NR_3," in *Protective Groups in Organic Chemistry*, J. F. W. McOmie, Ed., Plenum Press, New York and London, 1973, pp. 43–93; L. A. Carpino, *Acc. Chem. Res.*, **6,** 191–198 (1973); Y. Wolman, "Protection of the Amino Group," in *The Chemistry of the Amino Group*, S. Patai, Ed., Wiley-Interscience, New York, 1968, Vol. 4, pp. 669–699; E. Gross and J. Meienhofer, Eds., "The Peptides: Analysis, Synthesis, Biology, Vol. 3: Protection of Functional Groups in Peptide Synthesis," Academic Press, New York, 1981.
3. Work of N. L. Benoiton, reported by J. L. Fox, *Chem. Eng. News*, August 6, 1979, p. 20.

1. Methyl and Ethyl Carbamate: $CH_3OC(O)NR_2$ (Chart 8)

Formation

1. CH_3OCOCl, K_2CO_3, reflux 12 h.[1]
2. CO, O_2, MeOH, HCl, $PdCl_2$, $CuCl_2$.[2]
3.

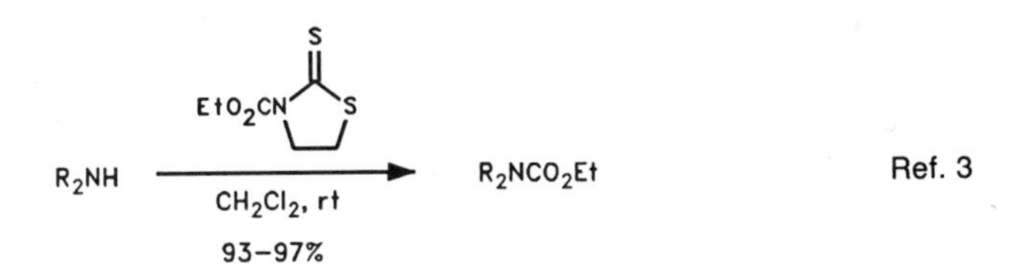

Cleavage

1. *n*-PrSLi, 0°, 8.5 h, 75–80% yield.[4]
2. Me_3SiI, 50°, 70% yield.[5,6]
3. KOH, H_2O, ethylene glycol, 100°, 12 h, 88% yield.[7]
4. HBr, AcOH, 25°, 18 h.[8]
5. $NaAlH_2(OCH_2CH_2OCH_3)_2$, benzene, rt, 80% yield.[9]
6. $Ba(OH)_2$, H_2O, MeOH, 110°, 12 h.[10]
7. NH_2NH_2–H_2O, KOH, 98% yield.[11]
8. Dimethyl sulfide, methanesulfonic acid, 5°, 58–100% yield.[12]

1. E. J. Corey, M. G. Bock, A. P. Kozikowski, A. V. Rama Rao, D. Floyd, and B. Lipshutz, *Tetrahedron Lett.*, 1051 (1978).
2. H. Alper and F. W. Hartstock, *J. Chem. Soc., Chem. Commun.*, 1141 (1985).
3. L. C. Chen and S. C. Yang, *J. Chin. Chem. Soc. (Tapei)*, **33,** 347 (1986).
4. E. J. Corey, L. O. Weigel, D. Floyd, and M. G. Bock, *J. Am. Chem. Soc.*, **100,** 2916 (1978).
5. R. S. Lott, V. S. Chauhan, and C. H. Stammer, *J. Chem. Soc., Chem. Commun.*, 495 (1979).
6. S. Raucher, B. L. Bray, and R. F. Lawrence, *J. Am. Chem. Soc.*, **109,** 442 (1987).
7. E. Wenkert, T. Hudlicky, and H. D. H. Showalter, *J. Am. Chem. Soc.*, **100,** 4893 (1978).
8. M. C. Wani, H. F. Campbell, G. A. Brine, J. A. Kepler, M. E. Wall, and S. G. Levine, *J. Am. Chem. Soc.*, **94,** 3631 (1972).
9. G. R. Lenz, *J. Org. Chem.*, **53,** 4447 (1988).
10. P. M. Wovkulich and M. R. Usković, *Tetrahedron*, **41,** 3455 (1985).

11. T. Shono, Y. Matsumura, K. Uchida, K. Tsubata, and A. Makino, *J. Org. Chem.*, **49,** 300 (1984).
12. H. Irie, H. Nakanishi, N. Fujii, Y. Mizuno, T. Fushimi, S. Funakoshi, and H. Yajima, *Chem. Lett.*, 705 (1980).

2. 9-Fluorenylmethyl Carbamate ($Fmoc–NR_2$) (Chart 8):

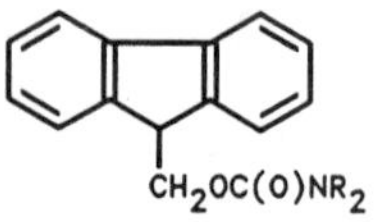

$CH_2OC(O)NR_2$

Some advantages of the Fmoc protective group are that it has excellent acid stability; thus BOC and benzyl-based groups can be removed in its presence. It is readily cleaved, nonhydrolytically, by simple amines, and the protected amine is liberated as its free base.[1] The Fmoc group is generally considered to be stable to hydrogenation conditions, but it has been shown that under some circumstances it can be cleaved with H_2/Pd–C, AcOH, MeOH, ($t_{1/2}$ = 3–33 h).[2]

Formation

1. Fmoc–Cl, $NaHCO_3$, aq. dioxane, 88–98% yield.[3] Diisopropylethylamine is reported to suppress dipeptide formation during Fmoc introduction with Fmoc–Cl.[4]
2. Fmoc–N_3, $NaHCO_3$, aq. dioxane, 88–98% yield.[3,5] This reagent reacts more slowly with amino acids than does the acid chloride. It is not the most safe method for Fmoc introduction because of the azide.
3. Fmoc–OBt (Bt = benzotriazol-1-yl).[6,7]
4. Fmoc–OSu (Su = succinimidyl), H_2O, CH_3CN.[6–9] The advantage of Fmoc–OSu is that little or no oligopeptides are formed when amino acid derivatives are prepared.[10]
5. Fmoc–OC_6F_5, $NaHCO_3$, H_2O, acetone, rt, 64–99% yield.[11]

Cleavage

1. The Fmoc group is cleaved under mild conditions with an amine base to afford the free amine and dibenzofulvene. The approximate half-lives for the deprotection of Fmoc–ValOH by a variety of amine bases in DMF are as follows:[10]

Amine	$t_{1/2}$
20% Piperidine	6 s
5% Piperidine	20 s
50% Morpholine	1 min
50% Dicyclohexylamine	35 min
10% *p*-Dimethylaminopyridine	85 min
50% Diisopropylethylamine	10.1 h

The half-lives shown in the table will vary depending on the structure of the Fmoc–amine derivative.

2. $Bu_4N^+F^-$, DMF, rt, 2 min.[12]
3. Piperazine attached to a polymer has also been used to cleave the Fmoc group.[13]

3. 9-(2-Sulfo)fluorenylmethyl Carbamate

Because of the electron-withdrawing sulfonic acid substituent, cleavage occurs under milder conditions than needed for the Fmoc group (0.1 *N* NH_4OH; 1% Na_2CO_3, 45 min).[14]

4. 9-(2,7-Dibromo)fluorenylmethyl Carbamate:

Br Br $CH_2OC(O)NR_2$

Because of the two electron-withdrawing bromine groups, pyridine can be used to cleave this derivative from its parent amine.[15]

1. L. A. Carpino, *Acc. Chem. Res.*, **20,** 401 (1987); L. A. Carpino, D. Sadat-Aalaee, and M. Beyermann, *J. Org. Chem.*, **55,** 1673 (1990).
2. E. Atherton, C. Bury, R. C. Sheppard, and B. J. Williams, *Tetrahedron Lett.*, 3041 (1979).
3. L. A. Carpino and G. Y. Han, *J. Org. Chem.*, **37,** 3404 (1972).
4. F. M. F. Chen and N. L. Benoiton, *Can. J. Chem.*, **65,** 1224 (1987).
5. M. Tessier, F. Albericio, E. Pedroso, A. Grandas, R. Eritja, E. Giralt, C. Granier, and J. Van Rietschoten, *Int. J. Pept. Protein Res.*, **22,** 125 (1983).
6. A. Paquet, *Can. J. Chem.*, **60,** 976 (1982).
7. G. F. Sigler, W. D. Fuller, N. C. Chaturvedi, M. Goodman, and M. Verlander, *Biopolymers*, **22,** 2157 (1983).
8. R. C. de L. Milton, E. Becker, S. C. F. Milton, J. E. H. Baxter, and J. F. Elsworth, *Int. J. Pept. Protein Res.*, **30,** 431 (1987).
9. L. Lapatsanis, G. Milias, K. Froussios, and M. Kolovos, *Synthesis*, 671 (1983).
10. For a review of the use of Fmoc protection in peptide synthesis, see: E. Atherton and R. C. Sheppard, "The Fluorenylmethoxycarbonyl Amino Protecting Group," in *The Peptides.*, S. Udenfriend and J. Meienhofer, Eds., Academic Press, Orlando, FL, 1987, Vol. 9, pp. 1–38.
11. I. Schoen and L. Kisfaludy, *Synthesis*, 303 (1986).
12. M. Ueki and M. Amemiya, *Tetrahedron Lett.*, **28,** 6617 (1987).
13. L. A. Carpino, E. M. E. Mansour, and J. Knapczyk, *J. Org. Chem.*, **48,** 666 (1983).
14. R. B. Merrifield and A. E. Bach, *J. Org. Chem.*, **43,** 4808 (1978).
15. L. A. Carpino, *J. Org. Chem.*, **45,** 4250 (1980).

5. 2,7-Di-*t*-butyl-[9-(10,10-dioxo-10,10,10,10-tetrahydrothioxanthyl)]methyl Carbamate (DBD–Tmoc–NR_2):

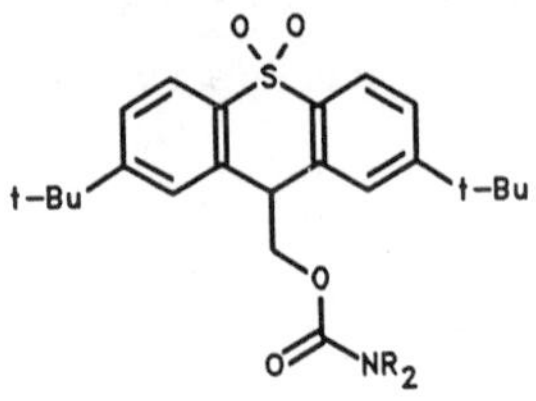

Formation

1. DBD–TmocCl, $NaHCO_3$, H_2O, dioxane.[1]

Cleavage[1]

The DBD–Tmoc group is stable to TFA and HBr/AcOH.

1. 50–75° in DMSO, 4.5–16 h, 100% yield.
2. Pd—C, HCO_2NH_4, MeOH.
3. Pyridine. The Fmoc group is stable to pyridine.

1. L. A. Carpino, H.-S. Gao, G.-S. Ti, and D. Segev, *J. Org. Chem.*, **54,** 5887 (1989).

6. 4-Methoxyphenacyl Carbamate (Phenoc–NR_2)

Formation/Cleavage[1]

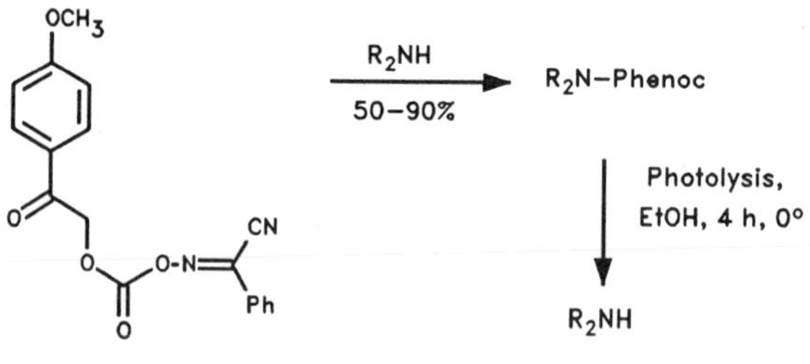

R_2NH, 50–90% → R_2N–Phenoc; Photolysis, EtOH, 4 h, 0° → R_2NH

This group is stable to 50% TFA/CH_2Cl_2, NaOH, and 20% piperidine/DMF.

1. G. Church, J.-M. Ferland, and J. Gauthier, *Tetrahedron Lett.*, **30,** 1901 (1989).

Substituted Ethyl Carbamates

7. 2,2,2-Trichloroethyl Carbamate (Troc–NR$_2$): $Cl_3CCH_2OC(O)NR_2$ (Chart 8)

Formation

1. Cl_3CCH_2OCOCl, Pyr or aq. NaOH, 25°, 12 h.[1,2]
2. Silylate with $Me_3SiN{=}C(OSiMe_3)CH_3$, then treat with Cl_3CCH_2OCOCl.[3]
3. Cl_3CCH_2OCO–O–succinimidyl, 1 *N* NaOH or 1 *N* Na_2CO_3, dioxane, 77–96% yield.[4,5] This method does not result in oligopeptide formation when used to prepare amino acid derivatives.
4. Treatment of a tertiary benzylamine also affords the Troc derivative with cleavage of the benzyl group (Cl_3CCH_2OCOCl, CH_3CN, 93% yield).[6]
5. 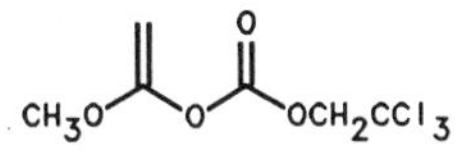 CH_3O–C(O)–O–C(O)–OCH_2CCl_3 CH_2Cl_2, rt, 3.5 h, 90–97% yield.[7]

Cleavage

1. Zn, THF, H_2O, pH 4.2, 30 min, 86% yield or pH 5.5–7.2, 18 h, 96% yield.[8] Under these conditions the Troc group can be cleaved in the presence of the BOC, benzyl, and trifluoroacetamido groups and these groups can in turn be cleaved individually in the presence of a Troc group.[9]
2. Electrolysis at a Hg cathode, 1.7 V (SCE), DMF, >72% yield.[10]
3. Electrolysis, −1.7 V, 0.1 *M* $LiClO_4$, 85% yield.[11]
4. Zn–Pb couple, 4 : 1 THF/1 *M* NH_4OAc.[12]
5. Cd, AcOH.[13] These conditions were reported to be superior to the use of Zn/AcOH. The authors also reported that the Troc group is not stable to hydrogenation with Pd–C (TsOH, DMF, H_2), but is stable to hydrogenation with Ru–C or Pt–C.
6. Cobalt(I) phthalocyanine.[14]
7.

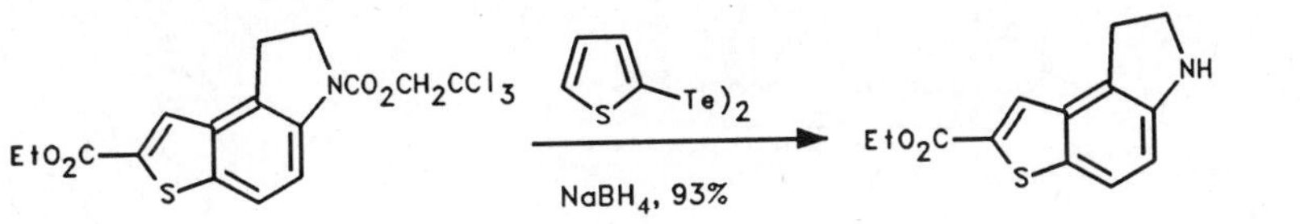

Ref. 15

1. T. B. Windholz and D. B. R. Johnston, *Tetrahedron Lett.*, 2555 (1967).
2. J. F. Carson, *Synthesis*, 268 (1981).

3. S. Raucher and D. S. Jones, *Synth. Commun.*, **15,** 1025 (1985).
4. A. Paquet, *Can. J. Chem.*, **60,** 976 (1982).
5. L. Lapatsanis, G. Milias, K. Froussios, and M. Kolovos, *Synthesis*, 671 (1983).
6. V. H. Rawal, R. J. Jones, and M. P. Cava, *J. Org. Chem.*, **52,** 19 (1987).
7. Y. Kita, J.-i. Haruta, H. Yasuda, K. Fukunaga, Y. Shirouchi, and Y. Tamura, *J. Org. Chem.*, **47,** 2697 (1982).
8. G. Just and K. Grozinger, *Synthesis*, 457 (1976).
9. R. J. Bergeron and J. S. McManis, *J. Org. Chem.*, **53,** 3108 (1988).
10. L. Van Hijfte and R. D. Little, *J. Org. Chem.*, **50,** 3940 (1985).
11. M. F. Semmelhack and G. E. Heinsohn, *J. Am. Chem. Soc.*, **94,** 5139 (1972).
12. L. E. Overman and R. L. Freerks, *J. Org. Chem.*, **46,** 2833 (1981).
13. G. Hancock, I. J. Galpin, and B. A. Morgan, *Tetrahedron Lett.*, **23,** 249 (1982).
14. H. Eckert and I. Ugi, *Liebigs Ann. Chem.*, 278 (1979).
15. M. V. Lakshmikantham, Y. A. Jackson, R. J. Jones, G. J. O'Malley, K. Ravichandran, and M. P. Cava, *Tetrahedron Lett.*, **27,** 4687 (1986).

8. 2-Trimethylsilylethyl Carbamate (Teoc–NR$_2$): $(CH_3)_3SiCH_2CH_2OC(O)NR_2$ (Chart 8)

Formation

1. Teoc–O–succinimidyl, $NaHCO_3$ or TEA, dioxane, H_2O, rt, overnight, 43–96% yield.[1,2] The use of Teoc–OSu for the protection of amino acids proceeds without oligopeptide formation. Teoc–O–benzotriazolyl was also examined, but was inferior to the succinimide derivative.
2. Teoc–OC_6H_4–4-NO_2, NaOH, *t*-BuOH, 66–89% yield.[3,4]
3. Teoc–Cl or Teoc–N_3.[5]
4. The Teoc derivative can be prepared by cleavage of an N–Bn bond with Teoc–Cl in THF. This is a general method for removal of benzyl groups from nitrogen.[6] Methyl and ethyl groups are also cleaved, but more slowly (24 h vs. 4 h) and in lower yield.

CO_2Et / N–Bn → ($TMSCH_2CH_2OCOCl$, THF, 85%) → CO_2Et / N–Teoc → ($Bu_4N^+F^-$, 97%) → CO_2Et / N–H

Cleavage

1. $Bu_4N^+F^-$, $KF \cdot 2H_2O$, CH_3CN, 50°, 8 h, 93% yield or 28°, 70 h, 93% yield.[7]
2. CF_3COOH.[5]
3. $ZnCl_2$, CH_3NO_2 or $ZnCl_2$, CF_3CH_2OH.[3] These conditions cause partial BOC

cleavage. The BOC group can be removed in the presence of a Teoc group with TsOH.[4]

1. R. E. Shute and D. H. Rich, *Synthesis*, 346 (1987).
2. A. Paquet, *Can. J. Chem.*, **60,** 976 (1982).
3. E. Wuensch, L. Moroder, and O. Keller, *Hoppe-Seyler's Z. Physiol. Chem.*, **362,** 1289 (1981).
4. A. Rosowsky and J. E. Wright, *J. Org. Chem.*, **48,** 1539 (1983).
5. L. A. Carpino, J.-H. Tsao, H. Ringsdorf, E. Fell, and G. Hettrich, *J. Chem. Soc., Chem. Commun.*, 358 (1978).
6. A. L. Campbell, D. R. Pilipauskas, I. K. Khanna, and R. A. Rhodes, *Tetrahedron Lett.*, **28,** 2331 (1987).
7. L. A. Carpino and A. C. Sau, *J. Chem. Soc., Chem. Commun.*, 514 (1979).

9. 2-Phenylethyl Carbamate (hZ–NR_2): $R_2NCO_2CH_2CH_2Ph$

The 2-phenylethyl carbamate ("homo Z" = homobenzyloxycarbonyl derivative) is prepared from the chloroformate, and can be cleaved with H_2/Pd–C if the catalyst is freshly prepared [$Pd(OAc)_2$, HCO_2NH_4]. This derivative is stable to CF_3COOH, HBr/AcOH, HCl/Et_2O, and normal hydrogenation with Pd/C (1 atm). Hydrogenolysis of the hZ group is slower than the Fmoc group, which is slower than the Z group (Cbz).[1]

1. L. A. Carpino and A. Tunga, *J. Org. Chem.*, **51,** 1930 (1986).

10. 1-(1-Adamantyl)-1-methylethyl Carbamate (Adpoc–NR_2): 1-(Adamantyl)$C(Me)_2OC(O)NR_2$

The Adpoc derivative is cleaved by CF_3COOH (0°, 4–5 min) 10^3 times faster than the *t*-BOC derivative.[1]

1. H. Kalbacher and W. Voelter, *Angew. Chem., Int. Ed. Engl.*, **17,** 944 (1978); *idem, J. Chem. Soc., Chem. Commun.*, 1265 (1980).

11. 1,1-Dimethyl-2-haloethyl Carbamate: $XCH_2C(CH_3)_2OC(O)NR_2$, X = Br, Cl (Chart 8)

Formation[1]

1. $XCH_2C(CH_3)_2OCOCl$, THF, Et_3N, H_2O, $CHCl_3$, 0°, 1.5 h (X = Br, 41–79% yield; X = Cl, 60–86% yield). These halo-substituted *t*-butyl chloroformates are more stable than an unsubstituted *t*-butyl chloroformate.

Cleavage[1]

1. CH_3OH, reflux, 1 h.
2. $BF_3 \cdot Et_2O$, CF_3COOH, 25°.
3. 4 *N* HBr, AcOH, 25°, 1 h.
4. Na, NH_3.

12. 1,1-Dimethyl-2,2-dibromoethyl Carbamate (DB-*t*-BOC–NR_2): $Br_2CHC(CH_3)_2OC(O)NR_2$

The DB-*t*-BOC group is introduced with the chloroformate and can be cleaved solvolytically in hot ethanol or by HBr/AcOH. It is stable to CF_3COOH, 24 h; HCl, $MeNO_2$, 24 h; HCl, AcOH, 24 h; HBr, $MeNO_2$, 5 h.[2]

13. 1,1-Dimethyl-2,2,2-trichloroethyl Carbamate (TCBOC–NR_2): $Cl_3CC(CH_3)_2OC(O)NR_2$

The TCBOC group is stable to the alkaline hydrolysis of methyl esters and to the acidic hydrolysis of *t*-butyl esters. It is rapidly cleaved by the supernucleophile lithium cobalt(I)phthalocyanine, by zinc in acetic acid,[3] and by cobalt phthalocyanine (0.1 eq., $NaBH_4$, EtOH, 77–90% yield).[4]

1. T. Ohnishi, H. Sugano, and M. Miyoshi, *Bull. Chem. Soc. Jpn.*, **45,** 2603 (1972).
2. L. A. Carpino, N. W. Rice, E. M. E. Mansour, and S. A. Triolo, *J. Org. Chem.*, **49,** 836 (1984).
3. H. Eckert, M. Listl, and I. Ugi, *Angew. Chem., Int. Ed. Engl.*, **17,** 361 (1978).
4. H. Eckert and Y. Kiesel, *Synthesis*, 947 (1980).

14. 1-Methyl-1-(4-biphenylyl)ethyl Carbamate (Bpoc–NR_2): *p*-$PhC_6H_4C(Me)_2OC(O)NR_2$ (Chart 8)

Formation

1. Bpoc–N_3, 35–80% yield.[1]

Cleavage

1. This derivative is readily cleaved by acidic hydrolysis (dil. CF_3COOH, CH_2Cl_2, 10 min, quant.). It is cleaved 3000 times faster than the *t*-BOC derivative because of stabilization of the cation by the biphenyl group.[1] BnSH

was found to be the most effective scavenger for $PhC_6H_4C^+Me_2$ when deblocking is performed in 0.5% TFA/CH_2Cl_2.[2]

2. Tetrazole, trifluoroethanol, 24 h, 95% yield.[3] These conditions will also cleave the *N*-trityl group. If deprotection is performed in the presence of an acylating agent, acylation proceeds directly.
3. *N*-Hydroxybenzotriazole, trifluoroethanol, rt.[4] Trityl and Nps (2-nitrophenylsulfenyl) groups are also cleaved under these conditions.

1. R. S. Feinberg and R. B. Merrifield, *Tetrahedron*, **28**, 5865 (1972).
2. D. S. Kemp, N. Fotouhi, J. G. Boyd, R. I. Carey, C. Ashton, and J. Hoare, *Int. J. Pept. Protein Res.*, **31**, 359 (1988).
3. M. Bodansky, A. Bodansky, M. Casaretto, and H. Zahn, *Int. J. Pept. Protein Res.*, **26**, 550 (1985).
4. M. Bodansky, M. A. Bednarek, and A. Bodansky, *Int. J. Pept. Protein Res.*, **20**, 387 (1982).

15. 1-(3,5-Di-*t*-butylphenyl)-1-methylethyl Carbamate (*t*-Bumeoc–NR_2):

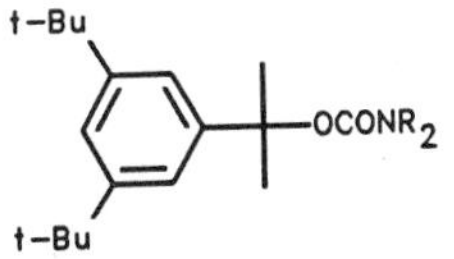

t-Bu

$OCONR_2$

t-Bu

Formation[1]

The *t*-Bumeoc adduct is prepared from the acid fluoride or the mixed carbonate in dioxane, H_2O, NaOH.

Cleavage[1]

Cleavage occurs with acid. The following tables give relative rate data that are useful for comparing other, more commonly employed, derivatives of phenylalanine (Phe).

Half-life of *t*-Bumeoc–Phe–OH with Different Acids

Acid	Half-life (min)	Complete Cleavage (min)
3% TFA/CH_2Cl_2	0.07	0.6
80% $AcOH/H_2O$	2.1	18.8
$AcOH/HCO_2H/H_2O$ (7:1:2)	22.0	167.0

Comparison of Cleavage Rates for Various Carbamate Protective Groups

Group	k^a_{rel}	k^b_{rel}
BOC	1	1
Ppoc[c]	700	750
Adpoc[d]	2400	600
Bpoc[e]	2800	2000
t-Bumeoc	4000	8000

[a] 80% $AcOH/H_2O$
[b] $AcOH/HCO_2H/H_2O$ (7:1:2)
[c] Ppoc = 2-Triphenylphosphonioisopropyl
[d] Adpoc = 1-Methyl-1-(1-adamantyl)ethyl
[e] Bpoc = 1-Methyl-1-(4-biphenyl)ethyl

1. W. Voelter and J. Mueller, *Liebigs Ann. Chem.*, 248 (1983).

16. 2-(2′- and 4′-Pyridyl)ethyl Carbamate (Pyoc–NR_2)

Formation/Cleavage[1,2]

$H_2NCH(R)CO_2Na$ → (4-pyridyl-$CH_2CH_2OCO_2C_6H_4$-4-NO_2, pH 11, 76–82%) → Pyoc-NHCH(R)CO_2H

Pyoc-NHCH(R)CO_2H → (1. MeI; 2. Morpholine; 90–95%) → $H_2NCH(R)CO_2Na$

The Pyoc derivative is not affected by H_2/Pd–C or TFA.

1. H. Kunz and S. Birnbach, *Tetrahedron Lett.*, **25,** 3567 (1984).
2. H. Kunz and R. Barthels, *Angew. Chem., Int. Ed. Engl.*, **22,** 783 (1983).

17. 2-(*N*,*N*-Dicyclohexylcarboxamido)ethyl Carbamate: $(C_6H_{11})_2NC(O)CH_2CH_2OCONR_2$

Formation

1. $(C_6H_{11})_2NC(O)CH_2CH_2OCOCl$, diisopropylethylamine, CH_2Cl_2, 0°, 15 min.[1]

Cleavage

1. *t*-BuOK, *t*-BuOH, 18-crown-6, THF, 0°, 30 min, 100% yield. This protective group is stable to $LiAlH_4$; 3 *N* NaOH, MeOH, rt; H_2, RaNi, 1500 psi, 100°, EtOH; and TFA.[1]

1. T. Fukuyama, L. Li, A. A. Laird, and R. K. Frank, *J. Am. Chem. Soc.*, **109,** 1587 (1987).

18. *t*-Butyl Carbamate (BOC group): $(CH_3)_3COC(O)NR_2$ (Chart 8)

The BOC group is used extensively in peptide synthesis for amine protection.[1] It is not hydrolyzed under basic conditions and is inert to many other nucleophilic reagents.

Formation

1. $(BOC)_2O$, NaOH, H_2O, 25°, 10–30 min, 75–95% yield.[2,3] This is one of the more common methods for introduction of the BOC group. It has the advantage that the byproducts are innocuous and are easily removed.
2. BOC-ON=C(CN)Ph, Et_3N, 25°, several hours, 72–100% yield.[4]
3. $BOC\text{-}ONH_2$.[5] This reagent reacts with amines 1.5–2.5 times faster than $(BOC)_2O$. Hydroxylamine can be used catalytically in the presence of $(BOC)_2O$ to generate this reagent *in situ*.
4. $BOC\text{-}OCH(Cl)CCl_3$ (1,2,2,2-tetrachloroethyl *tert*-butyl carbonate, BOC-OTCE), THF, K_2CO_3 or dioxane, H_2O, Et_3N, 60–91% yield.[6] This reagent is a cheap, distillable solid that has the effectiveness of $(BOC)_2O$.
5. t-BuO–C(O)–S–(2-pyridyl) $\xrightarrow{RNH_2}$ RNHBOC Ref. 7

6. $BOC\text{-}N_3$, DMSO, 25°.[8]
7. $BOC\text{-}OC_6H_4S^+Me_2\ MeSO_4^-$, H_2O.[9] This is a water-soluble reagent for the introduction of the BOC group.
8. BOC derivatives can be prepared directly from azides by hydrogenation in the presence of $(BOC)_2O$.[10]

TBDMSO–CH₂–CH(OH)–CH(N₃)–CO₂Me $\xrightarrow[86\%]{Pd\text{-}C,\ EtOAc,\ H_2,\ (BOC)_2O}$ TBDMSO–CH₂–CH(OH)–CH(NHBOC)–CO₂Me

9. Derivatization of cyclic urethanes with $(BOC)_2O$ makes the urethane carbonyl susceptible to hydrolysis under mild conditions and leaves the amine protected as a BOC derivative.[11]

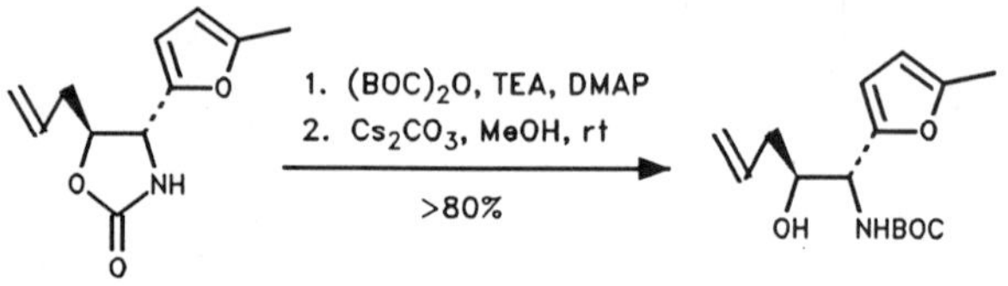

1. $(BOC)_2O$, TEA, DMAP
2. Cs_2CO_3, MeOH, rt

>80%

OH NHBOC

10. 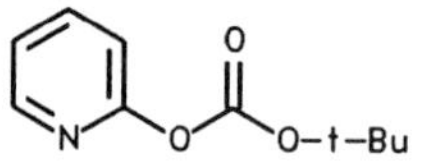 dioxane, H_2O, Et_3N, 70–94% yield.[12]

11. 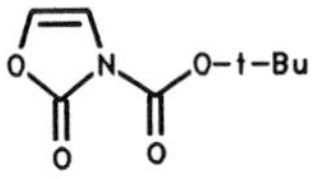 50% acetone, H_2O, DMAP, TEA, 85–95% yield.[13]

This method was also used to prepare the benzyl, methyl, ethyl, and *p*-methoxybenzyl derivatives. A polymeric version of the reagent was also described.

12. *t*-BuOCOF.[14, 15]

Cleavage

1. 3 *M* HCl, EtOAc, 25°, 30 min, 96% yield.[16] With MeOH as the solvent a diphenylmethyl ester is not affected.[17]
2. CF_3COOH, PhSH, 20°, 1 h, 100% yield.[18] Thiophenol is used to scavenge the liberated *t*-butyl cations, thus preventing alkylation of methionine or tryptophan. Other scavengers such as anisole, thioanisole, thiocresol, cresol, and dimethyl sulfide have also been used.[19]
3. Me_3SiI, $CHCl_3$ or CH_3CN, 25°, 6 min, 100% yield.[20, 21] Me_3SiI also cleaves carbamates, esters, ethers, and ketals under neutral, nonhydrolytic conditions. Some selectivity can be achieved by control of reaction conditions.
4. $AlCl_3$, $PhOCH_3$, CH_2Cl_2, CH_3NO_2, 0–25°, 2–5 h, 73–88% yield.[22]
5. The BOC group can be removed thermally, either neat (185°C, 20–30 min, 97% yield)[23] or in diphenyl ether[24].
6. Bromocatecholborane.[25]

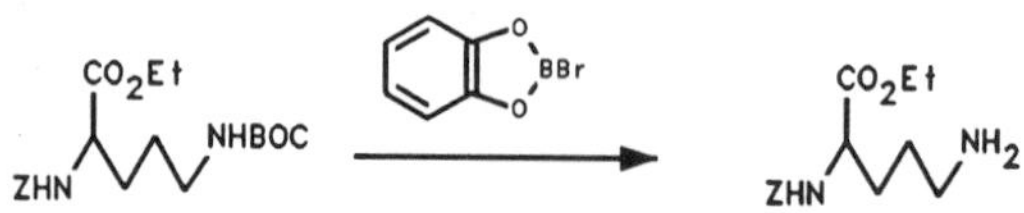

CO_2Et ZHN NHBOC → CO_2Et ZHN NH_2

7. Me_3SiCl, PhOH, CH_2Cl_2, 20 min, 100% yield.[26] Under these conditions benzyl groups are not cleaved.
8. Trimethylsilyl triflate (TMSOTf), $PhSCH_3$, CF_3COOH.[27] These conditions also cleave the following protective groups used in peptide synthesis: (MeO)Z–, Bn–, Ts–, $Cl_2C_6H_3CH_2$–, BOM (benzyloxymethyl)–, Mts–, MBS–, *t*-Bu–SR, Ad–SR, but not a BnSR, Acm, or $Arg(NO_2)$ group. The rate of cleavage is reported to be faster than with TfOH/TFA.
9. 10% H_2SO_4, dioxane.[28] These conditions are similar to the use of 50% TFA/CH_2Cl_2 and are considered safer for large-scale use instead of the use of the volatile, corrosive, and expensive TFA. The authors provide a comparison of many of the acidic methods for BOC cleavage.
10. The BOC group can be cleaved with TBDMSOTf and the intermediate silyl carbamate converted to other nitrogen protective groups.[29]

NHBOC, CO_2Me → 1. TBDMSOTf; 2. $Bu_4N^+F^-$, RX → $NHCO_2R$, CO_2Me

RX = MeI, 84%
RX = AllylBr, 82%
RX = BnBr, 88%

11. 0.05 *M* $MeSO_3H$, dioxane, CH_2Cl_2 (1 : 9).[30] This reagent also cleaves the Moz (4-methoxybenzyloxycarbonyl) group.

1. M. Bodanszky, *Principles of Peptide Chemistry*, Springer-Verlag, New York, 1984, p. 99.
2. D. S. Tarbell, Y. Yamamoto, and B. M. Pope, *Proc. Natl. Acad. Sci. (USA)*, **69**, 730 (1972).
3. E. Ponnusamy, U. Fotadar, A. Spisni, and D. Fiat, *Synthesis*, 48 (1986).
4. M. Itoh, D. Hagiwara, and T. Kamiya, *Bull. Chem. Soc. Jpn.*, **50**, 718 (1977).
5. R. B. Harris and I. B. Wilson, *Tetrahedron Lett.*, **24**, 231 (1983).
6. G. Barcelo, J. P. Senet, and G. Sennyey, *J. Org. Chem.*, **50**, 3951 (1985).
7. S. Kim, J. I. Lee, and K. Y. Yi, *Bull. Chem. Soc. Jpn.*, **58**, 3570 (1985).
8. J. B. Hansen, M. C. Nielsen, U. Ehrbar, and O. Buchardt, *Synthesis*, 404 (1982).
9. I. Azuse, H. Okai, K. Kouge, Y. Yamamoto, and T. Koizumi, *Chem. Express*, **3**, 45 (1988).
10. S. Saito, H. Nakajima, M. Inaba, and T. Moriwake, *Tetrahedron Lett.*, **30**, 837 (1989).
11. T. Ishizuka and T. Kunieda, *Tetrahedron Lett.*, **28**, 4185 (1987).
12. F. Effenberger and W. Brodt, *Chem. Ber.*, **118**, 468 (1985).
13. T. Kunieda, T. Higuchi, Y. Abe, and M. Hirobe, *Chem. Pharm. Bull.*, **32**, 2174 (1984).
14. L. A. Carpino, K. N. Parameswaran, R. K. Kirkley, J. W. Spiewak, and E. Schmitz, *J. Org. Chem.*, **35**, 3291 (1970).

15. For an improved preparation of this reagent, see: V. A. Dang, R. A. Olofson, P. R. Wolf, M. D. Piteau, and J.-P. G. Senet, *J. Org. Chem.*, **55,** 1847 (1990).
16. G. L. Stahl, R. Walter, and C. W. Smith, *J. Org. Chem.*, **43,** 2285 (1978).
17. Z. Tozuka, H. Takasugi, and T. Takaya, *J. Antibiotics*, **36,** 276 (1983).
18. B. F. Lundt, N. L. Johansen, A. Vølund, and J. Markussen, *Int. J. Pept. Protein Res.*, **12,** 258 (1978).
19. See, for example: M. Bodanszky and A. Bodanszky, *Int. J. Pept. Protein Res.*, **23,** 565 (1984); Y. Masui, N. Chino, and S. Sakakibara, *Bull. Chem. Soc. Jpn.*, **53,** 464 (1980).
20. R. S. Lott, V. S. Chauhan, and C. H. Stammer, *J. Chem. Soc., Chem. Commun.*, 495 (1979).
21. For a review on the use of Me_3SiI, see: G. A. Olah and S. C. Narang, *Tetrahedron*, **38,** 2225 (1982).
22. T. Tsuji, T. Kataoka, M. Yoshioka, Y. Sendo, Y. Nishitani, S. Hirai, T. Maeda, and W. Nagata, *Tetrahedron Lett.*, 2793 (1979).
23. V. H. Rawal, R. J. Jones, and M. P. Cava, *J. Org. Chem.*, **52,** 19 (1987).
24. H. H. Wasserman, G. D. Berger, and K. R. Cho, *Tetrahedron Lett.*, **23,** 465 (1982).
25. R. K. Boeckman, Jr., and J. C. Potenza, *Tetrahedron Lett.*, **26,** 1411 (1985).
26. E. Kaiser, Sr., J. P. Tam, T. M. Kubiak, and R. B. Merrifield, *Tetrahedron Lett.*, **29,** 303 (1988).
27. N. Fujii, A. Otaka, O. Ikemura, K. Akaji, S. Funakoshi, Y. Hayashi, Y. Kuroda, and H. Yajima, *J. Chem. Soc., Chem. Commun.*, 274 (1987).
28. R. A. Houghten, A. Beckman, and J. M. Ostresh, *Int. J. Pept. Protein Res.*, **27,** 653 (1986).
29. M. Sakaitani and Y. Ohfune, *Tetrahedron Lett.*, **26,** 5543 (1985).
30. Y. Kiso, A. Nishitani, M. Shimokura, Y. Fujiwara, and T. Kimura, *Pept. Chem., 1987*, 291 (1988); *Chem. Abstr.*, **109,** 190837t (1988).

19. 1-Adamantyl Carbamate (Adoc–NR_2): 1-Adamantyl–OC(O)NR_2 (Chart 8)

The Adoc group is very similar to the *t*-BOC group in its sensitivity to acid, but often provides more crystalline derivatives of amino acids.

Formation

1. 1-Adoc–Cl, NaOH, 27–98% yield.[1]
2. 1-Adoc–O-2-pyridyl, 70–95% yield.[2]

Cleavage

1. CF_3CO_2H, 25°, 15 min, 100% yield.[1]

1. W. L. Haas, E. V. Krumkalns, and K. Gerzon, *J. Am. Chem. Soc.*, **88,** 1988 (1966).
2. F. Effenberger and W. Brodt, *Chem. Ber.*, **118,** 468 (1985).

20. Vinyl Carbamate (Voc–NR$_2$): $CH_2{=}CHOC(O)NR_2$ (Chart 8)

The olefin of the Voc group is very susceptible to electrophilic reagents and thus is readily cleaved by reaction with bromine or mercuric acetate.

Formation

1. $CH_2{=}CHOCOCl$, MgO, H_2O, dioxane, pH 9–10, 90% yield.[1]
2. $CH_2{=}CHOCOSPh$, Et_3N, dioxane or DMF, H_2O, 25°, 16 h, 50–80% yield.[2]

Cleavage

1. Anhydrous HCl, dioxane, 25°, 97% yield.[1]
2. HBr, AcOH, 94% yield.[1]
3. Br_2, CH_2Cl_2 then MeOH, 95% yield.[1]
4. $Hg(OAc)_2$, AcOH, H_2O, 25°, 97% yield.[1]

1. R. A. Olofson, Y. S. Yamamoto, and D. J. Wancowicz, *Tetrahedron Lett.*, 1563 (1977).
2. A. J. Duggan and F. E. Roberts, *Tetrahedron Lett.*, 595 (1979). For an improved preparation of this reagent, see: R. A. Olofson and J. Cuomo, *J. Org. Chem.*, **45,** 2538 (1980).

21. Allyl Carbamate (Alloc–NR$_2$): $CH_2{=}CHCH_2OC(O)NR_2$ (Chart 8)

Formation

1. $CH_2{=}CHCH_2OCOCl$, Pyr.[1]
2. $(CH_2{=}CHCH_2OCO)_2O$, dioxane, H_2O, reflux or CH_2Cl_2, 1 h, rt, 67–96% yield.[2]
3. $CH_2{=}CHCH_2OC(O)$–O-benzotriazolyl.[3]

Cleavage

1. $Ni(CO)_4$ (*Caution: Very toxic!*), DMF, H_2O (95:5), 55°, 4 h, 83–95% yield.[1]
2. $Pd(Ph_3P)_4$, Bu_3SnH, AcOH, 70–100% yield.[4]
3. $Pd(Ph_3P)_4$, dimedone, THF, 88–95% yield.[5] The catalyst is not poisoned by the presence of thioethers such as methionine. Diethyl malonate has also been used as a nucleophile to trap the π-allylpaladium intermediate and regenerate Pd(O).[6]
4. $Pd(Ph_3P)_2Cl_2$, Bu_3SnH, p-$NO_2C_6H_4OH$, CH_2Cl_2, 70–100% yield.[4,7] This

reaction works best in the presence of acids. AcOH and pyridinium acetate are also effective.

5. $Pd_2(dba)_3 \cdot CHCl_3$, [tris(dibenzylideneacetone)dipalladium(chloroform)], HCO_2H, 74–100% yield.[8]
6. The Alloc group can be converted to a silyl carbamate that is readily hydrolyzed.[9]

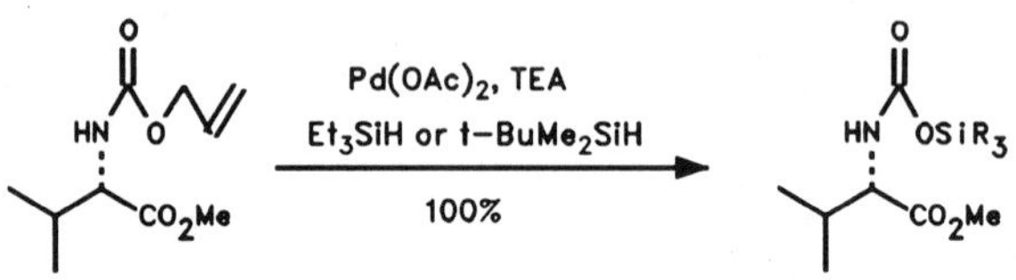

Pd(OAc)$_2$, TEA; Et$_3$SiH or t-BuMe$_2$SiH; 100%

7. $Pd(Ph_3P)_4$, 2-ethylhexanoic acid.[10]

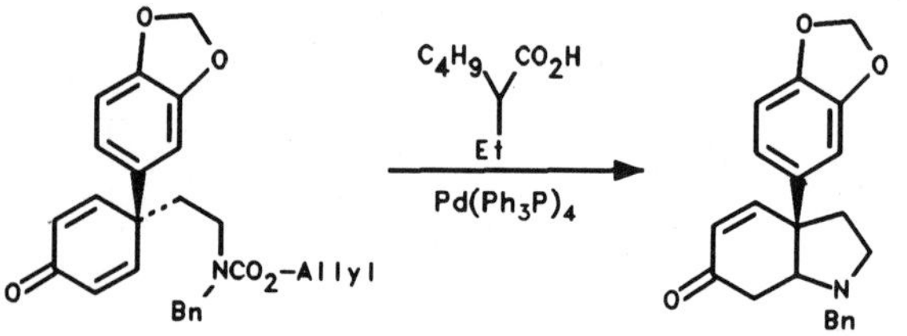

Ref. 11

22. 1-Isopropylallyl Carbamate (Ipaoc–NR$_2$):

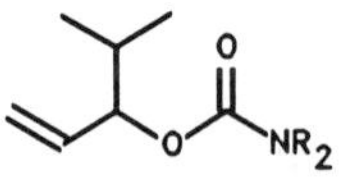

This group was developed to minimize the problem of nitrogen allylation during the deprotection step, because deprotection proceeds with β-hydride elimination. The derivative is stable to TFA and 6 *N* HCl.[12]

Formation/Cleavage

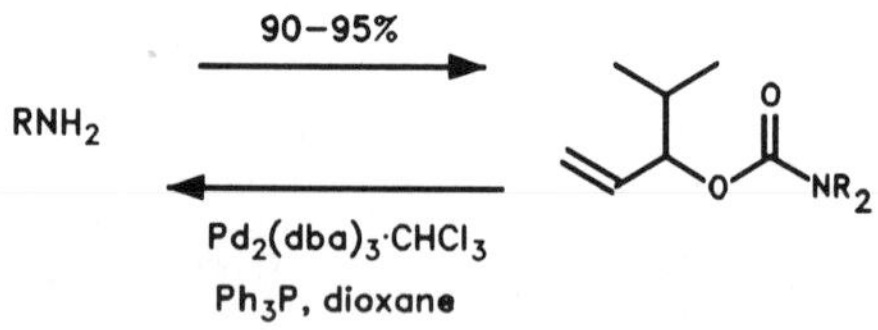

RNH$_2$ → (90–95%); ← Pd$_2$(dba)$_3$·CHCl$_3$, Ph$_3$P, dioxane

23. Cinnamyl Carbamate (Coc–NR$_2$): $PhCH{=}CHCH_2OC(O)NR_2$ (Chart 8)

Formation

1. $PhCH{=}CHCH_2OCO$-O-benzotriazolyl, Et_3N, dioxane or DMF, rt, 16 h, 71–100% yield.[13]

Cleavage

1. $Pd(Ph_3P)_4$, THF, Pyr, HCO_2H, heat, 4 min.[13]
2. $Hg(OAc)_2$, CH_3OH, HNO_3, 23°, 2–4 h, then KSCN, H_2O, 23°, 12–16 h.[14]

24. 4-Nitrocinnamyl Carbamate (Noc–NR_2):
$4\text{-}NO_2C_6H_4CH{=}CHCH_2OC(O)NR_2$

The Noc group, developed for amino acid protection, is introduced with the acid chloride (Et_3N, H_2O, dioxane, 2 h, 20°, 61–95% yield). It is cleaved with $Pd(Ph_3P)_4$ (THF, *N,N*-dimethylbarbituric acid, 8 h, 20°, 80% yield). It is not isomerized by Wilkinson's catalyst, thus allowing selective removal of the allyl ester group.[15]

1. E. J. Corey and J. W. Suggs, *J. Org. Chem.*, **38,** 3223 (1973).
2. G. Sennyey, G. Barcelo, and J.-P. Senet, *Tetrahedron Lett.*, **28,** 5809 (1987).
3. Y. Hayakawa, H. Kato, M. Uchiyama, H. Kajino, and R. Noyori, *J. Org. Chem.*, **51,** 2400 (1986).
4. O. Dangles, F. Guibé, G. Balavoine, S. Lavielle, and A. Marquet, *J. Org. Chem.*, **52,** 4984 (1987).
5. H. Kunz and C. Unverzagt, *Angew. Chem., Int. Ed. Engl.*, **23,** 436 (1984).
6. P. Boullanger and G. Descotes, *Tetrahedron Lett.*, **27,** 2599 (1986).
7. P. Four and F. Guibé, *Tetrahedron Lett.*, **23,** 1825 (1982).
8. I. Minami, Y. Ohashi, I. Shimizu, and J. Tsuji, *Tetrahedron Lett.*, **26,** 2449 (1985).
9. M. Sakaitani, N. Kurokawa, and Y. Ohfune, *Tetrahedron Lett.*, **27,** 3753 (1986).
10. S. F. Martin and C. L. Campbell, *J. Org. Chem.*, **53,** 3184 (1988).
11. P. D. Jeffrey and S. W. McCombie, *J. Org. Chem.*, **47,** 589 (1982).
12. I. Minami, M. Yukara, and J. Tsuji, *Tetrahedron Lett.*, **28,** 2737 (1987).
13. H. Kinoshita, K. Inomata, T. Kameda, and H. Kotake, *Chem. Lett.*, 515 (1985).
14. E. J. Corey and M. A. Tius, *Tetrahedron Lett.*, 2081 (1977).
15. H. Kunz and J. März, *Angew. Chem., Int. Ed. Engl.*, **27,** 1375 (1988).

25. 8-Quinolyl Carbamate (Chart 8):

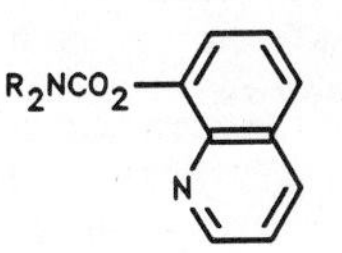

Formation/Cleavage[1]

An 8-quinolyl carbamate is cleaved under neutral conditions by Cu(II)- or Ni(II)-catalyzed hydrolysis.

1. E. J. Corey and R. L. Dawson, *J. Am. Chem. Soc.*, **84,** 4899 (1962).

26. *N*-Hydroxypiperidinyl Carbamate (Chart 8):

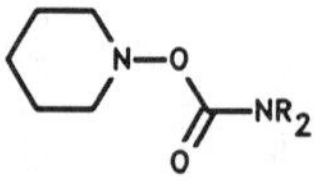

A piperidinyl carbamate, stable to aqueous alkali and to cold acid (30% HBr, 25°. several hours), is best cleaved by reduction.

Formation[1]

1. 1-Piperidinyl-OCOX (X = 2,4,5-trichlorophenyl, . . .), Et_3N, 55–85% yield.

Cleavage[1]

1. H_2, Pd–C, AcOH, 20°, 30 min, 95% yield.
2. Electrolysis, 200 mA, 1 *N* H_2SO_4, 20°, 90 min, 90–93% yield.
3. $Na_2S_2O_4$, AcOH, 20°, 5 min, 93% yield.
4. Zn, AcOH, 20°, 10 min, 94% yield.

1. D. Stevenson and G. T. Young, *J. Chem. Soc. C*, 2389 (1969).

27. Alkyldithio Carbamate: $R_2NCOSSR'$

Alkyldithio carbamates are prepared from the acid chloride (Et_3N, EtOAc, 0°) and amino acid, either free or as the *O*-silyl derivatives (70–88% yield).[1] The *N*-(*i*-propyldithio) carbamate has been used in the protection of proline during peptide synthesis.[2] Alkyldithio carbamates can be cleaved with thiols, NaOH, Ph_3P/TsOH. They are stable to acid. Cleavage rates are a function of the size of the alkyl group as illustrated in the table below.

Relative Rates[a] of Cleavage of Alkyldithio Carbamates

Alkyl Group (R′)	$HSCH_2CH_2OH$	NaOH
CH_3	100	100
Et	33	32
i-Pr	1.4	1.3
t-Bu	0.0002	—
Ph	460	500

[a]The rates were determined using the proline derivative as a substrate.[3]

1. E. Wünsch, L. Moroder, R. Nyfeler, and E. Jäger, *Hoppe-Seyler's Z. Physiol. Chem.*, **363,** 197 (1982).
2. F. Albericio and G. Barany, *Int. J. Pept. Protein Res.*, **30,** 177 (1987).
3. G. Barany, *Int. J. Pept. Protein Res.*, **19,** 321 (1982).

28. Benzyl Carbamate (Cbz- or Z-NR$_2$): PhCH$_2$OC(O)NR$_2$ (Chart 8)

Formation

1. PhCH$_2$OCOCl, Na$_2$CO$_3$, H$_2$O, 0°, 30 min, 72% yield.[1] Alpha-omega diamines can be protected somewhat selectively with this reagent at a pH between 3.5 and 4.5, but the selectivity decreases as the chain length increases [H$_2$N(CH$_2$)$_n$NH$_2$, n = 2, 71% mono; n = 7, 29% mono].[2]
2. (PhCH$_2$OCO)$_2$O, dioxane, H$_2$O, NaOH or Et$_3$N.[3,4] This reagent was reported to give better yields in preparing amino acid derivatives than when PhCH$_2$OCOCl was used. The reagent decomposes at 50°.
3. PhCH$_2$OCO$_2$–C(OMe)=CH$_2$, 90–98% yield.[5]
4. PhCH$_2$OCO$_2$–succinimidyl, >70% yield.[6] This reagent avoids the formation of amino acid dimers. It is a stable, easily handled solid.
5. PhCH$_2$OCO–benzotriazolyl, NaOH, dioxane, rt.[7]
6. PhCH$_2$OCOCN, CH$_2$Cl$_2$, CH$_3$CN or 1,2-dimethoxyethane.[8]
7. PhCH$_2$OCO–imidazolyl, 4-dimethylaminopyridine, 16 h, rt, 76% yield.[9] Two primary amines were protected in the presence of a secondary amine.
8.

ROH, Et$_3$N
dioxane
64–91%
NOCOCl
R'CH(NH$_2$)CO$_2$H
RO–C(O)–NH–CH(R')–CO$_2$H

ROH = t-BuOH, BnOH, FmOH, AdamantylOH, PhC$_6$H$_4$C(Me)$_2$OH, CH$_3$SO$_2$CH$_2$CH$_2$OH

This method is suitable for the preparation of BOC, Fmoc, Adoc, and Bpoc protected amino acids. The acid chloride is a stable, storable solid.[10]

Cleavage

1. H$_2$/Pd-C.[1] If hydrogenation is carried out in the presence of (BOC)$_2$O, the released amine is directly converted to the BOC derivative.[11]
2. H$_2$/Pd-C, NH$_3$, −33°, 3–8 h, quant.[12] When ammonia is used as the solvent, cysteine or methionine units in a peptide do not poison the catalyst.
3. Pd–C or Pd black, hydrogen donor, solvent, 25° or reflux in EtOH, 15 min–2 h, 80–100% yield. Several hydrogen donors, including cyclohex-

ene,[13] 1,4-cyclohexadiene,[14] formic acid,[15,16] *cis*-decalin,[17] and HCO_2NH_4,[18] have been used for catalytic transfer hydrogenation, which is in general a more rapid reaction than catalytic hydrogenation.

4. Et_3SiH, cat. Et_3N, cat. $PdCl_2$, reflux, 3 h, 80% yield.[19] If the reaction is performed in the presence of *t*-$BuMe_2SiH$, the *t*-butyldimethylsilyl carbamate can be isolated because of its greater stability.[20] *S*-Benzyl groups are stable to these conditions, but benzyl esters and benzyl ethers are cleaved.[19] A similar procedure has been published, but in this case the benzyl ether is stable to the cleavage conditions.[11]

BnO, NHZ, CO_2Me, OH → $Pd(OAc)_2$, Et_3SiH, TEA, $(BOC)_2O$, EtOH, rt, 30 h, 91% → BnO, NHBOC, CO_2Me, OH

5. Me_3SiI, CH_3CN, 25°, 6 min, 100% yield.[21,22]
6. Pd-Poly(ethylenimine), HCO_2H.[23] This catalyst system was reported to be better than Pd/C or Pd black for Z removal.
7. $AlCl_3$, $PhOCH_3$, 0–25°, 5 h, 73% yield.[24] These conditions are compatible with β-lactams.
8. BBr_3, CH_2Cl_2, −10°, 1 h → 25°, 2 h, 80–100% yield.[25] Benzyl carbamates of larger peptides can be cleaved by boron tribromide in trifluoroacetic acid, since the peptides are more soluble in acid than in methylene chloride.[26]
9. 253.7 nm, *hν*, 55°, 4 h, CH_3OH, H_2O, 70% yield.[27,28]
10. Electrolysis: −2.9 V, DMF, $R_4N^+X^-$, 70–80% yield[29] or Pd/graphite cathode, MeOH, AcOH, 2.5% $NaClO_4$ (0.5 mol/L), 99% yield.[30] Benzyl ethers and tosylates are stable to these conditions, but benzyl esters are cleaved.
11. Benzyl carbamates of pyrrole-type nitrogens can be cleaved with nucleophilic reagents such as hydrazine; hydrogenation and HF treatment are also effective.[31]
12. Benzyl carbamates are readily cleaved under strongly acidic conditions: HBr, AcOH;[32] 50% CF_3COOH (25°, 14 days, partially cleaved);[33] 70% HF, pyridine;[34] CF_3SO_3H;[35] FSO_3H,[36] or CH_3SO_3H.[36] In cleaving benzyl carbamates from peptides, 0.5 *M* 4-(methylmercapto)phenol in CF_3CO_2H has been recommended to suppress Bn^+ additions to aromatic amino acids.[37] To achieve deprotection via an S_N2 mechanism that also reduces the problem of Bn^+ addition, HF–Me_2S–*p*-cresol (25 : 65 : 10, v/v) has been recommended for peptide deprotection.[38]
13. Na/NH_3.[39]
14. 0.15 *M* $Ba(OH)_2$, heat, 40 h, 3 : 2 glyme/H_2O, 75% yield.[40]

Bu / NZ / OH → 0.15 M $Ba(OH)_2$, heat, 40 h, glyme, H_2O, 75% → Bu / NH / OH

In this case the following reagents failed to afford clean deprotection because of destruction of the acetylene: Me_3SiI, BBr_3, Me_2BBr, $BF_3/EtSH$, $AlCl_3/EtSH$, MeLi/LiBr, KOH/EtOH.

15. $BF_3 \cdot Et_2O$, CH_3SCH_3, CH_2Cl_2, 92% yield.[41]
16. 40% KOH, MeOH, H_2O, 85–94% yield.[42]
17. $K_3[Co(CN)_5]$, H_2, MeOH, 20°, 3 h.[43] Benzyl ethers are not cleaved under these conditions.
18. $LiBH_4$ or $NaBH_4$, Me_3SiCl, THF, 24 h, 88–95%.[44] This combination of reagents also reduces all functional groups that can normally be reduced with diborane.
19. Catecholborane halide cleaves benzyl carbamates in the presence of ethyl and benzyl esters and TBDMS ethers.[45]

1. M. Bergmann and L. Zervas, *Ber.*, **65,** 1192 (1932).
2. G. J. Atwell and W. A. Denny, *Synthesis*, 1032 (1984).
3. W. Graf, O. Keller, W. Keller, G. Wersin, and E. Wuensch, *Peptides 1986: Proceedings of the 19th European Peptide Symposium, Porto Carras, Chalkidiki, Greece, August 31-September 5, 1986*, D. Theodoropoulos, Ed., de Gruyter, New York, 1987, p. 73.
4. G. Sennyey, G. Barcelo, and J.-P. Senet, *Tetrahedron Lett.*, **27,** 5375 (1986).
5. Y. Kita, J. Haruta, H. Yasuda, K. Fukunaga, Y. Shirouchi, and Y. Tamura, *J. Org. Chem.*, **47,** 2697 (1982).
6. A. Paquet, *Can. J. Chem.*, **60,** 976 (1982).
7. E. Wuensch, W. Graf, O. Keller, W. Keller, and G. Wersin, *Synthesis*, 958 (1986).
8. S. Murahashi, T. Naota, and N. Nakajima, *Chem. Lett.*, 879 (1987).
9. S. K. Sharma, M. J. Miller, and S. M. Payne, *J. Med. Chem.*, **32,** 357 (1989).
10. P. Henklein, H.-U. Heyne, W.-R. Halatsch, and H. Niedrich, *Synthesis*, 166 (1987).
11. M. Sakaitani, K. Hori, and Y. Ohfune, *Tetrahedron Lett.*, **29,** 2983 (1988).
12. J. Meienhofer and K. Kuromizu, *Tetrahedron Lett.*, 3259 (1974).
13. A. E. Jackson and R. A. W. Johnstone, *Synthesis*, 685 (1976); G. M. Anantharamaiah and K. M. Sivanandaiah, *J. Chem. Soc., Perkin Trans. I*, 490 (1977).
14. A. M. Felix, E. P. Heimer, T. J. Lambros, C. Tzougraki, and J. Meienhofer, *J. Org. Chem.*, **43,** 4194 (1978).
15. K. M. Sivanandaiah and S. Gurusiddappa, *J. Chem. Res. Synop.*, 108 (1979); B. ElAmin, G. M. Anantharamaiah, G. P. Royer, and G. E. Means, *J. Org. Chem.*, **44,** 3442 (1979).
16. M. J. O. Anteunis, C. Becu, F. Becu, and M. F. Reyniers, *Bull. Soc. Chim. Belg.*, **96,** 775 (1987).

17. Y. Okada and N. Ohta, *Chem. Pharm. Bull.*, **30,** 581 (1982).
18. M. Makowski, B. Rzeszotarska, L. Smelka, and Z. Kubica, *Liebigs Ann. Chem.*, 1457 (1985).
19. L. Birkofer, E. Bierwirth, and A. Ritter, *Chem. Ber.*, **94,** 821 (1961).
20. M. Sakaitani, N. Kurokawa, and Y. Ohfune, *Tetrahedron Lett.*, **27,** 3753 (1986).
21. R. S. Lott, V. S. Chauhan, and C. H. Stammer, *J. Chem. Soc., Chem. Commun.*, 495 (1979).
22. M. Ihara, N. Taniguchi, K. Nogochi, K. Fujumoto, and T. Kametani, *J. Chem. Soc., Perkin Trans. I*, 1277 (1988).
23. D. R. Coleman and C. P. Royer, *J. Org. Chem.*, **45,** 2268 (1980).
24. T. Tsuji, T. Kataoka, M. Yoshioka, Y. Sendo, Y. Nishitani, S. Hirai, T. Maeda, and W. Nagata, *Tetrahedron Lett.*, 2793 (1979).
25. A. M. Felix, *J. Org. Chem.*, **39,** 1427 (1974).
26. J. Pless and W. Bauer, *Angew Chem., Int. Ed. Engl.*, **12,** 147 (1973).
27. S. Hanessian and R. Masse, *Carbohydr. Res.*, **54,** 142 (1977).
28. For a review of photochemically labile protective groups, see: V. N. R. Pillai, *Synthesis*, 1 (1980).
29. V. G. Mairanovsky, *Angew Chem., Int. Ed. Engl.*, **15,** 281 (1976).
30. M. A. Casadei and D. Pletcher, *Synthesis*, 1118 (1987).
31. M. Chorev and Y. S. Klausner, *J. Chem. Soc., Chem. Commun.*, 596 (1976).
32. D. Ben-Ishai and A. Berger, *J. Org. Chem.*, **17,** 1564 (1952).
33. A. R. Mitchell and R. B. Merrifield, *J. Org. Chem.*, **41,** 2015 (1976).
34. S. Matsuura, C.-H. Niu, and J. S. Cohen, *J. Chem. Soc., Chem. Commun.*, 451 (1976).
35. H. Yajima, N. Fujii, H. Ogawa, and H. Kawatani, *J. Chem. Soc., Chem. Commun.*, 107 (1974).
36. H. Yajima, H. Ogawa, and H. Sakurai, *J. Chem. Soc., Chem. Commun.*, 909 (1977).
37. M. Bodanszky and A. Bodanszky, *Int. J. Pept. Protein Res.*, **23,** 287 (1984).
38. J. P. Tam, W. F. Heath, and R. B. Merrifield, *J. Am. Chem. Soc.*, **105,** 6442 (1983).
39. I. Schon, T. Szirtes, T. Uberhardt, A. Rill, A. Csehi, and B. Hegedus, *Int. J. Pept. Protein Res.*, **22,** 92 (1983).
40. L. E. Overman and M. J. Sharp, *Tetrahedron Lett.*, **29,** 901 (1988).
41. I. H. Sanchez, F. J. López, J. J. Soria, M. I. Larraza, and H. J. Flores, *J. Am. Chem. Soc.*, **105,** 7640 (1983).
42. S. R. Angle and D. O. Arnaiz, *Tetrahedron Lett.*, **30,** 515 (1989).
43. G. Losse and H. R. Stiehl, *Z. Chem.*, **21,** 188 (1981).
44. A. Giannis and K. Sandhoff, *Angew. Chem., Int. Ed. Engl.*, **28,** 218 (1989).
45. R. K. Boeckman, Jr., and J. C. Potenza, *Tetrahedron Lett.*, **26,** 1411 (1985).

29. *p*-Methoxybenzyl Carbamate (Moz–NR_2): *p*-$MeOC_6H_4CH_2OC(O)NR_2$

Formation

1. Moz–ON=C(CN)Ph, H_2O, Et_3N, rt, 6 h, 90% yield.[1]

Cleavage

The Moz group is more readily cleaved by acid than is the benzyloxycarbonyl or BOC group.[2,3] The section on benzyl carbamates should be consulted since many of the methods for formation and cleavage should be applicable to the Moz group as well.

1. TsOH, CH_3CN, acetone, rt.[4,5]

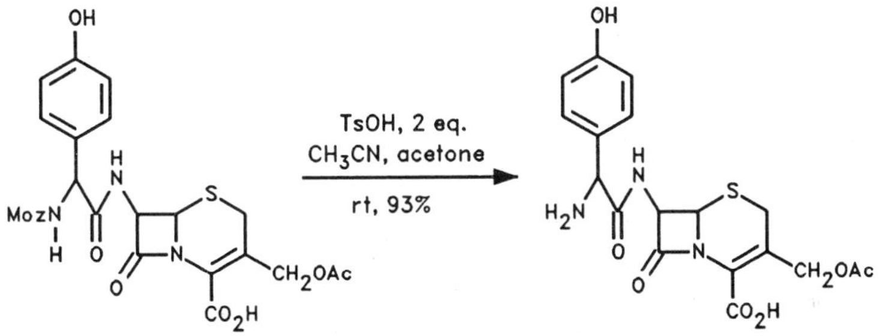

TsOH, 2 eq.
CH_3CN, acetone
rt, 93%

Ref. 5

2. 10% CF_3COOH, CH_2Cl_2, 100% yield.[1,2]

1. S.-T. Chen, S.-H. Wu, and K.-T. Wang, *Synthesis*, 36 (1989).
2. F. Weygand and K. Hunger, *Chem. Ber.*, **95,** 1 (1962).
3. S. S. Wang, S. T. Chen, K. T. Wang, and R. B. Merrifield, *Int. J. Pept. Protein Res.*, **30,** 662 (1987).
4. H. Yajima, H. Ogawa, N. Fujii, and S. Funakoshi, *Chem. Pharm. Bull.*, **25,** 740 (1977).
5. H. Yamada, H. Tobiki, N. Tanno, H. Suzuki, K. Jimpo, S. Ueda, and T. Nakagome, *Bull. Chem. Soc. Jpn.*, **57,** 3333 (1984).

30. *p*-Nitrobenzyl Carbamate: *p*-$NO_2C_6H_4CH_2OC(O)NR_2$ (Chart 8)

Formation[1]

1. *p*-$NO_2C_6H_4CH_2OCOCl$, base, 0°, 1.5 h, 78% yield.

Cleavage

1. H_2/Pd–C, 10 h, 87% yield.[1] A nitrobenzyl carbamate is more readily cleaved by hydrogenolysis than a benzyl carbamate; it is more stable to acid-catalyzed hydrolysis than is a benzyl carbamate, and therefore selective cleavage is possible.
2. 4 *N* HBr, AcOH, 60°, 2 h, 68% yield.[1]
3. $Na_2S_2O_4$, NaOH.[2]
4. Electrolysis, −1.2 V, DMF, $R_4N^+X^-$.[3]

1. J. E. Shields and F. H. Carpenter, *J. Am. Chem. Soc.*, **83,** 3066 (1961); S. Hashiguchi, H. Natsugari, and M. Ochiai, *J. Chem. Soc., Perkin Trans. I*, 2345 (1988).
2. P. J. Romanovskis, P. Henklein, J. A. Benders, I. V. Siskov, and G. I. Chipens, in *5th All-Union Symposium on Protein and Peptide Chemistry and Physics, Abstracts*, Baku (Soviet Azerbaijan), 1980, p. 229 (in Russian); P. J. Romanovskis, I. V. Siskov, I. K. Liepkaula, E. A. Porunkevich, M. P. Ratkevich, A. A. Skujins, and G. I. Chipens, "Linear and Cyclic Analogs of ACTH Fragments: Synthesis and Biological Activity," in *Peptides: Synthesis, Structure, Function: Proceedings of the Seventh American Peptide Symposium*, University of Wisconsin, Madison, 1981, D. H. Rich and E. Gross, Eds., Pierce Chemical Co., Rockford, IL, 1981, pp. 229–232.
3. V. G. Mairanovsky, *Angew. Chem., Int. Ed. Engl.*, **15,** 281 (1976); H. L. S. Maia, M. J. Medeiros, M. I. Montenegro, and D. Pletcher, *Port. Electrochim. Acta*, **5,** 187 (1987); *Chem. Abstr.*, **109,** 118114n (1989).

Benzyl carbamates substituted with one or more halogens are much more stable to acidic hydrolysis than are the unsubstituted benzyl carbamates.[1,2] For example, the 2,4-dichlorobenzyl carbamate is 80 times more stable to acid than is the simple benzyl derivative.[3] Halobenzyl carbamates can also be cleaved by hydrogenolysis with Pd–C.[3] The following halobenzyl carbamates have been found to be useful when increased acid stability is required.

31. *p*-Bromobenzyl Carbamate[4]

32. *p*-Chlorobenzyl Carbamate[1,2]

33. 2,4-Dichlorobenzyl Carbamate[3] (Chart 8)

1. K. Noda, S. Terada, and N. Izamiya, *Bull. Chem. Soc. Jpn.*, **43,** 1883 (1970).
2. B. W. Erickson and R. B. Merrifield, *J. Am. Chem. Soc.*, **95,** 3757 (1973).
3. Y. S. Klausner and M. Chorev, *J. Chem. Soc., Perkin Trans. I*, 627 (1977).
4. D. M. Channing, P. B. Turner, and G. T. Young, *Nature*, **167,** 487 (1951).

34. 4-Methylsulfinylbenzyl Carbamate (Msz–NR_2): $CH_3S(O)C_6H_4CH_2OCONR_2$

The Msz group is stable to TFA/anisole, NaOH, and hydrazine.

Formation

1. Msz–O-succinimidyl, CH_3CN, H_2O, Et_3N, 45% yield.[1]

Cleavage

1. $SiCl_4$, TFA, anisole.[1] $SiCl_4$ serves to reduce the sulfoxide prior to acid-catalyzed cleavage. Other sulfoxide reducing agents could probably be used.

1. Y. Kiso, T. Kimura, M. Yoshida, M. Shimokura, K. Akaji, and T. Mimoto, *J. Chem. Soc., Chem. Commun.*, 1511 (1989).

35. 9-Anthrylmethyl Carbamate: (Chart 8):

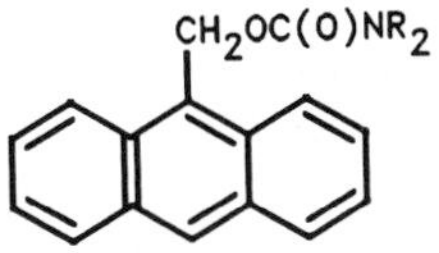

$CH_2OC(O)NR_2$

Formation

1. 9-Anthryl-$CH_2OCO_2C_6H_4$-*p*-NO_2, DMF, 25°, 86% yield.[1]

Cleavage[1]

1. CH_3SNa, DMF, −20°, 1–7 h, 77–91% yield or 25°, 4 min, 86% yield.
2. CF_3COOH, CH_2Cl_2, 0°, 5 min, 88–92% yield. The anthrylmethyl carbamate is stable to 0.01 *N* lithium hydroxide (25°, 6 h), to 0.1 *N* sulfuric acid (25°, 1 h), and to 1 *M* trifluoroacetic acid (25°, 1 h, dioxane).

1. N. Kornblum and A. Scott, *J. Org. Chem.*, **42,** 399 (1977).

36. Diphenylmethyl Carbamate: $Ph_2CHOC(O)NR_2$ (Chart 8)

The diphenylmethyl carbamate, prepared from the azidoformate, is readily cleaved by mild acid hydrolysis (1.7 *N* HCl, THF, 65°, 10 min, 100% yield).[1]

1. R. G. Hiskey and J. B. Adams, *J. Am. Chem. Soc.*, **87,** 3969 (1965).

Assisted Cleavage

Several protective groups have been prepared that rely on a β-elimination to effect cleavage. Often the protective group must first be activated to increase the acidity of the β-hydrogen. In general the derivatives are prepared by standard procedures, from either the chloroformate or mixed carbonate.

37. 2-Methylthioethyl Carbamate: $MeSCH_2CH_2OC(O)NR_2$

A 2-methylthioethyl carbamate is cleaved by 0.01 *N* NaOH after alkylation to $Me_2S^+CH_2CH_2$— or by 0.1 *N* NaOH after oxidation to the sulfone.[1]

38. 2-Methylsulfonylethyl Carbamate: $MeSO_2CH_2CH_2OC(O)NR_2$

This is the oxidized form of the methylthio derivative above. It is stable to catalytic hydrogenolysis and does not poison the catalyst. It is stable to liquid HF (30 min), but is cleaved in 5 s with 1 *N* NaOH.[2]

39. 2-(*p*-Toluenesulfonyl)ethyl Carbamate:
$4\text{-}CH_3\text{-}C_6H_4SO_2CH_2CH_2OC(O)NR_2$

This derivative is similar to the methylsulfonylethyl derivative. It is cleaved by 1 *M* NaOH, <1 h.[3] The related 4-chlorobenzenesulfonylethyl carbamate has also been used as a protective group that can be cleaved with DBU or tetramethylguanidine.[4]

40. [2-(1,3-Dithianyl)]methyl Carbamate (Dmoc-NR$_2$): S S $CH_2OC(O)NR_2$

Cleavage occurs by prior activation with peracetic acid to the bissulfone followed by mild base treatment.[5]

41. 4-Methylthiophenyl Carbamate (Mtpc-NR$_2$): $4\text{-}MeSC_6H_4OC(O)NR_2$[6]

42. 2,4-Dimethylthiophenyl Carbamate (Bmpc-NR$_2$):
$2,4\text{-}(MeS)_2C_6H_3OC(O)NR_2$[7]

After activation with peracetic acid and base treatment these derivatives form the isocyanate, which can be trapped with water to effect hydrolysis or with an alcohol to form other carbamates.

43. 2-Phosphonioethyl Carbamate (Peoc-NR$_2$): $R_3P^+CH_2CH_2OC(O)NR'_2\ X^-$

This derivative is stable to trifluoroacetic acid; it is cleaved by mild bases (pH 8.4; 0.1 *N* NaOH, 1 min, 100% yield).[8]

44. 2-Triphenylphosphonioisopropyl Carbamate (Ppoc-NR$_2$):
$Ph_3P^+CH_2CH(CH_3)OC(O)NR_2\ X^-$

This derivative is similar to the Peoc group except that it is 4 times more stable to base and is not as susceptible to side reactions as is the Peoc group.[9]

45. 1,1-Dimethyl-2-cyanoethyl Carbamate: $(CN)CH_2C(CH_3)_2OC(O)NR_2$ (Chart 8)

This derivative is stable to trifluoroacetic acid and is cleaved by aqueous K_2CO_3 or Et_3N, 25°, 6 h, 90% yield.[10]

1. H. Kunz, *Chem. Ber.*, **109,** 3693 (1976).
2. G. I. Tesser and I. C. Balver-Geers, *Int. J. Pept. Protein Res.*, **7,** 295 (1975).
3. A. T. Kader and C. J. M. Sterling, *J. Chem. Soc.*, 258 (1964).
4. V. V. Samukov, A. N. Sabirov, and M. L. Troshkov, *Zh. Obshch. Khim.*, **58,** 1432 (1988); *Chem. Abstr.*, **110,** 76008u (1989).
5. H. Kunz and R. Barthels, *Chem. Ber.*, **115,** 833 (1982); R. Barthels and H. Kunz, *Angew Chem., Int. Ed. Engl.*, **21,** 292 (1982).
6. H. Kunz and K. Lorenz, *Angew Chem., Int. Ed. Engl.*, **19,** 932 (1980).
7. H. Kunz and H.-J. Lasowski, *Angew. Chem., Int. Ed. Engl.*, **25,** 170 (1986).
8. H. Kunz, *Angew. Chem., Int. Ed. Engl.*, **17,** 67 (1978).
9. H. Kunz and G. Schaumloeffel, *Liebigs Ann. Chem.*, 1784 (1985).
10. E. Wünsch and R. Spangenberg, *Chem. Ber.*, **104,** 2427 (1971).

A series of carbamates have been prepared that are cleaved by liberation of a phenol, which, when treated with base, cleaves the carbamate by quinone methide formation through a 1,6-elimination.[1]

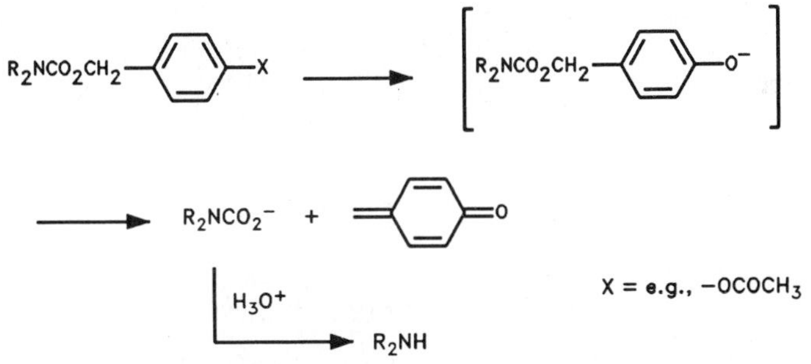

$R_2NCO_2CH_2$–C_6H_4–X → [$R_2NCO_2CH_2$–C_6H_4–O^-] → $R_2NCO_2^-$ + (quinone methide)

$R_2NCO_2^-$ —H_3O^+→ R_2NH

X = e.g., $-OCOCH_3$

46. *m*-Chloro-*p*-acyloxybenzyl Carbamate:[2,3]

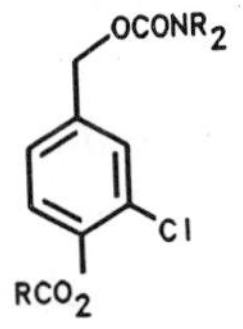

$OCONR_2$, Cl, RCO_2

Cleavage

1. $NaHCO_3/Na_2CO_3$ or H_2O_2/NH_3, $NaHSO_3$, 1 h.
2. 0.1 *N* NaOH, 10 min, 100% yield.

3. H_2/Pd–C.
4. HBr, AcOH.

47. *p*-(Dihydroxyboryl)benzyl Carbamate (Dobz–NR_2):

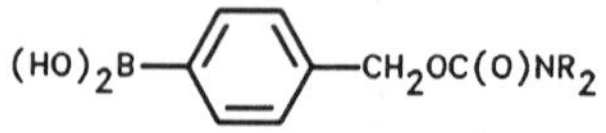

$(HO)_2B-C_6H_4-CH_2OC(O)NR_2$

Formation[4]

1. 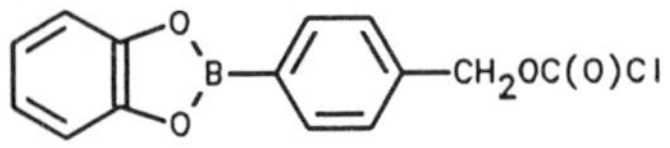 [catechol boronate]$-C_6H_4-CH_2OC(O)Cl$ aq. base; aq. acid

Cleavage[4]

1. H_2O_2, pH 9.5, 25°, 5 min, 90% yield.
2. H_2, Pd–C.
3. HBr, AcOH.

48. 5-Benzisoxazolylmethyl Carbamate (Bic–NR_2) (Chart 8):

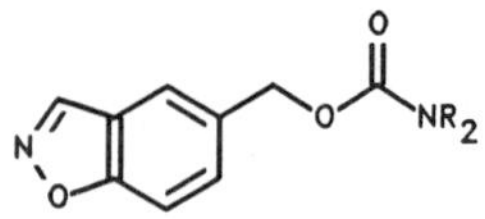

Formation

1. $ClCO_2CH_2$-5-benzisoxazole, pH 8.5–9.0, CH_3CN, 0°, 1 h, 63% yield.[5]

Cleavage[5]

1. Et_3N, CH_3CN or DMF, 25°, 30 min; Na_2SO_3, EtOH, H_2O, 40°, 3 h, pH 7, 92% yield or CF_3COOH, 90 min, 95% yield.
2. H_2, Pd–C.
3. HBr, AcOH. This derivative is stable to trifluoroacetic acid.

49. 2-(Trifluoromethyl)-6-chromonylmethyl Carbamate (Tcroc–NR_2):[6,7]

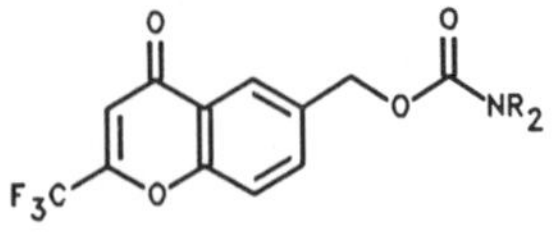

Cleavage

1. $PrNH_2$ or hydrazine.

The Tcroc group resists cleavage by CF_3COOH.

1. M. Wakselman, *Nouv. J. Chim.*, **7,** 439 (1983).
2. M. Wakselman and E. G.-Jampel, *J. Chem. Soc., Chem. Commun.*, 593 (1973).
3. G. Le Corre, E. G.-Jampel, and M. Wakselman, *Tetrahedron*, **34,** 3105 (1978).
4. D. S. Kemp and D. C. Roberts, *Tetrahedron Lett.*, 4629 (1975).
5. D. S. Kemp and C. F. Hoyng, *Tetrahedron Lett.*, 4625 (1975).
6. D. S. Kemp and G. Hanson, *J. Org. Chem.*, **46,** 4971 (1981).
7. D. S. Kemp, D. R. Bolin, and M. E. Parham, *Tetrahedron Lett.*, 4575 (1981).

Photolytic Cleavage

The following carbamates can be cleaved by photolysis.[1] They can be prepared from either the chloroformate or the mixed carbonate.

50. *m*-Nitrophenyl Carbamate[2]

51. 3,5-Dimethoxybenzyl Carbamate[3]

52. *o*-Nitrobenzyl Carbamate[4]

53. 3,4-Dimethoxy-6-nitrobenzyl Carbamate[4] (Chart 8)

54. Phenyl(*o*-nitrophenyl)methyl Carbamate[5]

1. For a review of photochemically labile protective groups, see: V. N. R. Pillai, *Synthesis*, 1 (1980).
2. Th. Wieland, Ch. Lamperstorfer, and Ch. Birr, *Makromol. Chem.*, **92,** 277 (1966).
3. J. W. Chamberlin, *J. Org. Chem.*, **31,** 1658 (1966).
4. B. Amit, U. Zehavi, and A. Patchornik, *J. Org. Chem.*, **39,** 192 (1974).
5. J. A. Baltrop, P. J. Plant, and P. Schofield, *J. Chem. Soc., Chem. Commun.*, 822 (1966).

Urea-Type Derivatives

55. Phenothiazinyl-(10)-carbonyl Derivative:

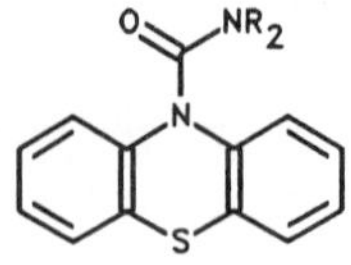

This derivative is prepared in 51–82% yield and cleaved with $Ba(OH)_2$ or NaOH in 52–96% yield after oxidation of the sulfur with hydrogen peroxide. It is stable to CF_3COOH and NaOH.[1]

1. J. Gante, W. Hechler, and R. Weitzel in *Peptides 1986: Proceedings of the 19th European Peptide Symposium, Porto Carras, Chalkidiki, Greece, August 31–September 5, 1986*, D. Theodoropoulos, Ed., de Gruyter, Berlin, 1987, pp. 87–90.

56. *N'-p*-Toluenesulfonylaminocarbonyl Derivative:
$R_2NCONHSO_2C_6H_4$-*p*-CH_3

This sulfonyl urea, prepared from an amino acid and *p*-tosyl isocyanate in 20–80% yield, is cleaved by alcohols (95% aq. EtOH, *n*-PrOH, or *n*-BuOH, 100°, 1 h, 95% yield). It is stable to dilute base, to acids (HBr/AcOH or cold CF_3CO_2H), and to hydrazine.[1]

1. B. Weinstein, T. N.-S. Ho, R. T. Fukura, and E. C. Angell, *Synth. Commun.*, **6,** 17 (1976).

57. *N'*-Phenylaminothiocarbonyl Derivative: $R_2NCSNHC_6H_5$ (Chart 8)

This thiourea, prepared from an amino acid and phenyl isothiocyanate,[1] is cleaved by anhydrous trifluoroacetic acid (an N–$COCF_3$ group is stable),[2] and by oxidation (*m*-$ClC_6H_4CO_3H$, 0°, 1.5 h, 73% yield; H_2O_2/AcOH, 80°, 80 min, 44% yield).[3]

1. P. Edman, *Acta Chem. Scand.*, **4,** 277 (1950).
2. F. Borrás and R. E. Offord, *Nature*, **227,** 716 (1970).
3. J. Kollonitsch, A. Hajós and V. Gábor, *Chem Ber.*, **89,** 2288 (1956).

Miscellaneous Carbamates

The following carbamates have seen little use since the preparation of the first edition of this book; they are listed here for completeness. For the most part they are variations of the BOC and benzyl carbamates, with the exception of the azo derivatives, which are highly colored. The differences between them are largely in the strength of the acid required for their cleavage.

58. *t*-Amyl Carbamate[1]

59. *S*-Benzyl Thiocarbamate[2] (Chart 8)

60. *p*-Cyanobenzyl Carbamate[3]

61. Cyclobutyl Carbamate[4] (Chart 8). The half-life for cleavage in neat CF_3CO_2H is > 300 min.

62. Cyclohexyl Carbamate[5]

63. Cyclopentyl Carbamate[5]

64. Cyclopropylmethyl Carbamate[4]

65. *p*-Decyloxybenzyl Carbamate[6]

66. Diisopropylmethyl Carbamate[5]

67. 2,2-Dimethoxycarbonylvinyl Carbamate[7]

68. *o*-(*N*,*N*-Dimethylcarboxamido)benzyl Carbamate[8]

69. 1,1-Dimethyl-3-(*N*,*N*-dimethylcarboxamido)propyl Carbamate[8]

70. 1,1-Dimethylpropynyl Carbamate[9] (Chart 8)

71. Di(2-pyridyl)methyl Carbamate[8]

72. 2-Furanylmethyl Carbamate[10]

73. 2-Iodoethyl Carbamate[11]

74. Isobornyl Carbamate[12]

75. Isobutyl Carbamate[13]

76. Isonicotinyl Carbamate[14] (Chart 8)

77. *p*-(*p*′-Methoxyphenylazo)benzyl Carbamate[15]

78. 1-Methylcyclobutyl Carbamate[4] (Chart 8)

79. 1-Methylcyclohexyl Carbamate[4] (Chart 8). The half-life for cleavage in neat CF_3CO_2H is 2 min and 180 min in formic acid.

80. 1-Methyl-1-cyclopropylmethyl Carbamate[5]

81. 1-Methyl-1-(3,5-dimethoxyphenyl)ethyl Carbamate[16]

82. 1-Methyl-1-(*p*-phenylazophenyl)ethyl Carbamate[17]

83. 1-Methyl-1-phenylethyl Carbamate[18] (Chart 8)

84. 1-Methyl-1-(4-pyridyl)ethyl Carbamate[8]

85. Phenyl Carbamate[19]

86. *p*-(Phenylazo)benzyl Carbamate[15]

87. 2,4,6-Tri-*t*-butylphenyl Carbamate[20]

88. 4-(Trimethylammonium)benzyl Carbamate[21]

89. 2,4,6-Trimethylbenzyl Carbamate[22]

1. S. Sakakibara, I. Honda, K. Takada, M. Miyoshi, T. Ohnishi, and K. Okumura, *Bull. Chem. Soc. Jpn.*, **42,** 809 (1969); S. Matsuura, C.-H. Niu, and J. S. Cohen, *J. Chem. Soc., Chem. Commun.*, 451 (1976).
2. H. B. Milne, S. L. Razniak, R. P. Bayer, and D. W. Fish, *J. Am. Chem. Soc.*, **82,** 4582 (1960).
3. K. Noda, S. Terada, and N. Izumiya, *Bull. Chem. Soc. Jpn.*, **43,** 1883 (1970).
4. S. F. Brady, R. Hirschmann, and D. F. Veber, *J. Org. Chem.*, **42,** 143 (1973).
5. F. C. McKay and N. F. Albertson, *J. Am. Chem. Soc.*, **79,** 4686 (1957).
6. H. Brechbühler, H. Büchi, E. Hatz, J. Schreiber, and A. Eschenmoser, *Helv. Chim. Acta*, **48,** 1746 (1965).
7. A. Gomez-Sanchez, P. B. Moya, and J. Bellanato, *Carbohydr. Res.*, **135,** 101 (1984).
8. S. Coyle, O. Keller, and G. T. Young, *J. Chem. Soc., Perkin Trans. I*, 1459 (1979).
9. G. L. Southard, B. R. Zaborowsky, and J. M. Pettee, *J. Am. Chem. Soc.*, **93,** 3302 (1971).
10. G. Losse and K. Neubert, *Tetrahedron Lett.*, 1267 (1970).
11. J. Grimshaw, *J. Chem. Soc.*, 7136 (1965).
12. M. Fujino, S. Shinagawa, O. Nishimura, and T. Fukuda, *Chem. Pharm. Bull.*, **20,** 1017 (1972).
13. R. L. Letsinger and P. S. Miller, *J. Am. Chem. Soc.*, **91,** 3356 (1969).
14. D. F. Veber, W. J. Paleveda, Y. C. Lee, and R. Hirschmann, *J. Org. Chem.*, **42,** 3286 (1977).
15. R. Schwyzer, P. Sieber, and K. Zatsko, *Helv. Chim. Acta*, **41,** 491 (1958).
16. C. Birr, W. Lochinger, G. Stahnke, and P. Lang, *Justus Liebigs Ann. Chem.*, **763,** 162 (1972).
17. A. T.-Kyi and R. Schwyzer, *Helv. Chim. Acta*, **59,** 1642 (1976).
18. B. E. B. Sandberg and U. Ragnarsson, *Int. J. Pept. Protein Res.*, **6,** 111 (1974); H. Franzen and U. Ragnarsson, *Acta Chem. Scand. B*, **33,** 690 (1979).
19. J. D. Hobson and J. G. McCluskey, *J. Chem. Soc. C*, 2015 (1967); R. W. Adamiak and J. Stawinski, *Tetrahedron Lett.*, 1935 (1977).
20. D. Seebach and T. Hassel, *Angew. Chem., Int. Ed. Engl.*, **17,** 274 (1978).
21. Y. Zhang, X. Wang, L. Li, and P. Zhang, *Sci. Sin. Ser. B*, **29,** 1009 (1986); *Chem. Abstr.*, **108,** 6354p (1988).
22. Y. Isowa, M. Ohmori, M. Sato, and K. Mori, *Bull. Chem. Soc. Jpn.*, **50,** 2766 (1977).

AMIDES

Simple amides are generally prepared from the acid chloride or the anhydride. They are exceptionally stable to acidic or basic hydrolysis, and are classically hydrolyzed by brute force by heating in strongly acidic or basic solutions. Among simple amides, hydrolytic stability increases from formyl to acetyl to benzoyl. Lability of the haloacetyl derivatives to mild acid hydrolysis increases with substitution: acetyl < chloroacetyl < dichloroacetyl < trichloroacetyl < trifluoroacetyl.[1] It should be noted that amide hydrolysis under acidic or basic[2] conditions is *greatly* facilitated in the presence of a neighboring hydroxyl group that can participate in the hydrolysis.[3] Although a number of imaginative amide-derived protective groups have been developed, most are not commonly used because they contain other reactive functionality, are not commercially available, or because other more easily introduced and cleaved groups such as the BOC, Alloc, and Cbz groups serve adequately for amine protection. Amide derivatives of the nucleotides are not discussed in this section since their behavior is atypical of amides. They are generally more easily hydrolyzed than is the typical amide. Several review articles discuss amides as —NH protective groups.[4–7]

1. R. S. Goody and R. T. Walker, *Tetrahedron Lett.*, 289 (1967).
2. B. F. Cain, *J. Org. Chem.*, **41,** 2029 (1976).
3. See, for example: C. K. Lai, R. S. Buckanin, S. J. Chen, D. F. Zimmerman, F. T. Sher, and G. A. Berchtold, *J. Org. Chem.*, **47,** 2364 (1982); E. R. Koft, P. Dorff, and R. Kullnig, *J. Org. Chem.*, **54,** 2936 (1989).
4. E. Wünsch, "Blockierung und Schutz der α-Amino-Function," in *Methoden der Organischen Chemie (Houben-Weyl)*, Georg Thieme Verlag, Stuttgart, 1974, Vol. 15/1, pp. 164–203, 250–264.
5. J. W. Barton, "Protection of N—H Bonds and NR_3," in *Protective Groups in Organic Chemistry*, J. F. W. McOmie, Ed., Plenum Press, New York and London, 1973, pp. 46–56.
6. Y. Wolman, "Protection of the Amino Group," in *The Chemistry of the Amino Group*, S. Patai, Ed., Wiley-Interscience, New York, 1968, Vol. 4, pp. 669–682.
7. *The Peptides; Analysis, Synthesis, Biology*, Vol. 3: *Protection of Functional Groups in Peptide Synthesis*, E. Gross and J. Meienhofer, Eds., Academic Press, New York, 1981.

1. **Formamide:** R_2NCHO (Chart 9)

Formation

1. 98% HCO_2H, Ac_2O, 25°, 1 h, 78–90% yield.[1,2]
2. HCO_2H, DCC, Pyr, 0°, 4 h, 87–90% yield.[3] These conditions produce *N*-formyl derivatives of *t*-butyl amino acid esters with a minimum of racemization.

3. HCO_2H, $EtN{=}C{=}N(CH_2)_3NMe_2 \cdot HCl$, 0°, 15 min; then *N*-methylmorpholine, 5°, 20 h, 65–96% yield. This method can be used with amine hydrochlorides.[4]
4. C_6F_5OCHO, $CHCl_3$, rt, 5–30 min, 85–99% yield.[5]
5. 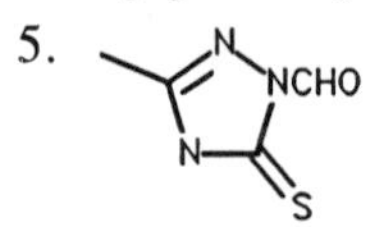

   This reagent also formylates alcohols in the presence of added base.[6]
6. *t*-$BuMe_2SiCl$, DMAP, Et_3N, DMF, 35–60°, 65–85% yield.[7]
7. DMF, silica gel, heat, 5 h, 100% yield,[8] or DMF, ZrO, heat, 5 h, 92% yield.[9]
8. HCO_2Et, heat.[10]

Cleavage

1. HCl, H_2O, dioxane, 25°, 48 h, or reflux, 1 h, 80–95% yield.[1]
2. Hydrazine, EtOH, 60°, 4 h, 60–80% yield.[11]
3. H_2/Pd–C, THF, HCl, 25°, 5–7 h, quant.[12]
4. 15% H_2O_2, H_2O, 60°, 2 h, 80% yield.[13]
5. AcCl, $PhCH_2OH$, 20°, 24 h, or 60°, 3 h, good yields.[14]
6. $h\nu$, 254 nm, CH_3CN, 100% yield.[15]
7. NaOH, H_2O, reflux, 18 h, 85% yield.[16]

1. J. C. Sheehan and D.-D. H. Yang, *J. Am. Chem. Soc.*, **80,** 1154 (1958).
2. E. G. E. Jahngen, Jr., and E. F. Rossomando, *Synth. Commun.*, **12,** 601 (1982).
3. M. Waki and J. Meienhofer, *J. Org. Chem.*, **42,** 2019 (1977).
4. F. M. F. Chen and N. L. Benoiton, *Synthesis*, 709 (1979).
5. L. Kisfaludy and L. Ötvös, Jr., *Synthesis*, 510 (1987).
6. H. Yazawa and S. Goto, *Tetrahedron Lett*, **26,** 3703 (1985).
7. S. W. Djurić, *J. Org. Chem.*, **49,** 1311 (1984).
8. European Patent to Japan Tobacco Inc., EP 271093, June 15, 1988.
9. K. Takahashi, M. Shibagaki, and H. Matsushita, *Agric. Biol. Chem.*, **52,** 853 (1988).
10. H. Schmidhammer and A. Brossi, *Can. J. Chem.*, **60,** 3055 (1982).
11. R. Geiger and W. Siedel, *Chem. Ber.*, **101,** 3386 (1968).
12. G. Losse and D. Nadolski, *J. Prakt. Chem.*, **24,** 118 (1964).
13. G. Losse and W. Zönnchen, *Justus Liebigs Ann. Chem.*, **636,** 140 (1960).
14. J. O. Thomas, *Tetrahedron Lett.*, 335 (1967).
15. B. K. Barnett and T. D. Roberts, *J. Chem. Soc., Chem. Commun.*, 758 (1972).
16. U. Hengartner, A. D. Batcho, J. F. Blount, W. Leimgruber, M. E. Larscheid, and J. W. Scott, *J. Org. Chem.*, **44,** 3748 (1979).

2. Acetamide: R_2NAc (Chart 9)

Formation

The simplest method for acetamide preparation involves reaction of the amine with acetic anhydride or acetyl chloride with or without added base. Some other methods are listed below.

1. C_6F_5OAc, DMF, 25°, 1–12 h, 78–91% yield.[1] These conditions allow selective acylation of amines in the presence of alcohols. If triethylamine is used in place of DMF, alcohols are also acylated (75–85% yield).
2. Ac_2O, 18-crown-6, Et_3N, 98% yield.[2] The crown ether forms a complex with a primary amine, thus allowing selective acylation of a secondary amine.
3. $AcOC_6H_4$-*p*-NO_2, pH 11.[3]

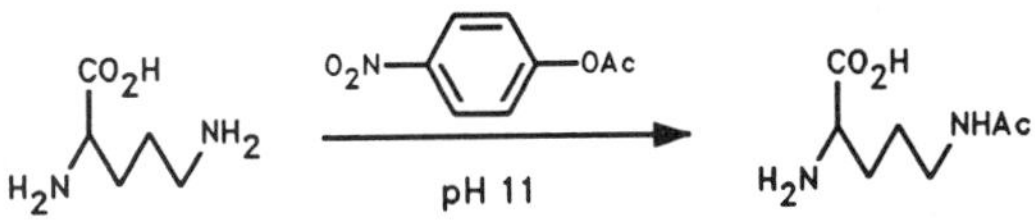

CO_2H / H_2N / NH_2 —[$O_2N-C_6H_4-OAc$, pH 11]→ CO_2H / H_2N / NHAc

Cleavage

1. 1.2 *N* HCl, reflux, 9 h, 61–77% yield.[4]
2. 85% hydrazine, 70°, 15 h, 68% yield.[5]
3. $Et_3O^+BF_4^-$, CH_2Cl_2, 25°, 1–2 h, 90% yield, then aq. $NaHCO_3$, satisfactory yields.[6]
4. Hog kidney acylase, pH 7, H_2O, 36°, 35 h.[7] In this case deprotection also proceeds with resolution since only one enantiomer is cleaved.

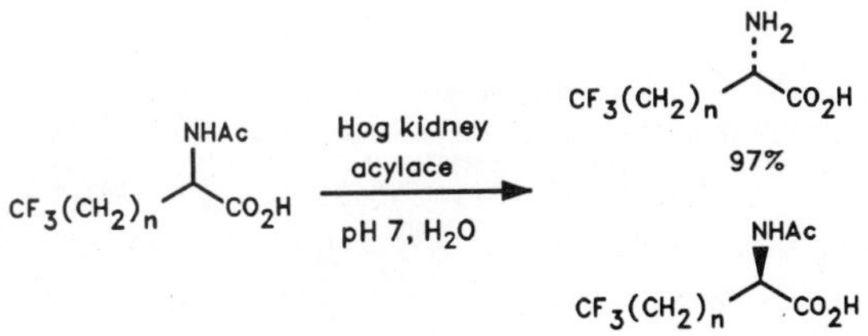

$CF_3(CH_2)_n$CH(NHAc)CO_2H —[Hog kidney acylace, pH 7, H_2O]→ $CF_3(CH_2)_n$CH(NH_2)CO_2H 97% + $CF_3(CH_2)_n$CH(NHAc)CO_2H

5. Simple amides that are difficult to cleave can first be converted to a BOC derivative by an exchange process that relies on the reduced electrophilicity of the carbamate as well as its increased steric bulk.[8–10]

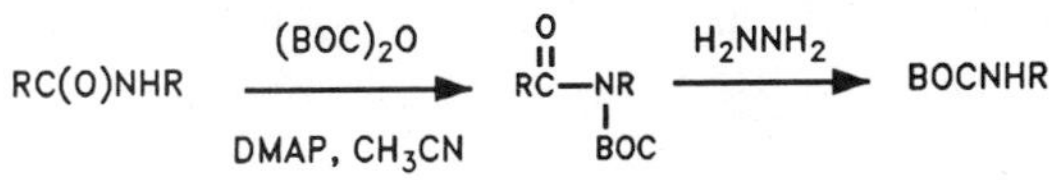

RC(O)NHR —[$(BOC)_2O$, DMAP, CH_3CN]→ RC(O)—N(BOC)R —[H_2NNH_2]→ BOCNHR

6. Na, BuOH, 120°, 62% yield.[11]
7. Ca, NH_3, DME, EtOH, 4 h, 96% yield.[12]

1. L. Kisfaludy, T. Mohacsi, M. Low, and F. Drexler, *J. Org. Chem.*, **44,** 654 (1979).
2. A. G. M. Barrett and J. C. A. Lana, *J. Chem. Soc., Chem. Commun.*, 471 (1978).
3. F. Kanai, T. Kaneko, H. Morishima, K. Isshiki, T. Takita, T. Takeuchi, and H. Umezawa, *J. Antibiotics (Tokyo)*, **38,** 39 (1985).
4. G. A. Dilbeck, L. Field, A. A. Gallo, and R. J. Gargiulo, *J. Org. Chem.*, **43,** 4593 (1978).
5. D. D. Keith, J. A. Tortora, and R. Yang, *J. Org. Chem.*, **43,** 3711 (1978).
6. S. Hanessian, *Tetrahedron Lett.*, 1549 (1967).
7. T. Tsushima, K. Kawada, S. Ishihara, N. Uchida, O. Shiratori, J. Higaki, and M. Hirata, *Tetrahedron*, **44,** 5375 (1988).
8. L. Grehn, K. Gunnarsson, and U. Ragnarsson, *J. Chem. Soc., Chem. Commun.*, 1317 (1985).
9. L. Grehn, K. Gunnarsson, and U. Ragnarsson, *Acta Chem. Scand. Ser B.*, **B40,** 745 (1986).
10. D. J. Kempf, *Tetrahedron Lett.*, **30,** 2029 (1989).
11. M. Obayashi and M. Schlosser, *Chem. Lett.*, 1715 (1985).
12. A. J. Pearson and D. C. Rees, *J. Am. Chem. Soc.*, **104,** 1118 (1982).

3. Chloroacetamide: R_2NCOCH_2Cl (Chart 9)

Monochloroacetamides are cleaved (by "assisted removal") by reagents that contain two nucleophilic groups (e.g., *o*-phenylenediamine,[1] thiourea,[2,3] 1-piperidinethiocarboxamide,[4] 3-nitropyridine-2-thione,[5] 2-aminothiophenol[6]):

R_2NCOCH_2Cl + H_2N, H_2N (o-phenylenediamine) —LiOH, H_2O / 100°, 1 h→ [R_2N, O, NH, NH_2 intermediate] → R_2NH (65–75%) + (H, N, N, H, O quinoxalinone)

The chloroacetamide can also be cleaved by first converting it to the pyridiniumacetamide (Pyr, 90°, 1 h, 70–90% yield) followed by mild basic hydrolysis (0.1 *N* NaOH, 25°).[7]

1. R. W. Holley and A. D. Holley, *J. Am. Chem. Soc.*, **74,** 3069 (1952).
2. M. Masaki, T. Kitahara, H. Kurita, and M. Ohta, *J. Am. Chem. Soc.*, **90,** 4508 (1968).
3. J. E. Baldwin, M. Otsuka, and P. M. Wallace, *Tetrahedron*, **42,** 3097 (1986); T. Allmendinger, G. Rihs, and H. Wetter, *Helv. Chim. Acta*, **71,** 395 (1988).
4. W. Steglich and H.-G. Batz, *Angew Chem., Int. Ed. Engl.*, **10,** 75 (1971)
5. K. Undheim and P. E. Fjeldstad, *J. Chem. Soc., Perkin Trans. I*, 829 (1973).
6. J. D. Glass, M. Pelzig, and C. S. Pande, *Int. J. Pept. Protein Res.*, **13,** 28 (1979).
7. C. H. Gaozza, B. C. Cieri, and S. Lamdan, *J. Heterocycl. Chem.*, **8,** 1079 (1971).

4. Trichloroacetamide: $R_2NCOCCl_3$ (Chart 9)

Formation

1. $Cl_3CCOCCl_3$, hexane, 65°, 90 min, 65–97% yield.[1]

Cleavage

1. $NaBH_4$, EtOH, 1 h, 65% yield.[2]

1. B. Sukornick, *Org. Synth., Collect. Vol. V*, 1074 (1973).
2. F. Weygand and E. Frauendorfer, *Chem. Ber.*, **103,** 2437 (1970).

5. Trifluoroacetamide (TFA): R_2NCOCF_3 (Chart 9)

Formation

1. CF_3CO_2Et, Et_3N, CH_3OH, 25°, 15–45 h, 75–95% yield.[1] A polymeric version of this approach has also been developed.[2]
2. $(CF_3CO)_2O$, 18-crown-6, Et_3N, 95% yield.[3] Complex formation of a primary amine with 18-crown-6 allows selective acylation of a secondary amine.
3. CF_3COO-succinimidyl, CH_2Cl_2, 0°, 85% yield.[4] These conditions selectively introduced the TFA group onto a primary amine in the presence of a secondary amine.
4. $(CF_3CO)_2O$, Pyr, CH_2Cl_2.[5]

Cleavage

The trifluoroacetamide is one of the more easily cleaved amides.

1. K_2CO_3 or Na_2CO_3, MeOH, H_2O, rt, 55–95% yield.[4,6] Note that the trifluoroacetamide has been cleaved in the presence of a methyl ester, which illustrates the ease of hydrolysis of the trifluoroacetamide group.[7]

OBn MeO2C NHZ HO2C NHCOCF3 K_2CO_3, MeOH H_2O, 6 h, 86% HO2C NH2

2. NH_3, MeOH.[8]
3. 0.2 N $Ba(OH)_2$, CH_3OH, 25°, 2 h, 79% yield.[9]
4. $NaBH_4$, EtOH, 20°, or 60°, 1 h, 60–100% yield.[10]
5. $PhCH_2N^+Et_3\ OH^-$, CH_2Cl_2, −40°, 48 h.[5]

1. T. J. Curphey, *J. Org. Chem.*, **44,** 2805 (1979).
2. P. I. Svirskaya, C. C. Leznoff, and M. Steinman, *J. Org. Chem.*, **52,** 1362 (1987).
3. A. G. M. Barrett and J. C. A. Lana, *J. Chem. Soc., Chem. Commun.*, 471 (1978).
4. R. J. Bergeron and J. J. McManis, *J. Org. Chem.*, **53,** 3108 (1988).
5. S. G. Pyne, *Tetrahedron Lett.*, **28,** 4737 (1987).
6. H. Newman, *J. Org. Chem.*, **30,** 1287 (1965); J. Quick and C. Meltz, *J. Org. Chem.*, **44,** 573 (1979); M. A. Schwartz, B. F. Rose, and B. Vishnuvajjala, *J. Am. Chem. Soc.*, **95,** 612 (1973).
7. D. L. Boger and D. Yohannes, *J. Org. Chem.*, **54,** 2498 (1989).
8. M. Imazawa and F. Eckstein, *J. Org. Chem.*, **44,** 2039 (1979).
9. F. Weygand and W. Swodenk, *Chem. Ber.*, **90,** 639 (1957).
10. F. Weygand and E. Frauendorfer, *Chem. Ber.*, **103,** 2437 (1970).

6. **Phenylacetamide:** $R_2NCOCH_2C_6H_5$

This amide, readily formed from an amine and the anhydride,[1] is readily cleaved by penicillin acylase (pH 8.1, *N*-methylpyrrolidone, 65–95% yield). This deprotection procedure works on peptides[2] as well as on nonpeptide substrates.[3]

1. A. R. Jacobson, A. N. Makris, and L. M. Sayre, *J. Org. Chem.*, **52,** 2592 (1987).
2. For a review on the use of enzymes in protective group manipulation in peptide chemistry, see: J. D. Glass, in *The Peptides*, S. Undenfriend and J. Meienhofer, Eds., Academic Press, Orlando, FL, 1987, Vol. 9, pp. 167–184.
3. H. Waldmann, *Tetrahedron Lett.*, **29,** 1131 (1988) and references cited therein.

7. **3-Phenylpropanamide:** $R_2NCOCH_2CH_2C_6H_5$ (Chart 9)

A 3-phenylpropanamide, prepared from a nucleoside, is hydrolyzed under mild conditions by α-chymotrypsin (37°, pH 7, 2–12 h).[1]

1. H. S. Sachdev and N. A. Starkovsky, *Tetrahedron Lett.*, 733 (1969).

8. Picolinamide: R_2NCO-2-pyridyl (Chart 9)

The picolinamide is prepared in 95% yield from picolinic acid/DCC and an amino acid, and is hydrolyzed in 75% yield by aqueous $Cu(OAc)_2$.[1]

9. 3-Pyridylcarboxamide: R_2NCO-3-pyridyl

The 3-pyridylcarboxamide, prepared from the anhydride (Pyr, 99% yield), is cleaved (55–86% yield) by basic hydrolysis (0.5 *M* NaOH, rt) after quaternization of the pyridine nitrogen with methyl iodide.[2]

1. A. K. Koul, B. Prashad, J. M. Bachhawat, N. S. Ramegowda, and N. K. Mathur, *Synth. Commun.*, **2,** 383 (1972).
2. S. Ushida, *Chem. Lett.*, 59 (1989).

10. *N*-Benzoylphenylalanyl Derivative: $R_2NCOCH(NHCOC_6H_5)CH_2C_6H_5$

This derivative, prepared from an amino acid and the acyl azide, is selectively cleaved in 80% yield by chymotrypsin.[1]

1. R. W. Holley, *J. Am. Chem. Soc.*, **77,** 2552 (1955).

11. Benzamide: $R_2NCOC_6H_5$ (Chart 9)

Formation

1. PhCOCl, Pyr, 0°, high yield.[1]
2. PhCOCN, CH_2Cl_2, −10°, 92% yield.[2] This reagent readily acylates amines in the presence of alcohols.
3. $PhCOCF(CF_3)_2$, $Me_2NCH_2CH_2NMe_2$ (TMEDA), 25°, 30 min, high yield.[3]
4. 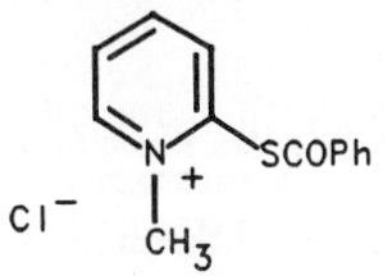 2-(SCOPh)-1-methylpyridinium Cl^-, aq. $NaHCO_3$ or aq. NaOH, good yields.[4]
5. $(PhCO)_2NOCH_3$, DMF, H_2O, or dioxane, 3–26 h, 66–89%.[5] The reagent is selective for primary amines.

Cleavage

1. 6 *N* HCl, reflux, 48 h or HBr, AcOH, 25°, 72 h, 80% yield.[6]
2. $(HF)_n$/Pyr, 25°, 60 min, 100% yield.[7] Polyhydrogen fluoride/pyridine cleaves most of the protective groups used in peptide synthesis.

3. Electrolysis, −2.3 V, $Me_4N^+X^-$, CH_3OH, 70 min, 60–90% yield.[8]
4. $(Me_2CHCH_2)_2AlH$, $PhCH_3$, −78°, 80% yield.[9] Since the *N*-benzoyl group in this substrate could not be removed by hydrolysis, a less selective reductive cleavage with diisobutylaluminum hydride was used.

1. E. White, *Org. Synth., Collect. Vol. V*, 336 (1973).
2. S.-I. Murahashi, T. Naota, and N. Nakajima, *Tetrahedron Lett.*, **26,** 925 (1985).
3. N. Ishikawa and S. Shin-ya, *Chem. Lett.*, 673 (1976).
4. M. Yamada, Y. Watabe, T. Sakakibara, and R. Sudoh, *J. Chem. Soc., Chem. Commun.*, 179 (1979).
5. Y. Kikugawa, K. Mitsui, T. Sakamoto, M. Kawase, and H. Tamiya, *Tetrahedron Lett.*, **31,** 243 (1990).
6. D. Ben-Ishai, J. Altman, and N. Peled, *Tetrahedron*, **33,** 2715 (1977); P. Hughes and J. Clardy, *J. Org. Chem.*, **53,** 4793 (1988).
7. S. Matsuura, C.-H. Niu, and J. S. Cohen, *J. Chem. Soc., Chem. Commun.*, 451 (1976).
8. L. Horner and H. Neumann, *Chem. Ber.*, **98,** 3462 (1965).
9. J. Gutzwiller and M. Uskokovic, *J. Am. Chem. Soc.*, **92,** 204 (1970).

12. *p*-Phenylbenzamide: $R_2NCOC_6H_4$-*p*-C_6H_5

The phenylbenzamide is prepared from the acid chloride in the presence of Et_3N (86% yield) and can be cleaved with 3% Na(Hg) (MeOH, 25°, 4 h, 81% yield).[1] Most amides react only slowly with Na(Hg). Phenylbenzamides are generally crystalline compounds, an aid in purification.[2]

1. R. B. Woodward and 48 co-workers, *J. Am. Chem. Soc.*, **103,** 3210 (1981).
2. R. M. Scribner, *Tetrahedron Lett.*, 3853 (1976).

Assisted Cleavage

A series of amides has been prepared as protective groups that are cleaved by intramolecular cyclization after activation, by reduction of a nitro group, or by activation by other chemical means. These groups have not found much use since the first edition of this volume and are therefore only listed for completeness. The concept is generalized in the following scheme:

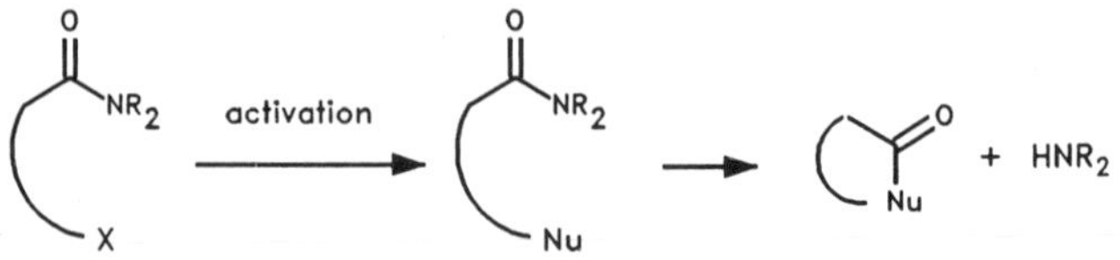

O
NR_2
X
activation
O
NR_2
Nu
O
Nu
+ HNR_2

13. ***o*-Nitrophenylacetamide**[1] (Chart 9)

14. ***o*-Nitrophenoxyacetamide**[2] (Chart 9)

15. **Acetoacetamide**[3] (Chart 9)

16. **(*N'*-Dithiobenzyloxycarbonylamino)acetamide**[4]

17. **3-(*p*-Hydroxyphenyl)propanamide**[5] (Chart 9)

18. **3-(*o*-Nitrophenyl)propanamide**[6]

19. **2-Methyl-2-(*o*-nitrophenoxy)propanamide**[1,7] (Chart 9)

20. **2-Methyl-2-(*o*-phenylazophenoxy)propanamide**[8] (Chart 9)

21. **4-Chlorobutanamide**[9] (Chart 9)

22. **3-Methyl-3-nitrobutanamide**[10]

23. ***o*-Nitrocinnamide**[11] (Chart 9)

24. ***N*-Acetylmethionine Derivative**[12] (Chart 9)

25. ***o*-Nitrobenzamide**[13]

26. ***o*-(Benzoyloxymethyl)benzamide**[14]

1. F. Cuiban, *Rev. Roum. Chim.*, **18,** 449 (1973).
2. R. W. Holley and A. D. Holley, *J. Am. Chem. Soc.*, **74,** 3069 (1952).
3. C. Di Bello, F. Filira, V. Giormani, and F. D'Angeli, *J. Chem. Soc. C*, 350 (1969).
4. F. E. Roberts, *Tetrahedron Lett.*, 325 (1979).
5. G. L. Schmir and L. A. Cohen, *J. Am. Chem. Soc.*, **83,** 723 (1961); L. Farber and L. A. Cohen, *Biochemistry*, **5,** 1027 (1966).
6. I. D. Entwistle, *Tetrahedron Lett.*, 555 (1979).
7. C. A. Panetta, *J. Org. Chem.*, **34,** 2773 (1969).
8. C. A. Panetta and A.-U. Rahman, *J. Org. Chem.*, **36,** 2250 (1971).
9. H. Peter, M. Brugger, J. Schreiber, and A. Eschenmoser, *Helv. Chim. Acta*, **46,** 577 (1963).
10. T.-L. Ho, *Synth. Commun.*, **10,** 469 (1980).
11. G. Just and G. Rosebery, *Synth. Commun.*, **3,** 447 (1973).
12. W. B. Lawson, E. Gross, C. M. Foltz and B. Witkop, *J. Am. Chem. Soc.*, **84,** 1715 (1962).
13. A. K. Koul, J. M. Bachhawat, B. Rashad, N. S. Ramegowda, A. K. Mathur, and N. K. Mathur, *Tetrahedron*, **29,** 625 (1973).
14. B. F. Cain, *J. Org. Chem.*, **41,** 2029 (1976).

A number of protective groups have been developed that simultaneously protect both sites of a primary nitrogen. These may prove to be useful for cases where acidic hydrogens on nitrogen cannot be tolerated.

27. 4,5-Diphenyl-3-oxazolin-2-one (Chart 8):

Formation[1]

1. DMF, 0.5 h; CF_3CO_2H, 67–85% yield.

Cleavage[1]

1. H_2/Pd–C, aq. HCl, 25°, 12 h, quant.
2. Na/NH_3, 75–85% yield.
3. *m*-$ClC_6H_4CO_3H$, then water, 70% yield.
4. O_2, photolysis, −30°, then Zn, AcOH, quant.[2]

Cyclic Imide Derivatives

28. *N*-Phthalimide (Chart 9):

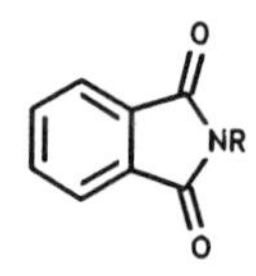

Formation

1. Phthalic anhydride, $CHCl_3$, 70°, 4 h, 85–93% yield.[3]
2. *o*-(CH_3OOC)C_6H_4COCl, Et_3N, THF, 0°, 2 h, 90–95% yield.[4]
3. Phthalimide–NCO_2Et, aq. Na_2CO_3, 25°, 10–15 min, 85–95% yield.[5] This reagent can be used to protect selectively primary amines in the presence of secondary amines.[6]

Cleavage

1. Hydrazine, EtOH, 25°, 12 h; H_3O^+, 76% yield.[3]
2. $PhNHNH_2$, *n*-Bu_3N, reflux, 2 h, 83% yield.[7]
3. $Na_2S \cdot H_2O$, THF, 68–90% yield; DCC($-H_2O$), 67–97% yield; hydrazine; dil. HCl, 55–95% yield.[8] This method is used to cleave *N*-phthalimido penicillins; hydrazine attacks an intermediate phthalisoimide instead of the azetidinone ring.

4. $NaBH_4$, 2-propanol, H_2O (6 : 1); AcOH, pH 5, 80°, 5–8 h.[9,10] This method was reported to be superior in cases where hydrazine proved to be inefficient.
5. $MeNH_2$, EtOH, rt, 5 min, then heat, 2.5 h, 89% yield.[11]

29. *N*-Dithiasuccinimide (Dts–NR) (Chart 9):

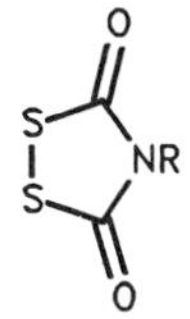

Formation

1. $EtOCS_2CH_2CO_2H$ or $EtOCS_2CSOEt$; ClSCOCl, 0–45°, 70–90% yield.[12,13]
2. PEG(2000)-$OCS_2CH_2CONH_2$; TMSNH(CO)NHTMS; ClCOSCl.[12]

Cleavage

The Dts group is cleaved by treatment with a thiol and base, e.g., $HOCH_2CH_2SH$, Et_3N, 25°, 5 min, $HSCH_2C(O)NHMe$, Pyr, 5 min.[14]

The Dts group—stable to acidic cleavage of *t*-butyl carbamates (12 *N* HCl, AcOH, reflux; HBr, AcOH), to mild base ($NaHCO_3$), and to photolytic cleavage of *o*-nitrobenzyl carbamates—can be used in orthogonal schemes for protection of peptides.[14] Merrifield defines an orthogonal system as a set of completely independent classes of protective groups wherein each class of protective groups can be removed in any order and in the presence of all other classes.

30. *N*-2,3,-Diphenylmaleimide:

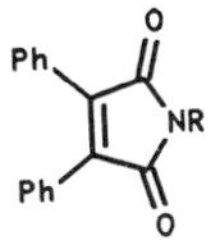

The diphenylmaleimide is prepared from the anhydride, 33–87% yield, and cleaved by hydrazinolysis, 65–75% yield. It is stable to acid (HBr, AcOH, 48 h) and to mercuric cyanide. It is colored and easily located during chromatography, and has been prepared to protect steroidal amines and amino sugars.[14]

31. *N*-2,5-Dimethylpyrrole:

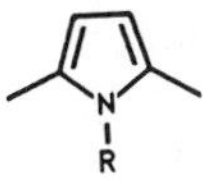

Formation

1. Acetylacetone, AcOH, 88% yield.[15]

Cleavage

1. $H_2NOH \cdot HCl$, EtOH, H_2O, 73% yield.[15,16]
2. Ozone.[17]

32. *N*-1,1,4,4-Tetramethyldisilylazacyclopentane Adduct (STABASE)

Formation/Cleavage[18–21]

$$\mathrm{RNH_2} \xrightarrow[\text{TEA, } CH_2Cl_2,\ 85\text{–}95\%]{ClMe_2SiCH_2CH_2SiMe_2Cl} \text{R–N}(\text{SiMe}_2\text{CH}_2\text{CH}_2\text{SiMe}_2)$$

The amine adducts are stable to the following reagents: *n*-BuLi (THF, −25°), *s*-BuLi (Et_2O, −25°); lithium diisopropylamide; saturated aqueous ammonium chloride; H_2O; MeOH; 2 *N* $NaHCO_3$; pyridinium dichromate, CH_2Cl_2; $KF \cdot 2H_2O$, THF, H_2O; saturated aqueous sodium dihydrogen phosphate. The derivative is not stable to strong acid or base; to pyridinium chlorochromate, CH_2Cl_2; or to $NaBH_4$, EtOH.

34. 5-Substituted 1,3-Dimethyl-1,3,5-triazacyclohexan-2-one

35. 5-Substituted 1,3-Dibenzyl-1,3,5-triazacyclohexan-2-one

Formation/Cleavage

$$\mathrm{R^1NH_2} + \mathrm{R^2HN{-}CO{-}NHR^2} \underset{\text{aqueous } NH_4Cl,\ 70°,\ 1\text{–}3\text{ h},\ 84\text{–}92\%}{\overset{70\text{–}110°,\ 56\text{–}95\%}{\rightleftarrows}} \text{1,3-di-}R^2\text{-5-}R^1\text{-1,3,5-triazacyclohexan-2-one}$$

R^2 = methyl or benzyl

The triazone is stable to $LiAlH_4$; $PtO_2/H_2/EtOH$, 48 h; Pd-black/H_2/THF, 1 h; *n*-BuLi/THF, −40°, 30 min; PhMgBr/THF −78°, 30 min; Wittig reagents; DIBAL/THF, rt, 3 h; $LiBH_4$/THF, 40°; acylation, silylation, and anhydrous acids ($TiCl_4$, CH_2Cl_2, −78°, 30 min; TsOH, toluene, 12 h; neat CF_3CO_2H, 15 min). Extended exposure (48 h) of a triazone to neat CF_3CO_2H results in cleavage.[22]

36. 1-Substituted 3,5-Dinitro-4-pyridone

Formation/Cleavage[23]

$$RNH_2 \xrightarrow[72\text{–}100\%]{\text{4-}NO_2\text{-}C_6H_4\text{-(3,5-dinitro-4-pyridone)},\ H_2O,\ Pyr,\ rt,\ 2\text{–}24\ h} \text{1-R-3,5-dinitro-4-pyridone}$$

$$\text{1-R-3,5-dinitro-4-pyridone} \xrightarrow[83\text{–}97\%]{MeNH_2,\ PrNH_2,\ \text{or hexyl}NH_2;\ H_2O,\ Pyr,\ 0.5\text{–}2\ h} RNH_2$$

1. J. C. Sheehan and F. S. Guziec, *J. Org. Chem.*, **38,** 3034 (1973).
2. F. S. Guziec, Jr., and E. T. Tewes, *J. Heterocycl. Chem.*, **17,** 1807 (1980).
3. T. Sasaki, K. Minamoto and H. Itoh, *J. Org. Chem.*, **43,** 2320 (1978).
4. D. A. Hoogwater, D. N. Reinhoudt, T. S. Lie, J. J. Gunneweg, and H. C. Beyerman, *Recl. Trav. Chim. Pays-Bas*, **92,** 819 (1973).
5. G. H. L. Nefkins, G. I. Tesser, and R. J. F. Nivard, *Recl. Trav. Chim. Pays-Bas*, **79,** 688 (1960); C. R. McArthur, P. M. Worster, and A. U. Okon, *Synth. Commun.*, **13,** 311 (1983).
6. G. Sosnovsky and J. Lukszo, *Z. Naturforsch. B*, **41B,** 122 (1986).
7. I. Schumann and R. A. Boissonnas, *Helv. Chim. Acta*, **35,** 2235 (1952).
8. S. Kukolja and S. R. Lammert, *J. Am. Chem. Soc.*, **97,** 5582 (1975).
9. F. Dasgupta and P. J. Garegg, *J. Carbohydr. Chem.*, **7,** 701 (1988).
10. J. O. Osby, M. G. Martin, and B. Ganem, *Tetrahedron Lett.*, **25,** 2093 (1984).
11. M. S. Motawia, J. Wengel, A. E. S. Abdel-Megid, and E. B. Pedersen, *Synthesis*, 384 (1989).
12. S. Zalipsky, F. Albericio, U. Slomczynska, and G. Barany, *Int. J. Pept. Protein Res.*, **30,** 748 (1987).
13. U. Zehavi, *J. Org. Chem.*, **42,** 2819 (1977).
14. G. Barany and R. B. Merrifield, *J. Am. Chem. Soc.*, **99,** 7363 (1977); *idem*, **102,** 3084 (1980).
15. S. P. Bruekelman, S. E. Leach, G. D. Meakins, and M. D. Tirel, *J. Chem. Soc., Perkin Trans. I*, 2801 (1984).
16. S. P. Breukelman, G. D. Meakins, and M. D. Tirel, *J. Chem. Soc., Chem. Commun.*, 800 (1982).
17. C. Kashima, T. Maruyama, Y. Fujioka, and K. Harada, *J. Chem. Soc., Perkin Trans. I*, 1041 (1989).
18. S. Djuric, J. Venit, and P. Magnus, *Tetrahedron Lett.*, **22,** 1787 (1981).

19. T. Högberg, P. Ström, and U. H. Lindberg, *Acta Chem. Scand., Ser. B.*, **B39,** 414 (1985).
20. T. L. Guggenheim, *Tetrahedron Lett.*, **25,** 1253 (1984).
21. M. J. Sofia, P. K. Chakravarty, and J. A. Katzenellenbogen, *J. Org. Chem.*, **48,** 3318 (1983).
22. S. Knapp, J. J. Hale, M. Bastos, and F. S. Gibson, *Tetrahedron Lett.*, **31,** 2109 (1990).
23. E. Matsumura, H. Kobayashi, T. Nishikawa, M. Ariga, Y. Tohda, and T. Kawashima, *Bull. Chem. Soc. Jpn.*, **57,** 1961 (1984); E. Matsumura, M. Ariga, Y. Tohda, and T. Kawashima, *Tetrahedron Lett.*, **22,** 757 (1981).

SPECIAL —NH PROTECTIVE GROUPS

N-Alkyl and *N*-Aryl Amines

1. *N*-Methylamine: CH_3NR_2

The cleavage of a methylamine can be accomplished photochemically in the presence of an electron acceptor such as 9,10-dicyanoanthracene.[1,2]

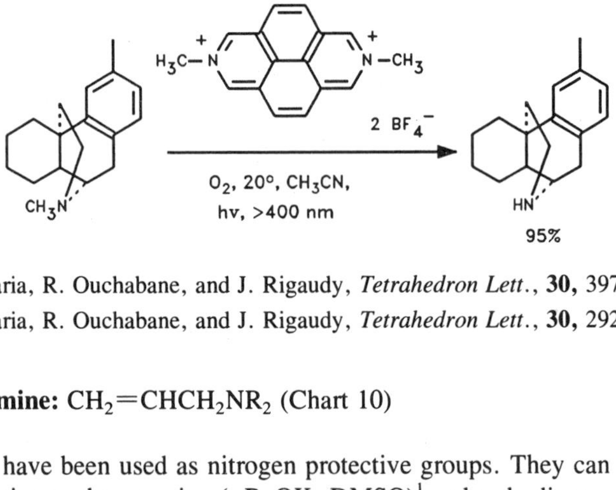

H_3C-N^+ ... N^+-CH_3, 2 BF_4^-

O_2, 20°, CH_3CN, hv, >400 nm

CH_3N → HN 95%

1. J. Santamaria, R. Ouchabane, and J. Rigaudy, *Tetrahedron Lett.*, **30,** 3977 (1989).
2. J. Santamaria, R. Ouchabane, and J. Rigaudy, *Tetrahedron Lett.*, **30,** 2927 (1989).

2. *N*-Allylamine: $CH_2=CHCH_2NR_2$ (Chart 10)

Allylamines have been used as nitrogen protective groups. They can be removed by isomerization to the enamine (*t*-BuOK, DMSO)[1] or by rhodium-catalyzed isomerization.[2]

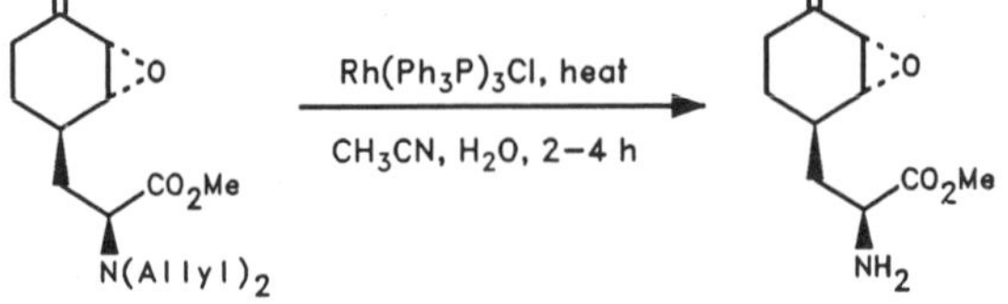

O, O, CO_2Me, $N(Allyl)_2$

$Rh(Ph_3P)_3Cl$, heat

CH_3CN, H_2O, 2–4 h

O, O, CO_2Me, NH_2

1. R. Gigg and R. Conant, *J. Carbohydr. Chem.*, **1,** 331 (1983).
2. B. C. Laguzza and B. Ganem, *Tetrahedron Lett.*, **22,** 1483 (1981).

3. *N*-[2-(Trimethylsilyl)ethoxy]methylamine (SEM–NR_2): $(CH_3)_3SiCH_2CH_2OCH_2$–NR_2

The SEM derivative of a secondary aromatic amine, prepared from SEMCl (NaH, DMF, 0°, 100% yield), can be cleaved with HCl (EtOH, >88% yield).[1]

1. Z. Zeng and S. C. Zimmerman, *Tetrahedron Lett.*, **29**, 5123 (1988).

4. *N*-3-Acetoxypropylamine: $R_2NCH_2CH_2CH_2OCOCH_3$ (Chart 10)

Formation

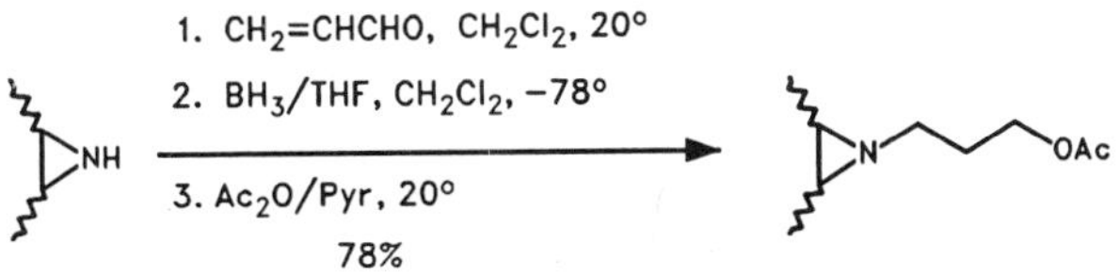

1. CH_2=CHCHO, CH_2Cl_2, 20°
2. BH_3/THF, CH_2Cl_2, −78°
3. Ac_2O/Pyr, 20°

78%

Cleavage

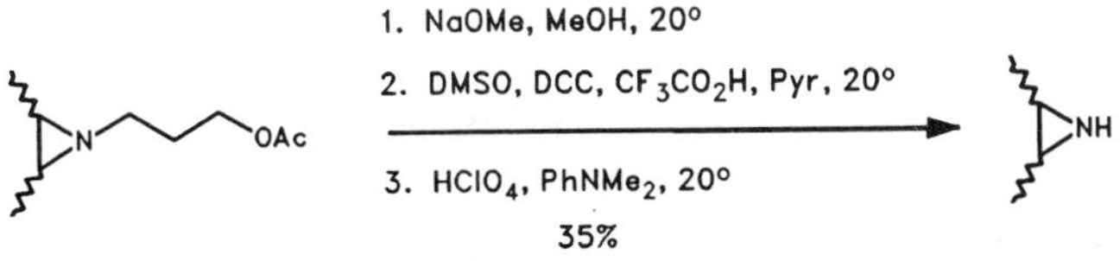

1. NaOMe, MeOH, 20°
2. DMSO, DCC, CF_3CO_2H, Pyr, 20°
3. $HClO_4$, $PhNMe_2$, 20°

35%

A 3-acetoxypropyl group was used to protect an aziridine —NH group during the synthesis of mitomycins A and C; acetyl, benzoyl, ethoxycarbonyl, and methoxymethyl groups were unsatisfactory.[1]

1. T. Fukuyama, F. Nakatsubo, A. J. Cocuzza, and Y. Kishi, *Tetrahedron Lett.*, 4295 (1977).

5. *N*-(1-Isopropyl-4-nitro-2-oxo-3-pyrrolin-3-yl)amine

Formation/Cleavage[1]

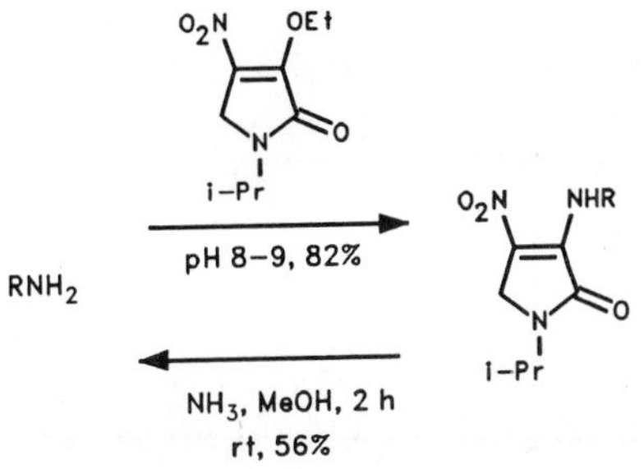

RNH_2 → pH 8–9, 82% (with 1-i-Pr-3-OEt-4-O_2N-pyrrolin-2-one)

← NH_3, MeOH, 2 h, rt, 56%

1. P. L. Southwick, G. K. Chin, M. A. Koshute, J. R. Miller, K. E. Niemela, C. A. Siegel, R. T. Nolte, and W. E. Brown, *J. Org. Chem.*, **49,** 1130 (1984).

6. Quaternary Ammonium Salts: $R_3N^+CH_3\ I^-$ (Chart 10)

Formation

1. CH_3I, CH_3OH, $KHCO_3$, 20°, 24 h, 85–95% yield.[1] These salts are generally used to protect tertiary amines during oxidation reactions. The conditions cited above form quaternary salts from primary, secondary, or tertiary amines, including amino acids, in the presence of hydroxyl or phenol groups.[1]

Cleavage

1. PhSNa, 2-butanone, reflux, 24–36 h, 85% yield.[2]

1. F. C. M. Chen and N. L. Benoiton, *Can. J. Chem.*, **54,** 3310 (1976).
2. M. Shamma, N. C. Deno, and J. F. Remar, *Tetrahedron Lett.*, 1375 (1966).

7. *N*-Benzylamine (R_2N–Bn): R_2NCH_2Ph (Chart 10)

Formation

1. BnCl, aq. K_2CO_3, reflux, 30 min; H_2, Pd–C, 77% yield.[1]

H_2N–CH(R)–CO_2H → (BnCl, K_2CO_3, H_2O, reflux, 30 min, 85%) BnHN–CH(R)–CO_2H + Bn_2N–CH(R)–CO_2H; Bn_2N–CH(R)–CO_2H → (H_2, Pd–C, 92%) BnHN–CH(R)–CO_2H

2. BnBr, EtOH, Na_2CO_3, H_2O, CH_2Cl_2, reflux.[2]
3. BnBr, Et_3N, CH_3CN.[3]

Examples 2 and 3 above produce dibenzyl derivatives from primary amines.

4. $PhCHN_2$, HBF_4, −40°, CH_2Cl_2, 57–68% yield.[4]

Cleavage

1. Pd–C, 4.4% HCOOH, CH_3OH, 25°, 10 h, 80–90% yield.[3,5] The cleavage of benzylamines with H_2/Pd–C is often very slow.[6] Note in example 1

above that one of the benzyl groups can be selectively removed from a dibenzyl derivative.

2. Pd–C, ROH, HCO_2NH_4,[7] hydrazine or sodium hypophosphite, 42–91% yield.[8] 2-Benzylaminopyridine and benzyladenine were stable to these reaction conditions. Lower yields occurred because of the water solubility of the product, thus hampering isolation.
3. Na, NH_3, excellent yields.[9]
4. $h\nu$, 405 nm ($CuSO_4$: NH_3 solution filter), CH_3CN, H_2O, 9,10-dicyanoanthracene, 6–10 h, 78–90% yield.[10] Benzyl groups as well as other alkyl groups can be converted to various carbamates as a variation of the von Braun reaction.[11] These can then be cleaved by conditions that are outlined in the section on carbamates.
5. CCl_3CH_2OCOCl, CH_3CN, 93%.[12]
6. $Me_3SiCH_2CH_2OCOCl$, THF, −50°, then 25°, overnight, 78–91% yield.[13]
7. α-Chloroethyl chloroformate; NaOH.[14,15]
8. Vinyl chloroformate is reported to be the best reagent for dealkylation of alkyl amines.[16]

Oxidative methods include the following:

9. RuO_4, NH_3, H_2O, 70% yield.[17]
10. *m*-Chloroperoxybenzoic acid followed by $FeCl_2$, −10°, 6–80% yield.[18]

1. L. Velluz, G. Amiard, and R. Heymes, *Bull. Soc. Chim. Fr.*, 1012 (1954).
2. N. Yamazaki, and C. Kibayashi, *J. Am. Chem. Soc.*, **111,** 1397 (1989).
3. B. D. Gray and P. W. Jeffs, *J. Chem. Soc., Chem. Commun.*, 1329 (1987).
4. L. J. Liotta and B. Ganem, *Tetrahedron Lett.*, **30,** 4759 (1989).
5. B. ElAmin, G. M. Anantharamaiah, G. P. Royer, and G. E. Means, *J. Org. Chem.*, **44,** 3442 (1979).
6. W. H. Hartung and R. Simonoff, *Org. React.*, **VII,** 263 (1953).
7. S. Ram and L. D. Spicer, *Tetrahedron Lett.*, **28,** 515 (1987); *idem, Synth. Commun.*, **17,** 415 (1987).
8. B. M. Adger, C. O'Farrell, N. J. Lewis, and M. B. Mitchell, *Synthesis*, 53 (1987).
9. V. du Vigneaud and O. K. Behrens, *J. Biol. Chem.*, **117,** 27 (1937).
10. G. Pandey and K. S. Rani, *Tetrahedron Lett.*, **29,** 4157 (1988).
11. H. A. Hageman, *Org. React.*, **VII,** 198 (1953).
12. V. H. Rawal, R. J. Jones and M. P. Cava, *J. Org. Chem.*, **52,** 19 (1987).
13. A. L. Campbell, D. R. Pilipauskas, I. K. Khanna, and R. A. Rhodes, *Tetrahedron Lett.*, **28,** 2331 (1987).
14. R. A. Olofson, J. T. Martz, J.-P. Senet, M. Piteau, and T. Malfroot, *J. Org. Chem.*, **49,** 2081 (1984).
15. P. DeShong and D. A. Kell, *Tetrahedron Lett.*, **27,** 3979 (1986).

16. R. A. Olofson, R. C. Schnur, L. Bunes, and J. P. Pepe, *Tetrahedron Lett.*, 1567 (1977). For a review, see: J. H. Cooley and E. J. Evain, *Synthesis*, 1 (1989).
17. X. Gao and R. A. Jones, *J. Am. Chem. Soc.*, **109,** 1275 (1987).
18. T. Monković, H. Wong, and C. Bachand, *Synthesis*, 770 (1985).

8. *N*-Di(4-methoxyphenyl)methylamine: $(4\text{-MeOC}_6H_4)_2CHNR_2$ (Chart 10)

This derivative has been used to protect the amines of amino acids [$(4\text{-MeOC}_6H_4)_2CHCl$, Et_3N, 0° → 20°, 20 h, 67% yield]. It is easily cleaved with 80% AcOH (80°, 5 min, 73% yield).[1]

9. *N*-5-Dibenzosuberylamine:

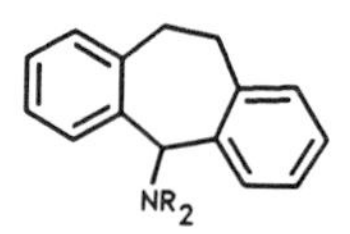

The dibenzosuberylamine is prepared in quantitative yield from an amine or amino acid and suberyl chloride; this chloride has also been used to protect hydroxyl, thiol, and carboxyl groups. Although the dibenzosuberylamine is stable to 5 *N* HCl/dioxane (22°, 16 h) and to refluxing HBr (1 h), it is completely cleaved by some acids (HCOOH, CH_2Cl_2, 22°, 2 h; CF_3COOH, CH_2Cl_2, 22°, 0.5 h; BBr_3, CH_2Cl_2, 22°, 0.5 h; 4 *N* HBr, AcOH, 22°, 1 h; 60% AcOH, reflux, 1 h) and by reduction (H_2, Pd–C, CH_3OH, 22°, 1 h, 100% cleaved).[2]

10. *N*-Triphenylmethylamine (Tr–NR$_2$): Ph_3CNR_2 (Chart 10)

The bulky triphenylmethyl group has been used to protect a variety of amines such as amino acids, penicillins, and cephalosporins. Esters of *N*-trityl α-amino acids are shielded from hydrolysis and require forcing conditions for cleavage. The α-proton is also shielded from deprotonation, which means that esters elsewhere in the molecule can be selectively deprotonated.

Formation

1. TrCl, Et_3N, 25°, 4 h.[3]
2. TrBr, $CHCl_3$, DMF, rt, 0.5–1 h; Et_3N, rt, 50 min.[4] These conditions also lead to tritylation of carboxyl groups in the amino acids, but they can be selectively hydrolyzed. This method was considered to be an improvement over the standard methods of *N*-tritylation of amino acids.
3. (i) Silylation of $-CO_2H$ with Me_3SiCl, Et_3N; (ii) TrCl, Et_3N; (iii) MeOH, 65–92% yield.[5] To effect *N*-tritylation of serine, Me_2SiCl_2 should be used in the silylation step.

Cleavage

1. HCl, acetone, 25°, 3 h, 80% yield.[1]
2. H_2, Pd black, EtOH, 45°, 92% yield.[6]
3. Na, NH_3.[7]
4. Hydroxybenzotriazole (HOBT), trifluorethanol, rt.[8]

11. *N*-[(4-Methoxyphenyl)diphenylmethyl]amine (MMTr–NR_2): (4-CH_3O–C_6H_4)(C_6H_5)$_2$C–NR_2 (Chart 10)

The MMTr derivative is easily prepared from amino acids, and is readily cleaved by acid hydrolysis (5% CCl_3CO_2H, 4°, 5 min, 100% yield).[9]

1. R. W. Hanson and H. D. Law, *J. Chem. Soc.*, 7285 (1965).
2. J. Pless, *Helv. Chim. Acta*, **59,** 499 (1976).
3. H. E. Applegate, C. M. Cimarusti, J. E. Dolfini, P. T. Funke, W. H. Koster, M. S. Puar, W. A. Slusarchyk, and M. G. Young, *J. Org. Chem.*, **44,** 811 (1979).
4. M. Mutter and R. Hersperger, *Synthesis*, 198 (1989).
5. K. Barlos, D. Papaioannou, and D. Theodoropoulos, *J. Org. Chem.*, **47,** 1324 (1982).
6. L. Zervas and D. M. Theodoropoulos, *J. Am. Chem. Soc.*, **78,** 1359 (1956).
7. H. Nesvadba and Y. Roth, *Monatsh. Chem.*, **98,** 1432 (1967).
8. M. Bodansky, M. A. Bednarek, and A. Bodansky, *Int. J. Pept. Protein Res.*, **20,** 387 (1982).
9. Y. Lapidot, N. de Groot, M. Weiss, R. Peled, and Y. Wolman, *Biochim. Biophys. Acta*, **138,** 241 (1967).

12. *N*-9-Phenylfluorenylamine (PhF–NR_2): 9-(C_6H_5)–($C_{13}H_8$)–NR_2

Formation

1. 9-PhF–Br, $Pb(NO_3)_2$, CH_3CN, rt, 28 h, >80% yield.[1,2]

Cleavage

This group was reported to be 6000 times more stable to acid than the trityl group because of destabilization of the cation by the fluorenyl group.[3]

1. CF_3COOH, CH_3CN, H_2O, 0°, 1 h → rt, 1 h.

1. P. L. Feldman and H. Rapoport, *J. Org. Chem.*, **51,** 3882 (1986).
2. B. D. Christie and H. Rapoport, *J. Org. Chem.*, **50,** 1239 (1985).
3. R. Bolton, N. B. Chapman, and J. Shorter, *J. Chem. Soc.*, 1895 (1964).

13. *N*-2,7-Dichloro-9-fluorenylmethyleneamine: 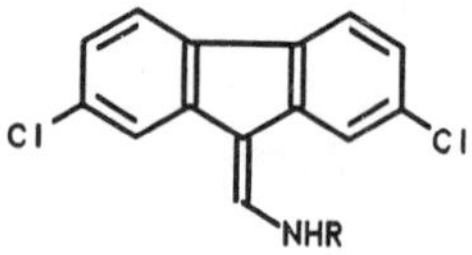
Cl
Cl
NHR

Formation/Cleavage[1]

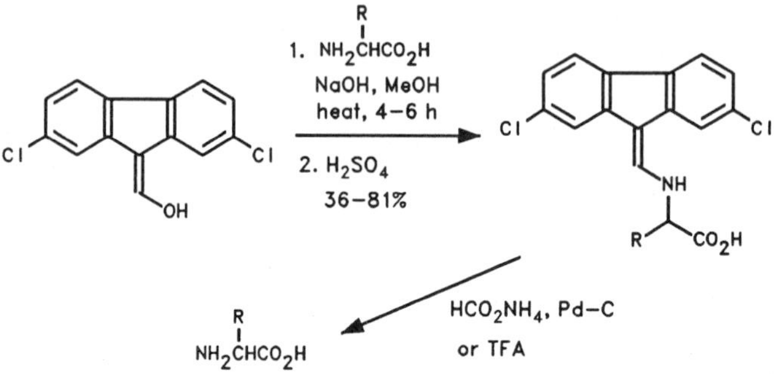
Cl
Cl
OH
R
1. NH2CHCO2H
NaOH, MeOH
heat, 4–6 h
2. H2SO4
36–81%
Cl
Cl
NH
R
CO2H
HCO2NH4, Pd–C
or TFA
R
NH2CHCO2H

1. L. A. Carpino, H. G. Chao, and J.-H. Tien, *J. Org. Chem.*, **54**, 4302 (1989).

14. *N*-Ferrocenylmethylamine (Fcm–NR_2): $C_{10}H_{10}FeCH_2NR_2$

The Fcm derivative is prepared from amino acids on treatment with formylferrocene and Pd-phthalocyanine by reductive alkylation (60–89% yield). It is cleaved with 2-thionaphthol/CF_3COOH. Its primary advantage is its color, making it easily detected.[1]

1. H. Eckert and C. Seidel, *Angew. Chem., Int. Ed. Engl.*, **25**, 159 (1986).

15. *N*-2-Picolylamine *N'*-Oxide: R_2NCH_2-2-pyridyl *N*-Oxide (Chart 10)

N-2-Picolylamine *N'*-oxide, used in oligonucleotide syntheses, is cleaved by acetic anhydride at 22°, followed by methanolic ammonia (85–95% yield).[1]

1. Y. Mizuno, T. Endo, T. Miyaoka, and K. Ikeda, *J. Org. Chem.*, **39**, 1250 (1974).

Imine Derivatives

A number of imine derivatives have been prepared as amine protective groups, but few of these have seen extensive use. The most widely used are the benzylidene and diphenylmethylene derivatives. The less used derivatives are listed, for completeness, with their references at the end of this section.

16. *N*-1,1-Dimethylthiomethyleneamine: $(MeS)_2C{=}NR$

This group was used to protect the nitrogen of glycine in a synthesis of amino acids.[1]

Formation

1. CS_2, TEA, $CHCl_3$, 20–40°, 1 h; MeI, reflux, 1 h, 77% yield.[2]

Cleavage

1. H_2O_2, HCO_2H, TsOH, 0° → 20°, 90% yield.[2]
2. HCl, H_2O, THF, rt, 100% yield.[2,3]

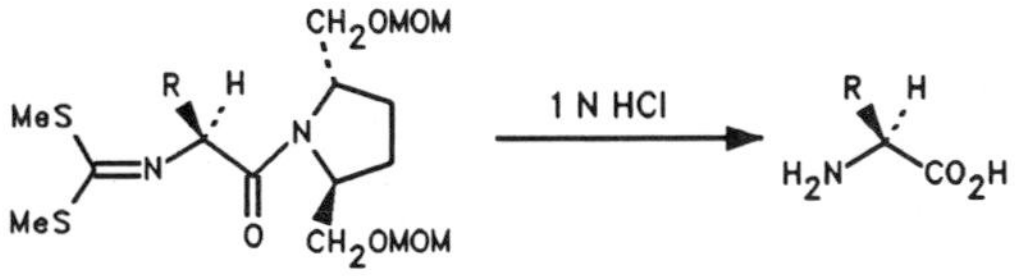

17. *N*-Benzylideneamine: RN=CHPh (Chart 10)

Most applications of this derivative have been for the preparation and modification of amino acids, although some applications in the area of carbohydrates have been reported. The derivative is stable to *n*-butyllithium and lithium diisopropylamide.

Formation

1. PhCHO, Et_3N, 80–90% yield.[4]
2. PhCHO, Na_2SO_4, benzene, rt, 99% yield.[5]

Cleavage

1. 1 *N* HCl, 25°, 1 h.[1,6]
2. H_2, Pd–C, CH_3OH.[7]
3. Hydrazine, EtOH, reflux, 6 h, 70% yield.[8]
4. Girard-T reagent, >75% yield.[9]

18. *N*-*p*-Methoxybenzylideneamine: $4\text{-}MeOC_6H_4CH{=}NR$

The *N*-*p*-methoxybenzylideneamine has been used to protect glucosamines.

Formation

1. 4-$MeOC_6H_4CHO$, benzene, pyridine, heat, >72% yield.[10]

Cleavage

1. MeOH, 10% aq. AcOH, $TsNHNH_2$, >81% yield.[7,11]
2. 5 *N* HCl.[12]

19. *N*-Diphenylmethyleneamine: $RN{=}CPh_2$

The derivative of glycine, prepared from benzophenone (cat. $BF_3 \cdot Et_2O$, xylene, reflux, 82% yield), has found considerable use in the preparation of amino acids. It can also be prepared by an exchange reaction with benzophenoneimine ($Ph_2C{=}NH$, CH_2Cl_2, rt).[13]

Cleavage

1. Concd. HCl, reflux, 6 h or aq. citric acid, 12 h.[14]
2. H_2, Pd–C, MeOH, rt, 14 h, 90% yield.[15]
3. NH_2OH, 3 min, pH 4–6.[16]

20. *N*-[(2-Pyridyl)mesityl]methyleneamine: $(C_5H_4N)(Me_3C_6H_2)C{=}NR$[17]

The imine, prepared from an amine and $(C_5H_4N)(Me_3C_6H_2)CO$ ($TiCl_4$, toluene, reflux, 12 h; NaOH, 80% yield), can be cleaved with concd. HCl (reflux). The protective group was used to direct α-alkylation of amines.

21. *N*-(*N*′,*N*′-Dimethylaminomethylene)amine: $RN{=}CHN(CH_3)_2$[18,19]

22. *N*,*N*′-Isopropylidenediamine[20] (Chart 10):

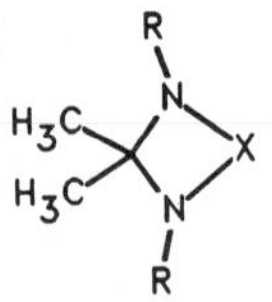

R
N
H_3C X
H_3C
N
R

23. *N*-*p*-Nitrobenzylideneamine: 4-$NO_2C_6H_4CH{=}NR$[21] (Chart 10)

24. *N*-Salicylideneamine: 2-HO–$C_6H_4CH{=}NR$[22] (Chart 10)

25. *N*-5-Chlorosalicylideneamine: 2-HO-5-ClC_6H_3CH=NR[23]

26. *N*-(5-Chloro-2-hydroxyphenyl)phenylmethyleneamine: RN=C(Ph)C_6H_3-2-OH-5-Cl[24,25]

27. *N*-Cyclohexylideneamine: C_6H_{11}N=CHR[26]

This imine is stable to the Fe(acac)$_3$-catalyzed Grignard coupling of aryl halides.

1. S. Ikegami, T. Hayama, T. Katsuki, and M. Yamaguchi, *Tetrahedron Lett.*, **27,** 3403 (1986); S. Ikegama, H. Uchiyama, T. Hayama, T. Katsuki, and M. Yamaguchi, *Tetrahedron*, **44,** 5333 (1988).
2. D. Hoppe and L. Beckmann, *Liebigs Ann. Chem.*, 2066 (1979).
3. W. Oppolzer, R. Moretti, and S. Thomi, *Tetrahedron Lett.*, **30,** 6009 (1989).
4. P. Bey and J. P. Vevert, *Tetrahedron Lett.*, 1455 (1977).
5. B. W. Metcalf and P. Casara, *Tetrahedron Lett.*, 3337 (1975).
6. D. Ferroud, J. P. Genet, and R. Kiolle, *Tetrahedron Lett.*, **27,** 23 (1986).
7. R. A. Lucas, D. F. Dickel, R. L. Dziemian, M. J. Ceglowski, B. L. Hensle, and H. B. MacPhillamy, *J. Am. Chem. Soc.*, **82,** 5688 (1960).
8. G. W. J. Fleet and I. Fleming. *J. Chem. Soc. C*, 1758 (1969).
9. T. Watanabe, S. Sugawara, and T. Miyadera, *Chem. Pharm Bull.*, **30,** 2579 (1982).
10. D. R. Mootoo and B. Fraser-Reid, *Tetrahedron Lett.*, **30,** 2363 (1989).
11. F. Baumberger, A. Vasella, and R. Schauer, *Helv. Chim. Acta*, 71, 429 (1988).
12. M. Bergmann and L. Zervas, *Ber.*, **64,** 975 (1931).
13. T. Hvidt, W. A. Szarek, and D. B. Maclean, *Can. J. Chem.*, **66,** 779 (1988).
14. M. J. O'Donnell, J. M. Boniece, and S. E. Earp, *Tetrahedron Lett.*, 2641 (1978).
15. L. Wessjohann, G. McGaffin, and A. de Meijere, *Synthesis*, 359 (1989).
16. K.-J. Fasth, G. Antoni, and B. Langström, *J. Chem. Soc., Perkin Trans. I*, 3081 (1988).
17. J. M. Hornback and B. Murugaverl, *Tetrahedron Lett.*, **30,** 5853 (1989).
18. J. Zemlicka, S. Chládek, A. Holy, and J. Smrt, *Collect. Czech. Chem. Commun.*, **31,** 3198 (1966).
19. J. J. Fitt and H. W. Gschwend, *J. Org. Chem.*, **42,** 2639 (1977).
20. P. M. Hardy and D. J. Samworth, *J. Chem. Soc., Perkin Trans. I*, 1954 (1977).
21. J. L. Douglas, D. E. Horning, and T. T. Conway, *Can. J. Chem.*, **56,** 2879 (1978).
22. J. N. Williams and R. M. Jacobs, *Biochem. Biophys. Res. Commun.*, **22,** 695 (1966).
23. J. C. Sheehan and V. J. Grenada, *J. Am. Chem. Soc.*, **84,** 2417 (1962).
24. B. Halpern and A. P. Hope, *Aust. J. Chem.*, **27,** 2047 (1974).
25. A. Abdipranoto, A. P. Hope, and B. Halpern, *Aust. J. Chem.*, **30,** 2711 (1977).
26. L. N. Pridgen, L. Snyder, and J. Prol, Jr., *J. Org. Chem.*, **54,** 1523 (1989).

Enamine Derivative

28. *N*-(5,5,-Dimethyl-3-oxo-1-cyclohexenyl)amine (Chart 10):

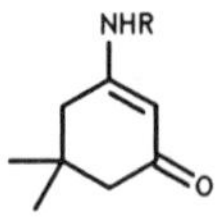

This vinylogous amide has been prepared in 70% yield to protect amino acid esters.[1] It is cleaved by treatment with either aqueous bromine[1] or nitrous acid (90% yield).[2]

1. B. Halpern and L. B. James, *Aust. J. Chem.*, **17,** 1282 (1964).
2. B. Halpern and A. D. Cross, *Chem. Ind. (London)*, 1183 (1965).

N-Hetero Atom Derivatives

Six categories of *N*-hetero atom derivatives are considered: N-M (M = boron, copper), N-N (e.g., *N*-nitro, *N*-nitroso), *N*-oxides (used to protect teriary amines), N-P (e.g., phosphinamides, phosphonamides), N-SiR_3 (R = CH_3), and N-S (e.g., sulfonamides, sulfenamides).

N-METAL DERIVATIVES

29. *N*-Borane Derivatives: $R_3N^+BH_3^-$

Aminoboranes can be prepared from diborane to protect a tertiary amine during oxidation; they are cleaved by refluxing in ethanol[1] or methanolic sodium carbonate.[2]

30. *N*-Diphenylborinic Acid Derivative

Formation/Cleavage[3,4]

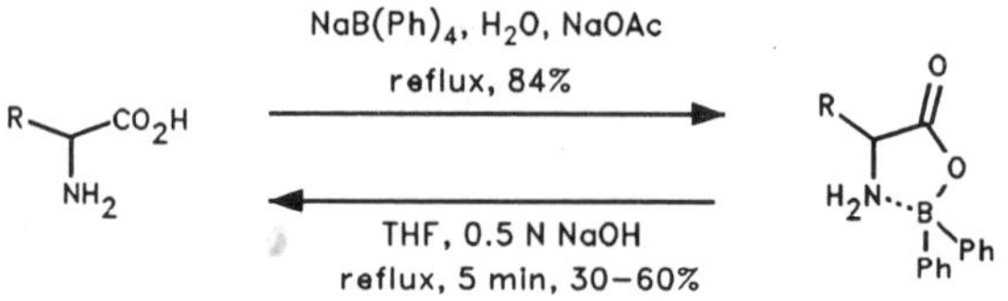

This derivative is stable to acetic acid and CF_3CO_2H.[4] The related diethylborinic acid derivative has been prepared from triethylborane (THF, reflux).[5]

1. A. Picot and X. Lusinchi, *Bull. Soc. Chim. Fr.*, 1227 (1977).
2. M. A. Schwartz, B. F. Rose, and B. Vishnuvajjala, *J. Am. Chem. Soc.*, **95,** 612 (1973).
3. I. Staatz, U. H. Granzer, A. Blume, and H. J. Roth, *Liebigs Ann. Chem.*, 127 (1989).
4. G. H. L. Nefkens and B. Zwanenburg, *Tetrahedron*, **39,** 2995 (1983).
5. F. Albericio, E. Nicolás, J. Rizo, M. Ruiz-Gayo, E. Pedroso, and E. Giralt, *Synthesis*, 119 (1990).

31. *N*-[Phenyl(pentacarbonylchromium- or -tungsten)carbenyl]amine

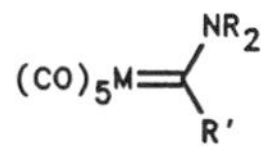

$(CO)_5M{=}C(NR_2)R'$

$R' = C_6H_5$ or CH_3; M = Cr or W

These transition metal carbenes, prepared in 66–97% yield from amino acid esters, are cleaved by acid hyrolysis (CF_3CO_2H, 20°, 80% yield; 80% AcOH; M = W, BBr_3, −25°).[1]

1. K. Weiss and E. O. Fischer, *Chem. Ber.*, **109,** 1868 (1976).

32. *N*-Copper or *N*-Zinc Chelate: $RNH_2 \cdots M \cdots OH$

M = Cu(II), Zn(II)

Formation/Cleavage

1. 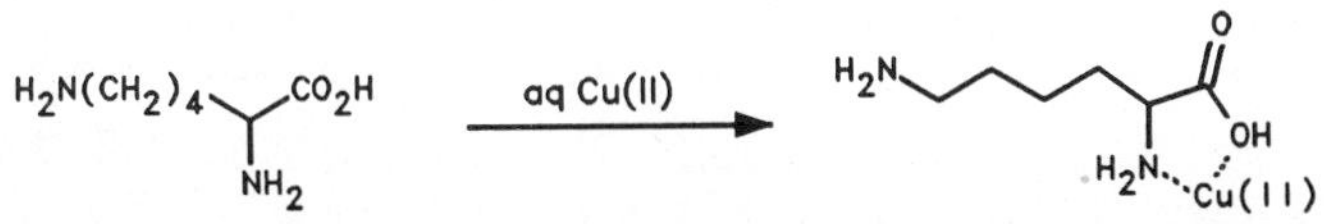 $H_2N(CH_2)_4CH(NH_2)CO_2H$ —aq Cu(II)→ $H_2N(CH_2)_4CH(NH_2)C(O)OH$ chelated to Cu(II)

   A copper chelate selectively protects the α-NH_2 group in lysine. The chelate is cleaved by 2 *N* HCl or by EDTA $(HO_2CCH_2)_2NCH_2CH_2N(CH_2CO_2H)_2$.[1]
2. In an aminoglycoside a vicinal amino hydroxy group can be protected as a Cu(II) chelate. After acylation of other amine groups, the chelate is cleaved by aqueous ammonia.[2] The copper chelate can also be cleaved with $Bu_2NC(S)NHBz$ (EtOH, reflux, 2 h).[3]
3. After examination of the complexing ability of Ca(II), Cr(III), Mn(II), Fe(III), Co(II), Ni(II), Cu(II), Zn(II), Ru(III), Ag(I), and Sn(IV), the authors decided that Zn(II) provides the best protection for vicinal amino hydroxy groups during trifluoroacetylation of other amino groups in the course of some syntheses of kanamycin derivatives.[4]

1. R. Ledger and F. H. C. Stewart, *Aust. J. Chem.*, **18,** 933 (1965).
2. S. Hanessian and G. Patil, *Tetrahedron Lett.*, 1035 (1978).
3. K. H. König, L. Kaul, M. Kuge, and M. Schuster, *Liebigs Ann. Chem.*, 1115 (1987).
4. T. Tsuchiya, Y. Takagi, and S. Umezawa, *Tetrahedron Lett.*, 4951 (1979).

N-N DERIVATIVES

33. *N*-Nitroamine: R_2NNO_2 (Chart 10)

Formation

$H_2NC(=NH)NH(CH_2)_3CH(NHPG)CO_2H$ $\xrightarrow[0°,\ 1\ h,\ 80\%]{HNO_3,\ H_2SO_4}$ $H_2NC(=NNO_2)NH(CH_2)_3CH(NHPG)CO_2H$

An *N*-nitro derivative is used primarily to protect the guanidino group in arginine; it is cleaved by reduction: H_2/Pd–C, AcOH/CH_3OH, ~80% yield;[1] 10% Pd–C/cyclohexadiene, 25°, 2 h, good yields;[2] Pd–C/4% HCO_2H–CH_3OH, 5 h, 100% yield;[3] $TiCl_3$/pH 6, 25°, 45 min, 70–98% yield;[4] $SnCl_2$/60% HCO_2H, 63% yield;[5] electrolysis, 1 *N* H_2SO_4, 1–6 h, 85–95% yield.[6]

1. K. Hofmann, W. D. Peckham, and A. Rheiner, *J. Am. Chem. Soc.*, **78,** 238 (1956).
2. A. M. Felix, E. P. Heimer, T. J. Lambros, C. Tzougraki, and J. Meienhofer, *J. Org. Chem.*, **43,** 4194 (1978).
3. B. ElAmin, G. M. Anantharamaiah, G. P. Royer, and G. E. Means, *J. Org. Chem.*, **44,** 3442 (1979).
4. R. M. Freidinger, R. Hirschmann, and D. F. Veber, *J. Org. Chem.*, **43,** 4800 (1978).
5. T. Hayakawa, Y. Fujiwara, and J. Noguchi, *Bull. Chem. Soc. Jpn.*, **40,** 1205 (1967).
6. P. M. Scopes, K. B. Walshaw, M. Welford, and G. T. Young, *J. Chem. Soc.*, 782 (1965).

34. *N*-Nitrosoamine: R_2NNO

N-Nitroso derivatives, prepared from secondary amines and nitrous acid, are cleaved by reduction (H_2/Raney Ni, EtOH, 28°, 3.5 h;[1] CuCl/concd. HCl[2]). Since many *N*-nitroso compounds are carcinogens, and because some racemization and cyclodehydration of *N*-nitroso derivatives of *N*-alkyl amino acids occur during peptide syntheses,[3,4] *N*-nitroso derivatives are of limited value as protective groups.

1. M. Harfenist and E. Magnein, *J. Am. Chem. Soc.*, **79,** 2215 (1957).
2. C. F. Koelsch, *J. Am. Chem. Soc.*, **68,** 146 (1946).
3. P. Quitt, R. O. Studer, and K. Vogler, *Helv. Chim. Acta*, **47,** 166 (1964).
4. F. H. C. Stewart, *Aust. J. Chem.*, **22,** 2451 (1969).

35. Amine *N*-Oxide: $R_3N \rightarrow O$ (Chart 10)

Amine oxides, prepared to protect tertiary amines during methylation[1,2] and to prevent their protonation in diazotized aminopyridines,[3] can be cleaved by reduction (e.g., SO_2/H_2O, 1 h, 22°, 63% yield;[1] H_2/Pd–C, AcOH, Ac_2O, 7 h, 91% yield;[2] Zn/HCl, 30% yield).[3] Photolytic reduction of an aromatic amine oxide has been reported [i.e., 4-nitropyridine *N*-oxide, 300 nm, $(MeO)_3PO/CH_2Cl_2$, 15 min, 85–95% yield].[4]

1. F. N. H. Chang, J. F. Oneto, P. P. T. Sah, B. M. Tolbert, and H. Rapoport, *J. Org. Chem.*, **15,** 634 (1950).
2. J. A. Berson and T. Cohen, *J. Org. Chem.*, **20,** 1461 (1955).
3. F. Koniuszy, P. F. Wiley, and K. Folkers, *J. Am. Chem. Soc.*, **71,** 875 (1949).
4. C. Kaneko, A. Yamamoto, and M. Gomi, *Heterocycles*, **12,** 227 (1979).

N-P DERIVATIVES

36. Diphenylphosphinamide (Dpp–NR_2): $Ph_2P(O)NR_2$ (Chart 10)

Phosphinamides are stable to catalytic hydrogenation, used to cleave benzyl-derived protective groups, and to hydrazine.[1] The rate of hydrolysis of phosphinamides is a function of the steric and electronic factors around the phosphorus.[2] This derivative has largely been used for the protection of amino acids and has seen little or no use in the general synthetic literature.

Formation

1. Ph_2POCl, *N*-methylmorpholine, 0°, 60–90% yield.[3]

Cleavage

The Dpp group is cleaved by the following acidic conditions: AcOH, HCOOH, H_2O, 24 h, 100% yield; 80% CF_3COOH, ~ quant; 0.4 *M* HCl, 90% CF_3CH_2OH, ~ quant; *p*-TsOH, H_2O–CH_3OH, ~ quant; 80% AcOH, 3 days, not completely cleaved.[2,3] The Dpp group is slightly less stable to acid than is the BOC group.[2,3]

37. Dimethylthiophosphinamide (Mpt–NR_2): $(CH_3)_2P(S)NR_2$ (Chart 10)

38. Diphenylthiophosphinamide (Ppt–NR_2): $Ph_2P(S)NR_2$

The Mpt and Ppt derivatives can be prepared from an amino acid and the thiophosphinyl chloride (Me_2PSCl or Ph_2PSCl, respectively, 41–78% yield, lysine gives 16% yield).[4] The Mpt group is cleaved with HCl or $Ph_3P\cdot HCl$[5] and is cleaved 60 times faster than the BOC group. The Ppt group is the more stable of the two groups.

39. Dialkyl Phosphoramidates: $(RO)_2P(O)NR_2$

Formation

1. $(EtO)_2POH$, CCl_4, aq. NaOH, $PhCH_2N^+Et_3\ Cl^-$, 0°, 1 h → 22°, 1 h, 75–90% yield.[6,7]
2. $(BuO)_2P(O)H$, Et_3N, CCl_4.[8]
3. $(i\text{-}PrO)_2P(O)Cl$, 73–93% yield.[9]

Cleavage

Phosphoramidates are cleaved with HCl saturated THF (70–94% yield). Their stability is dependent on the alkyl group, the methyl derivative being the least stable. They also have good stability to organic acids and Lewis acids.[9]

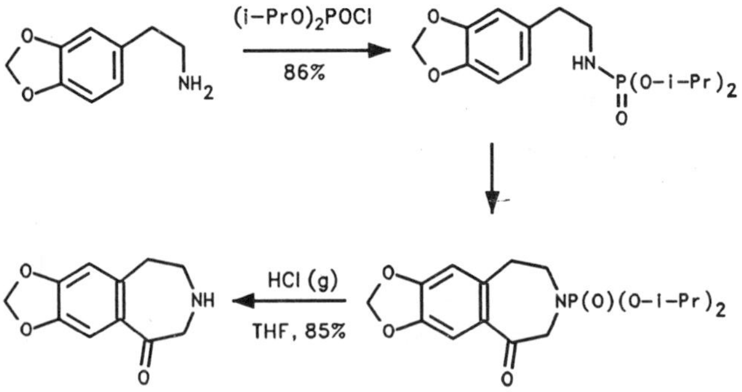

(i-PrO)$_2$POCl, 86%

HCl (g), THF, 85%

40. Dibenzyl Phosphoramidate: $(BnO)_2P(O)NR_2$

41. Diphenyl Phosphoramidate: $(PhO)_2P(O)NR_2$

Dibenzyl phosphoramidates have been prepared from amino acids and the phosphoryl chloride, $(BnO)_2P(O)Cl$.[10] A diphenyl phosphoramidate has been prepared from a glucosamine; it is converted by transesterification into a dibenzyl derivative to facilitate cleavage.[11]

1. G. W. Kenner, G. A. Moore, and R. Ramage, *Tetrahedron Lett.*, 3623 (1976).
2. R. Ramage, B. Atrash, D. Hopton, and M. J. Parrott, *J. Chem. Soc., Perkin Trans. I*, 1217 (1985).
3. R. Ramage, D. Hopton, M. J. Parrott, G. W. Kenner, and G. A. Moore, *J. Chem. Soc., Perkin Trans. I*, 1357 (1984).
4. S. Ikeda, F. Tonegawa, E. Shikano, K. Shinozaki, and M. Ueki, *Bull. Chem. Soc. Jpn.*, **52,** 1431 (1979).
5. M. Ueki, T. Inazu, and S. Ikeda, *Bull. Chem. Soc. Jpn.*, **52,** 2424 (1979).
6. A. Zwierzak, *Synthesis*, 507 (1975).

7. A. Zwierzak and K. Osowska, *Synthesis*, 223 (1984).
8. Y.-F. Zhao, S.-K. Xi, A.-T. Song, and G.-J. Ji, *J. Org. Chem.*, **49,** 4549 (1984).
9. Y. F. Zhao, G. J. Ji, S. K. Xi, H. G. Tang, A. T. Song, and S. Z. Wei, *Phosphorus Sulfur*, **18,** 155 (1983).
10. A. Cosmatos, I. Photaki, and L. Zervas, *Chem. Ber.*, **94,** 2644 (1961).
11. M. L. Wolfrom, P. J. Conigliaro, and E. J. Soltes, *J. Org. Chem.*, **32,** 653 (1967).

N-Si

For the most part silyl derivatives such as trimethylsilylamines have not been used extensively for amine protection because of their high reactivity to moisture, although they do provide satisfactory protection when prepared and used under anhydrous conditions.[1,2] The more stable and sterically demanding *t*-butyldiphenylsilyl group has been used to protect primary amines in the presence of secondary amines, thus allowing selective acylation or alkylation of the secondary amine.[3] Silylamines are reported not to be stable to oxidative conditions.[3] Silylamines are readily cleaved in the presence of silyl ethers.[4] For a more thorough discussion of silylating reagents, the section on alcohol protection should be consulted since many of the reagents described there will also silylate amines.

1. J. R. Pratt, W. D. Massey, F. H. Pinkerton, and S. F. Thames, *J. Org. Chem.*, **40,** 1090 (1975).
2. A. B. Smith, III, M. Visnick, J. N. Haseltine, and P. A. Sprengeler, *Tetrahedron*, **42,** 2957 (1986).
3. L. E. Overman, M. E. Okazaki, and P. Mishra, *Tetrahedron Lett.*, **27,** 4391 (1986).
4. T. P. Mawhinney and M. A. Madson, *J. Org. Chem.*, **47,** 3336 (1982).

N-S DERIVATIVES

N-SULFENYL DERIVATIVES

Sulfenamides, R_2NSR', prepared from an amine and a sulfenyl halide,[1,2] are readily cleaved by acid hydrolysis and have been used in syntheses of peptides, penicillins, and nucleosides. They are also cleaved by nucleophiles,[3] and by Raney nickel desulfurization.[4]

42. Benzenesulfenamide: $R_2NSC_6H_5$, **A** (Chart 10)

43. *o*-Nitrobenzenesulfenamide (Nps–NR₂): $R_2NSC_6H_4$-*o*-NO_2, **B** (Chart 10)

44. 2,4-Dinitrobenzenesulfenamide: $R_2NSC_6H_3$-2,4-$(NO_2)_2$, **C**

45. Pentachlorobenzenesulfenamide: $R_2NSC_6Cl_5$, **D**

Benzenesulfenamide, and a number of substituted benzenesulfenamides (compounds **B**, **C** and **D**) have been prepared to protect the 7-amino group in cephalosporins. They are cleaved by sodium iodide (CH_3OH, CH_2Cl_2, AcOH, 0°, 20 min, 53% yield from sulfenamide **B**.)[5] The *o*-nitrobenzenesulfenamide has been used for the protection of amino acids.[6,7] *o*-Nitrobenzenesulfenamides, **B**, are also cleaved by acidic hydrolysis (HCl/Et_2O or EtOH, 0°, 1 h, 95% yield),[8] by nucleophiles (13 reagents, 5 min–12 h, 90% cleaved;[3] PhSH or $HSCH_2CO_2H$, 22°, 1 h;[9] 2-mercaptopyridine/CH_2Cl_2, 1 min, 100% yield;[10] NH_4SCN, 2-methyl-1-indolylacetic acid[7]), and by catalytic desulfurization (Raney Ni/DMF, column, a few hours, satisfactory yields).[4] The 2,4-dinitrobenzenesulfenamide is cleaved with *p*-thiocresol/TsOH.[11]

46. 2-Nitro-4-methoxybenzenesulfenamide: $R_2NSC_6H_3$-2-NO_2-4-OCH_3

This sulfenamide, prepared from an amino acid, the sulfenyl chloride, and sodium bicarbonate, is cleaved by acid hydrolysis (HOAc/dioxane, 22°, 30 min, 95% yield).[12]

47. Triphenylmethylsulfenamide: $R_2NSC(C_6H_5)_3$

The tritylsulfenamide can be prepared from an amine and the sulfenyl chloride (Na_2CO_3, THF, H_2O or Pyr, CH_2Cl_2, 64–96% yield);[13] it is cleaved by hydrogen chloride in ether or ethanol (0°, 1 h, 90% yield),[8] $CuCl_2$ (THF, EtOH, 58–67% yield), Me_3SiI (77–96% yield),[13] I_2 (0.1 *M*, THF, collidine, H_2O, 97% yield).[14] The tritylsulfenamide is stable to 1 *N* HCl, base, $NaCNBH_3$, $LiAlH_4$, *m*-chloroperoxybenzoic acid, pyridinium chlorochromate, Jones reagent, Collins oxidation, and Moffat oxidation. The stability of this group is due largely to steric hindrance.

48. 3-Nitropyridinesulfenamide (Npys–NR_2)

This group, which is more stable than the 2-nitrobenzenesulfenamide, has been developed to protect amino acids. It is readily introduced with the sulfenyl chloride (52–74% yield) and is cleaved with triphenylphosphine or 2-thiopyridine *N*-oxide. It is stable to CF_3COOH but can be cleaved with 0.1 *M* HCl.[15]

1. For other methods of preparation, see F. A. Davis and U. K. Nadir, *Org. Prep. Proc. Int.*, **11,** 33 (1979).
2. For a review of sulfenamides, see: L. Craine and M. Raban, *Chem. Rev.*, **89,** 689 (1989).
3. W. Kessler and B. Iselin, *Helv. Chim. Acta*, **49,** 1330 (1966).
4. J. Meienhofer, *Nature*, **205,** 73 (1965).
5. T. Kobayashi, K. Iino, and T. Hiraoka, *J. Am. Chem. Soc.*, **99,** 5505 (1977).
6. S. Romani, G. Bovermann, L. Moroder, and E. Wünsch, *Synthesis*, 512 (1985).
7. I. F. Luescher and C. H. Schneider, *Helv. Chim. Acta*, **66,** 602 (1983).

8. L. Zervas, D. Borovas, and E. Gazis, *J. Am. Chem. Soc.*, **85,** 3660 (1963).
9. A. Fontana, F. Marchiori, L. Moroder, and E. Schoffone, *Tetrahedron Lett.*, 2985 (1966).
10. M. Stern, A. Warshawsky, and M. Fridkin, *Int. J. Pept. Protein Res.*, **13,** 315 (1979).
11. E. M. Gordon, M. A. Ondetti, J. Pluscec, C. M. Cimarusti, D. P. Bonner, and R. B. Sykes, *J. Am. Chem. Soc.*, **104,** 6053 (1982).
12. Y. Wolman, *Isr. J. Chem.*, **5,** 231 (1967).
13. B. P. Branchaud, *J. Org. Chem.*, **48,** 3538 (1983).
14. H. Takaku, K. Imai, and M. Nagai, *Chem. Lett.*, 857 (1988).
15. R. Matsueda and R. Walter, *Int. J. Pept. Protein Res.*, **16,** 392 (1980).

N-SULFONYL DERIVATIVES

Sulfonamides (R_2NSO_2R') are prepared from an amine and sulfonyl chloride in the presence of pyridine or aqueous base.[1] The sulfonamide is one of the most stable nitrogen protective groups. Arylsulfonamides are stable to alkaline hydrolysis and to catalytic reduction; they are cleaved by Na/NH_3,[2] Na/butanol,[3] sodium naphthalenide,[4] or sodium anthracenide,[5] and by refluxing in acid (48% HBr/cat. phenol).[6] Sulfonamides of less basic amines such as pyrroles and indoles are much easier to cleave than are those of the more basic alkyl amines. In fact, sulfonamides of the less basic amines (pyrroles, indoles, and imidazoles) can be cleaved by basic hydrolysis, which is almost impossible for the alkyl amines. Because of the inherent differences between the aromatic —NH group and simple aliphatic amines, the protection of these compounds (pyrroles, indoles, and imidazoles) will be described in a separate section. One appealing property of sulfonamides is that the derivatives are more crystalline than amides or carbamates.

1. E. Fischer and W. Lipschitz, *Ber.*, **48,** 360 (1915).
2. V. du Vigneaud and O. K. Behrens, *J. Biol. Chem.*, **117,** 27 (1937).
3. G. Wittig, W. Joos, and P. Rathfelder, *Justus Liebigs Ann. Chem.* **610,** 180 (1957).
4. S. Ji, L. B. Gortler, A. Waring, A. Battisti, S. Bank, W. D. Closson, and P. Wriede, *J. Am. Chem. Soc.*, **89,** 5311 (1967).
5. K. S. Quaal, S. Ji, Y. M. Kim, W. D. Closson, and J. A. Zubieta, *J. Org. Chem.*, **43,** 1311 (1978).
6. H. R. Synder and R. E. Heckert, *J. Am. Chem. Soc.*, **74,** 2006 (1952).

49. *p*-Toluenesulfonamide ($TsNR_2$): *p*-$CH_3C_6H_4SO_2NR_2$ (Chart 10)

50. Benzenesulfonamide: $PhSO_2NR_2$

Formation

Tosylates are generally formed from an amine and tosyl chloride in an inert solvent such as CH_2Cl_2 with an acid scavenger such as pyridine or triethylamine.

Cleavage

1. HBr, AcOH, 70°, 8 h, 45–50% yield.[1] During the synthesis of L-2-amino-3-oxalylaminopropionic acid, a neurotoxin, cleavage with Na/NH_3 or $[C_{10}H_8\cdot]^-$ Na^+ gave a complex mixture of products.
2. $NaAlH_2(OCH_2CH_2OCH_3)_2$, benzene or toluene, reflux, 20 h, 65–75% yield.[2] Note that $LiAlH_4$ does not cleave sulfonamides of primary amines; those from secondary amines must be heated to 120°.
3. Electrolysis, $Me_4N^+Cl^-$, 5°, 65–98% yield.[3-5]
4. Sodium naphthalenide.[6,7] This reagent has been used to remove the tosyl group from an amide.[8]
5. Li, NH_3, 75% yield[9] or Na, NH_3.[10,11]
6. 48% HBr, phenol, 30 min, heat, 85% yield.[5,12]
7. $HClO_4$, AcOH, 100°, 1 h, 30–75% yield.[13]
8. $h\nu$, Et_2O, 6–20 h, 85–90% yield.[14,15]
9. $h\nu$, EtOH, H_2O, $NaBH_4$, 1,2-dimethoxybenzene.[16] This is a photosensitized electron-transfer reaction. Other reductants such a hydrazine and $BH_3\cdot NH_3$ are also effective.

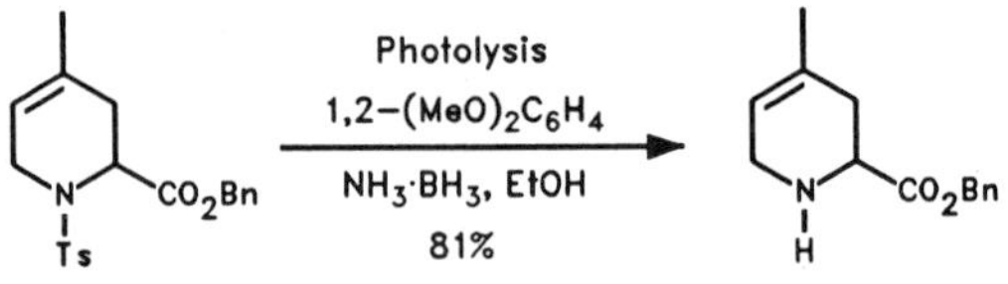

Photolysis
1,2-$(MeO)_2C_6H_4$
$NH_3\cdot BH_3$, EtOH
81%

10. Na(Hg), Na_2HPO_4.[17,18]

Na(Hg)
Na_2HPO_4, MeOH

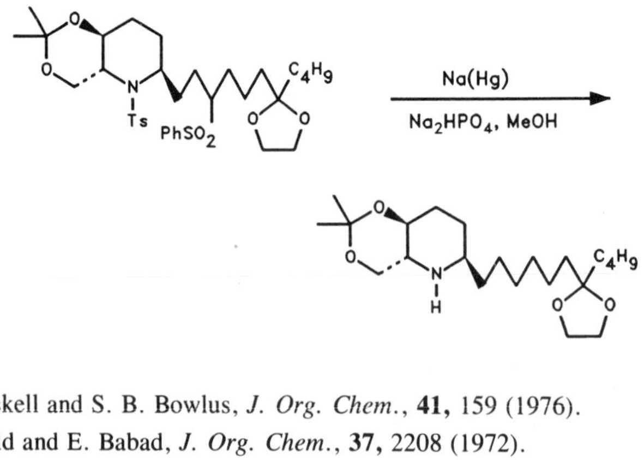

1. B. E. Haskell and S. B. Bowlus, *J. Org. Chem.*, **41,** 159 (1976).
2. E. H. Gold and E. Babad, *J. Org. Chem.*, **37,** 2208 (1972).
3. L. Horner and H. Neumann, *Chem. Ber.*, **98,** 3462 (1965).
4. T. Moriwake, S. Saito, H. Tamai, S. Fujita, and M. Inaba, *Heterocycles*, **23,** 2525 (1985).
5. R. C. Roemmele and H. Rapoport, *J. Org. Chem.*, **53,** 2367 (1988).

6. J. M. McIntosh and L. C. Matassa, *J. Org. Chem.*, **53,** 4452 (1988).
7. C. H. Heathcock, T. A. Blumenkopf, and K. M. Smith, *J. Org. Chem.*, **54,** 1548 (1989).
8. H. Nagashima, N. Ozaki, M. Washiyama, and K. Itoh, *Tetrahedron Lett.*, **26,** 657 (1985); J. R. Henry, L. R. Marcin, M. C. McIntosh, P. M. Scola, G. D. Harris, Jr., and S. M. Weinreb, *Tetrahedron Lett.*, **30,** 5709 (1989).
9. C. H. Heathcock, K. M. Smith, and T. A. Blumenkopf, *J. Am. Chem. Soc.*, **108,** 5022 (1986).
10. A. G. Schultz, P. J. McCloskey, and J. J. Court, *J. Am. Chem. Soc.*, **109,** 6493 (1987).
11. N. Yamazaki and C. Kibayashi, *J. Am. Chem. Soc.*, **111,** 1396 (1989).
12. R. S. Compagnone and H. Rapoport, *J. Org. Chem.*, **51,** 1713 (1986).
13. D. P. Kudav, S. P. Samant, and B. D. Hosangadi, *Synth. Commun.*, **17,** 1185 (1987).
14. A. Abad, D. Mellier, J. P. Pète, and C. Portella, *Tetrahedron Lett.*, 4555 (1971).
15. W. Yuan, K. Fearson, and M. H. Gelb, *J. Org. Chem.*, **54,** 906 (1989).
16. T. Hamada, A. Nishida, and O. Yonemitsu, *J. Am. Chem. Soc.*, **108,** 140 (1986).
17. T. N. Birkinshaw and A. B. Holmes, *Tetrahedron Lett.*, **28,** 813 (1987).
18. F. Chavez and A. D. Sherry, *J. Org. Chem.*, **54,** 2990 (1989).

51. 2,3,6-Trimethyl-4-methoxybenzenesulfonamide (Mtr–NR_2)[1]

52. 2,4,6-Trimethoxybenzenesulfonamide (Mtb–NR_2)[1] (Chart 10)

53. 2,6-Dimethyl-4-methoxybenzenesulfonamide (Mds–NR_2)[2]

54. Pentamethylbenzenesulfonamide (Pme–NR_2)[2]

55. 2,3,5,6-Tetramethyl-4-methoxybenzenesulfonamide (Mte–NR_2)[2]

56. 4-Methoxybenzenesulfonamide (Mbs–NR_2)[2]

57. 2,4,6-Trimethylbenzenesulfonamide (Mts–NR_2)[3]

58. 2,6-Dimethoxy-4-methylbenzenesulfonamide (iMds–NR_2)[3]

These sulfonamides have been used to protect the guanidino group of arginine.[4] Their acid stability as determined by TFA cleavage of the N^G-Arg derivative (25°, 60 min) is as follows: Mtr (52%) > Mds (22%) ≈ Mtb (20%) > Pme (2%) > Mte (1.6%) > Mts ≈ Mbs > iMbs. The Mtr group has been used to protect the ε-nitrogen of lysine. The percent (%) cleavage of Lys (Mtr) in various acids is as follows (MSA = methanesulfonic acid):[5]

	0.15 *M* MSA TFA, PhSMe (9:1) 20°	0.3 *M* MSA TFA, PhSMe (9:1) 20°	TFA PhSMe (9:1) 50°	HF PhSMe 0°	MSA PhSMe 20°	TFA 20°
1 h	80.7	95.1	15.1	3.6	2.3	0
2 h	91.9	99.3	33.6	—	—	0

The rate of cleavage is 4–5 times faster if dimethyl sulfide is included in the TFA/PhSMe mixture.[6]

59. 2,2,5,7,8-Pentamethylchroman-6-sulfonamide (Pmc–NR_2):

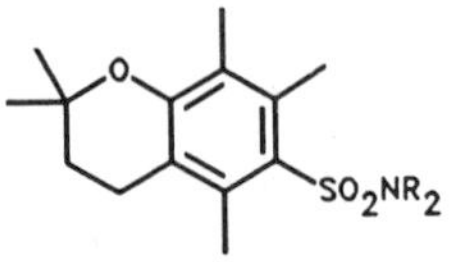

SO_2NR_2

This group was developed for the protection of N^G-Arg. It is effectively an analog of the Mtr group, but has the useful property that it is cleaved in TFA/PhSMe in only 20 min. The enhanced rate of cleavage is attributed to the forced overlap of the oxygen electrons with the incipient cation during cleavage. The Pmc group can also be cleaved with 50% TFA/CH_2Cl_2, which does not cleave the benzyl carbamate.[7,8]

1. E. Atherton, R. C. Sheppard, and J. D. Wade, *J. Chem. Soc., Chem. Commun.*, 1060 (1983).
2. M. Wakimasu, C. Kitada, and M. Fujino, *Chem. Pharm. Bull.*, **29**, 2592 (1981).
3. H. Yajima, K. Akaji, K. Mitani, N. Fujii, S. Funakoshi, H. Adachi, M. Oishi, and Y. Akazawa, *Int. J. Pept. Protein Res.*, **14**, 169 (1979).
4. M. Fujino, M. Wakimasu and C. Kitada, *Chem. Pharm. Bull.*, **29**, 2825 (1981); M. Fujino, O. Nishimura, M. Wakimasu, and C. Kitada, *J. Chem. Soc., Chem. Commun.*, 668 (1980).
5. M. Wakimasu, C. Kitada, and M. Fujino, *Chem. Pharm. Bull.*, **30**, 2766 (1982).
6. K. Saito, T. Higashijima, T. Miyazawa, M. Wakimasu, and M. Fujino, *Chem. Pharm. Bull.*, **32**, 2187 (1984).
7. R. Ramage and J. Green, *Tetrahedron Lett.*, **28**, 2287 (1987).
8. J. Green, O. M. Ogunjobi, R. Ramage, A. S. J. Stewart, S. McCurdy, and R. Noble, *Tetrahedron Lett.*, **29**, 4341 (1988).

60. Methanesulfonamide (Ms–NR_2): $CH_3SO_2NR_2$

The mesylate group, introduced with methanesulfonyl chloride, can be cleaved with lithium aluminum hydride and dissolving metal reduction (Na, *t*-BuOH, HMPT, NH_3, 64% yield).[1]

61. β-Trimethylsilylethanesulfonamide (SES–NR_2): $Me_3SiCH_2CH_2SO_2NR_2$

Formation

1. SES—Cl, Et_3N, DMF, 0°, 88–95% yield.[2]

Cleavage

The SES group is stable to TFA, hot 6 *M* HCl-THF, $LiBH_4$, CH_3CN, $BF_3 \cdot Et_2O$, 40% HF/EtOH.

1. DMF, CsF, 95°, 9–40 h, 80–93% yield.[2]
2. $Bu_4N^+F^-$, CH_3CN, reflux, >85% yield.[2,3]

62. 9-Anthracenesulfonamide

This group was used to protect the guanidine nitrogen of arginine. It is cleaved by hydrogenation (H_2, Pd–C, 24 h), SmI_2 (THF, *t*-BuOH), Al(Hg) (H_2O, pH 7), photolysis, TFA/anisole.[4]

63. 4-(4′,8′-Dimethoxynaphthylmethyl)benzenesulfonamide (DNMBS—NR_2):

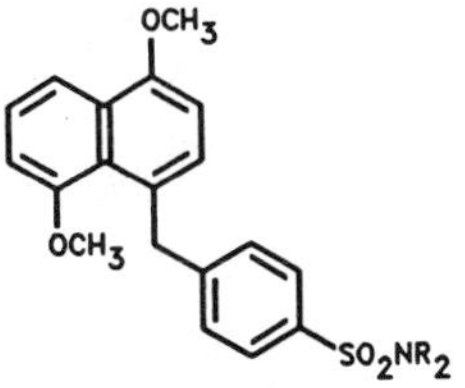

OCH_3

OCH_3

SO_2NR_2

The DNMBS derivative, readily prepared from an amine and the sulfonyl chloride, is efficiently ($\phi = 0.65$) cleaved photochemically ($h\nu$ >300 nm, EtOH, $NH_3 \cdot BH_3$, 77–91% yield).[5]

1. P. Merlin, J. C. Braekman, and D. Daloze, *Tetrahedron Lett.*, **29,** 1691 (1988).
2. S. M. Weinreb, D. M. Demko, T. A. Lessen, and J. P. Demers, *Tetrahedron Lett.*, **27,** 2099 (1986).
3. R. S. Garigipati and S. M. Weinreb, *J. Org. Chem.*, **53,** 4143 (1988).
4. H. B. Argens and D. S. Kemp, *Synthesis*, 32 (1988).
5. T. Hamada, A. Nishida, and O. Yonemitsu, *Tetrahedron Lett.*, **30,** 4241 (1989).

64. Benzylsulfonamide: $C_6H_5CH_2SO_2NR_2$ (Chart 10)

Benzylsulfonamides, prepared in 40–70% yield, are cleaved by reduction (Na, NH_3, 75% yield; H_2, Raney Ni, 65–85% yield, but not by H_2, PtO_2) and by acid hydrolysis (HBr or HI, slow).[1] They are also cleaved by photolysis (2–4 h, 40–90% yield).[2] The similar *p*-methylbenzylsulfonamide (PMS—NR_2) has been prepared to protect the ϵ-amino group in lysine; it is quantitatively cleaved by anhydrous hydrogen fluoride/anisole (−20°, 60 min).[3] Another example of this seldom used group is illustrated below.[4]

Formation

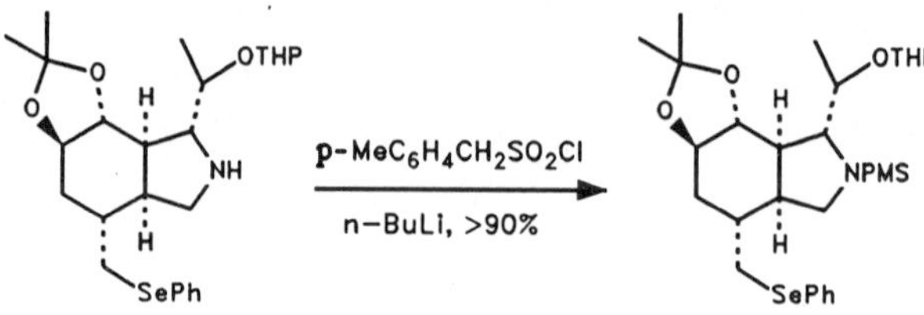
OTHP
NH
SePh
p-MeC6H4CH2SO2Cl
n-BuLi, >90%
NPMS

Cleavage

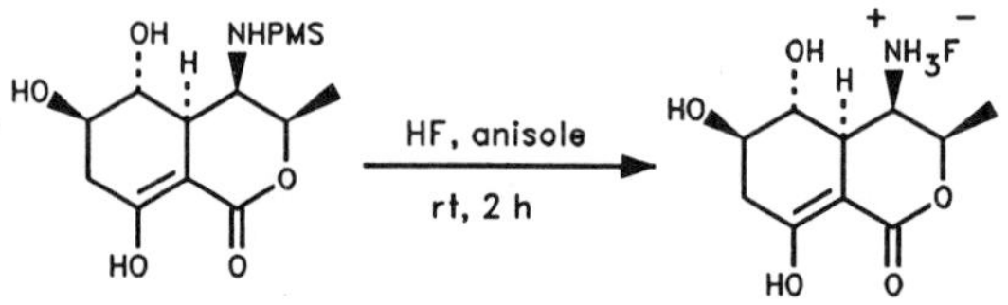
OH
NHPMS
HO
HF, anisole
rt, 2 h
NH3F

1. H. B. Milne and C.-H. Peng, *J. Am. Chem. Soc.*, **79,** 639, 645 (1957).
2. J. A. Pinock and A. Jurgens, *Tetrahedron Lett.*, 1029 (1979).
3. T. Fukuda, C. Kitada, and M. Fujino, *J. Chem. Soc., Chem. Commun.*, 220 (1978).
4. M. Yoshioka, H. Nakai, and M. Ohno, *J. Am. Chem. Soc.*, **106,** 1133 (1984).

65. Trifluoromethylsulfonamide: $R_2NSO_2CF_3$ (Chart 10)

A trifluoromethylsulfonamide can be prepared from a primary amine to allow monoalkylation of that amine.[1]

Formation

1. $(CF_3SO_2)_2O$, CH_2Cl_2, −78°, ~quant.[1]

Cleavage

1. $NaAlH_2(OCH_2CH_2OCH_3)_2$, benzene, reflux, a few minutes, 95% yield.[1]
2. 4-Br-$C_6H_4COCH_2Br$, K_2CO_3, acetone, 12 h; H_3O^+, 80% yield.[2]
3. $LiAlH_4$, Et_2O, reflux, 90–95% yield.[1]

1. J. B. Hendrickson and R. Bergeron, *Tetrahedron Lett.*, 3839 (1973).
2. J. B. Hendrickson, R. Bergeron, A. Giga, and D. Sternbach, *J. Am. Chem. Soc.*, **95,** 3412 (1973).

66. Phenacylsulfonamide: $R_2NSO_2CH_2COC_6H_5$ (Chart 10)

Like the trifluoromethylsulfonamides, phenacylsulfonamides are used to prevent dialkylation of primary amines. Phenacylsulfonamides are prepared in 91–94% yield from the sulfonyl chloride and cleaved in 66–77% yield by Zn/AcOH/trace HCl.[1]

1. J. B. Hendrickson and R. Bergeron, *Tetrahedron Lett.*, 345 (1970).

PROTECTION FOR IMIDAZOLES, PYRROLES, AND INDOLES

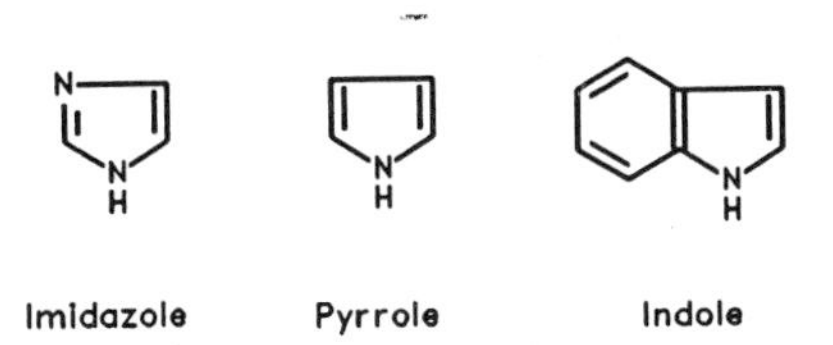

Protective group chemistry for these amines has been separated from the simple amines because chemically they behave quite differently with respect to protective group cleavage. The increased acidity of these aromatic amines makes it easier to cleave the various amide, carbamate, and sulfonamide groups that are used to protect this class. A similar situation arises in the deprotection of nucleoside bases (e.g., the isobutanamide is cleaved with methanolic ammonia[1]), again, because of the increased acidity of the NH group.

N-SULFONYL DERIVATIVES

1. *N,N*-Dimethylsulfonamide: $R_2N\text{-}SO_2NMe_2$

Formation

1. Imidazole, Me_2NSO_2Cl, Et_3N, PhH, 16 h, 95% yield.[2]

Cleavage

1. 2 *M* HCl, reflux, 4 h.[2,3]
2. 2% KOH, H_2O, reflux, 12 h, 64–92% yield.[3] This group is more stable to *n*-BuLi than is the benzyl group when used to protect imidazoles.

2. Mesitylenesulfonamide (Mts–NR$_2$): $R_2N–SO_2–C_6H_2–(2,4,6\text{-}CH_3)_3$

Formation/Cleavage[4]

$$\text{Z(OMe)Trp–OBn} \xrightarrow[\text{Cetyl(Me)}_3\text{N}^+\text{Cl}^-]{\text{MtsCl, NaOH, CH}_2\text{Cl}_2} \text{Z(OMe)Trp(N}^{in}\text{-Mts)–OBn}$$

The Mts group is stable to CF_3COOH, 1 *N* NaOH, hydrazine, 4 *N* HCl, 25% HBr–AcOH, and H_2–Pd, but is cleaved with 1 *M* CF_3SO_3H/CF_3COOH/thioanisole or CH_3SO_3H/CF_3COOH/thioanisole. Thioanisole is required to obtain clean conversions. The Mts group is not efficiently cleaved by HF.

3. *p*-Methoxyphenylsulfonamide (Mps–NR$_2$): $R_2N–SO_2–C_6H_4–4\text{-}OCH_3$

Formation

1. *p*-MeO–$C_6H_4SO_2Cl$, (Imidazole = His).[5,6]

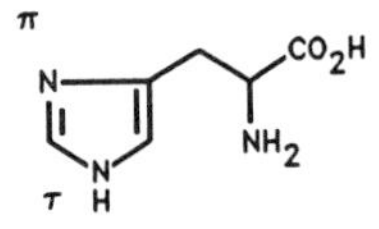

Histidine (His)

Cleavage

1. CF_3COOH, Me_2S, 40–60 min, 100% [Imidazole = His(Mps)].[7] This group is also cleaved with hydrazine, 1 *N* NaOH, HOBT, and HF.[7] The Mps group on histidine is stable to CF_3COOH/anisole and to 25% HBr/AcOH.

4. *p*-Toluenesulfonamide (TsNR$_2$): $R_2N–SO_2C_6H_4–4\text{-}CH_3$

Formation

1. Imidazole, *p*-toluenesulfonyl chloride, Et_3N.[8,9]

Cleavage

1. Ac_2O, pyridine; H_2O or trifluoroacetic anhydride, pyridine, 0.5–16 h, 95–100% yield, [Imidazole = His(Tos)].[5,9]
2. 1-Hydroxybenzotriazole (HOBT), THF, 1 h, [Imidazole = His(Tos)].[6]
3. Pyridine/HCl, DMF, [Imidazole = His(Tos)].[10]
4. HF, [Imidazole = His(Tos)].
5. CF_3CO_2H, Me_2S, 40–60 min, 100% yield, [Imidazole = His(Tos)].[11] The related phenylsulfonyl group has been used to protect pyrroles and indoles, and is cleaved with NaOH/H_2O/dioxane, rt, 2 h.[12,13]
6. KOH, MeOH, 98% yield (indole deprotection).[14,15]

CARBAMATES

5. 2,2,2-Trichloroethyl Carbamate (Troc–NR$_2$): $R_2NCO_2CH_2CCl_3$

Formation/Cleavage[16]

$$\text{BOC–TrpOBn} \xrightarrow{\text{TrocCl, NaOH, Bu}_4\text{NH}^+\text{SO}_4^-} \text{BOC–Trp(N}^{in}\text{-Troc)OBn}$$

The Troc group on tryptophan is stable to CF_3COOH, CF_3SO_3H, and H_2–Pd, but can be cleaved with 0.01 *M* NaOH/MeOH, hydrazine/MeOH/H_2O, Cd/AcOH/DMF. Cleavage with Zn/AcOH is only partially complete.

6. 2-(Trimethylsilyl)ethyl Carbamate (Teoc–NR$_2$): $R_2NCO_2CH_2CH_2Si(CH_3)_3$

The Teoc group is introduced onto pyrroles or indoles with 4-nitrophenyl 2-(trimethylsilyl)ethyl carbonate and NaH in 61–64% yield. The Teoc group can be removed with $Bu_4N^+F^-$ in CH_3CN.[17]

7. *t*-Butyl Carbamate (BOC–NR$_2$): $R_2N–CO_2–t–C_4H_9$

Formation

The BOC group has been introduced onto the imidazole nitrogen of histidine with BOCF (pH 7–8),[18] BOCN$_3$, MgO,[19] and $(BOC)_2O$.[17,20] It can be introduced onto pyrroles and indoles with phenyl *t*-butyl carbonate and NaH, 67–91% yield,[21] or with NaH, BOCN$_3$.[22]

Cleavage

The N^{im}-BOC group can be removed under the usual conditions for removing the BOC group: CF_3COOH and HF. It can also be removed with hydrazine and NH_3/MeOH. NaOMe/MeOH/THF has been used to remove the BOC group from pyrroles in 66–99% yield.[22] Thermolysis at 180° cleaves the BOC group from indoles and pyrroles in 92–99% yield.[23]

8. 1,1-Dimethyl-2,2,2-trichloroethyl Carbamate (TcBOC–NR$_2$): $R_2NCO_2C(CH_3)_2CCl_3$

Formation/Cleavage[24]

Forward: $Cl_3CC(Me)_2OCOCl$, TEA, CH_2Cl_2, 80%

Reverse: Zn, AcOH, MeOH, 20 min, 76%

Product N-substituent: $OC(Me)_2CCl_3$

9. 1-Adamantyl Carbamate (Adoc–NR$_2$): R$_2$NCO$_2$-1-adamantyl

Formation

1. AdocCl, histidine, NaOH, Na_2CO_3, H_2O, 86% yield; forms N$^{\alpha}$, N^{im}-Adoc–His(Adoc)OH.[25]

Cleavage

The Adoc group can be cleaved by the same methods used to cleave the BOC group. The Adoc group is somewhat more stable than is the BOC group to acid.

N-Alkyl and *N*-Aryl Derivatives

10. *N*-Vinylamine: $CH_2{=}CH{-}NR_2$

The vinyl group has been used to protect the nitrogen of benzimidazole during metalation with lithium diisopropylamide. It is introduced with vinyl acetate [$Hg(OAc)_2$, H_2SO_4, reflux, 24 h] and cleaved by ozonolysis (MeOH, −78°).[26]

11. *N*-2-Chloroethylamine: $R_2NCH_2CH_2Cl$

Formation/Cleavage[27]

$$\text{Pyrrole} \underset{\text{1. NaH, CH}_3\text{CN; 2. Hg(OAc)}_2\text{; 3. NaBH}_4}{\overset{\text{ClCH}_2\text{CH}_2\text{Cl, 50\% NaOH, Bu}_4\text{N}^+\text{I}^-\text{, >84\%}}{\rightleftarrows}} \text{C}_4\text{H}_4\text{NCH}_2\text{CH}_2\text{Cl}$$

12. *N*-(1-Ethoxy)ethylamine (EE–NR$_2$): $R_2NCH(OCH_2CH_3)CH_3$

Formation/Cleavage[28]

$$\text{Imidazole} \underset{1\ N\ \text{HCl, }72°}{\overset{\text{1. }n\text{-BuLi, }-10°\text{; 2. CH}_3\text{CH(Cl)OEt, }-20°\text{, 70–86\%}}{\rightleftarrows}} \text{Imidazole–EE}$$

13. *N*-2-(2′-Pyridyl)ethylamine: $R_2NCH_2CH_2$-2-(C_5H_4N)

14. *N*-2-(4′-Pyridyl)ethylamine: $R_2NCH_2CH_2$–4-(C_5H_4N)

Formation/Cleavage

2- or 4-vinylpyridine, AcOH, 22–94%[29]
or Na, 4-vinylpyridine[30]

$>$NH ⇄ $>NCH_2CH_2Pyr$

$AlCl_3$, $ClCH_2CH_2Cl$; NaOH, 18–93%
or MeI, acetone, 25°; NaOH[30]

A series of substituted benzimidazoles and pyrroles was protected and deprotected using this methodology.

N-Trialkylsilylamines

Pyrroles and indoles can be protected with the *t*-butyldimethylsilyl group by treatment with TBDMSCl and *n*-BuLi or NaH.[31] Triisopropylsilyl chloride (NaH, DMF, 0°–rt, 73% yield) has been used to protect the pyrrole nitrogen in order to direct electrophilic attack to the 3-position.[32] It has also been used to protect an indole.[33] This derivative can be prepared from the silyl chloride and K.[34] The silyl protective group is cleaved with $Bu_4N^+F^-$, THF, rt or with CF_3COOH.

15. *N-t*-Butyldimethylsilylamine (R_2N–TBDMS)

16. *N*-Triisopropylsilylamine (R_2N–TIPS)

17. *N*-Benzylamine (Bn–NR_2): $PhCH_2$–NR_2

Formation

1. BnCl, NH_3, Na.[35]

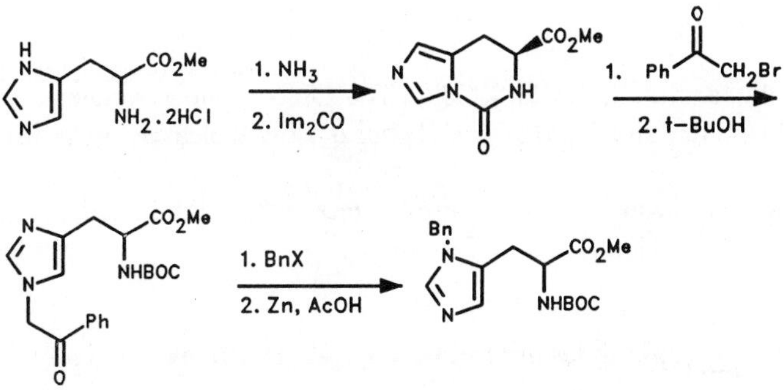

CO_2Me, $NH_2 \cdot 2HCl$ — 1. NH_3; 2. Im_2CO → CO_2Me, NH — 1. $PhCOCH_2Br$; 2. t-BuOH → CO_2Me, NHBOC, Ph — 1. BnX; 2. Zn, AcOH → Bn, CO_2Me, NHBOC

The following benzyl halides were used: $PhCH_2Br$, 82% yield; $PhCH(CH_3)Br$, 33% yield; $(Ph)_2CHBr$, 50% yield; 3,4-$(MeO)_2C_6H_3CH_2Cl$, 52% yield.[36]

Cleavage

1. Cyclohexadiene, Pd-black, 25°, 100% yield, [Imidazole = His(Bn)].[37]

With $H_2/Pd–C$, the normal conditions for benzyl group removal, it is difficult to remove the benzyl group on histidine without also causing reduction of other aromatic groups that may be present.[38]

18. *N-p*-Methoxybenzylamine (MPM)–NR_2): $R_2N–CH_2C_6H_4–4–OCH_3$

The MPM group was used in the preparation of a variety of triazoles. It is readily cleaved with CF_3COOH at 65° (52–100% yield).[39]

19. *N*-3,4-Dimethoxybenzylamine: 3,4-$(MeO)_2C_6H_3CH_2NR_2$

A 3,4-dimethoxybenzyl derivative, cleaved by acid (concd. H_2SO_4/anhyd. CF_3COOH, anisole), was used to protect a pyrrole —NH group during the synthesis of a tetrapyrrole pigment precursor. Neither an *N*-benzyl nor an *N-p*-methoxybenzyl derivative could be cleaved satisfactorily. Hydrogenolysis of the benzyl derivatives led to cyclohexyl compounds; acidic cleavage resulted in migration of the benzyl groups to the free α-position.[40]

20. *N*-2-Nitrobenzylamine (ONB–NR_2): $R_2N–CH_2C_6H_4–2–NO_2$ (Chart 10)

Formation

1. BOC—His(N^{im}Ag)OMe, 2-NO_2–$C_6H_4CH_2Br$, PhH, 4 h, reflux.[41]

Cleavage

1. *hν*, dioxane, 1 h, 100% yield.[41] The ONB group is stable to CF_3COOH, HCl–AcOH, and NaOH–MeOH, but is slowly cleaved by hydrogenation.

The related 4-nitrobenzyl group, used to protect a benzimidazole, can be cleaved with H_2O_2 (EtOH, NaOH, 50°, 72% yield).[42]

21. *N*-2,4-Dinitrophenylamine: 2,4-$(NO_2)_2$–$C_6H_3NR_2$ (Chart 10)

The dinitrophenyl group has been used to protect the imidazole —NH group in histidines (45% yield)[43] by reaction with 2,4-dinitrofluorobenzene and potassium carbonate. Imidazole —NH groups, but not α-amino acid groups, are quantitatively regenerated by reaction with 2-mercaptoethanol (22°, pH 8, 1 h).[44] The 2,4-

dinitrophenyl group on the N^{im} of histidine reduces racemization in peptide synthesis because of its electron-withdrawing character.[45]

22. *N*-Phenacylamine: $R_2NCH_2COC_6H_5$ (Chart 10)

The phenacyl group is stable to HBr–AcOH, CF_3COOH, and CF_3SO_3H.[46] It is used to protect the π-nitrogen in histidine in order to reduce racemization during peptide bond formation.[47]

1. TrCl, TEA 83%; 2. PhCOCH$_2$Br 93%; AcOH, H_2O 84%

23. *N*-Triphenylmethylamine (Tr–NR$_2$): R_2NCPh_3

24. *N*-Diphenylmethylamine (Dpm–NR$_2$): R_2NCHPh_2

Formation

1. BOC–His, TrCl, pyridine.[48]

Cleavage

The trityl group can be cleaved with HBr–AcOH, 2 h; CF_3COOH, 30 min; formic acid, 2 min; and by hydrogenation.[49] The trityl group in BOC—His(Tr)OH is stable to 1 *M* HCl/AcOH, rt, 20 h. The diphenylmethyl group was introduced in the same manner as the trityl group. It is more stable to acid than the trityl group, but not significantly.[48,49] The trityl group has also been used to protect simple imidazoles.[50]

The comparative stabilities of the N^α-Tr, N^{Im}-Tr, and N-BOC groups to various acidic conditions are as follows:[51]

	Cleavage (%)		
Cleavage Conditions	N^α-Tr	N^{Im}-Tr	N-BOC
5% HCO_2H, $ClCH_2CH_2Cl$, 8 min, 20°	100	1	0
$ClCH_2CH_2Cl$, MeOH, TEA, 5 min, 20°	100	<1	0
2.5 eq. HCl in 90% AcOH, 1 min, 20°	100	<1	<1
1 *N* HCl in 90% AcOH, 20 min, 20°	100	<1	100
90% AcOH, 1.5 h, 60°	100	100	<1
5% Pyr·HCl, in MeOH, 2 h, 60°	100	100	<1
95% TFA, 1 h, 20°	100	100	100

25. *N*-(Diphenyl-4-pyridylmethyl)amine (Dppm–NR$_2$): R$_2$N–C(Ph)$_2$(C$_5$H$_4$N) (Chart 10)

Formation

1. Ph$_2$(C$_5$H$_4$N)CCl, Et$_3$N, CHCl$_3$, (Z)- or (BOC)–HisOMe.[52,53]

Cleavage

The diphenyl-4-pyridylmethyl group is cleaved by Zn/AcOH, 1.5 h, 91% yield; H$_2$/Pd–C, 91% yield; or by electrolytic reduction, 2.5 h, 0°, 87% yield. The Dppm group is stable to trifluoroacetic acid.[52,54]

26. *N*-(*N'*, *N'*-Dimethyl)hydrazine: R$_2$N–NMe$_2$

The dimethylamino group can be cleaved from a pyrrole in low yield with chromous acetate.[55]

Amino Acetal Derivatives

27. *N*-Hydroxymethylamine: HOCH$_2$–NR$_2$

Formation/Cleavage[56]

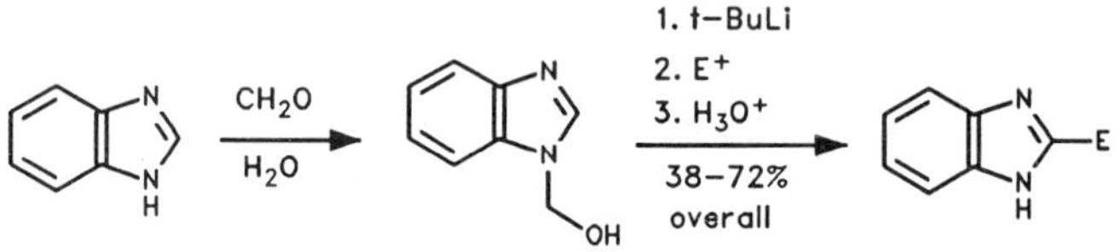

CH$_2$O, H$_2$O; 1. *t*–BuLi 2. E$^+$ 3. H$_3$O$^+$; 38–72% overall

28. *N*-Methoxymethylamine (MOM–NR$_2$): R$_2$NCH$_2$OCH$_3$ (Chart 10)

The MOM group is introduced onto an indole through the sodium salt (NaOH, DMSO, 0°, 0.5 h; MOMCl, 22°, 0.5 h, 90% yield). It is removed with BF$_3$·Et$_2$O (Ac$_2$O, LiBr, 20°, 48 h, 86% yield).[57]

29. *N*-Diethoxymethylamine: (EtO)$_2$CH–NR$_2$

Formation/Cleavage[58]

Imidazole —(EtO)$_3$CH, TsOH, 130° → (EtO)$_2$CH–imidazole; ← H$_3$O$^+$

30. *N*-2-Chloroethoxymethylamine: $R_2NCH_2OCH_2CH_2Cl$

This derivative has been prepared from an indole, the chloromethyl ether, and potassium hydride in 50% yield; it is cleaved in 84% yield by potassium cyanide/18-crown-6 in refluxing acetonitrile.[59]

31. *N*-[2-(Trimethylsilyl)ethoxy]methylamine (SEM–NR₂): $R_2NCH_2OCH_2CH_2Si(CH_3)_3$

Formation

1. Imidazole, indole, or pyrrole, NaH, SEMCl, 50–85% yield.[60–62]

Cleavage

1. 1 *M* $Bu_4N^+F^-$, THF, reflux, 45 min, 46–90% yield, or dilute HCl.[60,61]
2. $BF_3 \cdot Et_2O$; base.[63]
3. $Bu_4N^+F^-$, ethylenediamine (ethylenediamine was used as a formaldehyde scavenger), 45–98% yield.[63]
4. 3 *M* HCl, EtOH, reflux, 1 h, 95% yield.[64]

32. *N-t*-Butoxymethylamine (Bum–NR_2): $R_2NCH_2O\text{-}t\text{-}C_4H_9$

The Bum derivative has been used to protect the π-nitrogen of histidine to prevent racemization during peptide bond formation.[65]

33. *N*-Pivaloyloxymethylamine (POM–NR_2): $R_2NCH_2OCOC(CH_3)_3$ (Chart 10)

The POM group is introduced onto pyrroles and indoles by treatment with NaH, $(CH_3)_3CCO_2CH_2Cl$ in THF at room temperature in 65–78% yield.[66] It is removed by hydrolysis with MeOH, NaOH,[66] or NH_3, MeOH (25°, 4 h, 30–80% yield).[67]

34. *N*-Benzyloxymethylamine (BOM–NR_2): $R_2NCH_2OCH_2C_6H_5$ (Chart 10)

The BOM group is introduced onto an indole with the chloromethyl ether and sodium hydride in 80–90% yield. It is cleaved in 92% yield by catalytic reduction followed by basic hydrolysis,[68] or by CF_3COOH, HBr, or 6 *M* HCl at 110°.[69] It has been used to protect the π-nitrogen of histidine, preventing racemization during peptide bond formation.[70] It has also been used to protect the τ-nitrogen of histidine ($BnOCH_2Cl$, Et_2O; Et_3N, MeOH).[70]

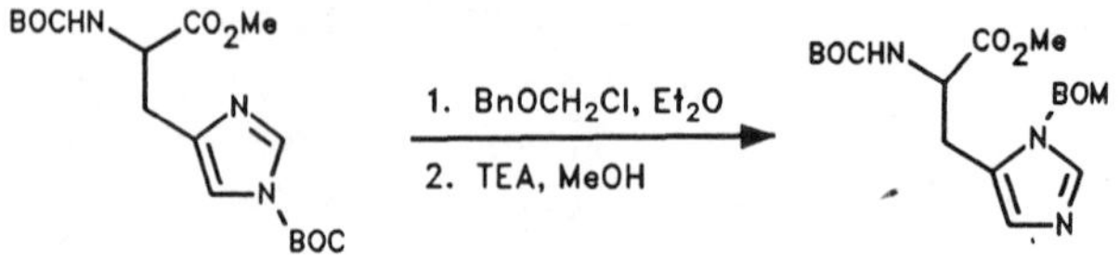

35. *N*-2-Tetrahydropyranylamine (THP–NR$_2$): R$_2$N-2-tetrahydropyranyl (Chart 10)

The THP derivative of the imidazole nitrogen in purines has been prepared by treatment with dihydropyran (TsOH, 55°, 1.5 h, 50–85% yield). It is cleaved by acid hydrolysis.[71]

Amides

36. Formamide: R$_2$N–CHO

Formation[72]***/Cleavage***[73]

$$\text{Tryptophan} \xrightarrow{\text{HCO}_2\text{H, HCl}} \text{Tryptophan(N}^{\text{im}}\text{-CHO)}$$

The formyl group is cleaved with HF/anisole/$(CH_2SH)_2$.[73] It is also cleaved at pH 9–10.[72]

37. Diphenylthiophosphinamide: $Ph_2P(S)$–NR$_2$

This group was used to protect the tryptophan nitrogen.

Formation

$Ph_2P(S)Cl$, $NaHSO_4$, NaOH, CH_2Cl_2, 0°, 88% yield.[74]

Cleavage

1. 0.25 *M* methanesulfonic acid, thioanisole in CF_3COOH, 0°, 90 min.[74]
2. 0.25 *M* trifluoromethanesulfonic acid, 0.25 *M* thioanisole in CF_3COOH, 0°, 50 min.[74]
3. 0.1 *M* $Bu_4N^+F^-$, DMSO or DMF, 25°, 10 min.[74,75]
4. 0.5 *M* KF, 18-crown-6, CH_3CN, 25°, 3 h.[74]

1. H. Büchi and H. G. Khorana, *J. Mol. Biol.*, **72,** 251 (1972).
2. D. J. Chadwick and R. I. Ngochindo, *J. Chem. Soc., Perkin Trans. I*, 481 (1984).

3. A. J. Carpenter and D. J. Chadwick, *Tetrahedron*, **42,** 2351 (1986).
4. N. Fujii, S. Futaki, K. Yasumura, and H. Yajima, *Chem. Pharm. Bull.*, **32,** 2660 (1984).
5. J. M. van der Eijk, R. J. M. Nolte, and J. W. Zwikker, *J. Org. Chem.*, **45,** 547 (1980).
6. T. Fujii and S. Sakakibara, *Bull. Chem. Soc. Jpn.*, **47,** 3146 (1974).
7. K. Kitagawa, K. Kitade, Y. Kiso, T. Akita, S. Funakoshi, N. Fujii, and H. Yajima, *J. Chem. Soc., Chem. Commun.*, 955 (1979).
8. S. Sakakibara and T. Fujii, *Bull. Chem. Soc. Jpn.*, **42,** 1466 (1969).
9. E. Wuensch, in *Methoden der Organischen Chemie* (*Houben-Weyl*), E. Mueller, Ed., Georg Thieme Verlag, Stuttgart, 1974, Vol. 15/1, p. 223.
10. H. C. Beyerman, J. Hirt, P. Kranenburg, J. L. M. Syrier, and A. Van Zon, *Recl. Trav. Chim. Pays-Bas*, **93,** 256 (1974).
11. K. Kitagawa, K. Kitade, Y. Kiso, T. Akita, S. Funakoshi, N. Fujii, and H. Yajima, *Chem. Pharm. Bull.*, **28,** 926 (1980).
12. J. Rokach, P. Hamel, M. Kakushima, and G. M. Smith, *Tetrahedron Lett.*, **22,** 4901 (1981).
13. W. A. Remers, R. H. Roth, G. J. Gibs, and M. J. Weiss, *J. Org. Chem.*, **36,** 1232 (1971).
14. A. P. Kozikowski and Y.-Y. Chen, *J. Org. Chem.*, **46,** 5248 (1981).
15. M. G. Saulnierand and G. W. Gribble, *J. Org. Chem.*, **47,** 2810 (1982).
16. Y. Kiso, M. Inai, K. Kitagawa, and T. Akita, *Chem. Lett.*, 739 (1983).
17. L. Grehn and U. Ragnarsson, *Angew. Chem., Int. Ed. Engl.*, **23,** 296 (1984).
18. E. Schnabel, H. Herzog, P. Hoffmann, E. Klauke, and I. Ugi, *Justus Liebigs Ann. Chem.*, **716,** 175 (1968); E. Schnabel, J. Stoltefuss, H. A. Offe, and E. Klauke, *Justus Liebigs Ann. Chem.*, **743,** 57 (1971).
19. M. Fridkin and H. J. Goren, *Can. J. Chem.*, **49,** 1578 (1971).
20. V. F. Pozdnev, *Zh. Obshch. Khim.*, **48,** 476 (1978); *Chem. Abstr.*, **89,** 24739m (1978).
21. D. Dhanak and C. B. Reese, *J. Chem. Soc., Perkin Trans. I*, 2181 (1986).
22. I. Hasan, E. R. Marinelli, L.-C. C. Lin, F. W. Fowler, and A. B. Levy, *J. Org. Chem.*, **46,** 157 (1981).
23. V. H. Rawal and M. P. Cava, *Tetrahedron Lett.*, **26,** 6141 (1985).
24. S. Raucher, J. E. Macdonald, and R. F. Lawrence, *J. Am. Chem. Soc.*, **103,** 2419 (1981).
25. W. L. Haas, E. V. Kromkalns, and K. Gerzon, *J. Am. Chem. Soc.*, **88,** 1988 (1966).
26. Y. L. Chen, K. G. Hedberg, and K. J. Guarino, *Tetrahedron Lett.*, **30,** 1067 (1989).
27. C. Gonzalez, R. Greenhouse, R. Tallabs, and J. M. Muchowski, *Can. J. Chem.*, **61,** 1697 (1983).
28. T. S. Manoharan and R. S. Brown, *J. Org. Chem.*, **53,** 1107 (1988).
29. M. Ichikawa, C. Yamamoto, and T. Hisano, *Chem. Pharm. Bull.*, **29,** 3042 (1981).
30. A. R. Katritzky, G. R. Khan, and C. M. Marson, *J. Heterocycl. Chem.*, **24,** 641 (1987).
31. B. H. Lipshutz, B. Huff, and W. Hagen, *Tetrahedron Lett.*, **29,** 3411 (1988).
32. J. M. Muchowski and D. R. Solas, *Tetrahedron Lett.*, **24,** 3455 (1983).
33. P. J. Beswick, C. S. Greenwood, T. J. Mowlem, G. Nechvatal, and D. A. Widdowson, *Tetrahedron*, **44,** 7325 (1988).

34. K. P. Stefan, W. Schuhmann, H. Parlar, and F. Korte, *Chem. Ber.*, **122,** 169 (1989).
35. V. du Vigneaud and O. K. Behrens, *J. Biol. Chem.*, **117,** 27 (1937).
36. C. J. Chivikas and J. C. Hodges, *J. Org. Chem.*, **52,** 3591 (1987).
37. A. M. Felix, E. P. Heimer, T. J. Lambros, C. Tzougraki, and J. Meienhofer, *J. Org. Chem.*, **43,** 4194 (1978).
38. E. C. Jorgensen, G. C. Windridge, and T. C. Lee, *J. Med. Chem.*, **13,** 352 (1970).
39. D. R. Buckle and C. J. M. Rockell, *J. Chem. Soc., Perkin Trans. I*, 627 (1982).
40. M. I. Jones, C. Froussios, and D. A. Evans, *J. Chem. Soc., Chem. Commun.*, 472 (1976).
41. S. M. Kalbag and R. W. Roeske, *J. Am. Chem. Soc.*, **97,** 440 (1975).
42. R. Balasuriya, S. J. Chandler, M. J. Cook, and D. J. Hardstone, *Tetrahedron Lett.*, **24,** 1385 (1983).
43. E. Siepmann and H. Zahn, *Biochim. Biophys. Acta*, **82,** 412 (1964).
44. S. Shaltiel, *Biochem. Biophys. Res. Commun.*, **29,** 178 (1967).
45. M. C. Lin, B. Gutte, D. G. Caldi, S. Moore, and R. B. Merrifield, *J. Biol. Chem.*, **247,** 4768 (1972).
46. A. R. Fletcher, J. H. Jones, W. I. Ramage, and A. V. Stachulski, in *Peptides 1978*, I. Z. Siemion and G. Kupryszeqski, Eds., Wroclaw University Press, Wroclaw, Poland, 1979, pp. 168–171.
47. A. R. Fletcher, J. H. Jones, W. I. Ramage, and A. V. Stachulski, *J. Chem. Soc., Perkin Trans. I*, 2261 (1979).
48. G. Losse and U. Krychowski, *J. Prakt. Chem.*, **312,** 1097 (1970).
49. G. Losse and U. Krychowski, *Tetrahedron Lett.*, 4121 (1971).
50. N. J. Curtis and R. S. Brown, *J. Org. Chem.*, **45,** 4038 (1980); K. L. Kirk, *J. Org. Chem.*, **43,** 4381 (1978); J. L. Kelley, C. A. Miller, and E. W. McLean, *J. Med. Chem.*, **20,** 721 (1977).
51. P. Sieber and B. Riniker, *Tetrahedron Lett.*, **28,** 6031 (1987).
52. S. Coyle and G. T. Young, *J. Chem. Soc., Chem. Commun.*, 980 (1976).
53. S. Coyle, O. Keller, and G. T. Young, *J. Chem. Soc., Chem. Commun.*, 939 (1975).
54. S. Coyle, A. Hallett, M. S. Munns, and G. T. Young, *J. Chem. Soc., Perkin Trans. I*, 522 (1981).
55. G. R. Martinez, P. A. Grieco, E. Williams, K.-i. Kanai, and C. V. Srinivasan, *J. Am. Chem. Soc.*, **104,** 1436 (1982).
56. A. R. Katritzky and K. Akutagawa, *J. Org. Chem.*, **54,** 2949 (1989).
57. R. J. Sundberg and H. F. Russell, *J. Org. Chem.*, **38,** 3324 (1973).
58. N. J. Curtis and R. S. Brown, *J. Org. Chem.*, **45,** 4038 (1980).
59. A. J. Hutchison and Y. Kishi, *J. Am. Chem. Soc.*, **101,** 6786 (1979).
60. J. P. Whitten, D. P. Matthews, and J. R. McCarthy, *J. Org. Chem.*, **51,** 1891 (1986).
61. B. H. Lipshutz, W. Vaccaro, and B. Huff, *Tetrahedron Lett.*, **27,** 4095 (1986).
62. M. P. Edwards, A. M. Doherty, S. V. Ley, and H. M. Organ, *Tetrahedron*, **42,** 3723 (1986).
63. J. M. Muchowski and D. R. Solas, *J. Org. Chem.*, **49,** 203 (1984).
64. D. P. Matthews, J. P. Whitten, and J. R. McCarthy, *J. Heterocycl. Chem.*, **24,** 689 (1987).

65. R. Colombo, F. Colombo, and J. H. Jones, *J. Chem. Soc., Chem. Commun.*, 292 (1984).
66. D. Dhanak and C. B. Reese, *J. Chem. Soc., Perkin Trans. I*, 2181 (1986).
67. M. Rasmussen and N. J. Leonard, *J. Am. Chem. Soc.*, **89,** 5439 (1967).
68. H. J. Anderson and J. K. Groves, *Tetrahedron Lett.*, 3165 (1971).
69. T. Brown, J. H. Jones, and J. D. Richards, *J. Chem. Soc., Perkin Trans. I*, 1553 (1982).
70. T. Brown and J. H. Jones, *J. Chem. Soc., Chem. Commun.*, 648 (1981).
71. R. K. Robins, E. F. Godefroi, E. C. Taylor, L. R. Lewis, and A. Jackson, *J. Am. Chem. Soc.*, **83,** 2574 (1961).
72. A. Previero, M. A. Coletti-Previero, and J. C. Cavadore, *Biochim. Biophys. Acta*, **147,** 453 (1967).
73. G. R. Matsueda, *Int. J. Pept. Protein Res.*, **20,** 26 (1982).
74. Y. Kiso, T. Kimura, M. Shimokura, and T. Narukami, *J. Chem. Soc., Chem. Commun.*, 287 (1988).
75. Y. Kiso, T. Kimura, Y. Fujiwara, M. Shimokura, and A. Nishitani, *Chem. Pharm. Bull.*, **36,** 5024 (1988).

PROTECTION FOR THE AMIDE —NH

Protection of the amide —NH is an area of protective group chemistry that has received little attention, and as a consequence few methods exist for amide —NH protection. Most of the cases found in the literature do not represent protective groups in the true sense, in that the protective group is often incorporated as a handle to introduce nitrogen into a molecule rather than installed to protect a nitrogen which at some later time is deblocked. For this reason many of the following examples deal primarily with removal rather than with both formation and cleavage.

AMIDES

1. Allylamide: $CH_2{=}CHCH_2$-NRCO-

The allyl group was used to protect the nitrogen in a β-lactam synthesis, but was removed in a four-step sequence. Whether a transition-metal-catalyzed isomerization to the enamide followed by hydrolysis is an effective cleavage procedure remains to be tested and warrants further study.[1]

2. Methoxymethylamide (MOM-NRCO-): CH_3OCH_2-NRCO-

Formation

1. MOMCl, *t*-BuOK, DMSO.[2]

Cleavage

1. BBr_3, 31% yield.[2]

3. Benzyloxymethylamide (BOM–NRCO–): $C_6H_5CH_2OCH_2$–NRCO–

The BOM group can be cleaved with $H_2/Pd(OH)_2$–C, MeOH, which also removes the BOM group from alcohols.[3]

4. *t*-Butyldimethylsiloxymethylamide: *t*-$C_4H_9(CH_3)_2SiOCH_2$–NRCO–

Formation

1. TBDMSOCH$_2$Cl, TEA, CH_2Cl_2, −78° → rt, 24 h, >89% yield.[4]

Cleavage

1. $Bu_4N^+F^-$, THF, rt, 30 min, 70% yield.[4]

5. Pyrrolidinomethylamide

Formation

1. HCHO, pyrrolidine, 93% yield.[5,6]

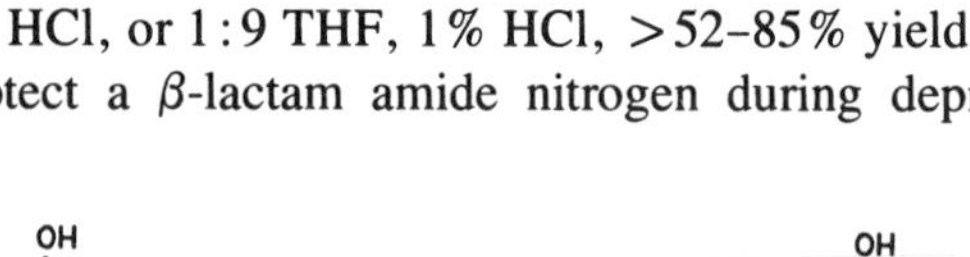

Pyrrolidine (NH), CH_2O, 93%

Cleavage

1. MeOH, 1% HCl, or 1:9 THF, 1% HCl, >52–85% yield.[6] This group was used to protect a β-lactam amide nitrogen during deprotonation of the α-position.

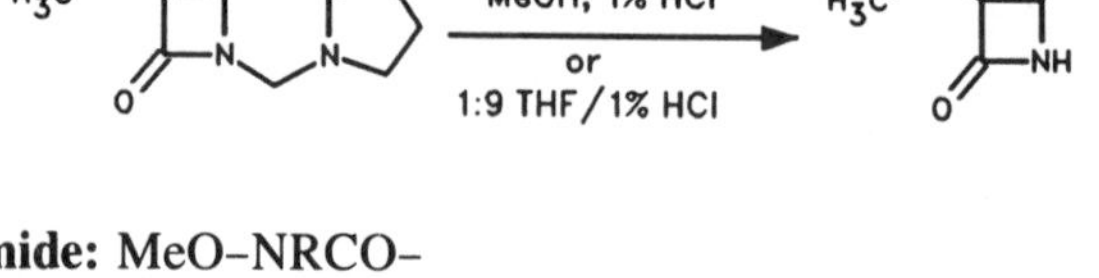

MeOH, 1% HCl or 1:9 THF/1% HCl

6. Methoxyamide: MeO–NRCO–

The methoxy group on a β-lactam nitrogen was cleaved by reduction with Li ($EtNH_2$, *t*-BuOH, THF, −40°, 71% yield). A benzyloxy group was stable to these cleavage conditions.[7]

7. Benzyloxyamide (BnO–NRCO–): $C_6H_5CH_2O$–NRCO–

The benzyloxy group on a β-lactam nitrogen was cleaved by hydrogenolysis (H_2, Pd–C) or by $TiCl_3$ [MeOH, H_2O, $(NH_4)_2CO_3$, Na_2CO_3].[8]

8. Methylthioamide: MeS–NRCO–

Formation

1. LDA, HMPA, $CH_3SSO_2CH_3$, $-78° \rightarrow 0°$, 94% yield.[9]

Cleavage

1. 2-Pyridinethiol, Et_3N, CH_2Cl_2, 95% yield. The methylthioamide group is stable to 2.5 *N* NaOH, THF, H_2O and to 10% H_2SO_4, MeOH, H_2O.[9]

The section on sulfenamides should be consulted for a related approach to nitrogen protection. Some of the derivatives presented there may also be applicable to amides.

9. *N*-Triphenylmethylthioamide: Ph_3CS–NRCO–

Cleavage

1. Bu_3P, EtOH, THF, 115°, 48 h, 75% yield.[10]
2. Me_3SiI, CH_2Cl_2, 25°, 7 h, 81% yield.[10]
3. Li, NH_3.[10]
4. W2 Raney Ni.[10] Li/NH_3 and Raney Ni also cleave benzylic C—N bonds.

10. *t*-Butyldimethylsilylamide (TBDMS–NRCO–): t-$C_4H_9(CH_3)_2Si$–NRCO–

Formation

1. TBDMSCl, Et_3N, CH_2Cl_2, 98% yield.[11-13]

Cleavage

1. 1 *N* HCl, MeOH, rt, 91% yield.[14]

The TBDMS derivative of a β-lactam nitrogen is reported to be stable to lithium diisopropylamide, citric acid, Jones oxidation, and BH_3–diisopropylamine, but not to $Pb(OAc)_4$ oxidation.

11. 4-Methoxyphenylamide: 4-CH_3O–C_6H_4–NRCO–

This group has been used extensively in β-lactam syntheses, where it is used to introduce the nitrogen as *p*-anisidine.

Cleavage

1. Electrolysis, CH_3CN, H_2O, $LiClO_4$, 1.5 V, rt, 60–95% yield.[15] The released quinone is removed by forming the bisulfite adduct, which can be washed out with water.
2. Ceric ammonium nitrate, CH_3CN, H_2O, 0°, 95% yield.[16,17]
3. Ozonolysis, then reduction with $Na_2S_2O_4$ at 50°, 57% yield.[18] The 3,4-dimethoxyphenyl derivative was cleaved in 71% yield using these conditions. Ceric ammonium nitrate was reported not to work in this example.

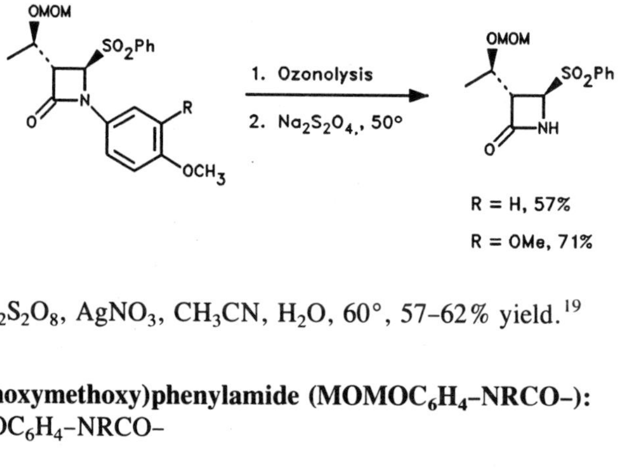

OMOM SO2Ph R OCH3 N O — 1. Ozonolysis 2. $Na_2S_2O_4$, 50° → OMOM SO2Ph NH O

R = H, 57%

R = OMe, 71%

4. $(NH_4)_2S_2O_8$, $AgNO_3$, CH_3CN, H_2O, 60°, 57–62% yield.[19]

12. 4-(Methoxymethoxy)phenylamide ($MOMOC_6H_4$–NRCO–):
4-$MeOCH_2OC_6H_4$–NRCO–

This group was developed for a case where direct oxidation of the methoxyphenyl group with CAN was not very efficient. Removal of the MOM group [HCl, $HC(OMe)_3$, MeOH], followed by oxidation with CAN, was reported to be more effective.[20]

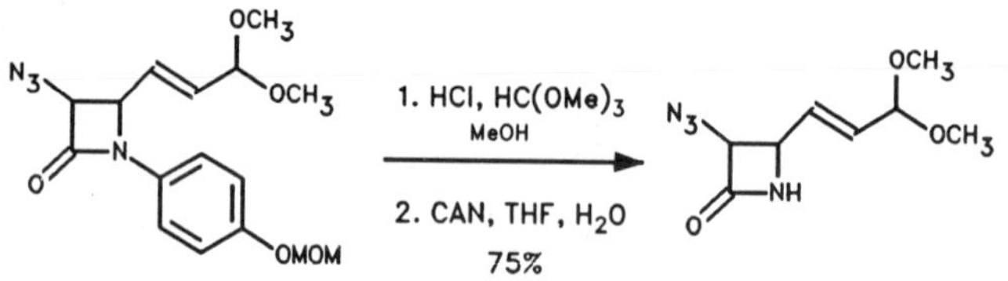

N_3 OCH3 OCH3 N O OMOM — 1. HCl, $HC(OMe)_3$ MeOH 2. CAN, THF, H_2O 75% → N_3 OCH3 OCH3 NH O

13. 2-Methoxy-1-naphthylamide: 2-CH_3O–$C_{10}H_6$–NRCO–

This group was removed from a cyclic urethane with CAN.[21]

14. Benzylamide (Bn–NRCO–): $C_6H_5CH_2$–NRCO–

Formation

1. BnCl, KH, THF, rt, 100% yield.[22]

Cleavage

1. H_2, Pd–C, AcOH, 2 days.[23] Debenzylation of a benzylacetamide by hydrogenolysis is much slower than hydrogenolysis of a benzyl oxygen bond. Hydroxyl groups protected with benzyl groups or benzylidene groups are readily cleaved without affecting amide benzyl groups.
2. Na or Li and ammonia, excellent yields.[24] A dissolving metal reduction can be effected without cleavage of a sulfur–carbon bond. Note also the unusual selectivity in the cleavage illustrated below. This was attributed to steric compression.[25] 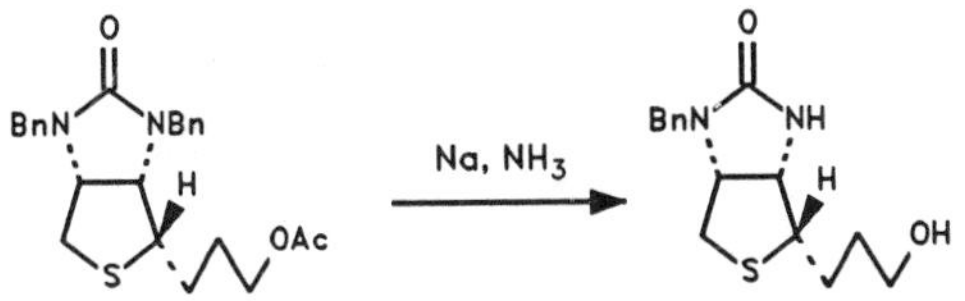
3. *t*-BuLi, THF, −78°; O_2 or MoOPH [oxodiperoxymolybdenum(hexamethylphosphorictriamide)(pyridine)], 30–68% yield.[26] This method uses the amide carbonyl to direct benzylic metalation.
4. *t*-BuOK, DMSO, O_2, 20°, 20 min.[27]
5. Sunlight, $FeCl_3$, H_2O, acetone, 21% yield.[28]
6. 95% HCO_2H, 50–60°, 74–91% yield.[29] This method was used to remove the α-methylbenzyl group from an amide.
7. Aqueous HBr, 85% yield.[30]
8. Orthophosphoric acid, phenol, 53% yield.[31]

Methods 7 and 8 were used to remove the benzyl group from a biotin precursor.

15. 4-Methoxybenzylamide (MPM–NRCO–): 4-$CH_3OC_6H_4CH_2$–NRCO–

Formation

1. NaH, 4-MeO-$C_6H_4CH_2Br$, DMF, rt, 12 h, 62% yield.[32]

Cleavage

1. Ceric ammonium nitrate (CAN), CH_3CN, H_2O, rt, 12 h, 96% yield.[33,34] Benzylamides are not cleaved under these conditions.

Some of the methods used to cleave the benzyl group should also be effective for cleavage of the MPM group.

16. 2,4-Dimethoxybenzylamide: 2,4-$(CH_3O)_2$–$C_6H_3CH_2$–NRCO–

Cleavage

1. TFA, 85% yield.[35,36]

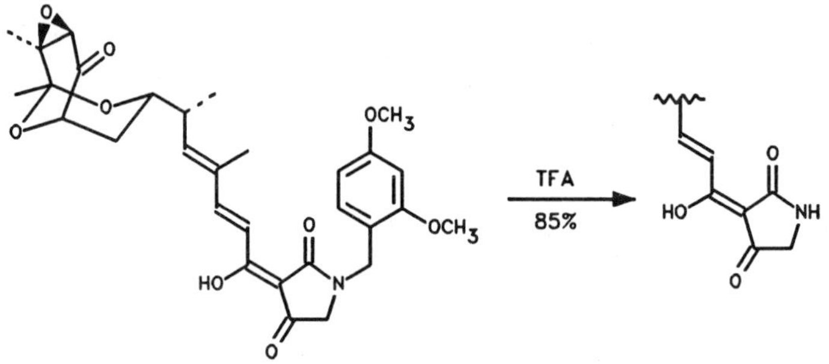

2. DDQ, $CHCl_3$, H_2O.[37]
3. Ceric ammonium nitrate, CH_3CN, H_2O, 78% yield.[38]
4. The related 3,4-dimethoxybenzyl group has been cleaved from an amide with Na/NH_3, 82% yield.[39]

17. Di(4-methoxyphenyl)methylamide (DAM–NRCO–):
(4-$MeOC_6H_4)_2$CH–NRCO–

The DAM group, used to protect the —NH group of a β-lactam, can be cleaved with ceric ammonium nitrate (H_2O, CH_3CN, 0°, 91% yield).[40]

18. Di(4-methoxyphenyl)phenylmethylamide (DMTr–NRCO–):
(4-$MeOC_6H_4)_2$PhC–NRCO–

The DMTr group was selectively introduced into a biotin derivative.[41]

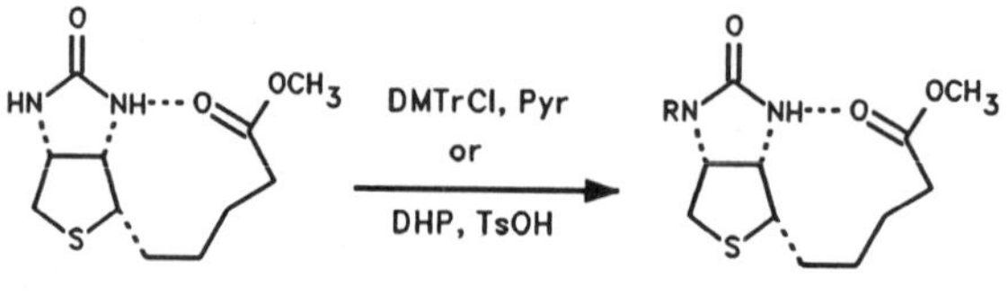

R = DMTr, 40%

R = THP, 45%

19. *N-t*-Butoxycarbonylamide (BOC–NRCO–): *t*-C_4H_9OCO–NRCO–

Formation

1. $(BOC)_2O$, Et_3N, DMAP, 25°, 15 h, 78–96% yield.[42,43]

Cleavage

It should be noted that when a BOC-protected amide is subjected to MeONa treatment the amide bond is cleaved in preference to the BOC group (85–96% yield) because of the difference in steric factors. The BOC group can be removed by the methods used to remove it from simple amines.

20. *N,O*-Isopropylidene Ketal:

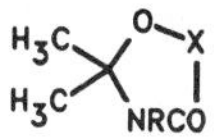

21. *N,O*-Benzylidene Acetal:

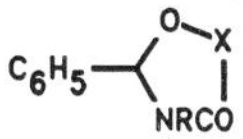

Isopropylidene[44,45] and benzylidene[46] groups have been used to protect simultaneously amide nitrogens and a neighboring hydroxyl. They can be removed by acid hydrolysis or hydrogenolysis (Pd–C, hydrazine, MeOH, 95% yield), respectively.

1. T. Fukuyama, A. A. Laird, and C. A. Schmidt, *Tetrahedron Lett.*, **25,** 4712 (1984).
2. G. W. Kirby, D. J. Robins, and W. M. Stark, *J. Chem. Soc., Chem. Commun.*, 812 (1983).
3. S. Hanessian, in *Trends in Synthetic Carbohydrate Chemistry, ACS Symposium Series No. 386*, D. Horton, L. D. Hawkins, and G. J. McGarvey, Eds., American Chemical Society, Washington, DC, 1989, p. 64.
4. T. Benneche, L. L. Gundersen, and K. Undheim, *Acta Chem. Scand., Ser. B,* **B42,** 384 (1988).
5. G. Cignarella, G. F. Cristiani, and E. Testa, *Justus Liebigs Ann. Chem.*, **661,** 181 (1963).
6. A. B. Hamlet and T. Durst, *Can. J. Chem.*, **61,** 411 (1983).
7. F. Shirai and T. Nakai, *Tetrahedron Lett.*, **29,** 6461 (1988).
8. P. G. Mattingly and M. J. Miller, *J. Org. Chem.*, **46,** 1557 (1981).
9. N. V. Shah and L. D. Cama, *Heterocycles*, **25,** 221 (1987).
10. D. A. Burnett, D. J. Hart, and J. Liu, *J. Org. Chem.*, **51,** 1929 (1986).
11. P. J. Reider and E. J. J. Grabowski, *Tetrahedron Lett.*, **23,** 2293 (1982).
12. H. Hiemstra, W. J. Klaver, and W. N. Speckamp, *Tetrahedron Lett.*, **27,** 1411 (1986).
13. D. J. Hart, C.-S. Lel, W. H. Pirkle, M. H. Hyon, and A. Tsipouras, *J. Am. Chem. Soc.*, **108,** 6054 (1986).
14. R. W. Ratcliffe, T. N. Salzmann, and B. G. Christensen, *Tetrahedron Lett.*, **21,** 31 (1980).

15. E. G. Corley, S. Karady, N. L. Abramson, D. Ellison, and L. M. Weinstock, *Tetrahedron Lett.*, **29,** 1497 (1988).
16. D. R. Kronenthal, C. Y. Han, and M. K. Taylor, *J. Org. Chem.*, **47,** 2765 (1982).
17. D.-C. Ha and D. J. Hart, *Tetrahedron Lett.*, **28,** 4489 (1987).
18. H. Yanagisawa, A. Ando, M. Shiozaki, and T. Hiraoka, *Tetrahedron Lett.*, **24,** 1037 (1983).
19. K. Bhattarai, G. Cainelli, and M. Panunzio, *Synlett*, 229 (1990).
20. T. Fukuyama, R. K. Frank, and C. F. Jewell, Jr., *J. Am. Chem. Soc.*, **102,** 2122 (1980).
21. B. M. Trost and A. A. Sudhakar, *J. Am. Chem. Soc*, **110,** 7933 (1988).
22. Y. Xia and A. P. Kozikowski, *J. Am. Chem. Soc.*, **111,** 4116 (1989).
23. R. Gigg and R. Conant, *Carbohydr. Res.*, **100,** C5 (1982).
24. T. Ohgi and S. M. Hecht, *J. Org. Chem.*, **46,** 1232 (1981); M. Y. Kim, J. E. Starrett, Jr., and S. M. Weinreb, *J. Org. Chem.*, **46,** 5383 (1981); S. Sugasawa and T. Fujii, *Chem. Pharm. Bull.*, **6,** 587 (1958); F. X. Webster, J. G. Millar, and R. M. Silverstein, *Tetrahedron Lett.*, **27,** 4941 (1986).
25. G. F. Field, *J. Org. Chem.*, **43,** 1084 (1978).
26. R. M. Williams and E. Kwast, *Tetrahedron Lett.*, **30,** 451 (1989).
27. R. Gigg and R. Conant, *J. Chem. Soc., Chem. Commun.*, 465 (1983).
28. M. Barbier, *Heterocycles*, **23,** 345 (1985).
29. J. E. Semple, P. C. Wang, Z. Lysenko, and M. M. Joullié, *J. Am. Chem. Soc.*, **102,** 7505 (1980).
30. E. G. Baggiolini, H. L. Lee, G. Pizzolato, and M. R. Uskokovic, *J. Am. Chem. Soc.*, **104,** 6460 (1982).
31. G. F. Field, W. J. Zally, L. H. Sternbach, and J. F. Blout, *J. Org. Chem.*, **41,** 3853 (1976).
32. M. Yamaura, T. Suzuki, H. Hashimoto, J. Yoshimura, and C. Shin, *Chem. Lett.*, 1547 (1984).
33. M. Yamaura, T. Suzuki, H. Hashimoto, J. Yoshimura, T. Okamoto, and C. Shin, *Bull. Chem. Soc. Jpn.*, **58,** 1413 (1985).
34. J. Yoshimura, M. Yamaura, T. Suzuki, and H. Hashimoto, *Chem. Lett.*, 1001 (1983).
35. R. H. Schlessinger, G. R. Bebernitz, P. Lin, and A. Y. Poss, *J. Am. Chem. Soc.*, **107,** 1777 (1985).
36. P. DeShong, S. Ramesh, V. Elango, and J. J. Perez, *J. Am. Chem. Soc.*, **107,** 5219 (1985).
37. S. Mori, H. Iwakura, and S. Takechi, *Tetrahedron Lett.*, **29,** 5391 (1988).
38. L. E. Overman and T. Osawa, *J. Am. Chem. Soc.*, **107,** 1698 (1985).
39. T. G. Back, K. Brunner, P. W. Codding, and A. W. Roszak, *Heterocycles*, **28,** 219 (1989).
40. T. Kawabata, Y. Kimura, Y. Ito, and S. Terashima, *Tetrahedron Lett.*, **27,** 6241 (1986).
41. A. M. Alves, D. Holland, and M. D. Edge, *Tetrahedron Lett.*, **30,** 3089 (1989).
42. D. L. Flynn, R. E. Zelle, and P. A. Grieco, *J. Org. Chem.*, **48,** 2424 (1983).

43. Y. Ohfune and M. Tomita, *J. Am. Chem. Soc.*, **104,** 3511 (1982).
44. D. Favara, A. Omodei-Salè, P. Consonni, and A. Depaoli, *Tetrahedron Lett.*, **23,** 3105 (1982).
45. F. A. Bouffard and B. G. Christensen, *J. Org. Chem.*, **46,** 2208 (1981).
46. Y. Hamada, A. Kawai, Y. Kohno, O. Hara, and T. Shioiro, *J. Am. Chem. Soc.*, **111,** 1524 (1989).

8

REACTIVITIES, REAGENTS, AND REACTIVITY CHARTS

REACTIVITIES

In the selection of a protective group it is of paramount importance to know the reactivity of the resulting protected functionality toward various reagents and reaction conditions. The number of reagents available to the organic chemist is large; approximately 8000 reagents are reviewed in the excellent series of books by the Fiesers.[a] In an effort to assess the effect of a wide variety of standard types of reagents and reaction conditions on the different possible protected functionalities, 108 prototype reagents have been selected and grouped into 16 categories:[b]

A. Aqueous
B. Nonaqueous Bases
C. Nonaqueous Nucleophiles
D. Organometallic
E. Catalytic Reduction
F. Acidic Reduction
G. Basic or Neutral Reduction
H. Hydride Reduction
I. Lewis Acids
J. Soft Acids
K. Radical Addition
L. Oxidizing Agents

M. Thermal Reactions
N. Carbenoids
O. Miscellaneous
P. Electrophiles

These 108 reagents are used in the Reactivity Charts that have been prepared for each class of protective groups. The reagents and some of their properties are described on the following pages.

REAGENTS

A. Aqueous

1. pH < 1, 100°	Refluxing HBr
2. pH < 1	1 *N* HCl
3. pH 1	0.1 *N* HCl
4. pH 2–4	0.01 *N* HCl; 1-0.01 *N* HOAc
5. pH 4–6	0.1 *N* H_3BO_3; phosphate buffer; HOAc–NaOAc
6. pH 6–8.5	H_2O
7. pH 8.5–10	0.1 *N* HCO_3^-; 0.1 *N* OAc^-; satd. $CaCO_3$
8. pH 10–12	0.1 *N* CO_3^{2-}; 1–0.01 *N* NH_4OH; 0.01 *N* NaOH; satd $Ca(OH)_2$
9. pH > 12	1–0.1 *N* NaOH
10. pH > 12, 150°	

B. Nonaqueous Bases

11. NaH	
12. $(C_6H_5)_3CNa$	$pK_a = 32$
13. $[C_{10}H_8]^- \cdot Na^+$	$pK_a \cong 37$
14. $CH_3SOCH_2^-Na^+$	$pK_a = 35$
15. $KO\text{-}t\text{-}C_4H_9$	$pK_a = 19$
16. $LiN(i\text{-}C_3H_7)_2$	(LDA) $pK_a = 36$
17. Pyridine; Et_3N	$pK_a = 5; 10$
18. $NaNH_2$; NaNHR	$pK_a = 36$

C. Nonaqueous Nucleophiles

19. $NaOCH_3/CH_3OH$, 25°	$pK_a = 16$
20. Enolate anion	$pK_a = 20$

21. NH_3; RNH_2; RNHOH	$pK_a = 10$
22. RS^-; N_3^-; SCN^-	
23. OAc^-; X^-	$pK_a = 4.5$
24. NaCN, pH 12	
25. HCN, cat. CN^-, pH 6	$pK_a = 9$. For cyanohydrin formation

D. Organometallic

26. RLi	
27. RMgX	
28. Organozinc	Reformatsky reaction. Similar: R_2Cu; R_2Cd
29. Organocopper	R_2CuLi
30. Wittig; ylide	Includes sulfur ylides

E. Catalytic Reduction

31. H_2/Raney Ni	
32. H_2/Pt, pH 2–4	
33. H_2/Pd–C	
34. H_2/Lindlar	
35. H_2/Rh–C or H_2/Rh–Al_2O_3	Avoids hydrogenolysis of benzyl ethers

F. Acidic Reduction

36. Zn/HCl
37. Zn/HOAc; $SnCl_2$/HCl
38. Cr(II), pH 5

G. Basic or Neutral Reduction

39. Na/*l* NH_3
40. Al(Hg)
41. $SnCl_2$/Py
42. H_2S or HSO_3^-

H. Hydride Reduction

43. $LiAlH_4$	
44. Li-*s*-Bu_3BH, $-50°$	Li-Selectride
45. $[(CH_3)_2CHCH(CH_3)]_2BH$	Disiamylborane
46. B_2H_6, 0°	
47. $NaBH_4$	

48. $Zn(BH_4)_2$ — Neutral reduction
49. $NaBH_3CN$, pH 4–6
50. $(i\text{-}C_4H_9)_2AlH$, $-60°$ — Dibal
51. $Li(O\text{-}t\text{-}C_4H_9)_3AlH$, 0°

I. Lewis Acids (Anhydrous conditions)

52. $AlCl_3$, 80°
53. $AlCl_3$, 25°
54. $SnCl_4$, 25°; $BF_3 \cdot Et_2O$
55. $LiClO_4$; $MgBr_2$ — For epoxide rearrangement
56. TsOH, 80° — Catalytic amount
57. TsOH, 0° — Catalytic amount

J. Soft Acids

58. Hg(II)
59. Ag(I)
60. Cu(II)/Py — For example, for Glaser coupling

K. Radical Addition

61. HBr/initiator — "Acidic" HX addition; acidity $\cong$ TsOH, 0°
62. HX/initiator — Neutral HX addition; X = P, S, Se, Si
63. NBS/CCl_4, $h\nu$ or heat — Allylic bromination
64. $CHBr_3$; $BrCCl_3$; $CCl_4/In\cdot$ — Carbon–halogen addition

L. Oxidizing Agents

65. OsO_4
66. $KMnO_4$, 0°, pH 7
67. O_3, $-50°$
68. RCO_3H, 0° — Epoxidation of olefins; prototype for H_2O_2/H^+
69. RCO_3H, 50° — Baeyer–Villiger oxidation of hindered ketones
70. CrO_3/Py — Collins oxidation
71. CrO_3, pH 1 — Jones oxidation
72. H_2O_2/OH^-, pH 10–12
73. Quinone — Dehydrogenation
74. $^1O_2$ — Singlet oxygen

75. CH_3SOCH_3, 100°	(DMSO); HCO_3^- may be added to maintain neutrality
76. NaOCl, pH 10	
77. Aq. NBS	Nonradical conditions
78. I_2	
79. C_6H_5SCl; C_6H_5SeX	
80. Cl_2; Br_2	
81. MnO_2/CH_2Cl_2	
82. $NaIO_4$, pH 5–8	
83. SeO_2, pH 2–4	
84. SeO_2/Py	In EtOH/cat. Py
85. $K_3Fe(CN)_6$, pH 7–10	Phenol coupling
86. Pb(IV), 25°	Glycol and α-hydroxy acid cleavage
87. Pb(IV), 80°	Oxidative decarboxylation
88. $Tl(NO_3)_3$, pH 2	Oxidative rearrangement of olefins

M. Thermal Reactions

89. 150°	Some Cope rearrangements and Cope eliminations
90. 250°	Claisen or Cope rearrangement
91. 350°	Ester cracking; Conia ''ene'' reaction

N. Carbenoids

92. $:CCl_2$	
93. $N_2CHCO_2C_2H_5$/Cu, 80°	
94. CH_2I_2/Zn–Cu	Simmons–Smith addition

O. Miscellaneous

95. *n*-Bu_3SnH/initiator	
96. $Ni(CO)_4$	
97. CH_2N_2	
98. $SOCl_2$	
99. Ac_2O, 25°	Acetylation
100. Ac_2O, 80°	Dehydration
101. DCC	Dicyclohexylcarbodiimide, $C_6H_{11}N{=}C{=}NC_6H_{11}$
102. CH_3I	
103. $(CH_3)_3O^+BF_4^-$	Or CH_3OSO_2F = Magic Methyl: **SEVERE POISON**

104. 1. LiN-i-Pr_2; 2. MeI — For C-alkylation
105. 1. K_2CO_3; 2. MeI — For *O*-alkylation

P. Electrophiles

106. RCHO
107. RCOCl
108. C^+ ion/olefin — For cation–olefin cyclization

REACTIVITY CHARTS

One requirement of a protective group is stability to a given reaction. The following charts were prepared as a guide to relative reactivities and thereby as an aid in the choice of a protective group. The reactivities in the charts were estimated by the individual and collective efforts of a group of synthetic chemists. *It is important to realize that not all the reactivities in the charts have been determined experimentally and considerable conjecture has been exercised.* For those cases in which a literature reference was available concerning the use of a protective group and one of the 108 prototype reagents, the reactivity is printed in italic type. However, an exhaustive search for such references has not been made; therefore the absence of italic type does not imply an experimentally unknown reactivity.

There are four levels of reactivity in the charts:

"H" (high) indicates that under the conditions of the prototype reagent the protective group is readily removed to regenerate the original functional group.

"M" (marginal) indicates that the stability of the protected functionality is marginal, and depends on the exact parameters of the reaction. The protective group may be stable, may be cleaved slowly, or may be unstable to the conditions. Relative rates are always important, as illustrated in the following example[e] (in which a monothioacetal is cleaved in the presence of a dithiane), and may have to be determined experimentally.

CH_3SCH_2O ... Me ... $OSiMe_2$-t-Bu ... S ... S ... HO

$HgCl_2/CaCO_3$[e]
CH_3CN-H_2O (4:1)
25°, 1–2 h

HO ... Me ... $OSiMe_2$-t-Bu ... S ... S ... HO

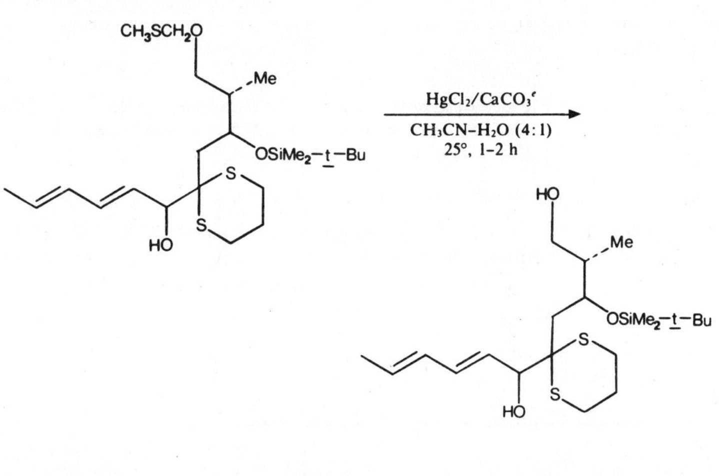

"L" (low) indicates that the protected functionality is stable under the reaction conditions.

"R" (reacts) indicates that the protected compound reacts readily, but that the original functional group is not restored. The protective group may be changed to a new protective group (eq. 1) or to a reactive intermediate (eq. 2), or the protective group may be unstable to the reaction conditions and react further (eq. 3).

$$(1)\ \mathrm{ROCOC_6H_4\text{-}}p\mathrm{\text{-}NO_2} \xrightarrow{\mathrm{H_2/Pd\text{-}C}} \mathrm{ROCOC_6H_4\text{-}}p\mathrm{\text{-}NH_2}$$

$$(2)\ \mathrm{RCONR'_2} \xrightarrow{\mathrm{Me_3O^+BF_4^-}} [\mathrm{R\underset{\displaystyle OMe}{\underset{|}{C}}{=}N^+R'_2\ BF_4^-}]$$

$$(3)\ \mathrm{RCH(OR')_2} \xrightarrow{\mathrm{pH} < 1,\ 100^\circ} [\mathrm{RCHO}] \longrightarrow \text{condensation products}$$

The reactivities in the charts refer *only* to the protected functionality, not to atoms adjacent to the functional group; for example, RCOOEt $\xrightarrow{\mathrm{LDA}}$: "L" (low) reactivity of PG(Et). However if the protected functionality is $R_2CHCOOEt$, this substrate obviously *will* react with LDA. Reactivity of the entire substrate must be evaluated by the chemist.

Five reagents [#25: HCN, pH 6; #88: $Tl(NO_3)_3$; #103: $Me_3O^+BF_4^-$; #104: LiN–*i*-Pr_2/MeI; and #105: K_2CO_3/MeI] were added after some of the charts had been completed; reactivities to these reagents are not included for all charts.

The number used to designate a protective group (PG) in a Reactivity Chart is the same as that used in the body of the text in the *first edition*.

Protective group numbers in the Reactivity Charts are not continuous, since not all of the protective groups described in the text are included in the charts. The protective groups that are included in the Reactivity Charts are in general those that have been used most widely; consequently, considerable experimental information is available for them.

The Reactivity Charts were prepared in collaboration with the following chemists, to whom we are most grateful: John O. Albright, Dale L. Boger, Dr. Daniel J. Brunelle, Dr. David A. Clark, Dr. Jagabandhu Das, Herbert Estreicher, Anthony L. Feliu, Dr. Frank W. Hobbs, Jr., Paul B. Hopkins, Dr. Spencer Knapp, Dr. Pierre Lavallée, John Munroe, Jay W. Ponder, Marcus A. Tius, Dr. David R. Williams, and Robert E. Wolf, Jr.

[a] L. F. Fieser and M. Fieser, *Reagents for Organic Synthesis*, Wiley-Interscience, New York, 1967, Vol. 1; M. Fieser and L. F. Fieser, Vols. 2–7, 1969–1979; M. Fieser, Vols. 8–15, 1980–1990.

[b] The categories and prototype reagents used in this study are an expansion of an earlier set of 11 categories and 60 prototype reagents,[c] originally compiled for use in LHASA[d] (Logic and Heuristics Applied to Synthetic Analysis), a long-term research program at Harvard University for Computer-Assisted Synthetic Analysis.

[c] E. J. Corey, H. W. Orf, and D. A. Pensak, *J. Am. Chem. Soc.*, **98,** 210 (1976).

[d] Selected references include: E. J. Corey, *Quart. Rev., Chem. Soc.*, **25,** 455 (1971); H. W. Orf, Ph.D. Thesis, Harvard University, 1976.

[e] E. J. Corey and M. G. Bock, *Tetrahedron Lett.*, 2643 (1975).

Reactivity Chart 1. Protection for the Hydroxyl Group: Ethers

1. Methyl Ether
2. Methoxymethyl Ether (MOM)
3. Methylthiomethyl Ether (MTM)
6. 2-Methoxyethoxymethyl Ether (MEM)
8. Bis(2-chloroethoxy)methyl Ether
9. Tetrahydropyranyl Ether (THP)
11. Tetrahydrothiopyranyl Ether
12. 4-Methoxytetrahydropyranyl Ether
13. 4-Methoxytetrahydrothiopyranyl Ether
15. Tetrahydrofuranyl Ether
16. Tetrahydrothiofuranyl Ether
17. 1-Ethoxyethyl Ether
18. 1-Methyl-1-methoxyethyl Ether
21. 2-(Phenylselenyl)ethyl Ether
22. *t*-Butyl Ether
23. Allyl Ether
26. Benzyl Ether
28. *o*-Nitrobenzyl Ether
35. Triphenylmethyl Ether
36. α-Naphthyldiphenylmethyl Ether
37. *p*-Methoxyphenyldiphenylmethyl Ether
41. 9-(9-Phenyl-10-oxo)anthryl Ether (Tritylone)
43. Trimethylsilyl Ether (TMS)
45. Isopropyldimethylsilyl Ether
46. *t*-Butyldimethylsilyl Ether (TBDMS)
48. *t*-Butyldiphenylsilyl Ether
51. Tribenzylsilyl Ether
53. Triisopropylsilyl Ether

(See chart, pp. 414–416)

Reactivity Chart 1. Protection for the Hydroxyl Group: Ethers

PG	pH<1, 100°	pH<1	pH 1	pH 2-4	pH 4-6	pH 6-8.5	pH 8.5-10	pH 10-12	pH>12	pH>12, 150°	NaH	Ph_3CNa	$(C_{10}H_8)^{\dot{-}}Na^+$	$MeSOCH_2^-Na^+$	KO-t-Bu	LiN-i-Pr_2	Py; R_3N	$NaNH_2$	NaOMe	Enolate	NH_3; RNH_2	RS^-; N_3^-; SCN^-	OAc^-; X^-	NaCN, pH 12	HCN, pH 6	RLi	RMgX	Organozinc	Organocopper	Wittig; ylide	H_2/Raney (Ni)	H_2/Pt pH 2-4	H_2/Pd	H_2/Lindlar	H_2/Rh	Zn/HCl	Zn/HOAc	Cr(II), pH 5
	1	2	3	4	5	6	7	8	9	10	11	12	13	14	15	16	17	18	19	20	21	22	23	24	25	26	27	28	29	30	31	32	33	34	35	36	37	38
	A. AQUEOUS										B. BASIC								C. NUCLEOPHILIC							D. ORGANOMET.					E. CAT. REDN.					F.		
1	H	M	L	L	L	L	L	L	L	L	L	L	L	L	L	L	L	L	L	L	L	L	L	L	L	L	L	L	L	L	L	L	L	L	L	L	L	L
2	H	H	H	M	L	L	L	L	L	M	L	L	L	L	L	L	L	L	L	L	L	L	L	L	L	L	L	L	L	L	L	M	L	L	L	H	M	L
3	H	H	M	L	L	L	L	L	L	M	L	M	R	M	L	M	L	L	L	L	L	L	L	L	L	L	L	L	L	L	R	R	R	L	R	R	R	M
6	H	H	L	L	L	L	L	L	L	L	L	L	L	L	L	L	L	L	L	L	L	L	L	L	L	L	L	L	L	L	L	L	L	L	L	M	L	L
8	H	H	H	H	M	L	L	L	M	H	L	R	R	R	R	R	L	R	R	R	M	R	L	R		R	R	M	R	L	R	R	R	L	L	H	H	M
9	H	H	H	H	M	L	L	L	L	L	L	L	L	L	L	L	L	L	L	L	L	L	L	L	M	L	L	L	L	L	L	H	L	L	L	H	H	M
11	H	H	H	H	L	L	L	L	L	M	L	M	R	R	L	M	L	L	L	L	L	L	L	L		L	L	L	L	L	R	R	R	L	R	H	H	M
12	H	H	H	H	M	L	L	L	L	L	L	L	L	L	L	L	L	L	L	L	L	L	L	L		L	L	L	L	L	L	H	L	L	L	H	H	M
13	H	H	H	H	M	L	L	L	L	M	L	L	M	L	L	L	L	L	L	L	L	L	L	L		M	L	L	L	L	R	R	R	L	R	H	H	M
15	H	H	H	H	H	L	L	L	L	L	L	L	L	L	L	L	L	L	L	L	L	L	L	L		L	L	L	L	L	L	H	L	L	L	H	H	H
16	H	H	H	M	L	L	L	L	L	M	L	M	R	R	L	M	L	L	L	L	L	L	L	L		L	L	L	L	L	R	R	R	L	R	H	H	M
17	H	H	H	H	L	L	L	L	L	L	L	L	L	L	L	L	L	L	L	L	L	L	L	L		L	L	L	L	L	L	H	L	L	L	H	H	L
18	H	H	H	H	M	L	L	L	L	L	L	L	L	L	L	L	L	L	L	L	L	L	L	L		L	L	L	L	L	L	H	L	L	L	H	H	M
21	H	H	M	L	L	L	L	L	L	M	L	R	R	R	M	R	L	R	M	L	L	R	L	L		R	L	L	L	L	R	R	R	L	R	H	H	M
22	H	H	L	L	L	L	L	L	L	L	L	L	L	L	L	L	L	L	L	L	L	L	L	L	L	L	L	L	L	L	L	L	L	L	L	L	L	L
23	H	H	L	L	L	L	L	L	M	R	L	L	L	L	R	L	L	L	R	L	L	L	L	L	L	L	L	L	L	L	R	R	R	L	R	L	L	L
26	H	H	L	L	L	L	L	L	L	L	L	L	R	L	L	L	L	L	L	L	L	L	L	L	L	L	L	L	L	L	H	H	H	L	L	L	L	L
28	H	H	L	L	L	L	L	L	L	H	R	M	R	R	L	R	L	R	L	L	L	L	L	L	L	M	R	R	R	L	H	H	H	L	R	R	R	R
35	H	H	H	H	L	L	L	L	L	L	L	L	H	L	L	L	L	L	L	L	L	L	L	L		L	L	L	L	L	H	H	H	L	L	H	H	L
36	H	H	H	H	L	L	L	L	L	M	L	L	H	L	L	L	L	L	L	L	L	L	L	L		L	L	L	L	L	H	H	H	L	L	H	H	L
37	H	H	H	H	M	L	L	L	L	L	L	L	H	L	L	L	L	L	L	L	L	L	L	L		L	L	L	L	L	H	H	H	L	L	H	H	M
41	H	H	L	L	L	L	L	L	L	L	L	L	R	L	L	L	L	L	L	M	R	L	L	L		R	R	R	L	R	R	R	R	L	R	R	R	R
43	H	H	H	H	H	H	H	H	H	H	H	H	H	H	L	L	L	H	H	L	H	H	L	H	H	H	H	L	L	L	L	H	H	L	L	H	H	H
45	H	H	H	H	M	L	L	L	H	H	L	L	L	L	L	L	L	L	L	L	L	L	L	L	L	H	H	L	L	H	M	H	M	L	L	H	H	M
46	H	H	H	H	L	L	L	L	H	H	L	L	L	L	L	L	L	L	L	L	L	L	L	L		L	L	L	L	L	L	H	L	L	L	H	L	L
48	H	M	M	L	L	L	L	L	H	H	L	L	L	L	L	L	L	L	L	L	L	L	L	L		L	L	L	L	L	L	L	L	L	L	M	L	L
51	H	H	H	H	L	L	L	L	H	H	L	L	H	L	L	L	L	L	L	L	L	L	L	L		L	L	L	L	L	L	H	L	L	L	H	H	L
53	H	H	H	H	L	L	L	L	H	H	L	L	L	L	L	L	L	L	L	L	L	L	L	L		L	L	L	L	L	L	H	L	L	L	H	H	L

Reactivity Chart 1. Protection for the Hydroxyl Group: Ethers (*Continued*)

PG	Na/NH_3	Al(Hg)	$SnCl_2/Py$	HSO_3^-; H_2S	$LiAlH_4$	Li-s-Bu_3BH	$(C_5H_{11})_2BH$	B_2H_6, 0°	$NaBH_4$	$Zn(BH_4)_2$	$NaBH_3CN$ pH 4-6	i-Bu_2AlH	$Li(OtBu)_3AlH$	$AlCl_3$, 80°	$AlCl_3$, 25°	$SnCl_4$; BF_3	$LiClO_4$; $MgBr_2$	TsOH, 80°	TsOH, 0°	Hg(II)	Ag(I)	Cu(II)/Py	HBr/In·	HX/In·	NBS/CCl_4	$Br_3CCl/In·$	OsO_4	$KMnO_4$, pH 7,0°	O_3, -50°	RCO_3H, 0°	RCO_3H, 50°	CrO_3/Py	CrO_3, pH 1	H_2O_2 pH 10-12	Quinone	$^1O_2$	DMSO, 100°	NaOCl pH 10	aq NBS
PG	39	40	41	42	43	44	45	46	47	48	49	50	51	52	53	54	55	56	57	58	59	60	61	62	63	64	65	66	67	68	69	70	71	72	73	74	75	76	77
	G.				H. HYDRIDE REDN.									I.						J.			K.				L. OXIDANTS												
1	L	L	L	L	L	L	L	L	L	L	L	L	L	H	H	H	L	L	L	L	L	L	L	L	M	L	L	L	L	L	L	L	L	L	L	L	L	L	L
2	L	L	L	L	L	L	L	L	L	L	L	L	L	H	H	H	L	H	L	L	L	L	H	L	R	L	L	L	R	L	M	L	L	L	L	L	L	L	R
3	R	L	M	L	L	L	L	L	L	L	L	L	L	H	H	M	M	M	L	H	H	L	M	L	M	M	L	R	R	R	R	L	H	L	L	R	M	R	R
6	L	L	L	L	L	L	L	L	L	L	L	L	L	H	H	H	H	H	L	L	L	L	H	L	R	L	L	L	R	L	L	L	M	L	L	L	L	L	M
8	R	M	M	L	M	L	L	L	L	L	L	M	L	H	H	H	L	H	M	L	R	L	H	L	R	R	L	L	R	L	M	L	H	L	L	L	L	L	M
9	L	L	L	L	L	L	L	L	L	L	L	L	L	H	H	H	M	H	L	L	L	L	H	L	M	L	L	L	H	L	H	L	H	L	L	L	L	L	L
11	R	L	L	L	L	L	L	L	L	L	L	L	L	H	M	M	L	H	L	H	H	L	M	L	M	L	L	R	R	R	R	L	H	L	L	R	M	R	R
12	L	L	L	L	L	L	L	L	L	L	M	L	L	H	H	H	L	H	L	L	L	L	H	L	M	L	L	L	R	L	H	L	H	L	L	L	L	L	R
13	R	L	L	L	L	L	L	L	L	L	M	L	L	H	H	H	L	H	L	L	L	L	H	L	M	L	L	R	R	R	R	L	H	L	L	R	M	R	R
15	L	L	L	L	L	L	L	L	L	L	H	L	L	H	H	H	M	H	L	L	L	L	H	L	M	L	L	L	R	L	H	L	H	L	L	L	L	L	R
16	R	L	L	L	L	L	L	L	L	L	L	L	L	H	M	M	L	H	L	H	H	L	M	L	M	L	L	R	R	R	R	L	H	L	L	R	M	R	R
17	L	L	L	L	L	L	L	L	L	L	L	L	L	H	H	H	L	H	L	L	L	L	M	L	R	L	L	L	R	L	H	L	H	L	L	L	L	L	M
18	L	L	L	L	L	L	L	L	L	L	M	L	L	H	H	H	M	H	M	L	L	L	H	L	M	L	L	L	L	H	R	L	R	L	L	L	L	L	M
21	R	M	L	L	L	L	L	L	L	L	L	L	L	H	M	M	L	M	L	L	L	L	R	L	R	L	R	R	R	R	R	R	R	R	L	R	M	R	R
22	L	L	L	L	L	L	L	L	L	L	L	L	L	H	H	H	L	H	L	L	L	L	M	L	L	L	L	L	L	L	L	L	L	L	L	L	L	L	L
23	H	L	L	L	L	L	R	R	L	L	L	R	L	H	H	H	L	L	L	R	L	L	R	R	R	R	R	R	R	R	R	L	M	L	L	R	L	L	R
26	H	L	L	L	L	L	L	L	L	L	L	L	L	H	H	H	L	L	L	L	L	L	R	L	R	L	L	L	L	L	L	L	L	L	L	L	L	L	L
28	R	R	L	L	R	L	L	L	L	L	L	L	L	H	L	L	L	L	L	L	L	L	R	L	R	L	L	L	L	L	L	L	L	L	L	L	L	L	L
35	H	M	H	L	L	L	L	L	L	L	M	L	L	H	H	H	L	H	H	L	L	L	H	L	L	L	L	L	L	L	H	L	H	L	L	L	L	L	L
36	H	M	L	L	L	L	L	L	L	L	M	L	L	H	H	H	L	H	H	L	L	L	H	L	L	L	L	L	L	L	H	L	H	L	L	L	L	L	L
37	H	M	L	L	L	L	L	L	L	L	M	L	L	H	H	H	L	H	H	L	L	L	H	L	L	L	L	L	L	M	H	L	H	L	L	L	L	L	L
41	R	R	L	R	R	R	R	L	R	R	L	R	M	H	H	H	M	H	L	L	L	L	H	L	L	L	L	L	L	L	R	L	L	L	L	L	L	L	L
43	H	L	M	H	H	H	L	H	H	H	H	H	H	H	M	M	L	H	L	L	L	L	H	M	R	M	L	H	L	L	H	L	H	H	L	L	H	H	H
45	L	L	L	L	L	L	L	L	L	L	M	L	L	H	L	L	L	H	M	L	L	L	H	L	M	L	L	L	L	L	H	L	H	L	L	L	L	L	L
46	L	L	L	L	L	L	L	L	L	L	L	L	L	H	M	M	L	H	L	L	L	L	H	L	M	L	L	L	L	L	L	L	L	L	L	L	L	L	L
48	L	L	L	L	L	L	L	L	L	L	L	L	L	H	L	L	L	M	L	L	L	L	L	L	M	L	L	L	L	L	L	L	M	L	L	L	L	L	L
51	R	R	L	L	L	L	L	L	L	L	L	L	L	H	L	L	L	H	L	L	L	L	R	L	R	L	L	L	L	L	L	L	H	L	L	L	L	L	L
53	L	L	L	L	L	L	L	L	L	L	L	L	L	H	M	M	L	H	L	L	L	L	H	L	M	L	L	L	L	L	L	L	H	L	L	L	L	L	L

Reactivity Chart 1. Protection for the Hydroxyl Group: Ethers (*Continued*)

PG	I_2	PhSeX; PhSCl	Br_2; Cl_2	MnO_2/CH_2Cl_2	$NaIO_4$ pH 5-8	SeO_2 pH 2-4	SeO_2/Py	$K_3Fe(CN)_6$, pH 8	Pb(IV), 25°	Pb(IV), 80°	$Tl(NO_3)_3$	150°	250°	350°	$:CCl_2$	N_2CHCO_2R/Cu	$CH_2I_2/Zn(Cu)$	$R_3SnH/In\cdot$	$Ni(CO)_4$	CH_2N_2	$SOCl_2$	Ac_2O, 25°	Ac_2O, 80°	DCC	MeI	$Me_3O^+BF_4^-$	1.LDA 2.MeI	1.K_2CO_3 2.MeI	RCHO	RCOCl	C^+/olefin
	78	79	80	81	82	83	84	85	86	87	88	89	90	91	92	93	94	95	96	97	98	99	100	101	102	103	104	105	106	107	108
	L. OXIDANTS											M.			N.			O. MISCELLANEOUS											P.		
1	*L*	*L*	M	*L*	L	L	L	L	L	L	*L*	L	*L*	L	L	L	L	*L*	L	*L*	*L*	*L*	L	*L*	*L*	L	*L*	L	L	*L*	*L*
2	*L*	L	R	L	*L*	H	L	L	L	H	H	L	L	R	*L*	L	L	*L*	L	*L*	*L*	*L*	*L*	L	L	M	L	L	L	*L*	H
3	L	L	R	L	R	H	M	L	R	R	*R*	L	*M*	R	*M*	R	L	M	L	*L*	L	*L*	*L*	L	*R*	R	M	R	L	L	M
6	L	L	M	L	L	L	L	L	L	M	M	*L*	L	R	*L*	*L*	L	*L*	L	*L*	L	L	L	L	*L*	M	L	L	L	L	H
8	L	L	M	L	L	H	L	L	M	H		L	L	R	L	M	R	R	L	L	L	L	M	L	L	M					
9	*L*	*L*	M	*L*	*L*	H	*L*	L	M	H	H	*L*	M	*H*	L	*L*	L	*L*	L	*L*	*L*	*L*	*L*	L	*L*	M	*L*	*L*	*L*	*L*	R
11	L	L	R	L	R	H	M	M	H	H		M	M	*H*	*M*	*M*	L	M	L	L	L	L	L	L	R	R	R	R			
12	*L*	L	R	L	L	H	L	L	H	H		L	M	*H*	L	L	L	L	L	L	L	L	M	*L*	L	M					
13	L	L	R	L	R	H	M	M	R	H		L	M	*H*	*M*	*M*	L	M	L	L	L	L	M	L	R	R	R	R			
15	L	L	R	L	M	H	L	L	H	H		L	M	R	L	L	L	L	L	L	L	L	H	L	L	M					
16	L	L	R	L	R	H	M	M	H	H		*R*	R	R	*M*	*M*	L	M	L	L	L	L	L	L	R	R	R	R			
17	L	L	M	L	*L*	H	L	L	M	H		L	L	R	*L*	L	L	L	L	L	L	L	L	L	L	M					
18	L	L	M	L	M	H	L	L	M	H		L	M	R	L	L	L	L	L	L	L	L	M	L	L	M					
21	R	L	R	*R*	*R*	R	*R*	L	*R*	*R*		R	R	R	M	M	*M*	*M*	*L*	*L*	L	*L*	*L*	L	R	R					
22	L	*L*	L	*L*	L	L	L	L	L	L	M	L	R	R	L	L	*L*	L	L	L	*L*	L	L	L	L	L	L	L	L	L	H
23	M	R	R	L	L	R	R	L	L	R	R	L	R	R	R	R	R	M	M	L	L	L	*L*	L	L	M	L	L	L	L	H
26	*L*	*L*	M	L	*L*	L	*L*	L	L	L	*L*	*L*	M	R	L	L	L	*L*	L	*L*	*L*	*L*	*L*	L	*L*	M	L	L	L	L	*L*
28	L	L	*M*	L	L	L	L	L	L	L	*L*	L	M	R	L	L	L	L	L	L	L	L	L	L	L	M	M	L	L	L	L
35	L	L	L	L	L	H	L	L	L	H		*L*	M	R	L	L	L	*L*	L	L	L	L	*M*	L	*L*	L				*L*	
36	L	L	L	L	L	H	L	L	L	H		L	M	R	L	L	L	L	L	L	L	L	L	L	L	L					
37	L	L	L	L	M	H	L	L	M	H		L	M	R	L	L	L	L	L	L	L	*L*	*M*	*L*	L	L					
41	L	L	L	L	L	L	L	L	L	L		L	R	R	L	L	L	M	L	L	L	L	*M*	L	L	L	*L*				
43	*L*	*L*	L	L	H	H	L	*H*	*H*	H	*H*	*L*	*L*	R	L	L	*L*	*L*	L	L	L	*M*	*H*	L	L	H	M	L	*L*	*L*	L
45	L	L	L	L	M	H	L	L	H	H		L	L	R	L	L	L	L	L	L	L	L	*H*	L	L	M					
46	*L*	*L*	L	*L*	L	H	L	L	M	*L*	M	*L*	*L*	R	L	L	L	L	L	L	L	L	*M*	L	L	M	L	L	*L*	*L*	*L*
48	L	L	L	L	L	L	L	L	L	M		L	L	R	L	L	L	L	L	L	L	L	*L*	L	L	L				*L*	
51	L	L	M	L	L	H	L	L	M	H		L	L	R	L	L	L	L	L	L	L	L	*M*	L	L	L					
53	L	L	L	L	L	H	L	L	M	H		L	L	R	L	L	L	L	L	L	L	L	*M*	L	L	M					

Reactivity Chart 2. Protection for the Hydroxyl Group: Esters

1. Formate Ester
3. Acetate Ester
6. Trichloroacetate Ester
10. Phenoxyacetate Ester
19. Isobutyrate Ester
22. Pivaloate Ester
23. Adamantoate Ester
27. Benzoate Ester
31. 2,4,6-Trimethylbenzoate (Mesitoate) Ester
34. Methyl Carbonate
36. 2,2,2-Trichloroethyl Carbonate
39. Allyl Carbonate
41. *p*-Nitrophenyl Carbonate
42. Benzyl Carbonate
46. *p*-Nitrobenzyl Carbonate
47. *S*-Benzyl Thiocarbonate
48. *N*-Phenylcarbamate
51. Nitrate Ester
53. 2,4-Dinitrophenylsulfenate Ester

(See chart, pp. 418–420)

Reactivity Chart 2. Protection for the Hydroxyl Group: Esters

PG	1 $pH<1$, 100°	2 $pH<1$	3 pH 1	4 pH 2–4	5 pH 4–6	6 pH 6–8.5	7 pH 8.5–10	8 pH 10–12	9 $pH>12$	10 $pH>12$, 150°	11 NaH	12 Ph_3CNa	13 $(C_{10}H_8)^{\dot{-}}Na^+$	14 $MeSOCH_2^-Na^+$	15 KO-t-Bu	16 LiN-i-Pr_2	17 Py; R_3N	18 $NaNH_2$
	A. AQUEOUS										B. BASIC							
1	H	H	H	M	L	L	*H*	*H*	*H*	H	H	L	*H*	*H*	*H*	H	*L*	*H*
3	H	M	L	*L*	L	L	M	*H*	*H*	H	R	H	*H*	*H*	*H*	H	L	*R*
6	H	M	L	L	L	L	*H*	*H*	H	H	L	H	*H*	*H*	*H*	L	*L*	L
10	H	M	L	L	L	L	M	*H*	H	H	M	M	*H*	*H*	*H*	H	*L*	*R*
19	H	M	L	L	L	L	*L*	M	*H*	H	H	H	*H*	*H*	*L*	M	*L*	*H*
22	H	M	L	L	L	L	L	L	M	H	L	L	*H*	*M*	*H*	L	*L*	*M*
23	H	H	L	L	L	L	L	*L*	M	H	L	L	*H*	*M*	*L*	L	*L*	*L*
27	H	M	L	*L*	L	L	L	M	*H*	H	L	*L*	*H*	*H*	*L*	*L*	*L*	*H*
31	H	M	L	L	L	L	L	L	L	H	*L*	L	*H*	*L*	*H*	L	*L*	*L*
34	H	M	L	L	L	L	L	*H*	H	H	L	L	*H*	*H*	*M*	L	*L*	*H*
36	H	*M*	*L*	L	L	L	M	H	H	H	L	H	*H*	*H*	*H*	H	*L*	*H*
39	H	M	L	L	L	L	L	H	H	H	L	H	*H*	*H*	*M*	H	*L*	*R*
41	H	M	L	L	L	L	H	H	H	H	L	L	*H*	*H*	*R*	L	*L*	*H*
42	H	M	L	L	L	L	L	H	H	H	L	H	*H*	*H*	*M*	H	*L*	*H*
46	H	M	L	*L*	L	L	L	H	H	H	L	H	*H*	*H*	*R*	H	*L*	*H*
47	H	*L*	L	L	L	L	L	H	H	H	L	H	*H*	*H*	*R*	H	*L*	*H*
48	H	L	*L*	L	L	L	L	L	M	H	R	H	*H*	*H*	*R*	R	*L*	*H*
51	H	H	M	*L*	L	L	L	L	H	H	L	L	*H*	*H*	*H*	M	*L*	*H*
53	H	M	M	*L*	L	L	L	H	*H*	H	L	L	*H*	*H*	*H*	H	*L*	*H*

PG	19 NaOMe	20 Enolate	21 NH_3; RNH_2	22 RS^-; N_3^-; SCN^-	23 OAc^-; X^-	24 NaCN, pH 12	25 HCN, pH 6	26 RLi	27 RMgX	28 Organozinc	29 Organocopper	30 Wittig; ylide
	C. NUCLEOPHILIC							D. ORGANOMET.				
1	*H*	*H*	H	H	*L*	M		H	H	M	M	H
3	*H*	*R*	M	L	*L*	*H*-	L	H	H	*L*	L	*L*
6	*H*	*R*	H	H	L	*R*		H	H	H	H	L
10	*H*	*R*	M	H	L	*L*		H	H	L	L	L
19	*M*	*M*	M	L	L	*L*		H	H	L	L	L
22	*M*	*M*	*H*	L	L	*L*	L	*H*	M	L	L	L
23	*L*	*L*	L	L	L	*L*		L	L	L	L	L
27	*M*	*M*	M	*L*	*L*	*L*	L	H	H	L	L	*L*
31	*L*	*L*	L	L	L	*L*		L	L	L	L	L
34	*M*	*M*	M	L	*L*	*L*	M	R	R	L	H	M
36	*R*	*R*	M	H	*L*	*R*		R	R	L	H	M
39	*M*	*M*	M	L	L	*L*		R	R	L	H	L
41	*R*	*R*	M	L	L	*M*		R	R	M	M	M
42	*M*	*M*	M	L	L	*L*	M	R	R	L	L	L
46	*R*	*R*	M	L	*L*	*L*		R	R	L	L	L
47	*R*	*R*	M	L	L	*L*		R	R	M	M	H
48	*M*	*L*	M	L	L	*L*		R	R	L	L	L
51	*H*	*H*	*H*	M	L	*L*		H	H	M	M	M
53	*H*	*H*	H	*H*	H	*H*		M	M	*H*	*H*	*H*

PG	31 H_2/Raney (Ni)	32 H_2/Pt pH 2–4	33 H_2/Pd	34 H_2/Lindlar	35 H_2/Rh	36 Zn/HCl	37 Zn/HOAc	38 Cr(II), pH 5
	E. CAT. REDN.					F.		
1	M	M	M	L	*L*	M	M	L
3	*L*	*L*	*L*	*L*	*L*	L	*L*	*L*
6	R	R	R	L	*R*	R	R	H
10	L	L	L	L	*L*	L	L	L
19	L	L	L	L	*L*	L	L	L
22	L	L	L	L	*L*	L	L	L
23	L	L	L	L	*L*	L	L	L
27	*L*	L	*L*	*L*	*L*	L	L	L
31	L	L	*L*	L	*L*	L	L	L
34	R	*L*	*L*	L	*L*	R	*H*	H
36	R	*L*	*L*	L	R	H	*H*	H
39	R	R	R	L	*R*	L	L	L
41	R	R	R	L	*R*	R	R	R
42	H	H	*H*	L	*M*	L	L	L
46	H	H	*H*	L	*R*	R	R	*R*
47	*R*	R	R	L	*H*	L	L	L
48	L	L	L	L	*L*	L	L	L
51	*H*	*H*	*H*	L	*H*	H	*H*	H
53	*H*	H	H	L	*H*	H	H	H

Reactivity Chart 2. Protection for the Hydroxyl Group: Esters (*Continued*)

	Na/NH_3	Al(Hg)	$SnCl_2/Py$	HSO_3^-; H_2S	$LiAlH_4$	Li-*s*-Bu_3BH	$(C_5H_{11})_2BH$	B_2H_6, 0°	$NaBH_4$	$Zn(BH_4)_2$	$NaBH_3CN$ pH 4-6	*i*-Bu_2AlH	Li(O*t*Bu)$_3$AlH	$AlCl_3$, 80°	$AlCl_3$, 25°	$SnCl_4$; BF_3	$LiClO_4$; $MgBr_2$	TsOH, 80°	TsOH, 0°	Hg(II)	Ag(I)	Cu(II)/Py	HBr/In·	HX/In·	NBS/CCl_4	Br_3CCl/In·	OsO_4	$KMnO_4$,pH 7,0°	O_3, -50°	RCO_3H, 0°	RCO_3H, 50°	CrO_3/Py	CrO_3, pH 1	H_2O_2 pH 10-12	Quinone	$^1O_2$	DMSO, 100°	NaOCl pH 10	aq NBS
PG	39	40	41	42	43	44	45	46	47	48	49	50	51	52	53	54	55	56	57	58	59	60	61	62	63	64	65	66	67	68	69	70	71	72	73	74	75	76	77
	G.				H. HYDRIDE REDN.									I.						J.			K.				L. OXIDANTS												
1	H	L	L	L	H	H	M	M	M	M	M	H	M	H	H	L	L	H	L	L	L	L	M	L	L	L	L	L	L	L	L	L	L	H	L	L	M	H	L
3	H	L	L	L	H	M	L	L	L	L	L	H	L	L	L	L	L	M	L	L	L	L	L	L	L	L	L	L	L	L	L	L	L	L	L	L	L	L	L
6	H	H	L	L	H	H	L	L	L	M	L	H	L	R	R	L	L	M	L	L	R	L	R	L	L	L	L	L	L	L	L	L	L	H	L	L	H	H	L
10	H	H	L	L	H	M	L	L	L	L	L	H	L	R	R	L	L	M	L	L	L	L	M	L	L	L	L	L	M	L	L	L	L	L	L	L	L	M	L
19	H	H	L	L	H	L	L	L	L	L	L	H	L	L	L	L	L	M	L	L	L	L	L	L	L	L	L	L	L	L	L	L	L	L	L	L	L	L	L
22	H	L	L	L	H	L	L	L	L	L	L	L	L	L	L	L	L	L	L	L	L	L	L	L	L	L	L	L	L	L	L	L	L	L	L	L	L	L	L
23	H	L	L	L	H	L	L	L	L	L	L	L	L	L	L	L	L	L	L	L	L	L	L	L	L	L	L	L	L	L	L	L	L	L	L	L	L	L	L
27	H	L	L	L	H	L	L	L	L	L	L	H	L	L	L	L	L	M	L	L	L	L	L	L	L	L	L	L	L	L	L	L	L	L	L	L	L	L	L
31	H	L	L	L	H	L	L	L	L	L	L	L	L	L	L	L	L	L	L	L	L	L	L	L	R	L	L	L	L	L	L	L	L	L	L	L	L	L	L
34	H	L	L	L	H	L	L	L	L	L	L	H	L	L	L	L	L	M	L	L	L	L	L	L	L	L	L	L	L	L	L	L	L	H	L	L	L	L	L
36	H	R	L	L	H	M	L	L	L	L	L	H	L	R	R	L	L	M	L	L	R	L	M	L	L	L	L	L	L	L	L	L	L	H	L	L	M	M	L
39	H	L	L	L	H	L	H	R	L	L	L	H	L	R	R	H	L	M	L	M	L	L	R	R	R	R	H	H	H	H	H	L	L	R	L	R	L	H	R
41	H	H	L	L	H	M	L	L	L	M	L	H	R	R	M	L	L	M	L	L	L	L	L	L	L	L	L	L	L	L	L	L	L	H	L	L	M	H	L
42	H	L	L	L	H	L	L	L	L	L	L	H	L	R	R	L	L	M	L	L	L	L	L	L	R	L	L	L	L	L	L	L	L	L	L	L	L	L	L
46	H	R	L	L	H	L	L	L	L	L	L	H	L	R	M	L	L	M	L	L	L	L	L	L	R	L	L	L	L	L	L	L	L	L	L	L	L	L	L
47	H	L	L	L	H	H	L	L	M	M	L	H	M	R	M	L	L	M	L	R	M	L	L	L	L	L	L	R	R	M	R	L	R	H	L	L	L	H	R
48	H	L	L	L	H	L	M	H	L	L	L	H	L	R	L	L	L	M	L	L	L	L	L	L	L	L	L	L	L	L	L	L	L	L	L	L	L	L	L
51	H	H	H	M	H	H	H	H	R	H	H	H	H	R	R	R	L	M	L	L	L	L	H	H	H	H	L	L	L	L	L	L	M	L	L	L	L	L	L
53	H	H	H	H	H	H	H	H	H	H	H	H	H	H	H	H	L	M	L	M	M	L	R	R	R	R	L	R	R	R	R	L	R	H	L	M	H	H	H

Reactivity Chart 2. Protection for the Hydroxyl Group: Esters (*Continued*)

PG	I_2	PhSeX; PhSCl	Br_2; Cl_2	MnO_2/CH_2Cl_2	$NaIO_4$ pH 5–8	SeO_2 pH 2–4	SeO_2/Py	$K_3Fe(CN)_6$, pH 8	Pb(IV), 25°	Pb(IV), 80°	$Tl(NO_3)_3$	150°	250°	350°	$:CCl_2$	N_2CHCO_2R/Cu	CH_2I_2/Zn(Cu)
	78	79	80	81	82	83	84	85	86	87	88	89	90	91	92	93	94
	L. OXIDANTS											M.			N.		
1	L	L	L	L	L	M	*M*	M	L	L		L	M	H	L	L	L
3	L	L	*L*	*L*	*L*	*L*	*L*	L	L	*L*	*L*	*L*	M	*H*	L	*L*	L
6	L	L	L	L	L	L	L	H	L	L		M	H	H	L	L	R
10	L	L	R	L	L	L	M	L	L	L		L	L	H	L	L	L
19	L	L	L	L	L	L	L	L	L	L		L	M	H	L	L	L
22	L	L	L	L	L	L	L	L	L	L	*L*	L	M	H	L	L	L
23	L	L	L	L	L	L	L	L	L	L		L	L	H	R	L	L
27	L	L	L	L	L	L	L	L	L	L	*L*	*L*	M	H	L	L	L
31	L	L	L	L	L	*L*	*L*	L	L	*L*		L	M	H	L	L	L
34	L	L	L	L	L	L	L	L	*L*	L	*L*	M	*H*	H	L	L	L
36	L	L	L	L	L	L	L	L	L	L	*L*	M	H	H	L	L	H
39	*M*	R	R	L	L	H	*H*	L	L	R		M	H	H	R	R	R
41	L	L	L	L	L	L	*M*	L	L	L		M	H	H	L	L	L
42	L	L	L	L	L	*M*	*M*	L	L	L	*L*	M	H	H	L	L	L
46	L	L	L	L	L	H	*H*	L	L	L		M	H	H	L	L	L
47	M	L	H	L	L	L	L	L	L	R		M	H	H	L	L	L
48	L	L	L	L	L	L	L	L	L	L		M	M	H	L	L	L
51	L	L	L	L	L	L	L	L	L	L		H	H	H	L	L	L
53	M	L	R	L	*M*	*M*	*M*	L	R	R		H	H	H	L	L	M

PG	R_3SnH/In·	$Ni(CO)_4$	CH_2N_2	$SOCl_2$	Ac_2O, 25°	Ac_2O, 80°	DCC	MeI	$Me_3O^+BF_4^-$	1.LDA 2.MeI	1.K_2CO_3 2.MeI	RCHO	RCOCl	C^+/olefin
	95	96	97	98	99	100	101	102	103	104	105	106	107	108
	O. MISCELLANEOUS											P.		
1	L	L	L	M	*L*	L	L	L						
3	*L*	L	*L*	*L*	*L*	*L*	*L*	*L*	*L*	R	M	L	L	L
6	*R*	*M*	*L*	L	L	L	L	L						
10	*L*	L	L	L	L	L	L	L						
19	L	L	L	L	L	L	L	L						
22	L	L	L	L	L	L	L	L	L	L	L	L	L	L
23	*L*	L	L	L	L	L	L	L						
27	*L*	L	*L*	L	*L*	L	L	L	L	L	L	L	L	L
31	L	L	*L*	L	*L*	L	L	L						
34	L	L	L	*L*	L	L	L	L	R	L	M	L	L	L
36	*R*	*M*	*L*	L	L	L	L	L	R	R	H	L	L	L
39	*H*	*H*	L	L	L	L	L	L						
41	*R*	L	L	L	L	L	L	L						
42	L	*L*	L	L	L	L	L	L	R	M	M	L	L	L
46	*H*	L	L	*L*	L	L	L	L						
47	*R*	L	*L*	L	*L*	*L*	L	L						
48	L	L	L	L	*R*	L	L	*L*		*R*	*L*			
51	H	L	L	H	*L*	L	*L*	L						
53	H	*M*	L	M	*L*	L	*L*	H						

Reactivity Chart 3. Protection for 1,2- and 1,3-Diols

1. Methylenedioxy Derivative
2. Ethylidene Acetal
6. Acetonide Derivative
11. Benzylidene Acetal
12. *p*-Methoxybenzylidene Acetal
18. Methoxymethylene Acetal
20. Dimethoxymethylenedioxy Derivative
28. Cyclic Carbonates
29. Cyclic Boronates

(See chart, pp. 422–424)

Reactivity Chart 3. Protection for 1,2- and 1,3-Diols

PG	pH<1, 100°	pH<1	pH 1	pH 2-4	pH 4-6	pH 6-8.5	pH 8.5-10	pH 10-12	pH>12	pH>12, 150°	NaH	Ph_3CNa	$(C_{10}H_8)^{\dot{-}}Na^+$	$MeSOCH_2^-Na^+$	KO-t-Bu	LiN-i-Pr_2	Py; R_3N	$NaNH_2$	NaOMe	Enolate	NH_3; RNH_2	RS^-; N_3^-; SCN^-	OAc^-; X^-	NaCN, pH 12	HCN, pH 6	RLi	RMgX	Organozinc	Organocopper	Wittig; ylide	H_2/Raney (Ni)	H_2/Pt pH 2-4	H_2/Pd	H_2/Lindlar	H_2/Rh	Zn/HCl	Zn/HOAc	Cr(II), pH 5
	1	2	3	4	5	6	7	8	9	10	11	12	13	14	15	16	17	18	19	20	21	22	23	24	25	26	27	28	29	30	31	32	33	34	35	36	37	38
	A. AQUEOUS										B. BASIC								C. NUCLEOPHILIC							D. ORGANOMET.					E. CAT. REDN.					F.		
1	H	H	L	*L*	L	L	L	L	L	L	L	L	L	L	L	L	L	L	L	L	L	L	L	L		L	L	L	L	L	L	L	L	L	L	L	L	L
2	H	H	H	M	L	L	L	L	L	L	L	L	L	L	L	L	L	L	L	L	L	L	L	*L*		L	L	L	L	L	L	M	L	L	L	H	M	L
6	H	H	*H*	M	L	L	L	*L*	*L*	*L*	*L*	L	*L*	L	*L*	L	*L*	L	*L*	*L*	*L*	*L*	*L*	*L*	L	L	*L*	L	*L*	*L*	*L*	M	*L*	L	L	H	M	*L*
11	H	H	H	*H*	L	L	L	L	*L*	M	*L*	L	R	L	L	L	*L*	L	*L*	L	*L*	L	*L*	L		L	L	L	*L*	L	H	H	*H*	L	L	H	H	L
12	H	H	H	H	M	L	L	L	L	M	L	L	R	L	L	L	L	L	L	L	L	L	L	L		L	L	L	L	L	H	H	H	L	L	H	H	M
18	H	H	H	*H*	*H*	L	L	L	L	M	L	L	L	M	L	L	L	L	L	L	L	L	L	L		M	H	H	L	L	L	H	L	L	L	H	H	H
20	H	H	H	H	H	L	L	L	L	*M*	L	L	L	*M*	*L*	L	L	L	L	*L*	L	*L*	L	L		*M*	*H*	H	L	*L*	L	H	L	L	L	H	H	H
28	H	*L*	*L*	L	L	L	L	*H*	*H*	H	L	L	*H*	*H*	*L*	L	*L*	*H*	*M*	*M*	M	M	M	*L*	L	H	H	L	L	L	L	L	*L*	L	*L*	L	L	L
29	H	H	H	H	H	*H*	H	H	H	H	L	L	*H*	*H*	*H*	L	L	*H*	*H*	*H*	*L*	*L*	*L*	*H*	M	H	H	H	H	H	L	H	L	L	*L*	H	H	M

Reactivity Chart 3. Protection for 1,2- and 1,3-Diols (*Continued*)

PG	Na/NH_3	Al(Hg)	$SnCl_2/Py$	HSO_3^-; H_2S	$LiAlH_4$	Li-s-Bu_3BH	$(C_5H_{11})_2BH$	B_2H_6, 0°	$NaBH_4$	$Zn(BH_4)_2$	$NaBH_3CN$ pH 4-6	i-Bu_2AlH	Li(OtBu)$_3$AlH	$AlCl_3$, 80°	$AlCl_3$, 25°	$SnCl_4$; BF_3	$LiClO_4$; $MgBr_2$	TsOH, 80°	TsOH, 0°
	39	40	41	42	43	44	45	46	47	48	49	50	51	52	53	54	55	56	57
	G.				H. HYDRIDE REDN.									I.					
1	L	L	L	L	L	L	L	L	L	L	*L*	L	L	H	H	*H*	L	M	L
2	L	L	L	L	L	L	L	L	*L*	L	*L*	L	L	H	H	H	L	M	L
6	*L*	*L*	L	L	*L*	L	*L*	L	*L*	L	*L*	*L*	L	H	H	H	L	M	L
11	*H*	L	L	L	L	L	L	L	*L*	L	*L*	L	L	H	H	*H*	L	H	*L*
12	H	L	L	L	L	L	L	L	L	L	*L*	L	L	H	H	H	L	M	L
18	L	L	L	L	R	L	L	L	L	L	*M*	L	*L*	H	H	H	M	*R*	H
20	L	L	L	L	R	L	L	L	L	L	*M*	L	*L*	H	H	H	M	R	H
28	H	L	L	*L*	H	*L*	*L*	L	L	*L*	L	H	L	H	H	L	L	M	L
29	H	M	L	*L*	H	*H*	*H*	H	H	*H*	H	H	H	H	H	*L*	L	*L*	L

PG	Hg(II)	Ag(I)	Cu(II)/Py	HBr/In·	HX/In·	NBS/CCl_4	Br_3CCl/In·	OsO_4	$KMnO_4$, pH 7, 0°	O_3, -50°	RCO_3H, 0°	RCO_3H, 50°	CrO_3/Py	CrO_3, pH 1	H_2O_2 pH 10-12	Quinone	$^1O_2$	DMSO, 100°	NaOCl pH 10	aq NBS
	58	59	60	61	62	63	64	65	66	67	68	69	70	71	72	73	74	75	76	77
	J.			K.				L. OXIDANTS												
1	L	L	L	R	L	L	L	L	L	R	L	L	L	L	L	L	L	L	L	*L*
2	L	L	L	M	L	L	L	L	L	R	L	*M*	L	H	L	L	L	L	L	*L*
6	*L*	*L*	L	M	L	*L*	L	*L*	L	*L*	*L*	*L*	*L*	H	*L*	*L*	L	*L*	L	*L*
11	*L*	L	L	H	L	*R*	L	L	L	*R*	L	H	*L*	H	L	L	L	*L*	L	*L*
12	L	L	L	H	L	R	L	L	L	R	M	H	L	H	L	L	L	L	L	*M*
18	L	L	L	H	L	H	L	L	M	L	H	H	L	H	L	L	L	L	L	*M*
20	L	L	L	H	L	L	L	L	M	L	H	H	L	H	L	L	L	L	L	*M*
28	L	L	L	L	L	L	L	L	*L*	L	*L*	L	*L*	L	H	L	L	L	M	L
29	H	H	L	L	L	L	L	L	H	L	*H*	*H*	L	H	H	L	L	*L*	H	H

Reactivity Chart 3. Protection for 1,2- and 1,3-Diols (*Continued*)

PG	I_2	PhSeX; PhSCl	Br_2; Cl_2	MnO_2/CH_2Cl_2	$NaIO_4$ pH 5-8	SeO_2 pH 2-4	SeO_2/Py	$K_3Fe(CN)_6$, pH 8	Pb(IV), 25°	Pb(IV), 80°	$Tl(NO_3)_3$	150°	250°	350°	$:CCl_2$	N_2CHCO_2R/Cu	$CH_2I_2/Zn(Cu)$
	78	79	80	81	82	83	84	85	86	87	88	89	90	91	92	93	94
	L. OXIDANTS											M.			N.		
1	L	L	*L*	L	L	L	L	L	L	L	L	L	L	M	L	L	L
2	L	L	L	L	*L*	M	L	L	L	M		L	L	M	L	L	L
6	*L*	L	*L*	*L*	*L*	M	L	L	*L*	*M*	H	*L*	*L*	L	L	L	L
11	L	L	R	L	*L*	H	L	L	L	M		L	M	H	L	L	L
12	L	L	R	L	L	H	L	L	L	M		L	M	H	*L*	*M*	L
18	L	L	L	L	M	H	L	L	*L*	*H*		L	M	H	L	L	L
20	L	L	L	L	M	R	L	L	*L*	*H*		M	H	H	L	L	L
28	L	L	*L*	L	L	L	L	L	*L*	L	L	L	M	H	L	L	L
29	L	L	*L*	*L*	*H*	H	L	H	H	H	H	L	H	H	L	L	L

PG	R_3SnH/In·	$Ni(CO)_4$	CH_2N_2	$SOCl_2$	Ac_2O, 25°	Ac_2O, 80°	DCC	MeI	$Me_3O^+BF_4^-$	1. LDA 2. MeI	1. K_2CO_3 2. MeI	RCHO	RCOCl	C^+/olefin
	95	96	97	98	99	100	101	102	103	104	105	106	107	108
	O. MISCELLANEOUS											P.		
1	L	L	*L*	L	L	*M*	L	L	M	L	L	L	L	H
2	L	L	*L*	L	*L*	*M*	L	L	M					
6	*L*	L	*L*	*L*	*L*	*M*	*L*	*L*	M	L	L	*L*	*L*	*H*
11	*L*	L	*L*	L	L	*M*	L	*L*	M				*L*	
12	L	L	L	L	L	*M*	L	L	R					
18	L	L	L	L	M	*R*	L	L	R					
20	L	L	L	L	M	*R*	L	L	R					
28	L	L	L	*L*	L	*L*	L	L	L	L	H	L	L	L
29	M	*L*	*L*	M	M	H	L	L	R	L	H	L	M	H

Reactivity Chart 4. Protection for Phenols and Catechols

PHENOLS

1. Methyl Ether
2. Methoxymethyl Ether
3. 2-Methoxyethoxymethyl Ether
4. Methylthiomethyl Ether
6. Phenacyl Ether
7. Allyl Ether
8. Cyclohexyl Ether
9. *t*-Butyl Ether
10. Benzyl Ether
11. *o*-Nitrobenzyl Ether
12. 9-Anthrylmethyl Ether
13. 4-Picolyl Ether
15. *t*-Butyldimethylsilyl Ether
16. Aryl Acetate
17. Aryl Pivaloate
18. Aryl Benzoate
19. Aryl 9-Fluorenecarboxylate
20. Aryl Methyl Carbonate
21. Aryl 2,2,2-Trichloroethyl Carbonate
22. Aryl Vinyl Carbonate
23. Aryl Benzyl Carbonate
25. Aryl Methanesulfonate

CATECHOLS

27. Methylenedioxy Derivative
28. Acetonide Derivative
30. Diphenylmethylenedioxy Derivative
31. Cyclic Borates
32. Cyclic Carbonates

(See chart, pp. 426–428)

Reactivity Chart 4. Protection for Phenols and Catechols

PG	pH<1, 100°	pH<1	pH 1	pH 2-4	pH 4-6	pH 6-8.5	pH 8.5-10	pH 10-12	pH>12	pH>12, 150°	NaH	Ph_3CNa	$(C_{10}H_8)^{\cdot -}Na^+$	$MeSOCH_2^-Na^+$	KO-t-Bu	LiN-i-Pr_2	Py; R_3N	$NaNH_2$	NaOMe	Enolate	NH_3; RNH_2	RS^-; N_3^-; SCN^-	OAc^-; X^-	NaCN, pH 12	HCN, pH 6	RLi	RMgX	Organozinc	Organocopper	Wittig; ylide	H_2/Raney (Ni)	H_2/Pt pH 2-4	H_2/Pd	H_2/Lindlar	H_2/Rh	Zn/HCl	Zn/HOAc	Cr(II), pH 5
	1	2	3	4	5	6	7	8	9	10	11	12	13	14	15	16	17	18	19	20	21	22	23	24	25	26	27	28	29	30	31	32	33	34	35	36	37	38
	A. AQUEOUS										B. BASIC								C. NUCLEOPHILIC							D. ORGANOMET.					E. CAT. REDN.					F.		
1	H	M	*L*	L	L	L	*L*	*L*	*L*	*L*	*L*	L	L	*L*	*L*	*L*	L	*L*	*L*	*L*	*L*	*H*	*L*	*H*	L	*L*	*L*	L	L	*L*	*L*	*L*	*L*	L	L	*L*	*L*	L
2	H	H	*H*	*M*	L	L	L	L	L	L	*L*	L	L	L	L	L	L	L	*L*	*L*	L	M	L	*L*	*L*	*L*	*L*	L	L	*L*	L	M	*L*	L	L	H	M	L
3	H	*H*	L	L	L	L	L	L	L	L	*L*	L	L	L	L	L	L	L	L	L	L	M	L	L	L	L	L	L	L	L	L	M	L	L	L	L	L	L
4	H	H	M	*M*	L	L	L	L	L	L	L	L	R	L	L	L	L	L	*L*	L	L	M	L	L	L	M	*L*	L	L	L	R	R	R	L	R	H	H	M
6	H	L	*L*	L	L	L	L	L	*L*	H	R	R	R	R	M	R	L	R	R	R	L	M	L	M	L	R	R	M	M	R	M	R	R	L	L	H	*H*	H
7	H	L	L	L	L	L	L	L	M	H	L	L	R	L	R	L	L	L	R	L	L	M	L	M	L	R	M	L	M	L	R	R	R	L	R	M	L	L
8	H	*H*	L	L	L	L	L	L	*L*	L	L	L	L	L	L	L	L	L	L	L	*L*	*L*	L	L	L	L	L	L	L	L	L	L	L	L	L	M	L	L
9	H	*H*	*H*	L	L	L	L	L	L	L	L	L	L	L	L	L	*L*	L	L	L	L	L	L	L	L	L	L	L	L	L	L	L	L	L	L	H	M	L
10	H	*H*	L	L	L	L	L	*L*	*L*	L	*L*	L	R	L	L	L	L	L	L	L	L	M	L	L	L	M	L	*L*	L	L	H	H	*H*	L	H	*L*	L	L
11	H	*L*	L	L	L	L	L	L	*L*	R	L	R	R	R	L	R	L	R	L	L	*L*	M	L	L	L	R	R	M	R	L	H	H	H	L	H	R	R	R
12	H	M	L	L	L	L	L	L	L	L	L	L	R	L	L	L	L	L	L	L	*L*	*H*	L	M	L	R	M	L	L	L	H	H	H	L	H	L	L	L
13	H	L	L	L	L	L	L	L	L	L	L	L	R	L	L	L	L	L	L	L	L	M	L	L	L	R	M	L	L	L	H	H	H	L	H	M	L	L
15	H	H	H	H	M	L	L	L	M	H	*L*	L	L	L	L	L	L	L	L	L	L	L	L	L	L	L	L	L	L	L	L	H	L	L	L	H	M	M
16	H	H	H	M	L	L	*H*	H	*H*	H	R	R	R	H	R	R	L	H	H	H	*H*	M	L	H	L	R	R	L	L	L	L	M	L	L	L	L	L	L
17	H	M	L	L	L	L	L	L	*M*	H	L	L	R	M	L	L	L	M	M	M	L	M	L	M	L	L	L	L	L	L	L	L	L	L	L	L	L	L
18	H	H	*H*	L	L	L	L	M	*H*	H	L	L	R	H	L	L	L	H	M	M	M	M	L	H	L	R	R	L	L	L	L	L	*L*	L	L	L	L	L
19	H	M	L	L	L	L	L	M	H	H	R	R	R	R	R	R	L	H	M	M	M	M	L	H	L	R	R	L	L	L	L	L	L	L	L	L	L	L
20	H	H	*H*	*H*	L	L	L	M	*H*	H	L	L	R	H	M	L	L	H	M	R	M	H	L	H	M	H	H	L	L	L	L	H	L	L	L	H	H	M
21	H	*M*	*L*	L	L	L	M	H	H	H	R	R	R	R	R	R	L	H	R	R	M	M	L	H	L	H	H	M	R	M	R	*L*	*L*	L	R	H	*H*	H
22	*H*	*M*	*L*	L	L	L	L	*H*	H	H	L	L	R	H	L	L	L	H	M	R	M	M	L	H	L	H	H	L	L	L	R	R	R	L	R	H	L	L
23	H	H	M	L	L	L	L	M	H	H	L	L	R	H	M	L	L	H	M	M	M	M	L	H	L	H	H	L	L	L	H	H	*H*	L	L	M	L	L
25	L	L	*L*	L	L	L	L	*L*	*H*	H	R	R	R	H	L	R	L	R	M	M	M	L	L	M	*L*	*H*	*M*	L	R	*M*	R	L	*L*	L	L	L	*L*	L
27	H	M	L	L	L	L	*L*	*L*	L	L	*L*	*L*	L	L	*L*	L	*L*	*L*	*L*	L	*L*	L	L	L	L	L	M	L	L	L	*L*	L	*L*	L	L	L	L	L
28	H	H	*H*	*M*	L	L	L	L	L	L	L	L	L	L	L	L	*L*	L	L	L	*L*	L	L	L	*L*	*L*	L	L	L	L	L	M	L	L	L	H	M	L
30	H	H	H	*H*	L	L	L	L	L	L	L	L	R	L	L	L	L	L	L	L	L	L	L	L	L	L	*L*	L	L	L	H	H	*H*	L	L	H	M	M
31	H	H	*H*	L	L	L	L	L	H	H	R	R	H	H	R	R	L	R	H	H	H	H	L	H	M	H	H	H	H	H	L	H	L	L	L	H	H	M
32	H	H	H	L	L	L	M	*H*	H	H	L	L	R	H	L	L	L	H	M	M	M	L	L	H	L	R	R	L	L	L	L	L	L	L	L	M	L	L

Reactivity Chart 4. Protection for Phenols and Catechols (*Continued*)

PG	Na/NH_3	Al(Hg)	$SnCl_2$/Py	HSO_3^-; H_2S	$LiAlH_4$	Li-s-Bu_3BH	$(C_5H_{11})_2BH$	B_2H_6, 0°	$NaBH_4$	$Zn(BH_4)_2$	$NaBH_3CN$ pH 4-6	i-Bu_2AlH	Li(OtBu)$_3AlH$	$AlCl_3$, 80°	$AlCl_3$, 25°	$SnCl_4$; BF_3	$LiClO_4$; $MgBr_2$	TsOH, 80°	TsOH, 0°
	39	40	41	42	43	44	45	46	47	48	49	50	51	52	53	54	55	56	57
	G.				H. HYDRIDE REDN.									I.					
1	R	L	L	L	L	L	L	L	L	L	L	L	L	H	M	L	L	L	L
2	R	L	L	L	L	L	L	L	L	L	L	L	L	H	H	H	M	H	L
3	R	L	L	L	L	L	L	L	L	L	L	L	L	H	M	L	L	H	L
4	R	L	L	L	L	L	L	L	L	L	L	M	L	H	H	M	L	M	L
6	R	R	L	L	R	R	R	R	R	R	R	R	R	H	H	L	L	M	L
7	R	L	L	L	L	L	R	R	L	L	L	L	L	H	H	L	L	L	L
8	R	L	L	L	L	L	L	L	L	L	L	L	L	M	L	L	L	L	L
9	R	L	L	L	L	L	L	L	L	L	L	L	L	H	H	H	L	H	L
10	R	L	L	L	L	L	L	L	L	L	L	L	L	H	M	L	L	L	L
11	R	R	L	L	R	R	L	L	L	L	L	R	L	H	M	L	L	L	L
12	R	L	L	L	L	L	L	L	L	L	L	L	L	H	M	L	L	M	L
13	R	L	L	L	L	L	L	L	L	L	L	L	L	H	H	L	L	M	L
15	R	L	L	L	L	L	L	L	L	L	M	M	L	H	M	L	L	H	L
16	R	L	L	L	H	H	L	L	M	L	L	H	M	R	R	M	L	H	L
17	R	L	L	L	H	L	L	L	L	L	L	M	L	R	L	L	L	M	L
18	R	L	L	L	H	M	L	L	L	L	L	H	M	R	R	M	L	H	L
19	R	L	L	L	H	H	L	L	L	L	L	H	M	H	H	L	L	H	L
20	R	L	L	L	H	L	L	L	L	L	M	H	M	H	H	L	L	H	M
21	R	M	L	L	H	M	L	L	M	L	L	H	M	H	H	L	L	H	L
22	R	L	L	L	H	L	R	R	L	L	L	H	M	H	H	L	L	H	L
23	R	L	L	L	H	L	L	L	L	L	L	H	M	H	H	L	L	H	L
25	R	L	L	L	L	L	L	L	L	L	L	L	L	H	L	L	L	L	L
27	R	L	L	L	L	L	L	L	L	L	L	L	L	H	H	L	L	M	L
28	R	L	L	L	L	L	L	L	L	L	M	L	L	H	H	M	L	M	L
30	R	M	L	L	L	L	L	L	L	L	L	L	L	H	H	M	L	M	L
31	R	M	L	L	H	H	H	H	H	H	H	H	H	H	H	H	L	H	L
32	R	L	L	L	H	L	L	L	L	L	L	H	M	H	H	L	L	H	L

PG	Hg(II)	Ag(I)	Cu(II)/Py	HBr/In·	HX/In·	NBS/CCl_4	Br_3CCl/In·	OsO_4	$KMnO_4$, pH 7, 0°	O_3, −50°	RCO_3H, 0°	RCO_3H, 50°	CrO_3/Py	CrO_3, pH 1	H_2O_2 pH 10-12	Quinone	$^1O_2$	DMSO, 100°	NaOCl pH 10	aq NBS
	58	59	60	61	62	63	64	65	66	67	68	69	70	71	72	73	74	75	76	77
	J.			K.				L. OXIDANTS												
1	L	L	L	L	L	L	L	L	L	L	L	L	L	L	L	L	L	L	L	L
2	L	L	L	H	L	R	L	L	L	L	L	M	L	R	L	L	L	L	L	L
3	L	L	L	M	L	R	L	L	L	L	L	L	L	M	L	L	L	L	L	L
4	H	M	L	H	L	R	L	L	R	R	R	R	L	R	L	L	R	L	R	R
6	L	L	L	L	L	L	L	L	L	L	L	R	L	L	L	L	L	L	L	L
7	R	L	L	R	R	R	R	R	R	R	R	R	L	L	L	L	R	L	L	R
8	L	L	L	M	L	L	L	L	L	L	L	L	L	L	L	L	L	L	L	L
9	L	L	L	M	L	L	L	L	L	L	L	L	L	R	L	L	L	L	L	L
10	L	L	L	R	R	R	R	L	L	L	L	M	L	L	L	L	L	L	L	L
11	L	L	L	M	M	M	M	L	L	L	L	M	L	L	L	L	L	L	L	L
12	L	L	L	R	R	R	R	L	L	L	L	M	L	L	L	L	L	L	L	L
13	L	L	L	R	R	R	R	L	R	R	R	R	L	M	L	L	L	L	L	L
15	L	L	L	H	L	L	L	L	L	L	L	M	L	R	L	L	L	L	L	L
16	L	L	L	L	L	L	L	L	L	L	L	L	L	L	H	L	L	L	H	L
17	L	L	L	L	L	L	L	L	L	L	L	L	L	L	L	L	L	L	L	L
18	L	L	L	L	L	L	L	L	L	L	L	L	L	L	H	L	L	L	M	L
19	L	L	L	R	L	R	L	L	L	L	L	L	L	L	M	L	L	L	M	L
20	L	L	L	H	L	L	L	L	L	L	L	L	L	H	M	L	L	L	M	L
21	L	R	L	M	L	L	L	L	L	L	L	L	L	L	R	L	L	L	R	L
22	H	L	L	R	R	R	R	R	R	R	R	R	L	R	R	L	R	L	R	R
23	L	L	L	M	L	R	L	L	L	L	L	M	L	M	M	L	L	L	L	L
25	L	L	L	L	L	L	L	L	L	L	L	L	L	L	L	L	L	L	L	L
27	L	L	L	M	L	R	L	L	L	L	L	L	L	M	L	L	L	L	L	L
28	L	L	L	H	L	L	L	L	L	L	L	L	L	R	L	L	L	L	L	L
30	L	L	L	H	L	L	L	L	L	L	L	L	L	R	L	L	L	L	L	L
31	L	L	L	H	L	L	L	L	L	L	L	R	L	R	L	L	L	L	L	L
32	L	L	L	H	L	L	L	L	L	L	L	L	L	R	R	L	L	L	M	L

Reactivity Chart 4. Protection for Phenols and Catechols (*Continued*)

PG	I_2	PhSeX; PhSCl	Br_2; Cl_2	MnO_2/CH_2Cl_2	$NaIO_4$ pH 5-8	SeO_2 pH 2-4	SeO_2/Py	$K_3Fe(CN)_6$, pH 8	Pb(IV), 25°	Pb(IV), 80°	$Tl(NO_3)_3$	150°	250°	350°	$:CCl_2$	N_2CHCO_2R/Cu	CH_2I_2/Zn(Cu)	R_3SnH/In·	$Ni(CO)_4$	CH_2N_2	$SOCl_2$	Ac_2O, 25°	Ac_2O, 80°	DCC	MeI	$Me_3O^+BF_4^-$	1.LDA 2.MeI	1.K_2CO_3 2.MeI	RCHO	RCOCl	C^+/olefin
	78	79	80	81	82	83	84	85	86	87	88	89	90	91	92	93	94	95	96	97	98	99	100	101	102	103	104	105	106	107	108
	L. OXIDANTS											M.			N.			O. MISCELLANEOUS											P.		
1	*L*	*L*	*L*	*L*	*L*	L	L	*L*	L	*L*	*L*	L	L	*L*	L	L	*L*	L	L	*L*	*L*	*L*	*L*	L	*L*	*L*	L	*L*	*L*	*L*	*L*
2	L	L	L	L	L	M	L	L	M	R	R	L	M	H	L	L	L	L	L	*L*	L	*L*	L	L	L	L	L	*L*	L	L	H
3	L	L	L	L	*L*	L	L	L	L	L	L	L	M	H	L	L	L	L	L	L	L	L	L	L	L	L	*L*	L	L	L	L
4	L	L	R	L	R	M	L	L	R	R	R	L	M	H	M	R	L	M	L	L	L	L	L	L	R	R	L	L	L	L	M
6	L	M	M	L	L	R	L	L	M	R	R	L	L	M	R	R	L	L	L	L	L	L	L	L	L	M	R	M	L	L	L
7	L	R	R	L	L	H	M	L	L	R	R	M	R	R	R	R	R	M	M	L	L	L	L	L	L	M	L	L	L	L	R
8	L	L	L	L	L	L	L	L	L	L	L	L	L	L	L	L	L	L	L	L	L	L	L	L	L	L	L	L	L	L	L
9	L	L	L	L	L	L	L	L	L	M	L	L	M	H	L	L	L	L	L	L	L	L	L	L	L	L	L	L	L	L	H
10	L	L	L	L	L	L	L	L	L	L	L	L	L	L	L	L	L	L	L	L	L	L	L	*L*	*L*	L	L	L	L	*L*	L
11	L	L	L	L	L	L	L	L	L	L	L	L	L	L	L	L	L	L	L	L	L	L	L	L	L	L	L	L	L	L	L
12	L	L	M	L	L	L	L	L	L	L	L	L	L	L	L	L	L	L	L	L	L	L	L	L	L	L	L	L	L	L	L
13	L	L	L	L	L	L	L	L	L	L	L	L	L	L	L	L	L	L	L	L	L	L	*L*	L	R	R	L	L	L	L	L
15	L	L	L	L	L	M	L	L	M	R	R	L	L	L	L	L	L	L	L	L	L	L	M	L	L	L	L	L	L	L	L
16	L	L	L	L	L	M	L	R	L	*L*	M	L	L	*L*	L	L	L	L	L	L	L	L	L	L	L	L	R	H	L	L	L
17	L	L	L	L	L	L	L	L	L	L	L	L	L	H	L	L	L	L	L	L	L	L	L	L	L	L	L	L	L	L	L
18	L	L	L	L	L	L	L	L	L	L	L	L	*L*	M	L	L	L	L	L	L	L	L	L	L	L	L	L	M	L	L	L
19	L	L	L	L	L	L	L	L	L	L	L	L	L	M	L	L	L	L	L	L	L	L	L	L	L	L	R	M	L	L	L
20	L	L	L	L	L	H	L	L	L	M	R	L	M	H	L	L	L	L	L	L	*L*	L	L	L	*L*	R	L	M	L	L	M
21	L	L	L	L	L	L	L	M	L	L	L	L	M	H	L	L	R	R	M	L	L	L	L	L	L	M	R	H	L	L	L
22	M	R	R	L	L	L	L	L	M	R	R	L	M	H	R	R	R	R	R	L	L	L	L	L	L	R	L	H	L	L	R
23	L	L	L	L	L	L	L	L	L	L	*R*	L	M	H	L	L	L	L	L	L	L	L	L	L	L	R	L	M	L	L	L
25	L	L	L	L	L	L	L	L	L	L	L	*L*	*L*	M	L	L	L	L	L	L	L	L	*L*	L	L	L	R	L	L	L	L
27	*L*	L	*L*	*L*	*L*	L	L	L	L	L	L	*L*	L	L	L	L	L	L	L	*L*	M	*L*	L	L	*L*	L	L	L	*L*	L	*L*
28	L	L	L	L	L	M	L	L	L	M	M	L	*L*	H	L	L	L	L	L	L	*L*	L	L	L	L	L	L	L	L	L	M
30	L	L	L	L	L	R	L	L	L	R	R	L	M	H	L	L	L	L	L	L	L	L	L	L	L	L	L	L	L	L	M
31	L	L	L	L	L	L	L	L	L	R	R	L	M	H	L	L	L	*L*	L	*L*	R	R	R	L	R	R	R	R	R	R	H
32	L	L	L	L	L	*L*	L	M	L	L	M	L	L	M	L	L	L	L	L	*L*	L	L	L	L	L	R	L	L	L	L	L

Reactivity Chart 5. Protection for the Carbonyl Group

1. Dimethyl Acetals and Ketals
3. Bis(2,2,2-trichloroethyl) Acetals and Ketals
5. 1,3-Dioxanes
6. 5-Methylene-1,3-dioxanes
7. 5,5-Dibromo-1,3-dioxanes
8. 1,3-Dioxolanes
9. 4-Bromomethyl-1,3-dioxolanes
10. 4-*o*-Nitrophenyl-1,3-dioxolanes
11. *S*,*S*′-Dimethyl Acetals and Ketals
19. 1,3-Dithianes
20. 1,3-Dithiolanes
24. 1,3-Oxathiolanes
26. *O*-Trimethylsilyl Cyanohydrins
29. *N*,*N*-Dimethylhydrazones
30. 2,4-Dinitrophenylhydrazones
33. *O*-Phenylthiomethyl Oximes
34. Substituted Methylene Derivatives
43. Bismethylenedioxy Derivatives

(See chart, pp. 430–432)

Reactivity Chart 5. Protection for the Carbonyl Group

PG	$pH<1$, 100°	$pH<1$	pH 1	pH 2-4	pH 4-6	pH 6-8.5	pH 8.5-10	pH 10-12	$pH>12$	$pH>12$, 150°	NaH	Ph_3CNa	$(C_{10}H_8)^{\dot{-}}Na^+$	$MeSOCH_2^-Na^+$	KO-t-Bu	LiN-i-Pr_2	Py; R_3N	$NaNH_2$
	1	2	3	4	5	6	7	8	9	10	11	12	13	14	15	16	17	18
	A. AQUEOUS										B. BASIC							
1	H	*H*	*H*	L	*L*	L	L	L	*L*	L	*L*	L	L	L	*L*	*L*	L	L
3	H	H	H	L	L	L	L	L	M	R	R	R	H	R	R	R	L	R
5	H	H	*H*	L	L	L	L	L	L	L	L	L	L	L	L	*L*	L	L
6	H	H	H	L	L	L	L	L	L	R	L	*L*	H	R	R	L	L	L
7	H	H	H	L	L	L	L	M	R	R	R	R	H	R	R	*R*	L	R
8	H	H	*H*	M	L	L	*L*	*L*	*L*	*L*	*L*	L	L	*L*	*L*	*L*	*L*	*L*
9	H	H	H	M	L	L	L	L	R	R	R	R	H	R	R	R	L	R
10	H	H	H	M	L	L	L	L	L	H	L	*M*	H	H	L	R	L	*M*
11	R	R	L	L	L	L	L	L	L	L	*L*	L	L	L	L	L	L	L
19	R	M	*L*	*L*	L	L	L	L	*L*	L	*L*	L	L	L	*L*	*L*	*L*	L
20	R	*R*	L	L	L	L	L	L	*L*	*H*	L	L	L	L	*L*	L	*L*	L
24	H	H	*H*	*H*	L	L	*L*	L	L	R	L	L	L	L	L	L	L	L
26	H	H	H	*H*	H	*H*	H	*H*	H	H	L	L	R	R	M	*L*	L	R
29	H	L	L	L	L	L	L	L	L	H	L	L	*M*	L	L	*R*	L	L
30	H	*L*	L	L	L	L	L	L	L	*H*	R	R	R	R	R	R	L	R
33	H	H	M	L	L	L	L	L	L	R	L	R	*H*	L	L	M	L	L
34	R	M	*L*	L	L	L	L	L	H	H	*H*	*H*	R	R	R	R	L	R
43	*H*	H	M	M	L	L	L	L	L	*L*	L	L	L	L	L	L	L	L

PG	NaOMe	Enolate	NH_3; RNH_2	RS^-; N_3^-; SCN^-	OAc^-; X^-	NaCN, pH 12	HCN, pH 6	RLi	RMgX	Organozinc	Organocopper	Wittig; ylide	H_2/Raney (Ni)	H_2/Pt pH 2-4	H_2/Pd	H_2/Lindlar	H_2/Rh	Zn/HCl	Zn/HOAc	Cr(II), pH 5
	19	20	21	22	23	24	25	26	27	28	29	30	31	32	33	34	35	36	37	38
	C. NUCLEOPHILIC							D. ORGANOMET.					E. CAT. REDN.					F.		
1	*L*	*L*	*L*	*L*	L	*L*		*L*	*L*	L	*L*	*L*	L	L	*L*	*L*	*L*	H	H	L
3	R	R	*M*	R	R	R		R	R	R	R	R	R	R	R	L	*M*	H	R	M
5	L	L	L	L	L	L		L	L	L	L	L	L	M	L	L	L	H	H	M
6	L	L	L	L	L	L		*L*	L	L	L	L	R	R	R	*L*	*R*	H	H	M
7	R	R	M	R	R	R		R	R	R	R	R	R	R	R	L	*M*	H	R	M
8	*L*	*L*	*L*	*L*	*L*	*L*		*L*	*L*	L	*L*	*L*	*L*	*R*	L	*L*	*L*	H	*L*	L
9	R	R	*L*	R	R	R		*L*	R	R	M	L	R	R	R	L	L	H	H	M
10	L	M	L	L	L	L		R	R	*M*	R	L	R	R	R	*M*	*R*	H	R	R
11	*L*	L	*L*	L	L	L		*L*	L	L	L	L	R	R	R	*H*	*R*	L	L	L
19	*L*	*L*	*L*	L	L	L		*L*	*L*	L	L	*L*	*R*	R	R	*H*	*R*	L	L	L
20	*L*	L	*L*	L	*L*	L		*L*	L	L	L	*L*	*R*	R	*L*	*H*	*R*	L	L	L
24	L	L	L	L	*L*	*L*		L	*L*	L	L	L	*R*	R	R	*H*	*R*	H	M	L
26	H	*L*	L	H	*L*	L		M	M	H	L	R	R	R	R	L	R	H	R	R
29	L	*L*	*R*	L	L	L		*L*	L	L	L	L	R	R	R	L	R	H	H	M
30	M	M	L	L	L	L		R	R	R	L	R	R	R	R	L	R	R	*R*	*R*
33	*L*	L	L	L	L	L		R	L	L	L	L	R	R	R	L	R	R	R	M
34	H	R	M	L	L	L		R	R	R	R	R	R	R	*L*	L	R	M	M	M
43	*L*	L	*L*	L	L	L		L	*L*	L	L	L	*L*	L	L	L	L	H	H	L

Reactivity Chart 5. Protection for the Carbonyl Group (*Continued*)

PG	Na/NH_3	Al(Hg)	$SnCl_2/Py$	HSO_3^-; H_2S	$LiAlH_4$	Li-s-Bu_3BH	$(C_5H_{11})_2BH$	B_2H_6, 0°	$NaBH_4$	$Zn(BH_4)_2$	$NaBH_3CN$ pH 4-6	i-Bu_2AlH	Li(OtBu)$_3$AlH	$AlCl_3$, 80°	$AlCl_3$, 25°	$SnCl_4$; BF_3	$LiClO_4$; $MgBr_2$	TsOH, 80°	TsOH, 0°
	39	40	41	42	43	44	45	46	47	48	49	50	51	52	53	54	55	56	57
	G.				H. HYDRIDE REDN.									I.					
1	L	L	L	L	L	L	L	M	L	L	L	L	L	H	R	H	L	H	M
3	R	M	L	L	R	L	L	M	L	L	L	L	L	H	H	H	L	H	M
5	L	L	L	L	L	L	L	M	L	L	L	L	L	H	H	H	L	L	L
6	R	L	L	L	L	L	R	R	L	L	L	L	L	H	H	H	L	H	M
7	R	H	M	L	R	L	L	L	L	L	L	M	M	H	H	H	L	L	L
8	L	L	L	L	L	L	L	M	L	L	L	L	L	H	H	H	L	L	L
9	L	H	L	L	M	L	L	M	L	L	L	L	L	H	H	H	L	L	L
10	R	R	L	L	R	R	L	M	L	L	L	L	L	H	H	H	L	L	L
11	R	L	L	L	L	L	L	L	L	L	L	L	L	H	L	L	L	L	L
19	R	L	L	L	L	L	L	L	L	L	L	L	L	H	L	L	L	L	L
20	R	L	L	L	L	L	L	L	L	L	L	L	L	H	L	L	L	L	L
24	R	L	L	L	L	L	L	L	L	L	L	L	L	H	H	M	L	H	L
26	R	R	M	L	R	R	R	R	M	M	L	R	R	H	H	H	M	H	H
29	L	L	R	L	R	L	L	R	L	L	R	R	L	H	L	L	L	L	L
30	R	R	M	L	R	R	L	L	L	M	R	R	M	L	L	L	L	L	L
33	R	R	M	L	L	L	L	R	L	L	R	R	M	H	M	L	L	L	L
34	R	R	M	L	R	R	R	R	M	L	R	R	M	M	L	L	L	L	L
43	L	L	L	L	L	L	L	L	L	L	L	L	L	R	R	M	M	L	L

PG	Hg(II)	Ag(I)	Cu(II)/Py	HBr/In·	HX/In·	NBS/CCl_4	Br_3CCl/In·	OsO_4	$KMnO_4$, pH 7, 0°	O_3, −50°	RCO_3H, 0°	RCO_3H, 50°	CrO_3/Py	CrO_3, pH 1	H_2O_2 pH 10-12	Quinone	$^1O_2$	DMSO, 100°	NaOCl pH 10	aq NBS
	58	59	60	61	62	63	64	65	66	67	68	69	70	71	72	73	74	75	76	77
	J.			K.				L. OXIDANTS												
1	L	L	L	H	L	L	L	L	L	L	L	L	L	H	L	L	L	L	L	L
3	L	R	L	R	R	R	R	L	L	L	L	L	L	H	L	L	L	H	L	L
5	L	L	L	H	L	L	L	L	L	L	L	L	L	M	L	L	L	L	L	L
6	R	L	L	H	R	H	R	R	R	R	R	R	L	H	L	L	M	L	L	R
7	L	R	L	H	H	H	H	L	L	L	L	L	L	H	L	L	L	H	L	L
8	L	L	L	H	L	L	L	L	L	L	L	L	L	H	L	L	L	L	L	L
9	L	R	L	H	H	L	L	L	L	L	L	L	L	L	L	L	L	R	L	L
10	L	L	L	H	H	L	L	L	L	L	L	L	L	H	L	L	L	L	L	L
11	H	H	H	H	L	L	L	L	R	R	R	R	L	M	L	L	R	M	R	R
19	H	H	H	H	L	L	L	L	R	R	R	R	L	M	L	L	R	M	R	R
20	H	H	H	H	L	L	L	L	R	R	R	R	L	M	L	L	R	M	R	R
24	H	H	H	H	L	L	L	L	R	R	R	R	L	H	L	L	R	M	R	R
26	L	L	L	R	L	H	L	L	R	H	R	R	L	R	R	L	L	R	R	R
29	L	L	R	R	M	L	M	L	H	H	H	H	H	H	H	L	H	L	H	H
30	L	L	H	R	L	L	L	L	L	H	H	H	L	L	L	L	M	L	H	R
33	R	R	M	R	L	R	L	L	R	H	R	R	L	R	L	L	R	M	M	R
34	L	L	L	R	R	R	R	R	R	H	R	R	L	M	L	L	M	M	M	M
43	L	L	L	M	L	L	L	L	L	L	L	L	L	L	L	L	L	L	L	L

Reactivity Chart 5. Protection for the Carbonyl Group (*Continued*)

PG	I_2 78	PhSeX; PhSCl 79	Br_2; Cl_2 80	MnO_2/CH_2Cl_2 81	$NaIO_4$ pH 5-8 82	SeO_2 pH 2-4 83	SeO_2/Py 84	$K_3Fe(CN)_6$, pH 8 85	Pb(IV), 25° 86	Pb(IV), 80° 87	$Tl(NO_3)_3$ 88	150° 89	250° 90	350° 91	:CCl_2 92	N_2CHCO_2R/Cu 93	CH_2I_2/Zn(Cu) 94
	L. OXIDANTS											M.			N.		
1	L	*L*	L	*L*	*L*	M	L	L	L	L	*L*	*L*	H	*H*	L	*L*	L
3	L	L	L	L	L	M	L	L	L	L		*L*	M	R	*L*	*M*	H
5	L	L	R	L	L	M	L	L	*L*	L		L	*L*	R	L	*L*	L
6	R	R	R	L	*R*	M	R	L	R	R		L	L	R	R	R	R
7	L	L	R	L	L	M	L	L	L	*L*		L	L	R	*L*	*M*	H
8	*L*	*L*	*R*	L	*L*	H	*L*	*L*	*L*	L		*L*	*L*	R	*L*	*L*	*L*
9	L	L	R	L	M	H	L	L	L	*L*		L	L	R	*L*	*L*	H
10	L	L	R	L	M	H	L	L	L	M		L	L	R	*L*	*L*	L
11	*H*	L	R	M	*R*	M	L	L	R	R		L	*L*	R	M	M	L
19	L	*L*	R	M	R	M	L	L	*L*	M		L	L	R	M	M	L
20	*H*	*L*	R	M	R	M	L	*L*	R	R		*L*	*L*	R	M	M	L
24	L	L	R	L	R	M	L	L	R	R		*L*	*L*	R	M	M	L
26	L	L	R	R	R	R	L	R	R	R		L	M	R	L	R	L
29	L	L	*H*	*M*	*H*	*L*	*M*	L	*R*	R		L	R	R	R	R	R
30	L	L	*L*	M	H	*L*	*L*	L	R	R		L	M	R	R	R	R
33	L	L	R	L	R	M	L	L	R	R		L	R	R	R	R	R
34	L	L	R	R	L	L	L	L	L	L		L	M	R	R	R	R
43	L	L	*L*	L	*L*	L	L	L	*L*	M		L	L	L	L	L	L

PG	$R_3SnH/In\cdot$ 95	$Ni(CO)_4$ 96	CH_2N_2 97	$SOCl_2$ 98	Ac_2O, 25° 99	Ac_2O, 80° 100	DCC 101	MeI 102	$Me_3O^+BF_4^-$ 103	1. LDA 2. MeI 104	1. K_2CO_3 2. MeI 105	RCHO 106	RCOCl 107	C^+/olefin 108
	O. MISCELLANEOUS											P.		
1	L	L	L	L	*L*	L	L	*L*	M			*L*	*L*	M
3	R	M	L	L	L	L	L	L	M			L	L	M
5	L	L	L	L	L	L	L	L	*L*			L	*L*	*L*
6	L	R	L	L	L	L	L	L	M			L	L	*R*
7	R	M	L	L	L	L	L	L	M			L	L	M
8	L	L	*L*	*L*	*L*	*L*	L	*L*	*L*			L	L	*R*
9	R	L	L	L	L	L	L	L	M			L	L	M
10	R	L	L	L	L	L	L	L	M			L	L	M
11	M	L	L	L	L	L	L	R	*R*			L	L	M
19	M	L	L	L	*L*	L	L	*H*	*L*			L	*L*	M
20	M	L	*L*	L	L	L	L	*H*	*R*			*L*	*L*	*L*
24	M	L	L	L	L	M	L	*R*	R			L	L	M
26	L	L	L	L	*R*	R	L	L	M			L	L	*R*
29	L	L	L	L	L	*L*	L	*R*	R			*L*	L	M
30	L	L	L	L	*L*	L	L	M	R			L	L	L
33	M	L	L	L	L	L	L	R	R			L	L	M
34	R	L	L	L	L	*L*	L	L	M			L	L	*R*
43	*L*	L	*L*	*L*	L	M	L	L	M			L	L	L

Reactivity Chart 6. Protection for the Carboxyl Group

ESTERS

1. Methyl Ester
2. Methoxymethyl Ester
3. Methylthiomethyl Ester
4. Tetrahydropyranyl Ester
7. Benzyloxymethyl Ester
8. Phenacyl Ester
13. *N*-Phthalimidomethyl Ester
15. 2,2,2-Trichloroethyl Ester
16. 2-Haloethyl Ester
21. 2-(*p*-Toluenesulfonyl)ethyl Ester
23. *t*-Butyl Ester
27. Cinnamyl Ester
30. Benzyl Ester
31. Triphenylmethyl Ester
33. Bis(*o*-nitrophenyl)methyl Ester
34. 9-Anthrylmethyl Ester
35. 2-(9,10-Dioxo)anthrylmethyl Ester
42. Piperonyl Ester
45. Trimethylsilyl Ester
47. *t*-Butyldimethylsilyl Ester
50. *S*-*t*-Butyl Ester
59. 2-Alkyl-1,3-oxazolines

AMIDES AND HYDRAZIDES

64. *N*,*N*-Dimethylamide
68. *N*-7-Nitroindoylamide
71. Hydrazides
72. *N*-Phenylhydrazide
73. *N*,*N*′-Diisopropylhydrazide

(See chart, pp. 434–436)

Reactivity Chart 6. Protection for the Carboxyl Group

PG	pH<1, 100°	pH<1	pH 1	pH 2-4	pH 4-6	pH 6-8.5	pH 8.5-10	pH 10-12	pH>12	pH>12, 150°	NaH	Ph_3CNa	$(C_{10}H_8)^{\cdot -}Na^+$	$MeSOCH_2^-Na^+$	KO-t-Bu	LiN-i-Pr_2	Py; R_3N	$NaNH_2$	NaOMe	Enolate	NH_3; RNH_2	RS^-; N_3^-; SCN^-	OAc^-; X^-	NaCN, pH 12	HCN, pH 6	RLi	RMgX	Organozinc	Organocopper	Wittig; ylide	H_2/Raney (Ni)	H_2/Pt pH 2-4	H_2/Pd	H_2/Lindlar	H_2/Rh	Zn/HCl	Zn/HOAc	Cr(II), pH 5
	1	2	3	4	5	6	7	8	9	10	11	12	13	14	15	16	17	18	19	20	21	22	23	24	25	26	27	28	29	30	31	32	33	34	35	36	37	38
	A. AQUEOUS										B. BASIC								C. NUCLEOPHILIC							D. ORGANOMET.					E. CAT. REDN.					F.		
1	H	H	L	L	L	L	L	M	H	H	L	L	R	R	L	L	L	L	L	R	M	H	L	L		R	R	L	L	L	L	L	L	L	L	L	L	L
2	H	H	H	M	L	L	L	L	H	H	L	L	R	R	L	L	L	L	R	R	M	L	L	L		R	R	L	L	L	L	L	L	L	L	H	M	L
3	H	M	L	L	L	L	L	L	M	H	L	L	R	R	L	L	L	L	R	R	M	M	L	L		R	R	L	L	L	R	R	R	L	R	H	M	M
4	H	H	H	H	L	L	L	M	H	H	L	L	R	R	L	L	L	L	R	R	M	L	L	L		R	R	L	L	L	L	H	L	L	L	H	H	L
7	H	H	H	M	L	L	L	L	H	H	L	L	R	R	L	L	L	L	R	R	M	L	L	L		R	R	L	L	L	H	H	H	L	M	H	M	L
8	H	L	L	L	L	L	L	M	H	H	R	R	R	R	M	R	L	R	R	R	M	H	L	H		R	R	R	L	R	H	H	H	L	M	H	H	H
13	H	H	L	L	L	L	H	H	H	H	L	L	R	R	L	L	L	L	R	R	H	L	L	L		R	R	L	L	L	L	M	L	L	M	H	R	L
15	R	L	L	L	L	L	M	H	H	H	R	R	R	R	R	L	L	R	L	R	M	M	L	M		R	R	M	R	L	R	R	L	L	L	H	H	M
16	H	H	M	L	L	L	L	H	H	H	R	R	R	R	R	R	R	R	R	R	M	R	L	H		R	R	R	R	R	R	R	R	L	L	M	M	M
21	H	H	L	L	L	L	M	H	H	H	R	R	R	R	R	R	R	R	R	R	M	R	M	H		R	R	L	L	H	R	L	L	L	L	M	M	M
23	H	H	H	H	L	L	L	L	H	H	L	L	R	R	L	L	L	L	L	L	L	L	L	H		R	R	L	L	L	L	L	L	L	L	H	L	L
27	H	H	H	L	L	L	L	H	H	H	L	L	R	R	L	L	L	L	R	R	M	L	L	L		R	R	L	H	L	R	R	R	L	R	M	L	L
30	H	M	L	L	L	L	L	H	H	H	L	L	R	R	L	L	L	L	R	R	M	M	L	L		R	R	L	L	L	H	H	H	L	L	L	L	L
31	H	H	H	H	M	L	L	M	H	H	L	L	R	R	L	L	L	L	R	R	L	L	L	L		R	R	L	L	L	H	H	H	L	M	H	H	M
33	H	M	L	L	L	L	L	M	H	H	L	L	R	R	L	L	L	L	R	R	M	L	L	L		R	R	L	L	L	R	R	R	L	R	R	R	R
34	H	H	M	L	L	L	L	L	H	H	L	L	R	R	L	L	L	L	R	R	M	H	L	L		R	R	L	L	L	H	H	H	L	M	M	L	L
35	H	H	L	L	L	L	L	M	H	H	L	L	R	R	L	L	L	L	R	R	M	M	L	L		R	R	M	L	M	H	H	H	L	M	R	R	L
42	H	H	H	L	L	L	L	M	H	H	L	L	R	R	L	L	L	L	R	R	M	M	L	L		R	R	L	L	L	H	H	H	L	M	H	M	L
45	H	H	H	H	H	H	H	H	H	H	L	L	R	R	L	L	L	L	R	R	H	H	H	H		R	R	H	H	H	H	H	H	H	H	H	H	H
47	H	H	H	H	H	M	H	H	H	H	L	L	R	R	L	L	L	L	R	R	M	L	L	L		R	R	L	L	L	L	H	L	L	L	H	H	H
50	H	H	M	L	L	L	L	L	M	H	L	L	R	R	L	L	L	L	R	R	R	R	M	H		R	R	R	R	L	R	R	R	L	R	H	L	L
59	H	H	M	L	L	L	L	L	M	H	L	L	M	L	L	L	L	L	L	L	L	R	L	L		L	L	L	L	L	R	R	R	L	R	R	R	R
64	H	L	L	L	L	L	L	L	L	H	L	L	M	M	L	L	L	L	L	L	L	L	L	L		R	R	L	L	L	M	L	L	L	L	L	L	L
68	H	M	L	L	L	L	L	M	H	H	L	L	R	H	H	L	L	L	L	M	L	L	L	L		R	R	L	L	L	R	R	R	R	R	R	R	R
71	H	H	M	L	L	L	L	L	M	H	R	R	R	R	R	R	L	R	R	L	L	L	L	L		R	R	L	L	L	H	H	M	L	L	M	L	L
72	H	H	M	L	L	L	L	L	M	H	R	R	R	R	L	R	L	R	R	L	L	L	L	L		R	R	L	L	L	H	H	M	L	L	M	L	L
73	H	H	L	L	L	L	L	L	L	H	L	L	H	R	L	R	L	R	L	L	L	L	L	L		R	R	L	L	L	H	H	M	L	L	M	L	L

Reactivity Chart 6. Protection for the Carboxyl Group (*Continued*)

	Na/NH_3	Al(Hg)	$SnCl_2/Py$	HSO_3^-; H_2S	$LiAlH_4$	Li-s-Bu_3BH	$(C_5H_{11})_2BH$	B_2H_6, 0°	$NaBH_4$	$Zn(BH_4)_2$	$NaBH_3CN$ pH 4-6	i-Bu_2AlH	Li(OtBu)$_3AlH$	$AlCl_3$, 80°	$AlCl_3$, 25°	$SnCl_4$; BF_3	$LiClO_4$; $MgBr_2$	TsOH, 80°	TsOH, 0°	Hg(II)	Ag(I)	Cu(II)/Py	HBr/In·	HX/In·	NBS/CCl_4	Br_3CCl/In·	OsO_4	$KMnO_4$, pH 7, 0°	O_3, -50°	RCO_3H, 0°	RCO_3H, 50°	CrO_3/Py	CrO_3, pH 1	H_2O_2 pH 10-12	Quinone	$^1O_2$	DMSO, 100°	NaOCl pH 10	aq NBS
PG	39	40	41	42	43	44	45	46	47	48	49	50	51	52	53	54	55	56	57	58	59	60	61	62	63	64	65	66	67	68	69	70	71	72	73	74	75	76	77
	G.				H. HYDRIDE REDN.									I.						J.			K.				L. OXIDANTS												
1	*R*	*L*	L	*L*	*R*	M	*L*	*L*	*L*	*L*	*L*	R	M	R	M	L	L	M	*L*	*L*	*L*	L	*L*	*L*	*L*	*L*	*L*	L	*L*	*L*	L	*L*	L	M	*L*	*L*	*L*	M	*L*
2	R	L	L	L	R	M	L	L	L	L	L	R	M	H	H	L	L	H	L	L	L	L	L	L	L	L	L	L	M	L	H	L	H	L	L	L	L	L	L
3	R	L	L	L	R	M	L	L	*L*	L	L	R	M	H	H	M	L	R	L	*H*	*H*	L	L	L	*H*	L	L	R	R	R	R	L	R	L	L	R	M	R	R
4	R	L	L	L	R	M	L	L	L	L	L	R	M	H	H	*H*	L	H	*M*	L	L	L	M	L	*M*	L	L	L	M	L	H	L	H	M	L	L	L	H	*L*
7	R	L	L	L	R	M	L	L	L	L	L	R	M	R	R	M	L	H	L	L	L	L	L	L	R	L	L	L	M	L	H	L	H	L	L	L	L	L	L
8	R	R	L	L	R	R	R	R	M	M	L	R	M	R	R	M	L	R	*L*	L	L	L	L	L	L	L	L	L	L	L	R	L	L	R	L	L	L	M	R
13	R	L	L	L	R	M	R	R	L	L	L	R	L	R	L	L	L	L	*L*	L	L	L	L	L	L	L	L	L	L	L	L	L	L	H	L	L	H	H	L
15	R	M	L	*L*	R	M	L	*L*	L	L	L	R	M	R	R	*L*	L	L	L	*L*	R	L	R	R	R	R	L	L	L	*L*	L	L	L	M	L	L	M	H	*L*
16	R	M	M	L	R	M	L	L	L	L	R	R	M	R	R	L	L	L	L	L	M	L	L	L	L	L	L	L	L	L	L	L	M	H	L	L	M	H	L
21	R	M	M	L	R	M	L	L	L	L	R	R	M	R	L	L	L	L	L	L	L	L	L	L	R	L	L	L	L	L	L	L	L	H	L	L	M	H	L
23	R	L	L	L	R	L	L	*L*	*L*	L	*L*	*L*	L	R	*H*	*H*	L	M	L	L	*L*	L	L	L	*L*	L	L	L	L	*L*	H	L	H	L	L	L	L	L	L
27	R	L	L	L	R	M	R	R	L	L	L	R	M	R	R	R	L	H	L	*R*	L	L	R	R	R	R	R	R	R	R	R	L	H	R	L	R	L	H	R
30	*H*	L	L	*L*	R	M	L	*L*	*L*	L	L	R	M	R	*R*	M	L	*L*	*L*	L	L	L	L	L	R	L	*L*	*L*	*L*	*L*	L	*L*	*L*	M	L	L	L	M	L
31	R	L	L	L	R	L	L	*L*	L	L	M	R	L	R	R	L	L	H	M	*L*	L	L	M	L	L	L	L	L	L	M	H	L	H	M	L	L	H	M	L
33	H	R	L	L	R	M	L	L	L	L	L	R	M	R	R	L	L	L	L	L	L	L	L	L	R	L	L	L	L	L	L	L	L	M	L	L	L	M	L
34	H	L	L	L	R	M	L	L	L	L	L	R	M	R	R	M	L	L	L	L	L	L	L	L	R	L	L	L	L	L	L	L	M	L	L	R	L	L	L
35	H	R	M	L	R	R	M	M	M	M	L	R	M	R	R	M	L	L	L	L	L	L	L	L	R	L	L	L	L	L	L	L	L	M	L	L	L	M	L
42	H	L	L	L	R	M	L	L	L	L	L	R	M	R	R	M	L	R	L	L	L	L	L	L	R	L	L	L	L	L	L	L	H	M	L	L	L	M	L
45	R	H	L	H	R	H	H	R	H	H	H	R	H	H	H	H	H	H	M	H	H	L	H	L	*L*	L	L	H	L	H	H	L	H	H	L	L	H	H	H
47	R	L	L	L	R	M	L	*L*	L	L	H	R	M	H	H	H	L	H	L	L	L	L	L	L	L	L	L	M	L	H	H	L	H	H	L	L	H	H	M
50	R	L	L	L	R	M	L	*L*	L	L	L	R	M	R	*M*	L	L	*H*	L	*R*	*R*	*R*	L	L	L	L	L	R	*R*	L	M	L	M	R	L	R	M	H	H
59	R	M	L	L	*L*	L	*M*	*L*	*L*	*L*	*M*	*M*	L	*R*	*L*	*L*	*L*	*R*	L	L	L	L	*L*	*L*	*L*	*L*	R	R	*L*	R	R	*L*	R	L	L	L	L	L	L
64	R	L	L	L	*R*	L	*R*	R	L	L	L	*R*	*L*	*L*	L	*L*	L	*L*	L	*L*	L	L	L	L	L	L	L	L	L	*L*	L	L	L	L	L	L	L	L	L
68	R	R	L	L	R	L	R	R	L	L	L	R	L	*M*	L	L	L	*L*	L	R	L	L	R	R	R	R	R	M	R	L	M	L	L	M	L	*R*	L	L	M
71	*L*	L	L	L	R	L	R	R	L	L	L	R	L	R	L	L	L	L	L	*R*	L	H	R	L	R	L	R	R	R	*R*	R	R	M	R	R	R	R	*R*	*H*
72	R	L	L	L	*R*	L	R	R	L	L	L	R	L	*M*	L	L	L	L	L	H	*L*	*H*	L	L	R	L	R	R	R	R	R	R	H	R	R	R	R	H	H
73	R	L	L	L	R	L	R	R	L	L	L	R	L	*M*	L	L	L	L	L	H	*L*	H	L	L	R	L	R	R	R	R	R	R	*H*	R	R	R	R	H	*H*

Reactivity Chart 6. Protection for the Carboxyl Group (*Continued*)

	I_2	PhSeX; PhSCl	Br_2; Cl_2	MnO_2/CH_2Cl_2	$NaIO_4$ pH 5-8	SeO_2 pH 2-4	SeO_2/Py	$K_3Fe(CN)_6$, pH 8	Pb(IV), 25°	Pb(IV), 80°	$Tl(NO_3)_3$	150°	250°	350°	$:CCl_2$	N_2CHCO_2R/Cu	$CH_2I_2/Zn(Cu)$	$R_3SnH/In\cdot$	$Ni(CO)_4$	CH_2N_2	$SOCl_2$	Ac_2O, 25°	Ac_2O, 80°	DCC	MeI	$Me_3O^+BF_4^-$	1.LDA 2.MeI	1.K_2CO_3 2.MeI	RCHO	RCOCl	C^+/olefin
PG	78	79	80	81	82	83	84	85	86	87	88	89	90	91	92	93	94	95	96	97	98	99	100	101	102	103	104	105	106	107	108
	L. OXIDANTS											M.			N.			O. MISCELLANEOUS											P.		
1	*L*	L	*L*	*L*	*L*	*L*	*L*	L	L	*L*		*L*	*L*	*L*	*L*	*L*	L	*L*	*L*	*L*	*L*	*L*	*L*	*L*	*L*	*L*			*L*	*L*	*L*
2	L	L	R	L	L	L	L	L	L	H		L	L	R	L	L	L	L	L	L	L	L	L	L	L	L			L	L	M
3	L	L	R	L	R	M	L	L	R	R		L	L	R	M	M	L	M	L	L	L	L	L	L	*R*	R			L	L	L
4	L	L	M	L	L	H	L	L	L	H		*M*	*H*	R	L	L	L	L	L	L	*M*	M	M	L	L	L			L	*L*	M
7	L	L	R	L	L	M	L	L	L	H		L	L	R	L	L	L	L	L	L	L	L	M	L	L	L			L	L	M
8	L	R	R	L	L	L	L	L	L	L		L	L	R	L	L	L	L	L	L	L	L	L	L	L	*L*			L	L	L
13	L	L	L	L	L	L	L	H	L	L		L	M	R	L	L	L	L	L	L	L	L	L	L	L	R			L	L	L
15	L	L	L	L	*L*	L	L	M	L	L		M	H	R	L	M	R	R	M	L	L	*L*	L	*L*	L	L			L	L	L
16	L	L	L	L	L	L	L	L	L	L		L	M	R	L	L	R	R	L	L	L	L	L	L	L	L			L	L	L
21	L	L	L	L	L	L	L	L	L	M		M	H	R	L	L	L	L	L	L	L	L	L	L	L	L			L	L	L
23	*L*	*L*	L	*L*	L	H	L	L	*L*	H		*M*	*H*	H	L	*L*	L	L	L	*L*	*L*	*L*	M	L	L	L			*L*	*L*	H
27	R	R	R	L	L	M	L	L	L	M		R	R	R	R	R	R	L	R	L	L	L	L	L	L	L			L	L	R
30	L	L	*L*	*L*	*L*	M	L	L	L	L		L	L	R	L	L	L	*L*	L	L	*L*	L	L	*L*	L	L			L	*L*	L
31	L	L	L	L	L	H	L	L	L	H		M	H	R	L	L	L	L	L	L	L	L	M	L	L	L			L	L	H
33	L	L	L	L	L	L	L	L	L	L		L	L	R	L	L	L	L	L	L	L	L	L	L	L	L			L	*L*	L
34	L	L	L	L	L	M	L	L	L	L		L	L	R	L	L	L	L	L	L	L	L	L	L	L	L			L	L	L
35	L	L	M	L	L	M	L	L	L	L		L	L	R	L	L	L	L	L	L	L	L	L	L	L	L			L	L	L
42	L	L	M	L	L	M	L	L	L	R		L	L	R	L	*M*	L	L	L	L	L	L	L	L	L	L			L	L	R
45	H	L	L	L	H	H	L	H	H	H		*L*	*L*	R	L	M	L	L	L	L	*L*	H	H	L	L	H			L	M	L
47	H	L	L	L	M	H	L	H	H	H		L	L	R	L	L	L	L	L	L	L	L	H	L	L	L			L	M	L
50	L	L	R	*L*	L	L	L	L	L	R		M	H	R	M	M	L	L	L	L	L	*L*	L	L	L	R			L	L	L
59	L	L	*R*	L	L	M	M	L	L	R		L	L	R	L	L	L	L	L	L	*L*	*L*	*L*	*L*	*R*	*R*			*L*	*R*	R
64	L	L	L	L	L	L	L	L	L	*L*		*L*	L	L	L	L	L	L	L	L	L	L	L	L	L	R			L	L	*L*
68	L	L	M	L	L	L	L	L	L	M		L	L	M	R	R	R	L	L	M	L	L	L	L	L	R			L	L	*M*
71	*R*	*R*	*R*	R	R	R	M	*R*	*H*	R		M	*R*	*R*	L	R	L	L	L	L	R	R	R	L	R	R			*R*	*R*	*R*
72	H	R	R	H	H	R	M	H	H	R		M	R	R	L	M	L	L	L	L	R	*R*	R	L	R	R			R	R	R
73	H	R	R	H	*H*	R	M	H	*H*	R		M	R	R	L	M	L	L	L	L	R	R	R	L	R	R			R	*R*	R

Reactivity Chart 7. Protection for the Thiol Group

1. *S*-Benzyl Thioether
3. *S-p*-Methoxybenzyl Thioether
5. *S-p*-Nitrobenzyl Thioether
6. *S*-4-Picolyl Thioether
7. *S*-2-Picolyl *N*-Oxide Thioether
8. *S*-9-Anthrylmethyl Thioether
9. *S*-Diphenylmethyl Thioether
10. *S*-Di(*p*-methoxyphenyl)methyl Thioether
12. *S*-Triphenylmethyl Thioether
15. *S*-2,4-Dinitrophenyl Thioether
16. *S-t*-Butyl Thioether
19. *S*-Isobutoxymethyl Monothioacetal
20. *S*-2-Tetrahydropyranyl Monothioacetal
23. *S*-Acetamidomethyl Aminothioacetal
25. *S*-Cyanomethyl Thioether
26. *S*-2-Nitro-1-phenylethyl Thioether
27. *S*-2,2-Bis(carboethoxy)ethyl Thioether
30. *S*-Benzoyl Derivative
36. *S*-(*N*-Ethylcarbamate)
38. *S*-Ethyl Disulfide

(See chart, pp. 438–440)

Reactivity Chart 7. Protection for the Thiol Group

PG	1 pH<1, 100°	2 pH<1	3 pH 1	4 pH 2-4	5 pH 4-6	6 pH 6-8.5	7 pH 8.5-10	8 pH 10-12	9 pH>12	10 pH>12, 150°	11 NaH	12 Ph_3CNa	13 $(C_{10}H_8)^{\dot{-}}Na^+$	14 $MeSOCH_2^-Na^+$	15 KO-t-Bu	16 LiN-i-Pr_2	17 Py; R_3N	18 $NaNH_2$	19 NaOMe	20 Enolate	21 NH_3; RNH_2	22 RS^-; N_3^-; SCN^-	23 OAc^-; X^-	24 NaCN, pH 12	25 HCN, pH 6	26 RLi	27 RMgX	28 Organozinc	29 Organocopper	30 Wittig; ylide	31 H_2/Raney (Ni)	32 H_2/Pt pH 2-4	33 H_2/Pd	34 H_2/Lindlar	35 H_2/Rh	36 Zn/HCl	37 Zn/HOAc	38 Cr(II), pH 5
	A. AQUEOUS										B. BASIC								C. NUCLEOPHILIC							D. ORGANOMET.					E. CAT. REDN.					F.		
1	H	*L*	L	*L*	L	L	L	L	L	M	L	R	R	R	*L*	R	L	L	L	L	L	*L*	L	L		*R*	R	L	L	L	R	R	R	*M*	R	L	L	L
3	H	*H*	*M*	*L*	L	L	L	L	*L*	M	L	R	R	R	L	R	L	L	L	L	L	L	L	L		R	R	L	L	L	R	R	R	*M*	*R*	M	L	L
5	H	L	L	L	L	L	L	L	M	H	M	R	R	R	L	R	L	R	M	L	L	M	L	M		R	R	L	R	L	R	R	*R*	*M*	R	R	R	R
6	H	*L*	L	L	L	L	L	L	*L*	M	L	R	R	R	L	R	L	L	L	L	L	*L*	L	L		R	R	L	L	L	R	R	R	*M*	R	L	L	L
7	H	*H*	*M*	L	L	L	L	L	L	M	L	R	R	R	L	R	L	L	L	L	L	L	L	L		R	R	L	L	L	R	R	R	*R*	R	R	R	R
8	H	*M*	L	L	L	L	L	L	L	M	L	R	R	R	L	R	L	L	L	L	L	*H*	L	L		R	R	L	L	L	R	R	R	*M*	R	L	L	L
9	*H*	*L*	L	L	L	L	L	L	L	M	L	R	R	R	L	R	L	*M*	*L*	L	L	M	L	L		R	R	L	L	L	R	R	R	*M*	R	L	L	L
10	H	H	L	L	L	L	L	L	L	M	L	R	R	R	L	R	L	*M*	L	L	L	M	L	L		R	R	L	L	L	R	R	R	*M*	R	L	L	L
12	H	*H*	*M*	*L*	L	L	L	L	L	M	L	*L*	R	L	L	*L*	L	L	*L*	L	L	*R*	L	L		L	L	L	L	L	R	R	R	*R*	R	M	L	L
15	H	L	L	L	L	M	H	H	H	H	*R*	*R*	R	R	*R*	R	L	*R*	*M*	M	M	H	L	*H*		R	R	L	R	L	R	R	R	*R*	R	R	R	R
16	H	*M*	*L*	L	L	L	L	*L*	*L*	M	L	L	L	L	L	*L*	*L*	L	*L*	*L*	*L*	L	L	L		*L*	L	L	L	L	R	R	M	*L*	M	L	L	L
19	H	H	L	L	L	L	L	L	M	H	L	L	R	H	L	L	L	L	L	L	L	H	L	L		L	L	L	L	L	R	R	R	L	R	H	L	L
20	H	H	L	L	L	L	L	L	M	H	L	L	R	H	L	L	L	L	L	L	L	R	L	L		L	L	L	L	L	R	R	R	L	R	H	L	L
23	*H*	*M*	*L*	*L*	*L*	*L*	*L*	*L*	*L*	*H*	R	R	R	R	L	R	L	R	R	L	*L*	L	L	L		R	R	L	L	L	R	R	R	*M*	R	L	*L*	L
25	H	R	L	L	L	L	L	M	M	H	R	R	R	R	M	R	L	R	R	L	*L*	*L*	L	L		R	R	L	L	L	R	R	R	*L*	R	R	R	L
26	H	*L*	L	L	L	L	H	H	*H*	R	H	H	R	H	H	H	H	H	H	R	H	M	*L*	*M*		R	R	L	R	L	R	R	R	*R*	R	R	R	R
27	H	*L*	L	L	L	L	L	H	*H*	R	H	H	R	H	H	H	L	H	R	R	R	M	L	*M*		R	R	*R*	R	L	R	R	R	*L*	R	L	L	L
30	H	L	L	L	L	L	M	H	H	H	L	L	H	H	H	H	H	H	H	H	H	H	L	H		H	H	H	L	H	R	R	R	L	R	H	L	L
36	H	L	L	L	L	M	H	H	H	H	R	R	R	H	H	R	H	H	H	H	H	H	L	L		H	H	L	L	L	R	R	R	L	R	H	L	L
38	H	L	L	L	L	L	L	L	H	H	R	R	R	H	R	R	L	R	R	R	L	M	L	H		H	H	H	M	H	R	R	R	H	R	H	H	H

Reactivity Chart 7. Protection for the Thiol Group (*Continued*)

PG	G.				H. HYDRIDE REDN.									I.					
	Na/NH_3	Al(Hg)	$SnCl_2/Py$	HSO_3^-; H_2S	$LiAlH_4$	Li-s-Bu_3BH	$(C_5H_{11})_2BH$	B_2H_6, 0°	$NaBH_4$	$Zn(BH_4)_2$	$NaBH_3CN$ pH 4-6	i-Bu_2AlH	Li(OtBu)$_3$AlH	$AlCl_3$, 80°	$AlCl_3$, 25°	$SnCl_4$; BF_3	$LiClO_4$; $MgBr_2$	TsOH, 80°	TsOH, 0°
	39	40	41	42	43	44	45	46	47	48	49	50	51	52	53	54	55	56	57
1	*H*	L	L	L	L	L	L	*L*	L	L	L	L	L	L	L	L	L	M	L
3	*H*	L	L	L	L	L	L	L	L	L	L	L	L	M	L	L	L	M	L
5	H	R	L	L	R	*M*	L	L	L	L	L	L	L	L	L	L	L	M	L
6	H	L	L	L	L	L	L	L	L	L	L	L	L	M	L	L	L	M	L
7	H	R	R	*R*	*R*	*R*	L	R	M	L	L	R	*R*	H	M	L	L	L	L
8	H	L	L	L	L	L	L	L	L	L	L	L	L	M	L	L	L	M	L
9	*H*	L	L	L	L	L	L	L	L	L	L	L	L	H	M	L	L	M	L
10	*H*	L	L	L	L	L	L	L	L	L	L	L	L	H	M	M	L	M	L
12	*H*	L	L	L	L	L	L	*H*	L	L	L	L	L	H	H	M	L	M	L
15	R	R	M	L	R	*M*	*M*	*M*	L	L	L	*R*	L	L	L	L	L	M	L
16	L	L	L	L	*L*	L	L	L	L	L	L	L	L	H	M	L	L	M	L
19	H	L	L	L	L	L	L	L	L	L	L	L	L	H	H	M	L	M	L
20	M	L	L	L	L	L	L	L	L	L	L	L	L	H	H	M	L	M	L
23	L	L	L	L	R	L	R	R	L	L	L	R	L	L	L	L	L	M	L
25	R	L	L	L	R	*M*	L	R	L	L	L	R	L	M	L	L	L	M	L
26	R	R	L	L	R	*M*	L	L	L	L	L	L	*R*	L	L	L	L	M	L
27	R	L	L	L	R	R	L	*L*	L	L	L	R	L	M	L	L	L	M	L
30	H	L	L	L	H	H	L	M	H	M	L	H	L	R	M	L	L	M	L
36	H	L	L	L	H	M	M	H	L	L	L	H	L	R	H	L	L	M	L
38	H	H	H	H	H	H	H	H	H	H	M	H	H	R	H	L	L	M	L

PG	J.			K.				L. OXIDANTS												
	Hg(II)	Ag(I)	Cu(II)/Py	HBr/In·	HX/In·	NBS/CCl_4	Br_3CCl/In·	OsO_4	$KMnO_4$, pH 7, 0°	O_3, -50°	RCO_3H, 0°	RCO_3H, 50°	CrO_3/Py	CrO_3, pH 1	H_2O_2 pH 10-12	Quinone	$^1O_2$	DMSO, 100°	NaOCl pH 10	aq NBS
	58	59	60	61	62	63	64	65	66	67	68	69	70	71	72	73	74	75	76	77
1	*L*	M	L	R	L	R	R	L	R	R	R	R	L	R	L	L	*R*	L	R	*R*
3	*R*	R	L	R	L	R	R	L	R	R	R	R	L	R	L	L	*R*	L	R	R
5	L	L	L	R	L	R	R	L	R	M	R	R	L	R	L	L	*R*	L	R	R
6	L	L	L	R	L	R	R	L	R	R	R	R	L	R	L	L	*M*	M	R	R
7	L	L	L	R	L	R	R	L	R	R	R	R	L	R	L	L	*M*	L	R	R
8	M	M	L	R	L	R	R	L	R	R	R	R	L	R	L	L	*R*	L	R	R
9	*M*	*M*	L	R	L	R	R	L	R	R	R	R	L	R	L	L	*R*	L	R	R
10	R	R	L	R	L	R	R	L	R	R	R	R	L	R	L	L	*R*	L	R	R
12	*R*	*R*	L	L	L	L	L	L	R	R	R	R	L	R	L	L	*M*	L	R	R
15	L	L	L	L	L	L	L	L	R	L	M	R	L	R	R	L	*M*	R	R	R
16	*R*	R	L	L	L	L	L	L	R	R	R	R	L	R	L	L	*M*	L	R	R
19	R	R	L	L	L	R	L	L	R	R	R	R	M	R	L	L	M	L	R	R
20	R	R	L	L	L	R	L	L	R	R	R	R	M	R	L	L	M	L	R	R
23	*R*	R	L	L	L	L	L	L	R	R	R	R	L	R	L	L	*R*	L	R	R
25	L	L	L	R	L	R	L	L	R	R	R	R	L	R	R	L	*R*	L	R	R
26	L	L	L	R	L	R	R	L	R	R	R	R	L	R	R	L	*R*	M	R	R
27	L	L	L	L	L	R	L	L	R	R	R	R	L	R	R	L	*M*	L	R	R
30	R	R	L	L	L	M	L	L	R	M	R	R	L	R	H	L	R	L	R	R
36	R	R	L	L	L	M	L	L	R	R	R	R	L	R	H	L	R	M	R	R
38	R	R	M	R	R	R	R	M	R	R	R	R	R	R	L	R	R	L	R	R

Reactivity Chart 7. Protection for the Thiol Group (*Continued*)

PG	I_2	PhSeX; PhSCl	Br_2; Cl_2	MnO_2/CH_2Cl_2	$NaIO_4$ pH 5–8	SeO_2 pH 2–4	SeO_2/Py	$K_3Fe(CN)_6$, pH 8	Pb(IV), 25°	Pb(IV), 80°	$Tl(NO_3)_3$	150°	250°	350°	$:CCl_2$	N_2CHCO_2R/Cu	CH_2I_2/Zn(Cu)	R_3SnH/In·	$Ni(CO)_4$	CH_2N_2	$SOCl_2$	Ac_2O, 25°	Ac_2O, 80°	DCC	MeI	$Me_3O^+BF_4^-$	1.LDA 2.MeI	1.K_2CO_3 2.MeI	RCHO	RCOCl	C^+/olefin
	78	79	80	81	82	83	84	85	86	87	88	89	90	91	92	93	94	95	96	97	98	99	100	101	102	103	104	105	106	107	108
	L. OXIDANTS											M.			N.			O. MISCELLANEOUS											P.		
1	*L*	L	R	M	R	R	M	L	R	R		L	L	M	M	R	L	*R*	L	L	L	L	L	*L*	R	R	R	R	*L*	*L*	M
3	M	L	R	M	R	R	M	L	R	R		L	L	M	M	R	L	R	L	L	L	L	L	*L*	R	R	R	R	L	L	M
5	R	L	R	L	R	R	L	L	R	R		L	L	M	M	R	L	R	L	L	L	L	L	L	M	M	R	M	L	L	L
6	R	L	R	L	R	R	M	L	R	R		L	L	M	M	R	L	R	L	*L*	L	L	L	*L*	R	R	R	R	L	L	L
7	R	L	R	L	R	R	M	L	R	R		L	L	M	R	R	L	R	L	L	R	*R*	*R*	L	R	R	R	R	L	H	L
8	R	L	R	M	R	R	R	L	R	R		L	L	M	M	R	L	R	L	L	L	L	L	L	R	R	R	R	L	L	M
9	*M*	L	R	M	R	R	R	L	R	R		L	L	M	M	R	L	R	L	L	L	L	L	L	R	R	R	R	L	L	M
10	M	L	R	M	R	R	R	L	R	R		L	L	M	M	R	L	R	L	L	L	L	L	L	R	R	R	R	L	L	M
12	*R*	L	M	L	R	L	L	L	R	R		L	L	M	M	R	L	L	L	L	L	L	L	L	L	L	L	L	L	L	M
15	R	L	M	L	L	M	L	R	R	R		L	L	M	M	R	L	R	L	L	L	L	L	L	L	L	R	L	L	L	L
16	*L*	L	M	L	R	L	L	L	R	R		L	L	R	M	R	L	L	L	L	L	*L*	L	L	M	M	M	M	*L*	L	M
19	R	L	R	L	R	M	L	L	R	R		L	M	R	M	R	L	M	L	L	L	L	L	L	R	R	M	R	L	L	M
20	R	L	R	L	R	M	L	L	R	R		L	M	R	M	R	L	M	L	L	L	L	L	L	R	R	M	R	L	L	M
23	*R*	L	R	L	R	M	L	L	R	R		L	L	M	M	R	L	L	L	L	L	L	M	L	R	R	R	R	L	L	L
25	R	L	R	L	R	M	L	M	R	R		L	L	M	M	R	L	*M*	L	L	L	L	L	L	R	R	R	R	L	L	L
26	R	L	R	L	R	*M*	*L*	*H*	R	R		M	H	H	M	R	L	R	L	L	L	L	L	L	R	R	R	R	L	L	L
27	R	L	R	L	R	M	L	H	R	R		M	H	H	M	R	L	L	L	L	L	L	L	L	R	R	R	R	L	L	L
30	L	L	R	L	L	L	L	H	L	M		M	R	R	M	M	L	L	L	L	L	L	L	L	L	R	L	L	L	L	L
36	L	L	R	L	R	L	L	H	L	M		R	R	R	M	M	L	R	L	L	L	L	M	L	L	R	L	L	L	L	L
38	R	L	R	R	R	R	R	R	R	R		H	H	R	M	R	R	R	R	L	L	L	L	L	M	R	R	M	L	L	R

Reactivity Chart 8. Protection for the Amino Group: Carbamates

1. Methyl Carbamate
5. 9-Fluorenylmethyl Carbamate
8. 2,2,2-Trichloroethyl Carbamate
11. 2-Trimethylsilylethyl Carbamate
16. 1,1-Dimethylpropynyl Carbamate
20. 1-Methyl-1-phenylethyl Carbamate
22. 1-Methyl-1-(4-biphenylyl)ethyl Carbamate
24. 1,1-Dimethyl-2-haloethyl Carbamate
26. 1,1-Dimethyl-2-cyanoethyl Carbamate
28. *t*-Butyl Carbamate
30. Cyclobutyl Carbamate
31. 1-Methylcyclobutyl Carbamate
35. 1-Adamantyl Carbamate
37. Vinyl Carbamate
38. Allyl Carbamate
39. Cinnamyl Carbamate
44. 8-Quinolyl Carbamate
45. *N*-Hydroxypiperidinyl Carbamate
47. 4,5-Diphenyl-3-oxazolin-2-one
48. Benzyl Carbamate
53. *p*-Nitrobenzyl Carbamate
55. 3,4-Dimethoxy-6-nitrobenzyl Carbamate
58. 2,4-Dichlorobenzyl Carbamate
65. 5-Benzisoxazolylmethyl Carbamate
66. 9-Anthrylmethyl Carbamate
67. Diphenylmethyl Carbamate
71. Isonicotinyl Carbamate
72. *S*-Benzyl Carbamate
75. *N*-(*N'*-Phenylaminothiocarbonyl) Derivative

(See chart, pp. 442–444)

Reactivity Chart 8. Protection for the Amino Group: Carbamates

PG	1 pH<1, 100°	2 pH<1	3 pH 1	4 pH 2-4	5 pH 4-6	6 pH 6-8.5	7 pH 8.5-10	8 pH 10-12	9 pH>12	10 pH>12, 150°	11 NaH	12 Ph_3CNa	13 $(C_{10}H_8)^{\cdot -}Na^+$	14 $MeSOCH_2^-Na^+$	15 KO-t-Bu	16 LiN-i-Pr_2	17 Py; R_3N	18 $NaNH_2$
	A. AQUEOUS										B. BASIC							
1	H	*H*	*L*	L	L	L	*L*	L	M	*H*	L	L	L	L	L	L	*L*	L
5	H	M	L	L	L	L	L	L	M	H	M	H	L	L	L	L	*M*	H
8	H	H	*L*	L	L	L	L	L	*M*	M	R	R	R	R	R	R	L	R
11	H	*H*	H	M	L	L	L	L	*L*	*M*	L	L	L	L	L	L	L	L
16	H	H	H	L	L	L	L	L	L	M	L	R	R	R	L	R	L	R
20	H	H	M	L	L	L	L	L	L	M	L	L	L	L	L	L	L	L
22	H	H	H	M	L	L	L	L	L	M	L	L	L	L	L	L	L	L
24	H	*H*	M	L	L	L	L	L	L	H	L	L	H	L	L	L	L	M
26	H	*L*	L	L	L	L	M	*H*	H	H	M	R	M	H	H	H	H	H
28	H	*H*	M	M	L	L	L	L	*L*	M	L	L	L	L	L	L	L	L
30	H	*H*	M	L	L	L	L	L	L	H	L	L	L	L	L	L	L	L
31	H	*H*	H	*M*	L	L	L	L	L	H	L	L	L	L	L	L	L	L
35	H	H	H	M	L	L	L	L	L	M	L	L	L	L	L	L	L	L
37	H	*H*	H	H	M	L	L	*L*	*L*	M	L	L	L	L	L	L	L	L
38	H	H	M	M	L	L	L	L	L	M	L	L	L	L	L	L	L	L
39	H	H	M	M	L	L	L	L	L	M	L	L	L	L	L	L	L	L
44	H	H	M	L	L	L	L	L	H	H	L	L	L	L	L	L	L	L
45	H	*L*	L	L	L	L	L	L	*L*	M	L	L	M	L	L	L	L	L
47	H	*L*	L	L	L	L	L	L	L	M	L	L	H	L	L	L	L	L
48	H	*H*	*L*	*L*	L	L	*L*	*L*	*L*	*H*	*L*	R	L	L	*L*	*L*	L	H
53	H	*H*	L	L	L	L	L	L	L	M	L	R	R	L	L	L	H	H
55	H	M	*L*	L	L	L	L	L	L	M	L	L	L	L	L	L	L	L
58	H	M	L	L	L	L	L	L	L	M	L	L	L	L	L	L	H	M
65	H	*H*	M	L	L	L	*L*	L	M	H	L	R	H	H	H	H	*R*	H
66	H	H	*L*	L	L	L	L	*L*	L	M	L	R	L	L	L	L	L	L
67	H	*H*	*H*	M	L	L	L	L	L	M	L	R	L	L	L	L	L	L
71	H	*L*	*L*	*L*	L	L	L	L	L	M	L	R	L	L	L	L	L	L
72	H	M	L	L	L	L	L	L	*H*	H	L	R	H	L	L	L	L	L
75	H	*H*	M	L	L	L	L	L	H	H	L	L	L	L	L	L	L	L

PG	19 NaOMe	20 Enolate	21 NH_3; RNH_2	22 RS^-; N_3^-; SCN^-	23 OAc^-; X^-	24 NaCN, pH 12	25 HCN, pH 6	26 RLi	27 RMgX	28 Organozinc	29 Organocopper	30 Wittig; ylide	31 H_2/Raney (Ni)	32 H_2/Pt pH 2-4	33 H_2/Pd	34 H_2/Lindlar	35 H_2/Rh	36 Zn/HCl	37 Zn/HOAc	38 Cr(II), pH 5
	C. NUCLEOPHILIC							D. ORGANOMET.					E. CAT. REDN.					F.		
1	L	L	*L*	H	*L*	L	*L*	H	*L*	L	*L*	*L*	L	L	L	L	L	L	L	L
5	M	L	*H*	L	L	M	L	H	H	M	L	L	L	L	*M*	L	R	L	L	L
8	*L*	R	M	M	L	L	L	H	H	R	R	R	L	L	*L*	*L*	L	*H*	*H*	H
11	L	L	*L*	L	L	L	L	H	H	L	L	L	L	M	*L*	L	L	M	L	L
16	L	L	L	L	L	L	L	H	H	R	M	R	H	H	*H*	*H*	R	H	H	L
20	L	L	L	L	L	L	L	H	H	L	L	L	M	H	L	L	R	H	H	L
22	L	L	L	L	L	L	L	H	H	L	L	L	M	H	L	L	R	H	H	H
24	R	R	M	L	L	L	L	H	H	M	R	M	M	R	R	L	L	H	H	H
26	H	H	H	H	M	H	L	H	H	H	M	M	R	R	R	L	R	R	R	L
28	L	L	*L*	*L*	L	L	L	H	H	L	L	L	L	H	*L*	*L*	L	H	H	M
30	L	L	L	L	L	L	L	H	H	L	L	L	L	M	L	L	L	H	M	L
31	L	L	L	L	L	L	L	H	H	L	L	L	L	M	L	L	L	H	M	L
35	L	L	L	L	L	L	L	H	H	L	L	L	L	M	L	L	L	H	H	L
37	L	L	L	L	L	L	L	H	H	L	L	L	H	H	H	L	R	H	H	M
38	L	L	H	L	L	L	L	H	H	L	L	L	H	H	H	L	R	H	H	L
39	L	L	H	L	L	L	L	H	H	L	L	L	H	H	H	L	R	H	H	L
44	L	L	L	L	L	L	L	H	H	L	L	L	L	L	L	L	L	H	H	L
45	L	L	L	L	L	L	L	H	H	L	L	L	L	*H*	H	L	H	H	*H*	L
47	M	M	L	L	L	L	L	H	H	L	L	L	R	H	*H*	L	R	L	L	L
48	L	L	*H*	*L*	L	L	L	H	H	L	L	L	H	*H*	*H*	L	R	M	*L*	L
53	M	M	H	L	L	L	L	H	H	L	R	L	H	H	*H*	L	R	R	R	R
55	M	M	M	L	L	L	L	H	H	L	R	L	H	H	H	L	R	R	R	R
58	M	M	H	L	L	L	L	H	H	M	L	L	H	H	*H*	L	R	L	L	L
65	H	H	H	H	L	M	L	H	H	L	L	L	H	H	*H*	L	R	H	M	L
66	L	L	*L*	*H*	L	L	L	H	H	L	L	L	H	H	H	L	R	M	M	L
67	L	L	H	L	L	L	L	H	H	M	L	L	H	H	H	L	R	H	M	L
71	L	L	M	L	L	L	L	H	H	L	L	L	H	H	*H*	L	R	*H*	*H*	L
72	H	M	H	L	L	M	L	H	H	L	L	L	*H*	R	H	L	R	L	L	L
75	L	L	L	L	L	L	L	H	H	M	L	L	R	R	R	L	R	H	H	R

Reactivity Chart 8. Protection for the Amino Group: Carbamates (*Continued*)

PG	Na/NH_3	Al(Hg)	$SnCl_2/Py$	HSO_3^-; H_2S	$LiAlH_4$	Li-*s*-Bu_3BH	$(C_5H_{11})_2BH$	B_2H_6, 0°	$NaBH_4$	$Zn(BH_4)_2$	$NaBH_3CN$ pH 4-6	*i*-Bu_2AlH	Li(O*t*Bu)$_3$AlH	$AlCl_3$, 80°	$AlCl_3$, 25°	$SnCl_4$; BF_3	$LiClO_4$; $MgBr_2$	TsOH, 80°	TsOH, 0°	Hg(II)	Ag(I)	Cu(II)/Py	HBr/In·	HX/In·	NBS/CCl_4	Br_3CCl/In·	OsO_4	$KMnO_4$, pH 7, 0°	O_3, -50°	RCO_3H, 0°	RCO_3H, 50°	CrO_3/Py	CrO_3, pH 1	H_2O_2 pH 10-12	Quinone	$^1O_2$	DMSO, 100°	NaOCl pH 10	aq NBS
	39	40	41	42	43	44	45	46	47	48	49	50	51	52	53	54	55	56	57	58	59	60	61	62	63	64	65	66	67	68	69	70	71	72	73	74	75	76	77
	G.				H. HYDRIDE REDN.									I.						J.			K.				L. OXIDANTS												
1	L	*L*	*L*	L	*R*	L	L	L	L	L	L	*R*	L	R	L	*L*	L	L	L	L	L	L	L	L	*L*	L	*L*	L	L	L	*L*	L	H	*L*	L	*L*	*L*	L	*L*
5	H	L	L	L	M	L	L	L	L	L	L	M	L	R	L	L	L	M	L	L	L	H	R	R	R	L	L	L	L	L	L	H	L	L	L	L	L	L	L
8	R	M	H	L	R	L	L	*L*	L	L	L	M	L	R	R	R	L	L	L	L	L	L	M	M	M	M	L	L	*L*	*L*	L	*L*	*L*	L	L	L	R	L	L
11	L	L	L	L	R	L	L	L	L	L	L	M	L	R	L	L	L	H	H	L	L	L	L	L	L	L	L	L	L	L	L	L	H	L	L	L	L	L	L
16	H	L	L	L	R	L	R	R	L	L	L	M	L	R	H	L	L	R	H	R	L	R	R	R	L	R	R	R	R	M	R	L	H	L	L	L	L	L	R
20	R	L	L	L	R	L	L	L	L	L	L	M	L	R	H	M	L	R	H	L	L	L	L	L	L	L	L	L	L	L	L	L	H	L	L	L	L	L	L
22	R	L	L	L	R	L	L	L	L	L	M	M	L	R	H	M	L	R	H	L	L	L	H	L	L	L	L	L	L	L	M	L	H	L	L	L	L	L	L
24	*H*	L	H	L	R	L	L	L	L	L	L	M	L	R	*H*	M	L	M	L	L	M	L	M	L	L	L	L	L	L	L	L	L	M	L	L	L	R	L	L
26	H	L	L	L	R	M	L	M	L	L	L	M	L	R	H	M	L	M	L	L	L	H	L	L	L	L	L	L	L	L	L	H	L	R	L	L	M	M	L
28	*L*	*L*	L	*L*	M	L	L	L	*L*	L	M	*L*	L	R	*H*	M	L	R	*H*	L	*L*	L	H	L	L	L	L	*L*	*L*	*L*	M	L	H	L	L	L	L	L	L
30	L	L	L	L	M	L	L	L	L	L	L	M	L	R	M	L	L	M	L	L	L	L	M	L	L	L	L	L	L	L	L	L	M	L	L	L	L	L	L
31	L	L	L	L	M	L	L	L	L	L	L	M	L	R	M	M	L	R	M	L	L	L	M	L	L	L	L	L	L	L	L	L	H	L	L	L	L	L	L
35	L	L	L	L	M	L	L	L	L	L	L	M	L	R	L	L	L	L	L	L	L	L	M	L	L	L	L	L	L	L	L	L	H	L	L	L	L	L	L
37	L	L	L	L	R	L	R	R	L	L	M	M	L	R	L	L	L	R	L	*R*	L	L	R	R	R	R	R	R	R	R	R	L	H	L	L	R	L	L	R
38	L	L	L	L	R	L	R	R	L	L	L	M	L	R	H	H	L	R	L	R	L	L	R	R	R	R	R	R	R	R	R	L	M	L	L	M	L	L	R
39	H	L	L	L	R	L	R	R	L	L	L	M	L	R	H	H	L	R	L	*R*	L	L	R	R	R	R	R	R	R	R	R	L	M	L	L	R	L	L	R
44	H	L	L	L	M	L	L	L	L	L	L	M	L	R	L	L	L	L	L	L	M	*H*	R	L	L	L	L	L	L	R	R	L	L	L	L	L	L	L	L
45	H	L	M	L	M	L	L	L	L	L	L	M	L	R	L	L	L	L	L	L	L	L	L	L	L	L	L	L	L	R	R	L	L	L	L	L	L	L	L
47	*H*	L	L	L	M	L	M	R	L	L	L	M	L	R	L	L	L	M	L	L	L	L	R	R	R	R	M	M	M	*H*	H	L	M	L	L	R	L	L	R
48	*H*	L	L	L	R	L	L	*L*	*L*	L	L	*L*	L	R	L	*L*	L	R	*L*	L	*L*	H	H	R	R	L	*L*	L	L	L	L	*L*	*L*	*L*	L	L	L	L	*L*
53	*H*	R	L	L	R	L	L	L	L	L	L	M	L	R	L	L	L	L	L	L	L	H	M	L	R	L	L	L	L	L	L	H	L	L	L	L	L	R	L
55	H	R	L	L	R	L	L	L	L	L	L	M	L	R	R	L	L	R	L	L	L	L	M	L	R	L	L	L	L	L	L	L	L	L	L	L	L	M	L
58	H	L	L	L	R	L	L	L	L	L	L	M	L	R	H	L	L	M	L	L	L	H	M	L	R	L	L	L	L	L	L	H	L	L	L	L	L	L	L
65	H	L	M	L	M	L	L	L	L	L	L	M	L	R	H	L	L	R	L	L	L	H	H	R	R	L	R	R	R	R	R	R	M	L	L	M	L	L	R
66	H	L	L	L	M	L	L	L	L	L	L	M	L	R	H	L	L	R	L	L	L	L	R	R	R	R	L	L	L	L	L	L	L	L	L	R	L	L	L
67	H	L	L	L	M	L	L	L	L	L	L	M	L	R	H	L	L	R	L	L	L	L	H	R	R	L	L	L	L	L	M	L	H	L	L	L	L	L	L
71	H	L	L	L	M	L	L	L	L	L	L	M	L	R	M	L	L	R	L	L	L	L	M	M	R	L	L	L	L	R	R	L	L	L	L	L	L	L	L
72	*H*	L	M	L	M	L	L	L	L	L	L	M	L	R	H	L	L	R	L	H	H	L	R	R	R	L	L	R	R	*H*	H	L	L	M	L	L	L	R	R
75	H	R	L	L	M	L	L	L	L	L	L	M	L	R	M	M	L	L	L	*R*	M	M	R	R	R	L	L	L	R	*H*	H	L	M	*H*	L	R	R	R	R

Reactivity Chart 8. Protection for the Amino Group: Carbamates (*Continued*)

PG	I_2	PhSeX; PhSCl	Br_2; Cl_2	MnO_2/CH_2Cl_2	$NaIO_4$ pH 5-8	SeO_2 pH 2-4	SeO_2/Py	$K_3Fe(CN)_6$, pH 8	Pb(IV), 25°	Pb(IV), 80°	$Tl(NO_3)_3$	150°	250°	350°	$:CCl_2$	N_2CHCO_2R/Cu	$CH_2I_2/Zn(Cu)$
	78	79	80	81	82	83	84	85	86	87	88	89	90	91	92	93	94
	L. OXIDANTS											M.			N.		
1	L	L	L	L	L	L	L	L	L	L	L	L	L	M	L	L	L
5	L	L	L	L	L	M	L	L	L	L	L	L	M	H	L	L	L
8	L	L	L	L	L	L	L	L	L	L	L	L	M	H	L	L	R
11	L	L	L	L	L	M	L	L	L	L	L	L	M	H	L	L	L
16	L	R	R	L	L	R	R	L	R	R	R	L	M	H	R	R	R
20	L	L	L	L	L	H	L	L	L	L	L	L	M	H	L	L	L
22	L	L	L	L	L	H	L	L	L	L	L	L	M	H	L	L	L
24	L	L	L	L	L	L	L	L	L	L	L	H	H	H	L	L	M
26	L	L	L	L	L	L	H	M	L	L	L	M	H	H	L	L	L
28	L	L	L	L	L	H	L	L	L	L	L	H	H	H	L	L	L
30	L	L	L	L	L	M	L	L	L	L	L	L	M	H	L	L	L
31	L	L	L	L	L	M	L	L	L	L	L	L	M	H	L	L	L
35	L	L	L	L	L	M	L	L	L	L	L	L	L	M	L	L	L
37	L	R	R	L	L	H	L	L	M	R	R	L	M	H	R	R	R
38	L	R	R	L	L	M	R	L	M	R	R	L	M	R	R	R	R
39	L	R	R	L	L	M	M	L	M	R	R	L	M	R	R	R	R
44	L	L	L	L	L	L	L	L	L	L	L	L	L	M	L	L	R
45	L	L	L	L	L	L	L	L	L	L	L	L	M	H	L	L	L
47	L	L	R	L	L	L	L	L	M	R	M	L	L	M	R	R	R
48	L	L	L	L	L	L	L	L	L	L	L	L	L	M	L	L	L
53	L	L	L	L	L	L	R	L	L	L	L	L	L	M	L	L	L
55	L	L	L	L	L	L	M	L	L	L	L	L	L	M	M	M	M
58	L	L	L	L	L	L	R	L	L	L	L	L	L	M	L	L	L
65	L	L	R	L	L	L	R	L	M	R	M	L	L	M	M	M	M
66	L	L	L	L	L	L	L	L	L	L	L	L	L	M	L	L	L
67	L	L	L	L	L	L	L	L	L	L	L	L	L	M	L	L	L
71	L	L	L	L	L	L	L	L	L	L	L	L	L	M	L	L	M
72	L	L	R	L	R	L	L	L	L	M	L	L	L	M	L	L	L
75	R	R	R	L	R	M	L	L	M	R	M	L	M	H	R	R	R

PG	$R_3SnH/In\cdot$	$Ni(CO)_4$	CH_2N_2	$SOCl_2$	Ac_2O, 25°	Ac_2O, 80°	DCC	MeI	$Me_3O^+BF_4^-$	1.LDA 2.MeI	1.K_2CO_3 2.MeI	RCHO	RCOCl	C^+/olefin
	95	96	97	98	99	100	101	102	103	104	105	106	107	108
	O. MISCELLANEOUS											P.		
1	L	L	L	L	L	L	L	L	R	L	L	L	L	L
5	L	L	L	L	L	L	L	L	R	L	L	L	L	L
8	R	M	L	L	L	L	L	L	R	L	L	L	L	R
11	L	L	L	L	L	L	L	L	R	L	L	L	L	L
16	R	R	L	L	L	L	L	L	R	R	L	L	L	R
20	L	L	L	L	L	L	L	L	R	L	L	L	L	M
22	L	L	L	L	L	L	L	L	R	L	L	L	L	M
24	R	L	L	L	L	L	L	L	R	L	L	L	L	H
26	L	L	L	L	L	L	L	L	R	R	H	L	L	M
28	L	L	L	L	L	L	L	L	R	L	L	L	L	H
30	L	L	L	L	L	L	L	L	R	L	L	L	L	M
31	L	L	L	L	L	L	L	L	R	L	L	L	L	M
35	L	L	L	L	L	L	L	L	R	L	L	L	L	H
37	R	H	L	L	L	L	L	L	R	L	L	L	L	R
38	R	H	L	L	L	L	L	L	R	L	L	L	L	R
39	R	H	L	L	L	L	L	L	R	L	L	L	L	R
44	L	H	L	L	L	L	L	R	R	L	R	L	L	M
45	L	L	L	L	L	L	L	R	R	L	R	L	L	L
47	R	R	L	L	L	L	L	L	R	L	L	L	L	L
48	L	L	L	L	L	L	L	L	R	L	L	L	L	M
53	R	L	L	L	L	L	L	L	R	L	L	L	L	M
55	R	L	L	L	L	L	L	L	R	L	L	L	L	M
58	R	L	L	L	L	L	L	L	R	L	L	L	L	L
65	M	L	L	L	L	L	L	R	R	L	L	L	L	H
66	L	L	L	L	L	L	L	L	R	L	L	L	L	M
67	L	L	L	L	L	L	L	L	R	L	L	L	L	H
71	L	R	L	L	L	L	L	R	R	L	R	L	L	L
72	R	L	L	L	L	L	L	L	R	L	L	L	L	L
75	R	L	R	L	L	M	L	R	R	R	R	L	L	M

Reactivity Chart 9. Protection for the Amino Group: Amides

1. *N*-Formyl
2. *N*-Acetyl
3. *N*-Chloroacetyl
5. *N*-Trichloroacetyl
6. *N*-Trifluoroacetyl
7. *N*-*o*-Nitrophenylacetyl
8. *N*-*o*-Nitrophenoxyacetyl
9. *N*-Acetoacetyl
12. *N*-3-Phenylpropionyl
13. *N*-3-(*p*-Hydroxyphenyl)propionyl
15. *N*-2-Methyl-2-(*o*-nitrophenoxy)propionyl
16. *N*-2-Methyl-2-(*o*-phenylazophenoxy)propionyl
17. *N*-4-Chlorobutyryl
19. *N*-*o*-Nitrocinnamoyl
20. *N*-Picolinoyl
21. *N*-(*N'*-Acetylmethionyl)
23. *N*-Benzoyl
29. *N*-Phthaloyl
31. *N*-Dithiasuccinoyl

(See chart, pp. 446–448)

Reactivity Chart 9. Protection for the Amino Group: Amides

	pH<1, 100°	pH<1	pH 1	pH 2-4	pH 4-6	pH 6-8.5	pH 8.5-10	pH 10-12	pH>12	pH>12, 150°	NaH	Ph_3CNa	$(C_{10}H_8)^{\cdot -}Na^+$	$MeSOCH_2^-Na^+$	KO-t-Bu	LiN-i-Pr_2	Py; R_3N	$NaNH_2$	NaOMe	Enolate	NH_3; RNH_2	RS^-; N_3^-; SCN^-	OAc^-; X^-	NaCN, pH 12	HCN, pH 6	RLi	RMgX	Organozinc	Organocopper	Wittig; ylide	H_2/Raney (Ni)	H_2/Pt pH 2-4	H_2/Pd	H_2/Lindlar	H_2/Rh	Zn/HCl	Zn/HOAc	Cr(II), pH 5
PG	1	2	3	4	5	6	7	8	9	10	11	12	13	14	15	16	17	18	19	20	21	22	23	24	25	26	27	28	29	30	31	32	33	34	35	36	37	38
	A. AQUEOUS										B. BASIC								C. NUCLEOPHILIC							D. ORGANOMET.					E. CAT. REDN.					F.		
1	H	*H*	*H*	*L*	L	L	L	M	H	*H*	*H*	L	R	L	L	L	L	L	L	L	*H*	L	L	M	L	H	H	L	L	L	L	*H*	L	L	L	H	L	L
2	*H*	M	L	*L*	L	L	L	*L*	M	H	*L*	R	R	R	L	*R*	*L*	R	*L*	L	*H*	*L*	*L*	L	L	H	M	L	L	L	*L*	L	*L*	L	L	M	L	L
3	*H*	M	L	L	L	L	L	L	*H*	H	M	R	R	R	L	R	*L*	R	R	L	H	R	L	R	L	H	M	M	R	L	M	R	R	L	L	H	H	*R*
5	H	L	L	L	L	L	M	H	H	H	L	L	R	L	L	R	*L*	L	R	R	H	M	L	R	L	H	M	M	R	M	R	R	R	L	R	H	H	M
6	H	*L*	*L*	*L*	L	L	M	*H*	*H*	H	L	L	R	L	L	L	*L*	L	R	R	*H*	M	L	R	L	H	M	L	M	M	R	R	*L*	L	L	H	H	M
7	H	L	L	L	L	L	L	L	M	H	M	R	R	R	L	R	L	R	L	L	H	L	L	L	L	H	H	M	L	L	R	*H*	R	L	R	H	H	R
8	H	L	L	L	L	L	L	L	M	H	M	R	R	R	L	L	L	R	M	M	M	M	L	M	L	H	R	L	L	L	R	R	*R*	L	R	H	H	R
9	H	*L*	L	L	L	L	L	L	H	H	R	R	R	R	R	R	L	R	R	R	*H*	L	L	L	R	H	R	R	R	R	M	R	L	L	M	L	L	L
12	H	L	L	L	L	L	L	L	M	H	M	R	R	R	L	R	L	R	L	L	M	L	L	L	L	H	M	L	L	L	L	L	L	L	R	L	L	L
13	H	L	L	L	L	L	L	L	M	H	R	R	R	R	R	R	L	R	R	L	M	L	L	L	L	H	R	L	L	L	L	L	L	L	R	L	L	L
15	H	L	L	L	L	L	L	L	M	H	L	L	R	L	L	L	L	L	L	L	M	L	L	L	L	H	R	L	L	L	R	*H*	R	L	R	H	H	R
16	H	L	L	L	L	L	L	L	H	H	L	L	R	M	L	H	L	L	L	L	M	L	L	L	L	H	R	R	L	L	R	H	R	R	R	H	H	H
17	H	L	L	L	L	L	L	L	*L*	H	M	R	R	R	R	R	L	R	R	L	M	R	L	R	L	H	M	L	R	L	M	R	R	L	L	L	L	L
19	H	L	L	L	L	L	L	L	M	H	L	L	R	R	L	L	L	L	M	R	M	M	L	M	L	H	R	R	L	L	R	*H*	R	L	R	H	H	R
20	H	L	L	L	L	L	L	L	M	H	L	L	*R*	L	L	L	L	L	L	L	H	L	L	L	L	H	M	L	L	L	L	L	L	L	R	L	L	L
21	H	L	L	L	L	L	L	L	M	H	R	R	R	R	R	R	L	R	L	L	H	L	L	L	L	H	R	L	L	L	R	R	R	R	R	L	L	L
23	H	*H*	L	L	L	L	L	*L*	*H*	H	L	L	*R*	L	*L*	*L*	L	L	L	L	*H*	L	L	M	L	H	M	L	L	L	L	L	*L*	L	R	L	L	L
29	*H*	L	*L*	*L*	L	L	L	L	*R*	H	L	*L*	*R*	L	*L*	L	L	L	*L*	*L*	*H*	*L*	L	L	L	H	M	*L*	L	L	L	L	*L*	L	R	L	L	L
31	M	*L*	L	L	L	L	*L*	M	H	H	L	L	R	R	L	L	L	L	H	H	H	*H*	L	H	L	H	R	R	M	H	R	R	R	R	R	R	R	R

Reactivity Chart 9. Protection for the Amino Group: Amides (*Continued*)

PG	Na/NH_3	Al(Hg)	$SnCl_2/Py$	HSO_3^-; H_2S	$LiAlH_4$	Li-s-Bu_3BH	$(C_5H_{11})_2BH$	B_2H_6, 0°	$NaBH_4$	$Zn(BH_4)_2$	$NaBH_3CN$ pH 4–6	i-Bu_2AlH	Li(OtBu)$_3$AlH	$AlCl_3$, 80°	$AlCl_3$, 25°	$SnCl_4$; BF_3	$LiClO_4$; $MgBr_2$	TsOH, 80°	TsOH, 0°
	39	40	41	42	43	44	45	46	47	48	49	50	51	52	53	54	55	56	57
	G.				H. HYDRIDE REDN.									I.					
1	R	L	L	L	*R*	L	H	R	L	L	L	H	L	L	L	L	L	L	L
2	R	L	L	L	*R*	L	H	R	*L*	L	*L*	H	L	L	*L*	*L*	L	L	L
3	R	M	L	L	R	M	H	R	L	L	L	H	L	M	M	L	L	L	L
5	R	M	L	L	H	M	H	R	*H*	M	M	H	M	M	M	L	L	L	L
6	R	L	L	L	H	M	H	R	*H*	M	M	H	M	L	*L*	L	L	*L*	L
7	R	R	L	L	R	M	H	R	L	L	L	H	L	L	L	L	L	L	L
8	R	R	L	L	R	M	H	R	L	L	L	H	L	L	L	L	L	L	L
9	R	L	L	L	R	R	H	R	R	R	R	H	R	R	R	M	L	L	L
12	R	L	L	L	R	L	H	R	L	L	L	H	L	L	L	L	L	L	L
13	R	L	L	L	R	L	H	R	L	L	L	H	L	L	L	L	L	L	L
15	R	R	L	L	R	M	H	R	L	L	L	H	L	L	L	L	L	L	L
16	R	*H*	R	L	R	L	H	R	R	R	L	H	M	R	L	L	L	L	L
17	R	L	L	L	R	M	H	R	L	L	L	H	L	M	M	L	L	L	L
19	R	R	L	L	R	M	H	R	L	L	L	H	L	M	M	L	L	L	L
20	R	L	L	L	R	L	H	R	L	L	L	H	L	L	L	L	L	L	L
21	R	L	L	L	R	L	H	R	L	L	L	H	L	L	L	L	L	L	L
23	R	*L*	L	L	*R*	L	H	R	*L*	L	L	*H*	L	L	L	L	L	*L*	L
29	*R*	L	L	L	R	L	H	R	*L*	L	L	H	L	L	L	L	L	L	L
31	R	R	R	L	R	R	H	R	R	L	L	H	R	R	R	L	L	R	L

PG	Hg(II)	Ag(I)	Cu(II)/Py	HBr/In·	HX/In·	NBS/CCl_4	Br_3CCl/In·	OsO_4	$KMnO_4$, pH 7, 0°	O_3, −50°	RCO_3H, 0°	RCO_3H, 50°	CrO_3/Py	CrO_3, pH 1	H_2O_2 pH 10–12	Quinone	$^1O_2$	DMSO, 100°	NaOCl pH 10	aq NBS
	58	59	60	61	62	63	64	65	66	67	68	69	70	71	72	73	74	75	76	77
	J.			K.				L. OXIDANTS												
1	*L*	L	L	L	L	L	L	L	L	*L*	L	M	L	H	M	L	L	L	L	L
2	*L*	L	L	L	L	L	L	L	L	L	L	*L*	L	*L*	*L*	L	*L*	L	L	L
3	L	H	L	L	L	L	L	L	L	L	L	L	L	L	L	L	L	R	L	L
5	L	H	L	R	R	R	R	L	L	L	L	L	L	L	L	L	L	L	R	L
6	L	M	L	L	L	L	L	L	L	L	*L*	R	L	L	L	L	L	L	R	L
7	L	L	L	L	L	M	L	L	L	L	L	L	L	L	L	L	L	L	L	L
8	L	L	L	L	L	L	L	L	L	L	L	L	L	L	L	L	L	L	L	L
9	L	L	*L*	L	L	L	L	L	*H*	L	L	M	L	*H*	L	L	L	L	R	R
12	L	L	L	L	L	R	L	L	L	L	L	L	L	L	L	L	L	L	L	L
13	L	L	L	R	R	R	R	L	R	L	L	L	M	R	M	M	L	L	R	*H*
15	L	L	L	L	L	L	L	L	L	L	L	L	L	L	L	L	L	L	L	L
16	L	R	L	R	R	R	R	L	L	R	M	R	L	R	R	L	R	L	L	L
17	L	*H*	L	L	L	L	L	L	L	L	L	L	L	L	L	L	L	R	L	L
19	L	L	L	R	R	R	R	R	R	R	L	M	L	L	R	L	L	L	L	R
20	H	L	H	L	L	R	L	L	L	L	M	R	M	M	L	L	L	L	L	L
21	L	L	L	L	L	L	L	L	R	R	R	R	L	R	L	L	R	L	R	R
23	*L*	L	L	L	L	L	L	*L*	L	L	L	L	*L*	L	L	L	L	*L*	L	L
29	L	L	*H*	L	L	L	L	L	L	*L*	L	*L*	L	L	L	L	L	L	L	L
31	R	M	L	R	R	R	R	M	R	R	R	R	L	R	R	R	R	R	R	R

Reactivity Chart 9. Protection for the Amino Group: Amides (*Continued*)

PG	78 I_2	79 PhSeX; PhSCl	80 Br_2; Cl_2	81 MnO_2/CH_2Cl_2	82 $NaIO_4$ pH 5-8	83 SeO_2 pH 2-4	84 SeO_2/Py	85 $K_3Fe(CN)_6$, pH 8	86 Pb(IV), 25°	87 Pb(IV), 80°	88 $Tl(NO_3)_3$	89 150°	90 250°	91 350°	92 $:CCl_2$	93 N_2CHCO_2R/Cu	94 $CH_2I_2/Zn(Cu)$
	L. OXIDANTS											M.			N.		
1	L	L	M	L	L	L	L	L	L	L	L	*L*	L	L	L	L	L
2	L	L	L	L	L	L	L	L	L	*L*	L	*L*	*L*	L	L	L	L
3	L	L	L	L	L	L	L	L	L	L	L	L	L	L	L	L	M
5	L	L	L	L	L	L	L	M	L	L	L	*L*	M	R	L	L	H
6	L	L	L	L	M	L	L	M	L	L	L	L	M	R	L	L	L
7	L	L	L	L	L	L	L	L	L	L	*L*	L	L	L	L	L	L
8	L	L	L	L	L	L	L	L	L	L	L	L	L	L	L	L	L
9	L	L	R	L	*M*	*H*	*M*	L	M	R	M	L	L	M	L	M	L
12	L	L	L	L	L	L	L	L	L	L	L	L	L	L	L	L	L
13	M	L	*H*	M	M	M	M	R	M	R	L	L	L	L	L	L	L
15	L	L	L	L	L	L	L	L	L	L	L	L	L	L	L	L	L
16	L	L	L	L	L	L	L	R	L	M	R	R	R	R	M	R	R
17	L	L	L	L	L	L	L	L	L	L	L	R	R	R	L	L	R
19	L	L	M	L	L	L	L	L	L	R	M	L	L	L	R	R	R
20	L	L	L	L	L	L	L	L	L	L	L	L	L	L	L	M	M
21	L	L	R	L	R	M	L	L	R	R	L	M	H	H	M	M	L
23	L	L	L	L	L	L	L	*L*	L	L	L	L	*L*	L	L	L	L
29	L	L	*L*	L	L	L	L	L	L	L	L	L	L	L	L	L	L
31	M	M	R	R	M	R	*M*	R	R	R	L	L	M	H	M	M	H

PG	95 $R_3SnH/In\cdot$	96 $Ni(CO)_4$	97 CH_2N_2	98 $SOCl_2$	99 Ac_2O, 25°	100 Ac_2O, 80°	101 DCC	102 MeI	103 $Me_3O^+BF_4^-$	104 1.LDA 2.MeI	105 1.K_2CO_3 2.MeI	106 RCHO	107 RCOCl	108 C^+/olefin
	O. MISCELLANEOUS											P.		
1	L	L	L	L	L	L	*L*	L	R	*R*	L	L	*H*	L
2	L	L	*L*	L	*L*	*L*	*L*	L	*R*	*R*	L	L	*L*	L
3	R	L	L	L	L	*L*	L	L	R	R	L	L	L	L
5	R	M	L	L	L	*L*	L	L	R	L	M	L	L	L
6	R	L	L	L	L	*L*	*L*	L	R	L	L	L	*L*	L
7	L	L	L	L	L	*L*	L	L	R	R	L	L	L	L
8	L	L	L	L	L	*L*	L	L	R	R	L	L	L	L
9	*R*	L	M	L	M	M	L	L	R	R	R	L	M	M
12	L	L	L	L	L	*L*	L	L	R	R	L	L	L	L
13	L	L	R	L	R	R	R	L	R	R	R	L	R	L
15	L	L	L	L	L	*L*	L	L	R	L	L	L	L	L
16	R	L	L	L	L	*L*	L	L	R	L	M	L	L	R
17	R	L	L	L	L	*L*	L	L	R	R	L	L	L	L
19	L	L	L	L	L	*L*	L	L	R	L	L	L	L	*M*
20	L	L	L	L	L	*L*	L	M	R	L	M	L	L	L
21	L	L	L	L	L	*L*	L	R	R	R	R	L	L	L
23	L	L	*L*	L	*L*	*L*	L	L	R	L	L	L	*L*	L
29	L	L	L	*L*	L	*L*	*L*	L	R	L	L	L	L	L
31	R	R	L	L	L	*L*	L	M	R	M	L	L	L	L

Reactivity Chart 10. Protection for the Amino Group: Special —NH Protective Groups

1. *N*-Allyl
2. *N*-Phenacyl
3. *N*-3-Acetoxypropyl
5. Quaternary Ammonium Salts
6. *N*-Methoxymethyl
8. *N*-Benzyloxymethyl
9. *N*-Pivaloyloxymethyl
12. *N*-Tetrahydropyranyl
13. *N*-2,4-Dinitrophenyl
14. *N*-Benzyl
16. *N*-*o*-Nitrobenzyl
17. *N*-Di(*p*-methoxyphenyl)methyl
18. *N*-Triphenylmethyl
19. *N*-(*p*-Methoxyphenyl)diphenylmethyl
20. *N*-Diphenyl-4-pyridylmethyl
21. *N*-2-Picolyl *N*′-Oxide
24. *N*,*N*′-Isopropylidene
25. *N*-Benzylidene
27. *N*-*p*-Nitrobenzylidene
28. *N*-Salicylidene
33. *N*-(5,5-Dimethyl-3-oxo-1-cyclohexenyl)
37. *N*-Nitro
39. *N*-Oxide
40. *N*-Diphenylphosphinyl
41. *N*-Dimethylthiophosphinyl
47. *N*-Benzenesulfenyl
48. *N*-*o*-Nitrobenzenesulfenyl
55. *N*-2,4,6-Trimethylbenzenesulfonyl
56. *N*-Toluenesulfonyl
57. *N*-Benzylsulfonyl
59. *N*-Trifluoromethylsulfonyl
60. *N*-Phenacylsulfonyl

(See chart, pp. 450–452)

Reactivity Chart 10. Protection for the Amino Group: Special —NH Protective Groups

PG	1 pH<1, 100°	2 pH<1	3 pH 1	4 pH 2-4	5 pH 4-6	6 pH 6-8.5	7 pH 8.5-10	8 pH 10-12	9 pH>12	10 pH>12, 150°	11 NaH	12 Ph_3CNa	13 $(C_{10}H_8)^{\bar{\cdot}}Na^+$	14 $MeSOCH_2^-Na^+$	15 KO-t-Bu	16 LiN-i-Pr_2	17 Py; R_3N	18 $NaNH_2$
	A. AQUEOUS										B. BASIC							
1	H	L	L	L	L	L	L	L	L	R	L	R	R	R	*R*	R	L	R
2	H	*L*	L	L	L	L	L	L	*L*	R	R	R	R	R	R	R	L	R
3	H	R	L	L	L	L	M	R	R	R	L	R	R	R	L	R	L	R
5	H	L	L	L	L	L	L	L	L	R	L	R	L	M	M	R	L	R
6	H	H	L	L	L	L	L	L	L	L	L	L	L	L	L	L	L	L
8	H	H	*L*	L	L	L	L	L	*L*	L	L	L	L	L	L	L	L	L
9	H	H	M	L	L	L	M	H	H	H	L	L	R	R	L	L	L	H
12	H	H	*H*	H	H	L	L	L	L	*L*	L	L	L	L	L	L	L	L
13	H	L	*L*	L	L	L	L	L	H	H	L	M	R	L	L	L	L	H
14	H	L	*L*	L	L	L	L	L	L	L	L	L	L	L	L	L	L	L
16	H	L	L	L	L	L	L	L	L	R	L	M	R	L	L	L	L	L
17	H	H	H	*H*	L	L	L	L	*L*	L	L	L	L	L	L	L	L	L
18	H	*H*	*H*	*H*	M	L	L	L	L	L	L	L	L	L	L	L	L	L
19	H	H	*H*	H	M	L	L	L	L	L	L	L	L	L	L	L	L	L
20	M	*L*	L	L	L	L	L	L	L	L	L	L	L	L	L	L	L	L
21	H	L	L	L	L	L	L	L	L	R	L	L	R	L	L	*H*	L	*H*
24	H	H	H	M	L	L	L	L	L	H	L	L	R	L	L	L	L	L
25	H	H	*H*	M	L	L	L	L	M	H	L	L	R	L	L	L	L	L
27	H	H	H	M	L	L	L	L	L	H	L	M	R	L	L	*L*	L	L
28	H	H	*H*	M	L	L	L	L	M	H	R	R	R	R	R	R	L	R
33	H	H	*L*	L	L	L	L	L	L	H	L	R	R	R	R	R	L	R
37	H	*L*	*L*	L	L	L	L	L	*L*	H	L	L	R	L	L	L	H	H
39	H	M	L	L	L	L	L	L	L	H	L	L	R	L	L	M	L	L
40	H	*H*	*H*	*M*	L	L	L	L	L	H	L	L	L	L	L	L	L	L
41	H	*H*	*H*	*L*	L	L	L	L	L	H	L	L	R	L	L	L	L	L
47	H	M	L	L	L	L	L	L	M	H	L	L	R	L	L	L	L	L
48	H	H	*H*	*H*	L	L	L	L	M	H	L	M	R	L	L	L	L	L
55	*H*	L	L	L	L	L	L	L	*L*	*L*	L	L	H	L	L	L	L	L
56	*H*	L	L	L	L	L	L	L	L	*L*	L	L	*H*	L	L	L	L	L
57	H	M	L	L	L	L	L	L	L	H	R	R	R	R	R	R	L	R
59	H	L	L	L	L	L	L	L	L	H	L	L	H	L	L	L	L	L
60	H	L	L	L	L	L	L	*L*	R	H	R	R	R	R	R	R	L	R

PG	19 NaOMe	20 Enolate	21 NH_3; RNH_2	22 RS^-; N_3^-; SCN^-	23 OAc^-; X^-	24 NaCN, pH 12	25 HCN, pH 6	26 RLi	27 RMgX	28 Organozinc	29 Organocopper	30 Wittig; ylide	31 H_2/Raney (Ni)	32 H_2/Pt pH 2-4	33 H_2/Pd	34 H_2/Lindlar	35 H_2/Rh	36 Zn/HCl	37 Zn/HOAc	38 Cr(II), pH 5
	C. NUCLEOPHILIC							D. ORGANOMET.					E. CAT. REDN.					F.		
1	L	L	L	L	L	L	L	L	L	L	L	L	R	R	H	L	R	L	L	L
2	M	M	*M*	L	L	R	R	R	R	M	L	R	L	*R*	R	L	R	H	*H*	L
3	*R*	R	R	L	L	L	L	R	R	L	L	L	L	L	*L*	L	L	L	L	L
5	L	L	L	*H*	L	M	L	R	R	*L*	L	R	L	L	L	L	L	H	L	L
6	L	L	L	L	L	L	L	L	L	L	L	L	L	L	L	L	L	L	L	L
8	L	L	L	L	L	L	L	L	L	*L*	L	L	H	H	*R*	L	L	L	L	L
9	H	M	*H*	L	L	L	L	R	R	L	L	L	L	L	L	L	L	M	L	L
12	L	L	L	L	L	L	L	L	L	L	L	L	L	H	L	L	L	H	H	H
13	M	M	*L*	*H*	L	R	L	R	R	M	M	L	R	R	R	L	R	R	R	R
14	L	L	L	L	L	L	L	L	L	L	L	L	H	M	M	L	*H*	L	L	L
16	L	L	L	L	L	R	L	H	H	M	M	L	H	H	H	L	R	R	R	R
17	L	L	L	L	L	L	L	L	L	L	L	L	H	H	H	L	L	H	H	L
18	L	L	L	L	L	L	L	L	L	L	L	L	H	H	H	L	L	H	H	L
19	L	L	L	L	L	L	L	L	L	L	L	L	H	H	H	L	L	H	H	L
20	L	L	L	L	L	L	L	R	R	L	L	L	H	H	*H*	L	L	H	*H*	L
21	L	L	L	L	L	L	L	R	R	L	L	L	R	R	R	R	R	R	R	R
24	L	L	H	L	L	L	R	R	R	R	M	R	R	R	R	L	R	R	R	R
25	L	L	*H*	L	L	L	R	R	R	R	M	R	*R*	R	*H*	L	R	R	R	R
27	L	L	H	L	L	R	R	R	R	R	M	M	R	R	R	L	R	R	R	R
28	R	R	H	*L*	L	R	R	R	R	R	R	R	*R*	R	R	L	R	R	R	R
33	H	R	H	R	L	R	L	R	R	R	M	M	R	R	R	L	R	R	R	R
37	H	H	R	*L*	*L*	*L*	*L*	H	H	H	M	L	R	H	*H*	L	H	H	H	H
39	L	L	L	L	L	L	L	R	R	L	L	L	H	*H*	*H*	H	H	*H*	H	H
40	L	L	*L*	L	L	L	L	L	L	L	L	L	L	M	L	L	L	H	M	L
41	L	L	L	L	L	L	L	M	M	M	L	L	R	R	R	L	R	H	L	L
47	L	R	H	H	*H*	H	H	L	L	L	L	L	R	R	R	L	R	L	L	L
48	L	R	H	*H*	H	H	*H*	R	R	M	M	L	*R*	R	R	L	R	H	H	R
55	L	L	*L*	L	L	L	L	H	H	H	L	L	L	L	L	L	L	H	L	L
56	L	L	*L*	*L*	L	L	L	H	H	H	L	L	*L*	*L*	L	L	L	*H*	*L*	L
57	R	R	L	R	L	L	L	H	H	H	L	L	*H*	H	H	L	L	H	M	M
59	L	L	L	L	L	L	L	H	H	H	H	L	L	L	L	L	L	L	L	M
60	R	R	R	R	L	R	R	H	H	H	R	R	H	H	H	L	R	H	*H*	M

Reactivity Chart 10. Protection for the Amino Group: Special —NH Protective Groups (*Continued*)

PG	Na/NH_3	Al(Hg)	$SnCl_2/Py$	HSO_3^-; H_2S	$LiAlH_4$	Li-s-Bu_3BH	$(C_5H_{11})_2BH$	B_2H_6, 0°	$NaBH_4$	$Zn(BH_4)_2$	$NaBH_3CN$ pH 4-6	i-Bu_2AlH	Li(OtBu)$_3$AlH	$AlCl_3$, 80°	$AlCl_3$, 25°	$SnCl_4$; BF_3	$LiClO_4$; $MgBr_2$	TsOH, 80°	TsOH, 0°	Hg(II)	Ag(I)	Cu(II)/Py	HBr/In·	HX/In·	NBS/CCl_4	Br_3CCl/In·	OsO_4	$KMnO_4$, pH 7,0°	O_3, -50°	RCO_3H, 0°	RCO_3H, 50°	CrO_3/Py	CrO_3, pH 1	H_2O_2 pH 10-12	Quinone	$^1O_2$	DMSO, 100°	NaOCl pH 10	aq NBS
	39	40	41	42	43	44	45	46	47	48	49	50	51	52	53	54	55	56	57	58	59	60	61	62	63	64	65	66	67	68	69	70	71	72	73	74	75	76	77
	G.				H. HYDRIDE REDN.									I.						J.			K.				L. OXIDANTS												
1	R	L	L	L	L	L	R	R	L	L	L	L	L	L	L	L	L	L	L	R	L	L	R	R	R	R	R	R	R	R	R	R	L	R	R	R	M	M	R
2	*R*	R	L	R	R	R	R	R	R	R	M	R	R	L	L	L	L	L	L	L	L	L	L	L	L	L	R	R	R	R	R	R	L	R	R	R	M	M	R
3	R	L	L	L	R	L	L	L	L	L	L	R	M	R	L	L	L	R	L	L	L	L	L	L	L	L	R	R	R	R	R	R	L	R	R	R	M	R	R
5	L	L	L	L	L	L	L	L	L	L	L	L	L	R	L	L	L	L	L	L	M	L	L	L	L	L	L	L	L	L	L	L	L	L	L	L	H	L	L
6	L	L	L	L	L	L	L	L	L	L	L	L	L	H	H	H	L	M	L	L	L	L	L	L	L	L	R	R	R	R	R	R	L	R	R	R	M	M	R
8	H	L	L	L	L	L	L	L	L	L	L	L	L	H	H	H	L	M	L	L	L	L	L	L	R	L	R	R	R	R	R	R	L	R	R	R	M	M	R
9	R	L	L	L	R	L	L	L	L	L	L	R	M	H	H	M	L	R	L	L	L	L	L	L	L	L	R	R	R	R	R	R	M	R	R	R	M	R	R
12	L	L	L	L	L	L	L	L	L	L	L	L	L	H	H	H	L	M	L	L	L	L	L	L	L	L	R	R	R	R	R	R	H	R	R	R	M	M	R
13	R	R	R	L	R	L	L	L	L	L	L	R	M	L	L	L	L	L	L	L	L	L	L	L	L	L	*L*	*M*	*M*	*M*	*H*	*L*	L	*L*	*L*	L	*L*	*M*	*L*
14	H	M	L	L	*L*	L	L	L	L	L	L	L	L	L	L	L	L	L	L	L	L	L	L	L	R	L	R	R	R	R	R	R	L	R	R	R	M	M	R
16	H	R	L	L	R	L	L	L	L	L	M	R	M	L	L	L	L	L	L	L	L	L	L	L	R	L	R	R	R	R	R	R	L	R	R	R	M	M	R
17	H	L	L	L	L	L	L	L	L	L	L	L	L	R	R	L	M	L	L	L	L	L	L	L	R	L	R	R	R	R	R	R	H	R	R	R	M	M	R
18	*H*	L	L	L	L	L	L	L	L	L	L	L	L	L	L	L	L	L	L	L	L	L	L	L	L	L	R	R	R	R	R	R	H	R	R	L	M	M	R
19	H	L	L	L	L	L	L	L	L	L	L	L	L	R	R	L	M	L	L	L	L	L	L	L	L	L	R	R	R	R	R	R	H	R	R	L	M	M	R
20	H	L	L	L	L	L	L	L	L	L	L	L	L	L	L	L	L	L	L	L	L	L	L	L	L	L	R	R	R	R	R	R	L	R	R	L	M	M	R
21	R	R	R	L	R	M	R	R	M	L	L	M	M	R	L	L	L	L	L	L	L	R	L	L	R	L	R	R	R	R	R	R	L	R	R	R	R	M	R
24	R	M	L	M	R	L	R	R	L	L	R	R	R	L	L	L	L	M	L	M	M	H	R	R	R	R	R	R	R	R	R	R	H	M	R	*L*	R	R	R
25	R	M	L	*R*	R	L	R	R	M	M	R	R	R	L	L	L	L	M	L	M	M	H	R	R	R	R	R	R	R	R	R	R	H	M	R	*L*	R	R	R
27	R	R	L	R	R	L	R	R	M	M	R	R	R	L	L	L	L	M	L	M	M	H	R	R	R	R	R	R	R	R	R	R	H	M	R	*L*	R	*R*	R
28	R	M	L	M	R	L	R	R	M	M	R	R	R	R	R	R	L	R	L	M	M	H	R	R	R	R	R	R	R	R	R	R	H	M	R	*M*	R	R	R
33	R	L	L	R	R	R	R	R	L	L	M	R	R	L	L	L	L	H	L	L	M	H	R	R	R	R	R	R	R	R	R	*L*	L	*L*	R	R	*L*	*L*	R
37	R	R	R	L	R	L	R	R	L	L	R	*L*	R	L	L	L	L	L	L	L	L	H	R	R	R	R	R	R	R	R	R	R	R	R	R	R	R	R	R
39	H	H	H	*H*	H	M	R	R	M	L	M	H	H	R	L	L	L	L	L	L	L	R	L	L	L	L	L	L	L	L	L	L	L	L	L	L	H	L	L
40	H	L	L	L	M	L	L	L	L	L	L	L	L	L	L	L	L	M	L	L	L	L	L	L	L	L	L	L	L	L	L	L	H	L	L	L	L	L	L
41	H	L	L	L	M	L	L	L	L	L	L	L	L	L	L	L	L	M	L	H	H	H	L	L	L	L	L	L	L	L	L	L	H	L	L	L	L	L	L
47	H	H	M	M	H	M	R	R	R	R	M	H	H	*H*	L	L	L	M	L	H	H	H	L	L	*R*	L	R	R	R	R	R	R	L	L	L	R	*L*	R	R
48	H	H	M	*H*	H	H	R	R	R	R	M	H	H	L	L	L	L	M	L	H	H	H	L	L	L	L	R	R	R	R	R	R	L	L	L	M	*L*	R	R
55	H	H	L	L	L	L	L	L	L	L	L	L	L	L	L	L	L	L	L	L	L	L	L	L	R	L	L	L	L	L	L	L	L	L	L	L	L	M	L
56	*H*	H	L	L	*L*	L	L	L	L	L	L	L	L	L	L	L	L	L	L	L	L	L	L	L	R	L	L	L	L	L	L	L	L	L	L	L	L	M	L
57	*H*	H	L	L	L	L	L	L	L	L	L	L	L	L	L	L	L	L	L	L	L	L	L	L	M	L	L	L	L	L	L	L	L	L	L	L	L	M	*L*
59	M	L	L	L	*R*	L	L	L	L	L	L	L	L	L	L	L	L	L	L	L	L	L	L	L	L	L	L	L	L	L	L	L	L	L	L	L	L	M	L
60	H	H	L	L	R	R	R	R	M	M	M	R	R	L	L	L	L	L	L	L	L	L	L	L	L	L	L	L	L	L	L	L	L	L	L	R	L	M	R

Reactivity Chart 10. Protection for the Amino Group: Special —NH Protective Groups (*Continued*)

PG	I_2	PhSeX; PhSCl	Br_2; Cl_2	MnO_2/CH_2Cl_2	$NaIO_4$ pH 5-8	SeO_2 pH 2-4	SeO_2/Py	$K_3Fe(CN)_6$, pH 8	Pb(IV), 25°	Pb(IV), 80°	$Tl(NO_3)_3$	150°	250°	350°	$:CCl_2$	N_2CHCO_2R/Cu	$CH_2I_2/Zn(Cu)$	$R_3SnH/In\cdot$	$Ni(CO)_4$	CH_2N_2	$SOCl_2$	Ac_2O, 25°	Ac_2O, 80°	DCC	MeI	$Me_3O^+BF_4^-$	1.LDA 2.MeI	1.K_2CO_3 2.MeI	RCHO	RCOCl	C^+/olefin
	78	79	80	81	82	83	84	85	86	87	88	89	90	91	92	93	94	95	96	97	98	99	100	101	102	103	104	105	106	107	108
	L. OXIDANTS											M.			N.			O. MISCELLANEOUS											P.		
1	L	R	R	R	R	R	R	R	R	R	R	L	L	M	R	R	R	L	R	L	L	L	L	L	R	R	L	R	L	*L*	R
2	L	M	R	R	R	R	R	R	R	R	R	L	L	M	L	L	L	R	L	R	L	L	L	*L*	R	R	R	R	L	L	R
3	L	L	R	R	R	M	L	R	R	R	R	L	L	R	L	L	L	L	L	L	L	L	L	L	R	R	M	R	L	L	R
5	L	L	L	L	L	L	L	L	L	L	L	R	R	R	L	L	L	L	L	L	L	L	L	L	L	L	L	L	L	*R*	L
6	L	L	R	R	R	M	L	R	R	R	R	L	L	M	L	L	L	L	L	L	L	L	L	L	R	R	L	R	L	L	R
8	L	L	R	R	R	R	M	R	R	R	R	L	L	M	L	L	L	L	L	L	L	L	*L*	L	R	R	L	R	L	L	R
9	L	L	R	R	R	M	L	R	R	R	R	L	L	H	L	L	L	L	L	L	L	L	L	L	R	R	L	R	L	L	R
12	L	L	R	R	R	M	L	R	R	R	R	L	L	H	L	L	L	L	L	L	L	L	L	L	R	R	L	R	L	L	R
13	L	L	*M*	*L*	*L*	*L*	L	*L*	*L*	*L*	*L*	L	M	H	L	L	L	L	L	L	L	L	L	L	M	R	L	M	L	L	R
14	L	L	R	R	R	R	M	R	R	R	R	L	L	M	L	L	L	L	L	*L*	L	L	L	L	R	R	*L*	R	L	L	R
16	L	L	R	R	R	M	L	R	R	R	R	L	L	R	L	L	L	L	L	L	L	L	L	L	M	R	L	M	L	L	R
17	L	L	R	R	R	R	M	R	R	R	R	L	L	M	L	L	L	L	L	L	L	L	L	L	R	R	L	R	L	L	R
18	L	L	R	R	R	M	L	R	R	R	R	L	L	M	L	L	L	L	L	L	L	L	L	L	R	R	L	R	L	L	R
19	L	L	R	R	R	M	L	R	R	R	R	L	L	M	L	L	L	L	L	L	L	L	L	L	R	R	L	R	L	L	R
20	L	L	R	R	R	M	L	R	R	R	R	L	L	M	L	L	L	L	M	L	L	R	R	L	R	R	L	R	L	L	R
21	L	L	R	R	R	R	M	R	R	R	R	R	R	R	L	L	L	M	M	M	H	*H*	H	L	R	R	L	R	L	R	R
24	R	R	R	R	R	R	R	R	R	R	R	L	M	H	R	R	R	L	L	L	R	R	R	*L*	M	R	L	M	L	L	R
25	R	R	R	R	R	R	M	R	R	R	R	L	M	H	R	R	R	R	L	L	L	*L*	H	L	M	R	*L*	M	L	L	R
27	R	R	R	R	R	M	M	R	R	R	R	L	M	H	R	R	R	R	L	L	L	M	R	L	M	R	L	M	L	L	R
28	R	R	R	R	R	R	M	R	R	R	R	L	M	H	R	R	R	R	L	M	R	R	R	R	M	R	R	R	L	R	R
33	L	*R*	R	L	R	R	R	M	R	R	R	L	L	M	R	R	R	L	L	L	L	M	H	*L*	L	R	R	L	L	L	R
37	R	R	R	R	R	R	R	R	R	R	R	L	R	R	R	R	R	R	R	R	L	M	H	*L*	M	R	R	M	R	R	R
39	L	L	L	L	L	L	L	L	L	L	L	H	R	R	L	L	L	L	L	L	H	H	H	L	L	R	L	L	L	R	R
40	L	L	L	L	L	M	L	L	L	L	L	L	L	M	L	L	L	L	L	L	L	L	L	L	R	R	L	R	L	L	M
41	L	L	L	L	R	M	L	L	L	L	L	L	L	M	L	L	L	L	L	L	L	L	L	L	R	R	L	R	L	L	M
47	R	L	R	*R*	R	R	L	R	R	R	R	L	M	H	*H*	L	L	*H*	L	L	*L*	M	R	L	R	R	R	R	L	L	R
48	R	L	R	R	R	M	L	M	R	R	R	L	M	R	L	L	L	L	L	L	*L*	R	R	L	R	R	R	R	L	L	R
55	L	L	L	L	L	M	L	L	L	L	L	L	M	H	L	L	L	L	L	L	L	L	L	L	L	R	R	R	L	L	M
56	L	L	L	L	L	M	L	L	L	L	L	L	M	H	L	L	L	L	L	L	*L*	L	L	*L*	L	R	R	R	L	L	M
57	L	R	*M*	L	L	R	M	L	L	L	L	L	M	H	L	L	L	L	L	L	L	L	L	L	L	R	R	M	L	L	M
59	L	L	L	L	L	L	L	L	L	L	L	L	M	R	L	L	L	L	L	L	L	L	L	L	L	R	R	R	L	L	M
60	L	R	R	L	L	R	R	L	L	L	L	L	M	H	L	L	L	R	L	R	R	R	R	L	M	R	R	R	L	L	M

INDEX